PHOTOLUMINESCENCE OF SOLUTIONS

Photoluminescence of Solutions

With Applications to Photochemistry and Analytical Chemistry

C. A. PARKER, Ph.D., D.Sc., F.R.M.S., F.R.I.C.

Royal Naval Scientific Service,
Admiralty Materials Laboratory,
Holton Heath, Poole, Dorset, England

ELSEVIER PUBLISHING COMPANY

Amsterdam – London – New York

1968

ELSEVIER PUBLISHING COMPANY
335 Jan van Galenstraat
P.O. Box 211, Amsterdam, The Netherlands

ELSEVIER PUBLISHING CO. LTD.
Barking, Essex, England

AMERICAN ELSEVIER PUBLISHING COMPANY, INC.
52 Vanderbilt Avenue
New York, New York 10017

Library of Congress Catalog Card Number: 68-15622

With 188 illustrations and 53 tables.

Printed in the Netherlands

To my wife JOAN

Preface

The enormous growth in the literature relating to the processes of light emission from molecules in solution is largely due to the development of highly sensitive and efficient spectrophotometers, which have been exploited by the chemist and the physicist to elucidate the mechanisms of photoluminescence, and to develop new methods of chemical analysis. Many of the important advances in the theory and practice of photoluminescence are still adequately treated only in original papers and reviews scattered widely in the literature. There is thus a need for a text book that will at the same time instruct the student in the basic principles, and serve the expert as a source of reference. This is the dual purpose of the present work. It is written from the points of view of the photochemist and the analytical chemist but it will be of value to all workers who have occasion to measure photoluminescence.

The book deals mainly with liquid solutions, both fluid and glassy, but some reference is made to solid solutions and adsorbates. Although the principles of gas phase and crystal luminescence are not specifically dealt with, much of the chapter on experimental methods is relevant to these fields also. The mathematics, although extensive in some parts of the book, consist of the simple algebraical manipulations associated with the kinetics of photochemical processes or the operation of optical instruments. The essential parts are developed from first principles and should therefore be readily assimilated by the newcomer to the field. Quantum mechanical calculations are excluded and the concepts of molecular orbitals, energy levels, electronic transitions, energy transfer, etc., are dealt with in a pictorial manner. Adequate references are given for the reader to follow up any particular aspect in more detail if desired.

I have devoted a considerable amount of space to the principles governing the operation of the various components of spectrofluorimeters and spectrophosphorimeters, and the methods by which these components may be combined to produce a complete instrument of good performance. The object is to provide the reader with sufficient information to set up an instrument for himself, or to judge the relative merits of complete commercial luminescence spectrometers for his particular purpose. I have deliberately avoided giving details of specific commercial instruments because these would soon become out-of-date, and the latest information is best obtained directly from the manufacturers themselves. A great deal of the chapter on experimental methods is a summary of my own experience together with selected ideas published by other workers, and its layout was to some extent influenced by the many questions on instrumentation that I have received during the last ten years.

Chapter 4 will be of particular interest to the photochemist and physicist, while chapter 5 is intended primarily for analytical chemists. Both groups will I hope find something of value in both chapters. On the one hand many analytical methods have evolved as a result of fundamental research into photoluminescence processes, and on the other hand the preparation and analysis of pure materials is of vital concern in many branches of photoluminescence research. The book is not intended to be a compendium of photoluminescence methods of analysis, but chapter 5, read in conjunction with appropriate sections in the other chapters, will give the reader a clear idea of what kinds of analytical problem can be solved by the measurement of photoluminescence, and how to set about applying such measurement to a practical job of analysis.

In some areas I have attempted to give a balanced historical account of important developments, but in other areas I have had to be selective. There is a wide scope for the choice of examples illustrating certain aspects of the subject and in many cases I have chosen examples from my own work. This is not to say that equally good examples could not have been found elsewhere, but simply that the number of examples had to be limited to keep the size of the book within reasonable bounds.

I have attempted to make the subject index as comprehensive as possible. For example it is intended to include every mention of a chemical compound, and although many of these are discussed only in passing, the literature citation in the text will enable the reader to locate the sources of more detailed information. In the subject index the page numbers in italics followed by the letter T or F indicate that the item is to be found in a table or figure on that page. Separate indexes give abbreviated descriptions of tables and figures. All cross references in the text are given as section numbers: thus a reference to section 5 E 3 indicates that the matter can be found in chapter 5, main section E, sub-section 3, the page number of which is to be found in the Table of Contents. Definitions of symbols are given in a separate list.

I am greatly indebted to Dr. T. C. J. Ovenston, the Superintendent of the Admiralty Materials Laboratory, for his continued help and encouragement during the many years that I have worked in his laboratory, and to Dr. E. J. Bowen, F.R.S., who first stimulated my interest in photochemistry and has since given me the benefit of his wide experience in the field. I should like to thank my colleagues who have been closely associated with the work on photoluminescence in this laboratory: Mr. W. J. Barnes, Mr. C. G. Hatchard, L.R.I.C., Miss Thelma A. Joyce and Mr. W. T. Rees, M.Sc , F.R.I.C.

I am grateful to the authors of the many papers from which I have drawn material or ideas in compiling this book. I acknowledge the permission of the following authorities to reproduce figures or other matter: Academic Press Inc., American Institute of Physics, The Chemical Society, The Council of the Faraday Society, The Council of the Royal Society, W. Heffer & Sons Ltd., The Institute of Physics and The Physical Society, Pergamon Press Ltd., Verlag Chemie GmbH, John Wiley & Sons Inc., and the Editors of *The Analyst*, *Analytical Chemistry*, *Applied Optics*, *Chemistry in Britain*, *Electronic Technology* (now incorporated into *Industrial Electronics*), *Journal de Chimie Physique*, *Nature* and *Zeitschrift für Physikalische Chemie*.

Poole, Dorset
April 1968

C. A. Parker

Contents

PREFACE VII

1. Basic Principles and Definitions 1

A. LUMINESCENCE AND PHOTOCHEMISTRY 1
1. Luminescence 1
2. Nature of Light 2
3. Photochemistry 4

B. FLUORESCENCE 7
1. Origin of Fluorescence 7
2. The Franck–Condon Principle 11
3. Energy Difference in the 0–0 Transitions 13
4. Quantitative Aspects of Light Absorption and Emission 15
5. Transition Probability and Lifetime 22
6. Types of Electronic Transition 28

C. PHOSPHORESCENCE AND DELAYED FLUORESCENCE 36
1. The Triplet State 36
2. Intersystem Crossing 38
3. Definition of Phosphorescence and Delayed Fluorescence 42
4. E-Type Delayed Fluorescence 44
5. P-Type Delayed Fluorescence 46
6. Recombination Photoluminescence 48

D. POLARISATION OF PHOTOLUMINESCENCE 51
1. Polarised Light 51
2. Degree of Polarisation 53
3. Polarisation of Luminescence from Oriented Molecules 53
4. Polarisation of Luminescence from Randomly Oriented Molecules 55
5. Polarisation Spectra 57
6. Rotational Depolarisation of Fluorescence 59
7. Depolarisation Caused by Energy Transfer 60

E. LIGHT SCATTERING 60
1. Kinds of Light Scattering 60
2. Rayleigh Scattering 61
3. Tyndall Scattering 63
4. Scattering from Large Particles 64
5. The Raman Effect 64
6. Polarisation of Raman Emission 65
7. Raman Emission of Solvents 66

2. Kinetics of Photoluminescence 69

A. TRANSITION RATES 69

B. FLUORESCENCE 71
1. Fluorescence Efficiency and Lifetime 71
2. The Stern–Volmer Equation and Fluorescence Quenching 72

3. Diffusion-Controlled Processes . . . 74
4. Fluorescence from Upper Excited States . . . 78
5. Temperature Dependence of Prompt Fluorescence . . . 81
6. Energy Transfer between Singlet States . . . 83
7. Intramolecular Energy Transfer . . . 85
C. KINETICS OF PHOSPHORESCENCE . . . 86
1. Triplet Formation Efficiency . . . 86
2. Phosphorescence Efficiency and Lifetime . . . 87
3. Temperature Dependence of Phosphorescence . . . 89
4. Triplet-to-Singlet Energy Transfer and Sensitised Phosphorescence . . . 92
D. KINETICS OF DELAYED FLUORESCENCE . . . 97
1. E-Type Delayed Fluorescence . . . 97
2. Triplet-to-Triplet Energy Transfer and P-Type Delayed Fluorescence . . . 99
3. P-Type Delayed Fluorescence in Rigid Media . . . 103
4. Annihilation-Controlled Delayed Fluorescence . . . 108
5. Sensitised Delayed Fluorescence . . . 114
6. Anti-Stokes Delayed Fluorescence and Mutual Sensitisation . . . 122
E. SUMMARY OF KINDS OF PHOTOLUMINESCENCE . . . 126

3. Apparatus and Experimental Methods . . . 128
A. GENERAL COMMENTS . . . 128
B. MONOCHROMATORS . . . 131
1. Working Principles . . . 131
2. Angular and Linear Dispersion . . . 133
3. Spectral Purity with a Continuous Source . . . 133
4. Spectral Purity with a Discontinuous Source . . . 136
5. Determination of Linear Dispersion . . . 139
6. Optimum Entrance Slit Illumination . . . 142
7. Light Gathering Power . . . 144
8. Comparison of LGP Values . . . 147
9. Grating Inefficiencies . . . 149
10. Scattered Light . . . 150
11. Geometry of Exit Beam and Specimen Illumination . . . 151
12. Resolving Power . . . 152
13. Choice Between Wavelength and Wavenumber . . . 153
14. Summary . . . 156
15. Precautions . . . 157
C. LIGHT SOURCES . . . 158
1. Measurement of Spectral Distribution . . . 158
2. Incandescent Sources . . . 169
3. Xenon Arc Lamps . . . 172
4. Mercury Lamps . . . 174
5. Low Pressure Mercury Lamps . . . 174
6. Medium Pressure Mercury Lamps . . . 176
7. High Pressure Mercury Lamps . . . 178
8. Lamps Containing Elements other than Mercury . . . 179
9. Minimum Usable Source Size . . . 179
10. Examples of Light Fluxes Obtained . . . 182
11. Precautions to be Observed in Using Ultra-violet Lamps . . . 184
D. FILTERS . . . 186

1. General Comments . . . 186
2. Filters for Isolating Exciting Light . . . 188
3. Isolation of Fluorescence or Phosphorescence . . . 191
4. Choice of Filters for Filter Fluorimetry . . . 192
E. DETECTION AND MEASUREMENT OF LIGHT BY PHYSICAL METHODS . . . 193
1. Types of Photo-Detector . . . 193
2. Thermopiles . . . 194
3. Vacuum Photocells . . . 197
4. Photomultiplier Tubes . . . 200
5. Fluorescence Quantum Counters . . . 204
F. MEASUREMENT OF LIGHT BY CHEMICAL METHODS . . . 208
1. Advantages of the Chemical Actinometer in Radiant Energy Measurements 208
2. The Ferri-oxalate Actinometer . . . 208
3. Preparation of Actinometer Liquid and Choice of Concentration and Optical Depth . . . 210
4. Preparation of Calibration Graph for Ferrous Iron . . . 212
5. General Procedure for Actinometry . . . 213
6. Comments on Procedure . . . 214
G. AMPLIFICATION AND RECORDING OF PHOTOMULTIPLIER OUTPUT . . . 215
1. Manual Operation . . . 215
2. Single Beam Recording . . . 216
3. Double Beam Recording . . . 217
4. Calibration of Sensitivity . . . 219
H. FACTORS AFFECTING THE CHOICE OF GEOMETRICAL ARRANGEMENT OF THE SPECIMEN 220
1. Inner Filter Effects and Quenching . . . 220
2. Inner Filter Effects with Right Angle Illumination . . . 222
3. Inner Filter Effects with Frontal Illumination . . . 226
4. The Method of In-Line Illumination . . . 229
5. Choice of Specimen Arrangement . . . 233
J. SPECIMEN CONTAINERS AND COMPARTMENTS . . . 234
1. Containers for Specimens at Room Temperature . . . 234
2. Cell for Vacuum De-aeration . . . 239
3. Measurements at Controlled Temperature . . . 241
K. CORRECTION OF SPECTRA . . . 246
1. Correction of Excitation Spectra . . . 246
2. Automatic Recording of Corrected Excitation Spectra . . . 249
3. Correction of Emission Spectra . . . 252
4. Automatic Recording of Corrected Emission Spectra . . . 258
L. MEASUREMENT OF FLUORESCENCE EFFICIENCY . . . 261
1. Absolute Quantum Efficiency of Fluorescence . . . 261
2. Determination of Relative Fluorescence Efficiencies . . . 262
3. Precautions . . . 263
4. Standard Fluorescent Substances . . . 265
M. MEASUREMENT OF FLUORESCENCE LIFETIME . . . 269
1. Phase Fluorometers . . . 269
2. Flash Fluorometers . . . 270
3. Indirect Method . . . 271
N. MEASUREMENT OF LONG-LIVED LUMINESCENCE . . . 272
1. The Spectrophosphorimeter . . . 272
2. Phosphorimeter Factor . . . 275

3. Method of Phasing Choppers and Determination of Chopper Leakage . . . 278
4. Determination of Quantum Efficiency of Phosphorescence and Delayed Fluorescence . . . 279
5. Determination of Rate of Light Absorption . . . 282
6. Measurement of Lifetime of Long-Lived Luminescence . . . 282
7. Purity of Materials . . . 286
8. Artifacts and Trivial Effects . . . 291
P. MEASUREMENT OF POLARISATION OF LUMINESCENCE . . . 292
1. Polarising Units . . . 293
2. Polarisation Filter Fluorimeters . . . 295
3. Polarisation of Fluorescence Excitation Spectrum . . . 298
4. Conversion of Conventional Spectrofluorimeter or Spectrophosphorimeter for Polarisation Measurements . . . 299
5. Automatic Recording of Fluorescence Polarisation Spectra . . . 301

4. Special Topics and Applications . . . 303
A. DETERMINATION OF PARAMETERS OF THE LOWEST EXCITED SINGLET STATE . . . 303
1. General Comments . . . 303
2. Fluorescence Lifetimes . . . 303
3. Intersystem Crossing Rates . . . 304
4. Determination of Triplet Formation Efficiencies by the Method of Medinger and Wilkinson . . . 305
5. Triplet Formation Efficiencies from Delayed Fluorescence Measurements . . . 309
6. Triplet Formation Efficiencies from Phosphorescence Measurements . . . 313
B. TRIPLET STATE PARAMETERS . . . 313
1. Rate of $T_1 \rightarrow S_0$ Intersystem Crossing . . . 313
2. Rate of $S_1 \leftarrow T_1$ Intersystem Crossing . . . 320
3. Probability of Triplet-to-Triplet Energy Transfer . . . 321
4. Determination of Triplet Energies . . . 323
5. Triplet Data from $T_1 \leftarrow S_0$ Excitation . . . 326
C. CHEMICAL EQUILIBRIA IN THE EXCITED STATE . . . 328
1. General Principles . . . 328
2. Relationship between Spectra and Equilibrium Constants . . . 332
3. pH-Dependence of Fluorescence Intensities . . . 334
4. Effect of Added Buffer . . . 337
5. Reaction in the Absence of Buffer . . . 339
6. Relationship between (pK − pK*) and Chemical Structure . . . 340
7. Isomerisation in the Excited State . . . 342
8. Protolytic Equilibria in the Triplet State . . . 343
D. EXCITED DIMERS AND PROMPT FLUORESCENCE . . . 344
1. Types of Dimers . . . 344
2. Stability of Excited Dimers . . . 346
3. The Excited Dimer of Pyrene . . . 347
4. Lifetime of the Pyrene Monomer and Excited Dimer . . . 350
5. Thermodynamic Data for Excited Dimers . . . 351
6. Excited Dimers of Other Compounds . . . 356
7. Factors Affecting Excited Dimer Formation . . . 359
8. Intramolecular Excited Dimer Formation . . . 361
9. Other Methods of Producing Excited Dimer Emission . . . 362
E. EXCITED DIMERS AND DELAYED FLUORESCENCE . . . 363

1. Pyrene Solutions at Room Temperature 363
2. Naphthalene in Ethanol at Low Temperature 366
3. Pyrene Solutions at Low Temperature 369
4. General Mechanism for Excitation of Monomer and Dimer Emission . . 371
F. EFFECTS OF SOLVENT 373
1. General Comments 373
2. Solvation by Polarisation 374
3. Hydrogen Bonding 374
4. Viscosity Effects 378
5. Resolution of Vibrational Bands and the Shpol'skii Effect 379
G. PHOTOLUMINESCENCE MEASUREMENTS IN THE STUDY OF IRREVERSIBLE PHOTOCHEMICAL REACTIONS 386
1. Analysis of Photochemical Products 386
2. Measurement of Fluorescence Quenching 387
3. Investigation of Photochemical Reactions in Rigid Media 388
4. Sensitised Delayed Fluorescence of Photochemical Reactants and Products . . 390
H. APPLICATIONS OF POLARISATION MEASUREMENTS 392
1. Determination of Rotational Relaxation Time 392
2. Determination of Fluorescence Lifetime and Mechanism of Fluorescence Quenching 394
3. Determination of Orientation 395
4. Energy Transfer 396

5. Application to Analytical Chemistry 397
A. GENERAL COMMENTS 397
1. Introduction 397
2. Photoluminescence in Relation to Other Analytical Methods 398
3. Classification of Photoluminescence Methods 399
B. SENSITIVITY 402
1. Meaning of the Term "Sensitivity" 402
2. Instrumental Sensitivity 404
3. Absolute Sensitivity 406
4. Method Sensitivity 409
C. FACTORS CONTRIBUTING TO THE LUMINESCENCE BLANK 411
1. Raman Spectrum of the Solvent 411
2. Luminescence of Cuvette and Surroundings 416
3. Scattered Light 418
4. Luminescence of Impurities in Solvents or Reagents 419
5. Luminescence of Other Components of the Specimen 422
6. Photo-decomposition 426
D. LUMINESCENCE OF ORGANIC COMPOUNDS IN RELATION TO MOLECULAR STRUCTURE 428
1. General Rules 428
2. Degree of π-Electron Conjugation 429
3. Nature of Lowest Excited Singlet State 431
4. Classification of Substituent Groups 434
5. Position and Vibrational Resolution of Fluorescence Emission Spectra . . 436
6. Heavy Atom Substitution 437
E. ANALYSIS OF ORGANIC MATERIALS 438
1. Criteria of Purity 438
2. Direct Analysis of Mixtures by Fluorescence Measurement 440

3. Direct Analysis of Mixtures by Phosphorescence Measurement 443
4. Direct Analysis of Mixtures by the Measurement of Sensitised Delayed Fluorescence 449
5. Other Methods Utilising Prompt Fluorescence in Fluid Solution 455
6. Fluorescence Quenching and Indirect Methods 460
7. Other Methods Based on Fluorescence and Phosphorescence in Glassy Media 460
8. Methods Based on Fluorescence of Solid Solutions 466
9. Fluorescence and Phosphorescence of Adsorbates 469
F. LUMINESCENCE OF INORGANIC AND INORGANIC–ORGANIC COMPOUNDS IN RELATION TO STRUCTURE 470
1. Inorganic Compounds 470
2. Organo-Metallic Complexes 473
3. Intramolecular Energy Transfer 478
4. Intermolecular Energy Transfer 481
5. Organic Compounds Containing Non-Metals 481
G. ANALYSIS FOR ELEMENTS AND INORGANIC COMPOUNDS 483
1. Application of Prompt Fluorescence in Fluid Solution 483
2. Methods Based on Intra- and Intermolecular Energy Transfer 491
3. Quenching Methods 493
4. Indirect Methods 496
5. Inorganic Substances in Crystalline or Glassy Media 497
H. FUTURE DEVELOPMENTS 499

DEFINITIONS OF SYMBOLS 502
LIST OF TEXTBOOKS 510
REFERENCES 511
INDEX OF TABLES 522
INDEX OF FIGURES 524
SUBJECT INDEX 529

Chapter 1

Basic Principles and Definitions

A. LUMINESCENCE AND PHOTOCHEMISTRY

1. Luminescence

Hot bodies that are self-luminous solely because of their high temperature are said to emit *incandescence*. All other forms of light emission are called *luminescence*. A system emitting luminescence is losing energy and, if the light emission is to continue indefinitely, some form of energy must be supplied from elsewhere. Most kinds of luminescence are classified according to the source from which this energy is derived. Thus the light from a gas discharge lamp or from a gallium arsenide laser is *electroluminescence* produced by the passage of an electric current through the ionised gas or the semiconductor p – n junction. The luminous dial of a clock emits *radioluminescence* excited by the high energy particles from the radioactive material incorporated in the phosphor coating. Energy from a chemical reaction may excite *chemiluminescence*, and chemiluminescent reactions that take place in living organisms give rise to *bioluminescence*, e.g., the light emitted by glow-worms, fireflies, and the so-called "phosphorescence" of the sea. *Thermoluminescence* is a special form of chemiluminescence arising from the chemical reaction between reactive species trapped in a rigid matrix and released by raising the temperature. Some other kinds of luminescence are *triboluminescence*, observed when certain crystals are crushed, and *sonoluminescence*, produced in liquids exposed to intense sound waves. For *photoluminescence*, the energy is provided by the absorption of infra-red, visible or ultra-violet light.

The study of most kinds of luminescence can provide some information about the chemical composition of the emitting system and the processes that take place after the absorption of the excitation energy, but photoluminescence is one of the most informative because it allows a greater degree of experimental control over the excitation process. By choice of the wavelength of the exciting light the energy can be directed to specific components of the system so that the ensuing processes are simpler than if the energy is fed to the system as a whole. The processes leading to the emission of photoluminescence are closely associated with those that result in photochemical reactions and the study of photoluminescence thus forms an integral part of the subject matter of photochemistry.

2. Nature of Light

To account for the observed modes of interaction of light with matter it is necessary to use two complementary and apparently contradictory models, by regarding light sometimes as a collection of particles and sometimes as a succession of waves. The wave model will explain phenomena such as reflection, refraction or diffraction, that do not involve the absorption of energy. According to the electromagnetic theory of Maxwell the displacements forming the waves are varying electric and magnetic fields, the electric and magnetic oscillations taking place in directions at right angles to one another and to the direction of travel. The interaction of light with matter is then interpreted in terms of interactions between the electric oscillations of the waves and the electrons of the atoms. The wave properties are characterised by a frequency ν, and a wavelength λ, which are related by the equation:

$$\lambda\nu = \mathbf{c} \tag{1}$$

where **c** is the constant velocity of light *in vacuo* (3×10^{10} cm sec^{-1}).

To describe processes involving the absorption or emission of light it is necessary to invoke the requirements of the Quantum Theory, the basis of which is that radiant energy can only be absorbed in definite units, or *quanta*. The energy, E, carried by one quantum is proportional to the frequency of oscillation, i.e.:

$$\mathrm{E} = \mathrm{h}\nu = \mathrm{hc}/\lambda \text{ ergs} \tag{2}$$

where h is Planck's constant (6.624×10^{-27} erg sec). The characteristics of a monochromatic beam of radiation can therefore be expressed not only by its wavelength or frequency, but also by the size of its quantum. When dealing with chemical reactions the most convenient unit of mass is the gram molecule, containing N single molecules ($\mathrm{N} = 6.023 \times 10^{23}$), and similarly in photochemistry the most convenient unit of radiant energy is equal to 6.023×10^{23} quanta. This unit is called the *einstein* and is equal to N$\mathrm{h}\nu$ ergs.

Thus, if a photochemical reaction proceeds in such a way that one molecule reacts for each quantum absorbed, the absorption of one einstein of light will cause the reaction of one mole.

In consequence of the Planck relationship (equation 2) the size of the einstein varies enormously from one end of the electromagnetic spectrum to the other, as indicated in Table 1. Thus one einstein of gamma rays is equivalent to about 10^7 kilogram calories, that of ultra-violet and visible light to about 10^2, that of the far infra-red to about 10^{-1}, while the einstein of the longer radio waves carries only about 10^{-8} kilogram calories of energy.

TABLE 1

APPROXIMATE SIZES OF QUANTA

Radiation	λ (cm) (typical values)	Wave-number (μm^{-1})	Size of quantum (electron volts)	Size of einstein (kilogram calories)	Absorption or emission of radiation involves
Gamma rays	10^{-10}	10^6	1.2×10^6	2.9×10^7	Nuclear reactions
X-rays	10^{-8}	10^4	1.2×10^4	2.9×10^5	Transitions of inner atomic electrons
Ultraviolet	10^{-5}	10^1	1.2×10^1	2.9×10^2	Transitions of outer atomic electrons
Visible	4×10^{-5}	2.5	3.1	7.1×10^1	
	8×10^{-5}	1.25	1.6	3.6×10^1	
Infrared	10^{-3}	10^{-1}	1.2×10^{-1}	2.9	Molecular vibrations
Far infrared	10^{-2}	10^{-2}	1.2×10^{-2}	2.9×10^{-1}	Molecular rotations
Radar	10^1	10^{-5}	1.2×10^{-5}	2.9×10^{-4}	Oscillation of mobile or free electrons
Long radio waves	10^5	10^{-9}	1.2×10^{-9}	2.9×10^{-8}	

Note on units

The following list gives the equivalent values of wavelength, wavenumber, frequency and quantum energy in various units for a chosen wavelength of near infra-red light:

wavelength λ	10^{-4} cm
	1.000 micrometre (μm)
	1000 nanometre (nm)
wavenumber $\bar{\nu}$	10^4 cm^{-1}
	1.000 μm^{-1}
frequency ν	2.998×10^{14} sec^{-1}
energy of einstein	28.57 kcal mole^{-1}
energy of quantum	1.986×10^{-12} erg
	1.240 electron volt
1 einstein sec^{-1} =	$1.196\times10^8/\lambda$ watt (λ in nm)

It is to be expected that these radiations of widely differing energies differ greatly in their mode of emission or absorption by matter, as indicated in column 6 of Table 1. This table also illustrates an alternative method of expressing frequency, viz., *wavenumber* (number of waves per unit length), for which the symbol $\bar{\nu}$ is used. It is the reciprocal of the wavelength and has the advantage over frequency (expressed in vibrations per second) of simpler arithmetical conversion from wavelength. In terms of wavenumber equation 2 becomes:

$$E = hc\bar{\nu} \text{ ergs} \tag{3}$$

when $\bar{\nu}$ is measured in cm^{-1}. Some alternative units of wavelength, wavenumber, and energy are included in the note in Table 1.

The ultra-violet and visible regions of the spectrum are of most interest to the photochemist. Absorption in these regions causes the excitation of the outermost electrons of the molecule, i.e., those responsible for chemical binding, and hence it may lead to chemical change. The same wavelength regions are of interest in the study of photoluminescence because this also requires prior excitation of one of the outermost electrons. Radiation of wavelengths longer than that of near infra-red (> 1.5 μm) is of less importance in photochemistry because chemical reactions requiring activation energies much less than that carried by quanta of near infra-red radiation (< 20 kcal) are likely to proceed rapidly in the dark at room temperature by thermal activation. At low temperatures longer wavelengths may be effective in certain systems.

3. Photochemistry

The action of sunlight in promoting chemical changes such as the bleaching of fabrics or the development of the green colour of plants is a matter of common experience. Indeed the utilisation of sunlight by plants involves photochemical reactions on which the whole of the food supplies of the animal world ultimately depend. The quantitative study of photochemical reactions first emerged with the formulation by Grotthus (1817) of what may be regarded as the first law of photochemistry, that is "Only that light which is absorbed by a system can cause chemical change". The second law of photochemistry was first expounded by Stark (1908) and later by Einstein (1912). This states that "One quantum of light is absorbed per molecule of absorbing and reacting substance that disappears". This law was originally deduced for extremely simple reactions and in the strict sense it applies only to the primary photochemical act, i.e. the production of the excited species immediately following the absorption act, because some of the excited molecules may revert to their original state by various processes including the emission of luminescence. Furthermore, even if all molecules undergo reaction, the primary products are frequently unstable and undergo a complex series of reactions to give the products finally observed. It was Einstein who emphasized the importance of determining the *quantum efficiency* when investigating the mechanism of photochemical reactions. The quantum efficiency may be defined in terms of the number of molecules of reactant that disappear, or the number of molecules of a particular product that are produced, per quantum of light absorbed, thus:

$$\text{quantum efficiency of decomposition} = \frac{\text{moles decomposed}}{\text{einsteins absorbed}} \qquad (4)$$

$$\text{quantum efficiency of formation of A} = \frac{\text{moles of A produced}}{\text{einsteins absorbed}} \qquad (5)$$

Similarly, the efficiency of photoluminescence may be expressed thus:

$$\text{quantum efficiency of photoluminescence} = \frac{\text{einsteins emitted}}{\text{einsteins absorbed}} \qquad (6)$$

Owing to the frequent complexity of the reactions following the primary photochemical act the quantum yield of a photochemical reaction can vary from a million (e.g., the explosive reaction between hydrogen and chlorine) to a small fraction (e.g. the bleaching of light-resistant dyestuffs). The problem confronting the photochemist thus involves recognition of the primary photochemical process, and elucidation of the subsequent reactions. In the investigation of the mechanism of photoluminescence the problems are similar, but usually less complex, because light emission generally arises from only two of the processes that take place after the primary act of light absorption, and the quantum efficiency of photoluminescence never exceeds unity.

Just as it is necessary to employ both a "wave model" and a "particle model" to explain all the properties of light, so also it is necessary to treat the electrons and nuclei of atoms both as electrically charged particles and as waves. The dual nature of the electrons bound in atoms or molecules is expressed mathematically in the wave equation of Schrödinger. This takes the form of a differential equation of which the particular solutions for any one system represent the energies of the various possible states, or energy levels, in which the system can exist. The absorption of a quantum of light by an atom or molecule can only occur if the size of the quantum is exactly equal to the difference between two existing energy levels of the system concerned. As a result of the absorption of the quantum the system is raised from the lower level (energy E_1) to the higher level (energy E_2). Thus for absorption to occur:

$$h\nu = E_2 - E_1 \tag{7}$$

The vibrational and rotational motions of a molecule are also quantised i.e., vibrational or rotational energy can be taken up or lost only in discrete quantum units. Thus the total energy of a particular state of a molecule may

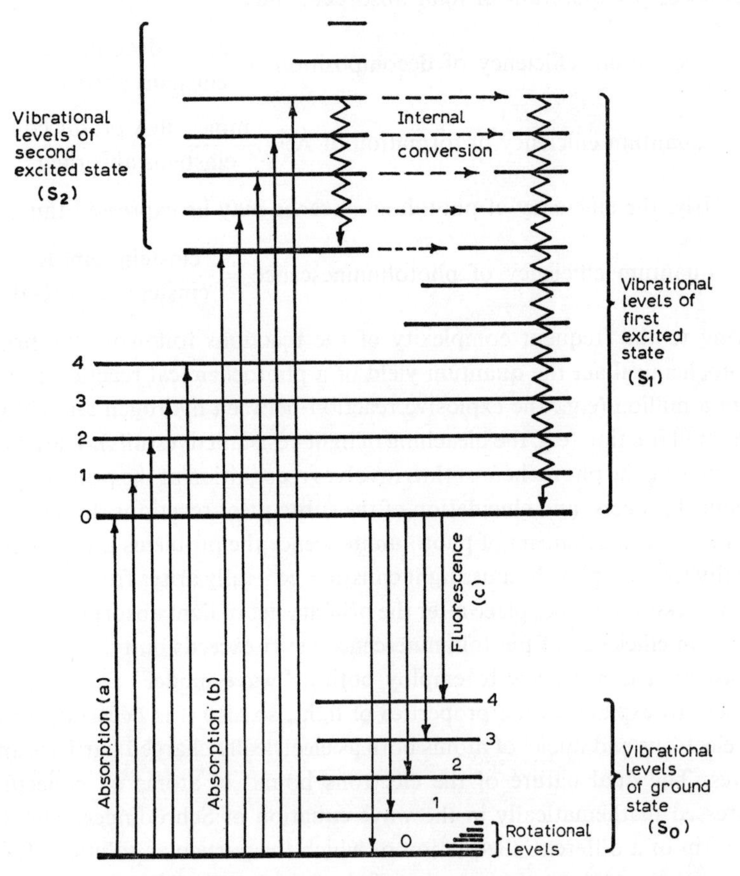

Fig. 1. Transitions giving rise to absorption and fluorescence emission spectra.

be represented as the sum of the electronic excitation energy, E_e, the vibrational energy, E_v, and the rotational energy, E_r, (as well as kinetic energy of translation which is not concerned in the primary act of light absorption) thus:

$$E = E_e + E_v + E_r \tag{8}$$

Equation 7 may thus be expanded as follows:

$$h\nu = \Delta E_e + \Delta E_v + \Delta E_r \tag{9}$$

i.e., the overall change of energy on absorption of a quantum of light may be represented as the sum of the changes in the number of electronic, vibrational and rotational quanta. The size of the vibrational quanta is less than that of the quanta required to excite an electron, and the size of the rotational quanta is smaller still. It is thus convenient to represent the various energy states available to the molecule by a simple energy level diagram like that shown in Fig. 1. Each electronic level is split into a series of vibrational levels, and each vibrational level is itself split into a series of closely spaced rotational levels. (In Fig. 1 only the rotational levels of the lowest vibrational level of the ground state are shown.) With this diagram we are now ready to give a simple qualitative description of the commonest form of photoluminescence, namely prompt fluorescence. The relationship between this and other forms of photoluminescence will be explained later.

B. FLUORESCENCE

1. Origin of Fluorescence

At room temperature most molecules are in the lowest vibrational level of the ground electronic state (level 0 of S_0 in Fig. 1), and from here transitions take place upwards on absorption of light (transitions *a* and *b* in Fig. 1). For all molecules in solution the rotational levels are so closely spaced that they cannot be distinguished spectroscopically and the vibrational levels in Fig. 1 should strictly be represented by bands containing the rotational levels. For some molecules the pattern of vibrational levels is comparatively simple and it appears in the absorption spectrum as a series of well separated maxima (see Fig. 2 curves A). However with many complex organic molecules the pattern of vibrational levels is so complex that all the transitions (*a*) in Fig. 1 to the various vibrational levels of the first electronic excited state give rise to one broad absorption band (e.g., curves B

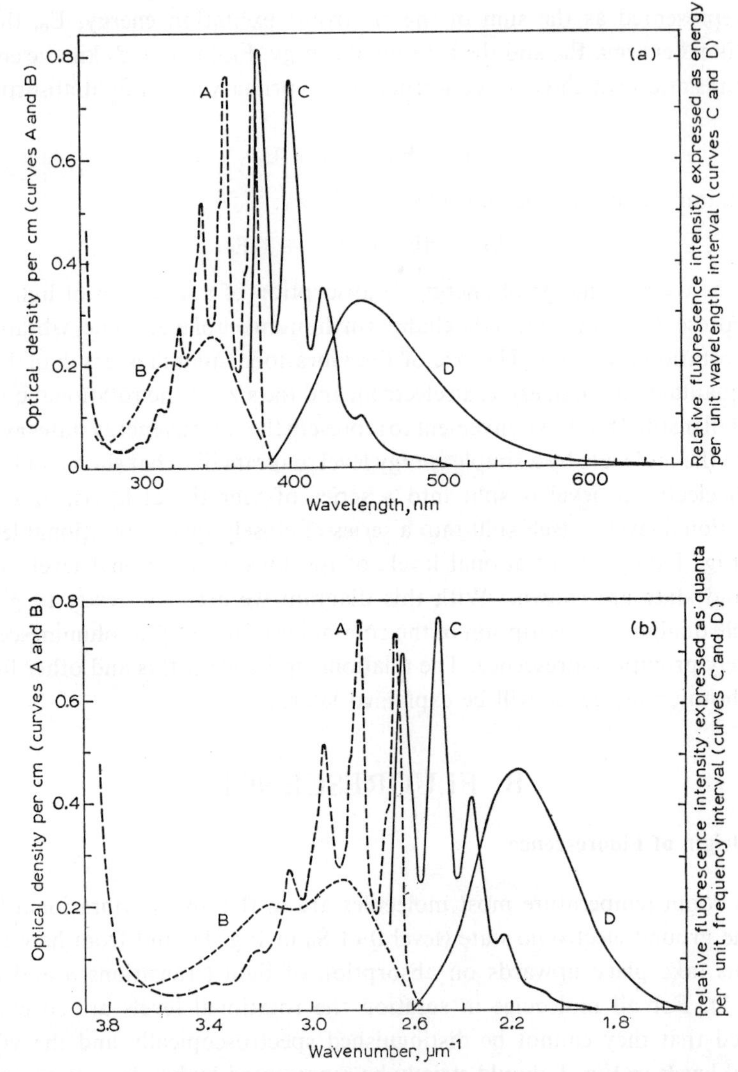

Fig. 2. Absorption and fluorescence emission spectra of anthracene and quinine bisulphate, illustrating the difference between plotting in units of (a) wavelength and energy and (b) wavenumber and quanta.

Curve A, absorption spectrum of anthracene in ethanol (17.2 μg per ml); curve B, absorption spectrum of quinine bisulphate in 0.1 N sulphuric acid (25 μg per ml); curve C, fluorescence emission spectrum of anthracene in ethanol (1.0 μg per ml); curve D, fluorescence emission spectrum of quinine bisulphate in 0.1 N sulphuric acid (1.0 μg per ml). Quartz prism spectrometers were used. The half-band width was 0.028 μm^{-1} (4.5 nm) at 2.5 μm^{-1} (400 nm) for fluorescence, and about 0.0005 μm^{-1} for absorption (from Parker and Rees[158]).

in Fig. 2), another broad absorption band appearing at shorter wavelengths (not shown in Fig. 2) corresponding to the transitions (*b* in Fig. 1) to the vibration–rotation levels of the second electronic excited state.

A molecule raised to an upper vibrational level of any excited state rapidly loses its excess of vibrational energy by collisions with surrounding molecules. This process is indicated by the wavy lines in Fig. 1. It is also found that almost all substances, when raised to electronic excited states higher than the first, undergo a process known as *internal conversion* whereby the molecule passes from a low vibrational level of the upper state to a high vibrational level of the lower state having the same total energy (see horizontal arrows in Fig. 1). Once internal conversion has occurred, the molecule again rapidly loses its excess vibrational energy by collision with solvent molecules. The net result of all these processes is that molecules raised to levels higher than the lowest vibrational level of the first excited state (i.e. level 0 of S_1) rapidly fall to the latter. Some substances may undergo photochemical reaction when raised to upper excited states but the processes leading to such reaction (e.g. dissociation) must take place rapidly to compete with the internal conversion and loss of vibrational energy by collisions. Similarly for light emission from these upper states to be appreciable, it also would have to be a much more rapid process than it normally is, and consequently in solution the observation of light emission due to transitions from upper excited states is very rare. Processes of light emission or chemical reaction by molecules in the lowest vibrational level of the first excited state need not be so rapid because internal conversion from here to the ground state is a relatively slow process and the lifetime of this state is typically 10^{-9} sec compared with about 10^{-12} sec for the upper excited states.

From level 0 of S_1 (Fig. 1) the molecule can return to any one of the vibration–rotation levels of the ground state with the emission of fluorescence (transitions *c* in Fig. 1). If all the molecules that originally absorb light return to the ground state in this way the solution will fluoresce with a quantum efficiency of unity. A proportion of the excited molecules may return to the ground state by other mechanisms, for example by conversion to the triplet state (which will be discussed later), or some may undergo photochemical reaction either in the first or a higher excited state. The quantum efficiency of fluorescence will then be less than unity and may be almost zero.

The fact that at room temperature absorption takes place almost exclusively from the lowest vibrational level of the ground state, while the emission takes place exclusively from the lowest vibrational level of the first excited state explains why only one transition, namely the 0–0 transition, is common to both the absorption and emission spectra. Thus all other transitions in absorption (transitions *a* in Fig. 1) require more energy than all the transitions observed in the emission spectrum (transitions *c* in Fig. 1). This means that the fluorescence emission spectrum overlaps the longest wavelength absorption band at the wavelength corresponding to the 0–0 transition and the rest of the emission spectrum lies immediately to the long wavelength side of the first absorption band. (With complex molecules even the 0–0 transitions do not coincide exactly: the 0–0 band in the fluorescence spectrum appears at very slightly lower energy than in the absorption spectrum—see Fig. 2 curves A and C. The reason for the non-coincidence of the 0–0 bands in absorption and emission is given in Section 1 B 3.) Frequently the emission spectrum is an approximate mirror-image of the first absorption band because the distribution of vibrational levels in the first excited state, which determines the shape of the first absorption band, is often similar to the distribution of vibrational levels in the ground state, which determines the shape of the fluorescence emission spectrum. Fig. 1 thus provides a representation of *Stokes Law*, which states that the wavelength of the fluorescence is always longer than that of the exciting light. However at room temperature a very small proportion of the molecules are thermally excited to the lower vibrational levels of the ground state (e.g. to level 1 of S_0) and can therefore undergo the transition $0 \leftarrow 1$ which would then be expected to appear in the absorption spectrum as a very weak band. In practice such a band is usually too weak to be observed in the absorption spectrum, but by taking special precautions it is possible to excite molecules at this wavelength and to observe emission corresponding to the $0 \rightarrow 0$ transition, i.e. of shorter wavelength than that of the exciting light. This is known as *anti-Stokes fluorescence*. An example is given in Chapter 5 where its importance in relation to some trace analytical work is described.

If the positions of the energy levels as indicated by the simplified diagram in Fig. 1 were the only factor governing light absorption and emission we might expect to find a series of equally intense absorption bands extending to short wavelengths and a series of equally intense fluorescence bands extending to longer wavelengths. In fact, as is indicated by the absorption and fluorescence spectra of anthracene shown in Fig. 2, the intensities of the bands vary considerably and to explain this we have to take into account the positions of the nuclei of the atoms in the molecule at the moment of light absorption or emission.

2. The Franck–Condon Principle

The Franck–Condon principle is based on the fact that light absorption takes place within about 10^{-15} sec, i.e. within the period of the vibration of the light wave. During this time the heavy nuclei of the atoms in the molecule do not appreciably change their positions and momenta and therefore the nuclear configuration and relative motions in the excited state immediately after absorption are identical with those in the ground state immediately before the light absorption occurred. To see how this affects the intensities of the various vibrational bands it is necessary to indicate the positions of the nuclei on a potential energy diagram. To represent the potential energy of a poly-atomic molecule as a function of its nuclear configuration requires a multi-dimensional "surface", but it is sufficient for our present purpose to consider a diatomic molecule and to plot potential energy against the inter-atomic distance as shown in Fig. 3. The curve MN represents the potential energy of the system in the ground state and the

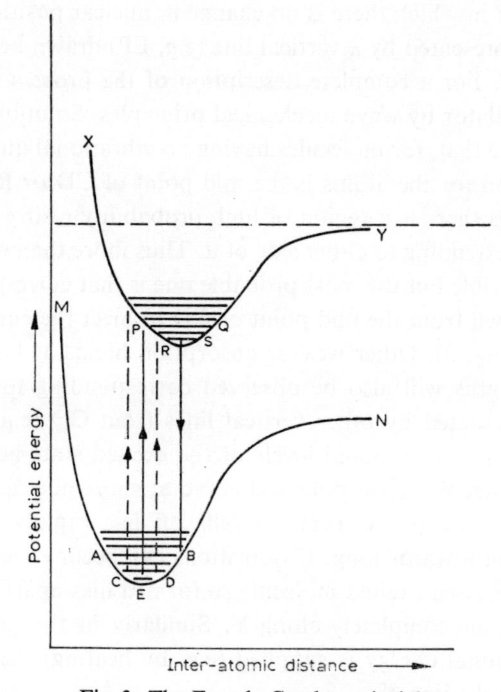

Fig. 3. The Franck–Condon principle

horizontal lines represent molecules in the ground state having various numbers of quanta of vibrational energy. Molecules with no vibrational quanta oscillate slightly about the mean separation distance (i.e. along the part of the curve CD). A molecule with three vibrational quanta oscillates along the path ACDB, and similarly for the other vibrational levels. All points on the straight line AB represent molecules having three vibrational quanta but with varying proportions of kinetic and potential energy, the sum of which is constant and equal to the height of the line AB above the minimum. The excited state is represented by a similar curve XY, situated higher up the diagram by a distance equal to the electronic excitation energy. The orbital occupied by the electron in the excited state is generally somewhat more extended in space than that occupied in the ground state and for this reason the equilibrium position in the excited state (mid point of RS) corresponds to a greater inter-atomic distance than that in the ground state (mid point of CD). At the moment of light absorption the molecule may have various combinations of kinetic and potential energy represented by points on the line CD. The Franck–Condon principle states that the most probable transition is that in which there is no change in nuclear position or momentum, i.e. it is represented by a vertical line (e.g. EP) drawn between the two electronic states. For a complete description of the process it is necessary to treat the oscillator by wave mechanical principles. Solutions of the wave equation indicate that, for molecules having no vibrational quanta, the most probable position for the atoms is the mid point of CD or RS. For higher vibrational levels there is a region of high probability near each end of the oscillation and extending to either side of it. Thus more than one absorption transition is possible but the most probable one is that corresponding to the vertical line drawn from the mid point of CD to meet the curve XY (this is the line EP in Fig. 3). Other weaker absorption bands at both longer and shorter wavelengths will also be observed corresponding to less probable transitions represented by other vertical lines from CD (e.g. those shown dotted in Fig. 3) to vibrational levels of the excited state adjacent to PQ. Because of the steeply rising potential curve at short nuclear distances the intensity of the absorption spectrum falls off less rapidly towards short wavelengths than towards long. If transitions can occur to levels above the line Y the compressed excited molecule so formed flies apart with sufficient energy to dissociate completely along Y. Similarly in the ground state, if sufficient vibrational energy is supplied (e.g. by heating) the molecule will dissociate along the line N.

After excitation to a vibrational level such as PQ the molecule rapidly loses excess vibrational energy and falls to the lowest vibrational level RS. From here it can undergo a transition back to the ground state with the emission of fluorescence. Again by the Franck–Condon principle the most intense emission band will be that corresponding to the vertical line drawn from the midpoint of RS to B. Weaker emission bands will also be observed corresponding to transitions to vibrational levels of the ground state adjacent to AB. The shape of the emission spectrum will depend on the relative positions of the minima of the upper and lower curves. If the positions of the minima differ by only a small amount the potential energy curve of the ground state will be steeper to the right of the vertical transition to B and the intensity will fall off less rapidly towards the long wavelengths. This is the situation with many compounds. If the positions of the minima differ by a large amount the curve for the ground state will be less steep to the right of the vertical transition and the fluorescence intensity will fall off less rapidly towards the short wavelengths. At an intermediate separation of the minima the intensity will fall off at about the same rate in both directions, i.e. an approximately symmetrical fluorescence emission spectrum will be observed.

3. Energy Difference in the O–O Transitions

The principles summarised in Fig. 1 provide an explanation of Stokes Law, i.e. that the fluorescence spectrum is situated to the long-wavelength side of the longest wavelength absorption band. The variation in the intensities of the various vibrational bands follows from the additional requirements of the Franck–Condon principle. We shall now explain why the 0–0 transitions in absorption and fluorescence do not coincide (see Fig. 4). In both the ground state and the excited state the molecule is solvated by dipolar attraction, due either to permanent dipoles in the solute and solvent or to induced dipoles in one or the other. Since the electron distribution in the excited state is different from that in the ground state, the permanent dipole moment and/or polarisability in the excited state will be different, and hence also the degree of solvation. In fluid solution at room temperature the solvent molecules will not have time to re-orient themselves during the process of light absorption and hence the molecule immediately after excitation finds itself in a state of solvation corresponding to a higher energy than its preferred equilibrium condition. The act of light absorption corresponds therefore to the transition from *a* to *b* in Fig. 4. The molecule will normally have

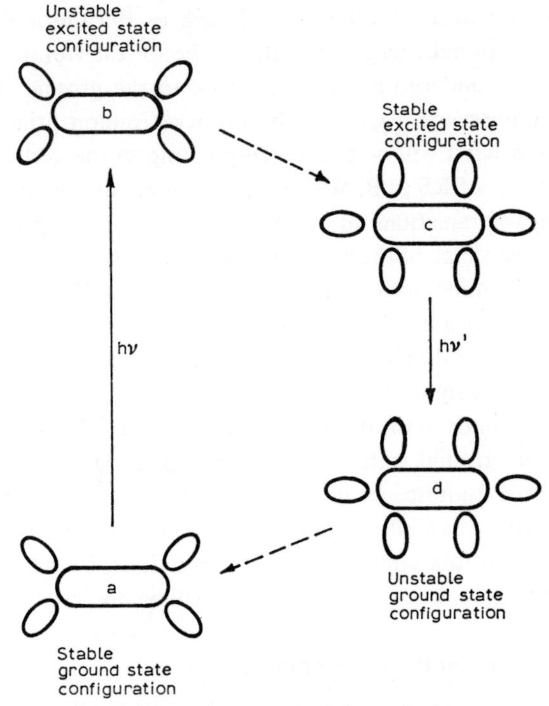

Fig. 4. Illustrating change of solvation after excitation or emission.
The large and small ovals representing solute and solvating solvent molecules are purely diagrammatic and are intended to represent a higher degree of solvation in the stable configuration of the excited state.

time to relax to the lower energy equilibrium configuration before emitting, i.e. the excited molecule will have time to relax to the state c in Fig. 4. Immediately after emission, the molecule, now in the ground state, finds itself in the state of solvation (d in Fig. 4) corresponding to the equilibrium configuration in the *excited* state. It thus has slightly more energy than the equilibrium configuration in the ground state and thus finally relaxes to the latter (state a in Fig. 4). It will be obvious that the process of light absorption will require slightly more energy than is released during the process of light emission, although both processes correspond to the 0–0 transition. The energy difference is dissipated as heat to the solvent during the two relaxation processes.

At sufficiently low temperature and high viscosity, the process $b \rightarrow c$ will not occur to an appreciable extent during the lifetime of the excited state, and emission will take place from *b*, or from some intermediate point between *b* and *c*, and the difference in energy between the 0–0 transitions will be reduced. At high temperatures the difference between configurations *b* and *c* is less pronounced and the difference between the energies of the 0–0 transitions is therefore less. The difference should thus pass through a maximum at intermediate temperatures and this has indeed been found to occur[1].

Since the energy difference between the 0–0 transitions depends on different degrees of solvation in the two states, it should be greatest for those molecules showing a large change in dipole moment on excitation to the first excited state. For a given substance it should also be greater in more polar solvents. There is in fact a relationship connecting the 0–0 band shift, the polarity of the solvent and the change in dipole moment of the solute on excitation. By measuring the 0–0 band shift in a series of solvents, the dipole moment in the excited state can be calculated[2,3].

It may be noted that since the electron in the excited state occupies a more extended orbital than in the ground state, the molecule in the excited state is expected to be more polarisable and hence, even in the absence of a change in permanent dipole moment on excitation, the degree of solvation in the excited state should increase slightly, and a small but appreciable 0–0 band separation should generally be observed.

4. Quantitative Aspects of Light Absorption and Emission

Consider a parallel beam of light of intensity I_0 directed on a parallel-sided specimen (Fig. 5). Part of the light (I_R and I_R') is reflected at each interface, part (I_S) is scattered within the medium, part (I_A) is absorbed, and part (I_T) is transmitted. Clearly,

$$I_0 = I_R + I_R' + I_S + I_A + I_T \tag{10}$$

The fraction reflected, with perpendicular illumination on an air/glass interface, is approximately 4% at each interface and is normally compensated for in the method of measuring absorption spectra. With clear solutions the proportion of light scattered is small and for the present therefore we shall consider only the light absorbed and transmitted. If the incident beam of light is monochromatic and the specimen is an assembly of similar absorbing molecules, it is observed that, provided the rate of light absorption is small compared with the number of molecules present, the fraction of light absorbed is independent of the intensity of the incident beam. In fact,

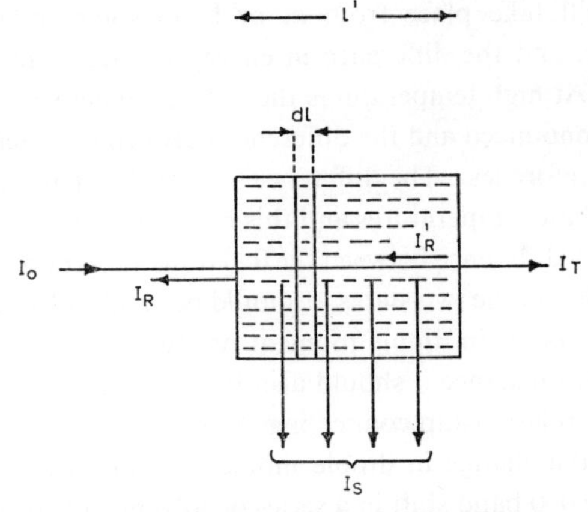

Fig. 5. Derivation of Beer–Lambert Law.

so long as the fraction of light absorbed is very small, each molecule behaves as though it had a specific cross sectional area, k, for that particular wavelength.

At some point within the medium bounded by the infinitely thin layer dl, let the light intensity be I. The intensity of light absorbed, dI_A, in traversing the thin layer is thus equal to the product of the incident intensity, I, on the layer, and the total effective cross-section of the molecules per sq.cm of the beam in the layer, i.e.

$$dI_A = -dI = Ikc'dl \tag{11}$$

where c′ is the number of molecules per cc and k is the effective cross-section of one molecule in sq.cm. Integration between the limits 0 and l′ leads to the familiar Beer–Lambert law:

$$I_T/I_0 = e^{-kc'l'} \tag{12}$$

in which the subscript "T" and the prime on l are usually omitted.

Naturally if the light intensity is so high that an appreciable fraction of the molecules are raised to an excited state (so as to acquire a different effective cross-section k), the above relation will not hold. However in aerated fluid solution at room temperature, where both singlet states and triplet states (see later) have very short lifetimes, it is only with extremely intense light that this occurs and the Beer–Lambert relationship is normally assumed to be universally obeyed. However it will be seen later that in a rigid medium at low temperature, where molecules in the triplet state often have long lifetimes, the accumulation of such long-lived molecules can be considerable even with beams of moderate intensity, and under these conditions the Beer–Lambert relationship will not hold unless the true concentrations and effective cross-sections of both the ground state molecules and the excited molecules are taken into account.

In equation 12 the concentration c′ refers to the number of molecules per cc, and k the effective cross-section of one molecule in square cm. A more practical form of equation 12 is as follows:

$$\log_{10}(I_0/I) = \varepsilon cl \qquad (13)$$

in which c is now measured in moles per litre and ε is known as the molecular extinction coefficient. k is simply related to ε by the expression:

$$k = 2{,}300\varepsilon/N \qquad (14)$$

where N is Avogadro's number, 6.023×10^{23}. With molecules having molecular weights of several hundreds, the maximum values observed for ε are about 10^5, and substitution in equation 14 indicates that the maximum effective cross-section of such a molecule is of the order 10^{-16} cm^2 i.e. the molecules behave as though they are opaque particles having dimensions of the order of one or two angstrom units (10^{-8} cm).

The quantity $\log_{10}(I_0/I)$ is known as the optical density or absorbance of the solution (see Table 2) and is proportional to the concentration of

TABLE 2

NOMENCLATURE AND SYMBOLS OF ABSORPTION SPECTROMETRY

Recommended[4]		Alternatives	
Incident radiant power	P_0	Intensity of incident light	I_0
Transmitted radiant power	P	Intensity of transmitted light	I
Absorptivity	a	Extinction coefficient	κ
Internal cell length	b	Optical depth	l
Absorbance	A	Optical density	D
Transmittance (P/P_0)	T	Transmission (I/I_0)	T
Molecular absorptivity	ε	Molecular extinction coefficient	ε

Note

ε refers to concentration expressed in moles per litre; κ (or a) to grams per litre. l is measured in cm. The relationship between the quantities is thus as follows:

$$D = \log_{10}(I_0/I) = \varepsilon\, cl = \log_{10}(1/T) = \kappa\, clM$$

where M is the molecular weight.

absorbing species. If more than one absorbing species is present the optical density (D_T) of the mixture can be simply related to the optical densities (D_A and D_B) that would be observed under the same conditions with each of the species A and B separately. Thus in equation 11, the total effective cross-section of the molecules per sq. cm of the layer dl is now $(k_A c_A' + k_B c_B')dl$ and substitution of this for $kc'dl$ leads to:

$$I/I_0 = \exp[-(k_A c_A' + k_B c_B')l] \tag{15}$$

or

$$D_T = \log_{10}(I_0/I) = \varepsilon_A c_A l + \varepsilon_B c_B l = D_A + D_B \tag{16}$$

The optical density of the mixture is thus equal to the sum of the optical densities of the separate components. The nomenclature and symbols of absorption spectroscopy are not yet completely standardised. Thus in the U.S.A. the recommendations of the Committee on Nomenclature in Applied Spectroscopy[4] are frequently adopted, while in the U.K. a different system is often used. The alternatives are summarised in Table 2.

Absorption spectra consist simply of a plot of ε, or of $\log \varepsilon$, against the wavelength or frequency of the light being absorbed. The question now arises how one should plot fluorescence emission spectra. Fluorescence emission spectra are normally plotted in one of two ways depending on whether it is desired to characterise the light by means of its wavelength or its frequency. In the first method the number of quanta emitted per second within a unit wavelength interval at the wavelength concerned (i.e. $d\mathbf{Q}/d\lambda$) is plotted against the wavelength λ. In the second method the number of quanta emitted per second within a unit frequency interval (i.e. $d\mathbf{Q}/d\nu$) is plotted against frequency. A third method (though generally less useful for photochemistry) is to plot *energy* emitted per second within unit wavelength interval (i.e. $dE/d\lambda$) against λ. The fluorescence is emitted from the specimen in all directions, but it is rarely necessary to plot the intensity scale in absolute units; relative values are all that are usually needed. The fluorescence emission spectra of solutions of anthracene and quinine bisulphate plotted in two of these ways, are shown in Fig. 2. With the first two methods of plotting, the area under the curve is proportional to the total rate of emission of fluorescence of all wavelengths, measured of course in quanta per unit time. With the third method, the area under the curve gives the total *energy* emitted per unit time. It is often necessary to convert wavelength units into frequency units and vice versa. The two quantities are related by equation 1. It is somewhat inconvenient to use frequency measured in vibrations per second, partly because of the large numbers involved, and partly because the conversion from wavelength requires the use of the factor, **c**, the velocity of light. The most commonly used unit of "frequency" is therefore *wavenumber* expressed as waves per cm. This is equal to the reciprocal of the wavelength expressed in cm, and for the spectral range of most interest in connection with photoluminescence (200–800 nanometres (nm) i.e. 200–800 millimicrons) the wavenumber range in cm^{-1} is 50,000–12,500. A still more convenient unit is the reciprocal micrometre (μm^{-1}) since the whole spectral range is then represented by 5.0–1.25 μm^{-1}.

Because the processes of absorption and emission are quantised, and the size of the quantum is proportional to frequency rather than wavelength, the spectra plotted in terms of frequency or wavenumber are theoretically more meaningful than those plotted in terms of wavelength. Plots showing a wide spectral range are also more conveniently handled on a frequency scale, where each spectral emission band is shown in its correct perspective in relation to others, while on a wavelength scale an undue amount of space is taken up by the spectral bands situated at long wavelengths. However, because automatic recording of spectra with a grating spectrometer can be most conveniently made with a linear wavelength scale, a scale in wavelengths is frequently used when a fully corrected spectrum is not needed. Both methods of plotting will be used in this book.

Because emission almost always takes place exclusively from the lowest vibrational level of the first excited state (see Fig. 1), the shape of the fluorescence emission spectrum is always the same, no matter what the wavelength of the exciting light. This is a most useful rule in practical spectrofluorimetry because if it is found that the shape of the emission spectrum of a solution does change when the wavelength of the exciting light is varied, the presence of more than one fluorescent component should be suspected. Admittedly there are some exceptions to this rule and these will be discussed later. The relationship between the intensity of fluorescence and the extinction coefficient can be simply derived from the Beer–Lambert Law (equation 12). The rate of emission of fluorescence, **Q**, is by definition (equation 6) equal to the rate of light absorption, measured in quanta per sec, multiplied by the quantum efficiency of fluorescence, ϕ_f, i.e.,

$$\mathbf{Q} = I_A\phi_f = (I_0 - I_T)\phi_f \tag{17}$$

and from equation 12 therefore

$$Q = I_0(1 - e^{-kc'l})\phi_f \qquad (18)$$

Expanding the exponential term we get:

$$Q = I_0(kc'l - (kc'l)^2/2 + (kc'l)^3/6 - \ldots)\phi_f \qquad (19)$$

Changing to concentration, c in moles per litre, and molecular extinction coefficient, (see equation 14), we have:

$$Q = I_0(2.3\varepsilon cl)(1 - 2.3\varepsilon cl/2 + (2.3\varepsilon cl)^2/6 - \ldots)\phi_f \qquad (20)$$

For weakly absorbing solutions for which the optical density, εcl, is small, the equation simplifies to:

$$Q = I_0(2.3\varepsilon cl)\phi_f \qquad (21)$$

The errors introduced by assuming the validity of equation 21 for solutions having various optical densities are shown in Table 3. The error can be regarded as an *inner filter effect*, i.e. the solution at the back of the cell is exposed to a lower intensity of exciting light than that at the front owing to absorption of a part of the exciting light by the intervening solution. Table 3 refers to measurements of the *total* fluorescence emitted in all directions. Generally only the fluorescence emitted within a comparatively small solid angle is collected and measured by the spectrometer, and the actual error introduced in such a measurement will depend on the precise geometry of the specimen with respect to the beam of exciting light and the direction from which the fluorescence is viewed. This will be dealt with in more detail in Chapter 3 under the heading of inner filter effects.

Equation 21 indicates that for a given substance in *dilute* solution contained in a particular cuvette, the intensity of the fluorescence observed is proportional to the product of concentration and intensity of the exciting light. Since with modern photomultiplier tubes exceedingly low intensities of light can be measured, then by using very high intensities of exciting light

TABLE 3

ERROR DUE TO INNER FILTER EFFECT

Optical density	$\frac{1}{2}(2.3\varepsilon cl)$	$\frac{1}{6}(2.3\varepsilon cl)^2$	% error in Q
0.001	0.0011	0.000001	+ 0.1
0.01	0.0115	0.0001	1.1
0.05	0.0575	0.0022	5.5
0.10	0.115	0.0088	10.6
0.20	0.230	0.035	20

(I_0), exceedingly low concentrations can be detected. Equation 21 thus illustrates why spectrofluorimetry is so much more sensitive than absorption spectrometry. Thus in absorption spectrometry the concentration is proportional log (I_0/I), and the instrumental factor governing the minimum detectable concentration is the minimum detectable difference between I_0 and I: to achieve high sensitivity I must be measured with a high degree of precision. We can make an estimate of the limiting sensitivity in absorption spectrophotometry as follows. With commercial instruments the *overall* precision with which log (I_0/I) can be measured, is certainly not better than 0.001 units, using a cuvette of 1 cm optical depth. For molecules of medium size the value of ε_{max} is rarely as great as 10^5. Substituting in equation 13 we find:

$$C_{min} > 10^{-3}/10^5 > 10^{-8}\text{ M} \tag{22}$$

On the other hand the instrumental sensitivity of a spectrofluorimeter is limited in principle only by the maximum intensity of exciting light available and not by the precision with which a light intensity can be measured. Under ideal conditions concentrations as low as 10^{-12} M can be detected.

The way in which the intensity of fluorescence changes with variation in the *wavelength* of the exciting light can also be deduced from equation 21. For a solution of a single solute at a constant concentration the fluorescence intensity is proportional to $I_0\varepsilon\phi_f$. Thus if the intensity of exciting light is kept constant as the wavelength is varied, the fluorescence intensity will be proportional to $\varepsilon\phi_f$. Plots of $\varepsilon\phi_f$ against wavelength or frequency of the exciting light are called *true fluorescence excitation spectra.* Now it so happens that for most complex molecules in solution, the fluorescence efficiency (ϕ_f) is approximately independent of the wavelength of the exciting light. Thus the true excitation spectrum of a dilute solution of a single absorbing solute will be proportional to ε, that is it will be a replica of the absorption spectrum of the compound (see for example Fig. 6). Spectrofluorimetry can thus be used to measure the absorption spectra of fluorescent compounds at concentrations far lower than would be needed to measure the absorption spectrum directly with an absorption spectrophotometer. As will be seen later, it has the advantage that the absorption spectrum of one component of a mixture of absorbing compounds can be picked out by tuning to the wavelength of the appropriate fluorescence emission band.

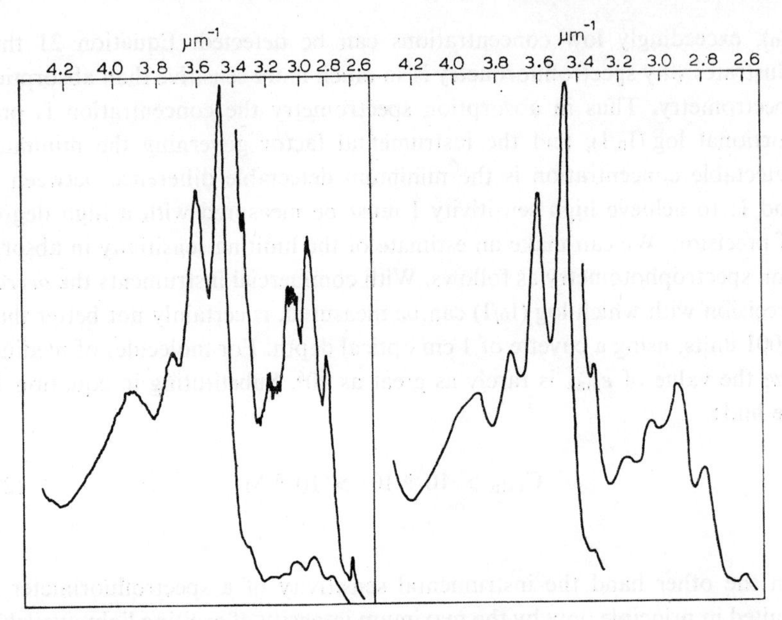

Fig. 6. Relationship between fluorescence excitation spectrum and absorption spectrum. 1,2-Benzanthracene in ethanol: left, excitation spectrum with fluorescence observed at 2.25–2.45 μm⁻¹; right, absorption spectrum (from Parker[143]).

5. Transition Probability and Lifetime

We have seen in Section 1 B 2 how the relative probabilities of transitions to various vibrational levels of an upper electronic excited state are reflected in the relative absorption intensities of these transitions. The absolute probabilities of the electronic transitions themselves depend on the types of the electronic states involved and determine the overall intensity of the corresponding absorption bands. Optical transitions that are probable in absorption are also probable in emission, and there exists a direct relationship between the probability of fluorescence emission on the one hand and molecular extinction coefficient of the corresponding absorption band on the other. Before discussing this relationship we must first consider the kinetics of the process of fluorescence emission. Suppose we have excited a number of molecules, n_0, to a particular electronic state, under conditions in which the only route for return to the ground state is that involving the emission of light. The probability that one of these molecules will emit light and return to the lower state is independent of the presence of the other excited molecules and thus at any time, t, the number of molecules emitting per second is proportional to the number, n, present at that time i.e.:

$$dn/dt = -k_f n \tag{23}$$

Integration leads to the equation:

$$n = n_0 e^{-k_f t} \tag{24}$$

The rate of emission of light, **Q**, is given by:

$$\mathbf{Q} = -dn/dt = kn_0 e^{-k_f t} = \mathbf{Q}_0 e^{-k_f t} \tag{25}$$

Thus the intensity of fluorescence decays exponentially. The mean *radiative lifetime* of the fluorescence, τ_r is defined by:

$$\tau_r = 1/k_f \tag{26}$$

and it is thus the time required for the intensity to fall to 1/e of its initial value. The parameter k_f is the first order rate constant for the process of fluorescence emission.

A variety of formulae have been derived to relate the radiative lifetime, τ_r, to the extinction coefficient. The simplest to use is that given by Bowen and Wokes[5]. Translated into units of μm^{-1} this takes the form:

$$1/\tau_r = 2900 n^2 \bar{\nu}_0^2 \int \varepsilon d\bar{\nu} \quad (\tau_r \text{ in seconds}) \tag{27}$$

where $\int \varepsilon d\bar{\nu}$ is the area under the curve of molecular extinction coefficient plotted against wavenumber (see Fig. 7), $\bar{\nu}_0$ is the wavenumber of the maximum of the absorption band, and n is the refractive index of the solvent. For many purposes this gives a sufficiently accurate value of τ_r. A somewhat more complex version is that recommended by Förster[6]:

$$1/\tau_r = 2900 n^2 \int \frac{(2\bar{\nu}' - \bar{\nu})^3}{\bar{\nu}} \varepsilon d\bar{\nu} \tag{28}$$

in which $\bar{\nu}'$ is the wavenumber of the mirror symmetry point between the absorption and fluorescence bands (see Fig. 7). Förster's formula corrects for the asymmetry of the absorption band but assumes a mirror image relationship with the fluorescence band. Birks and Dyson[7] later introduced a more refined equation which is valid when there is no mirror image relationship, and also takes into account the dispersion of the solvent. It does however require integrations to be carried out over the fluorescence

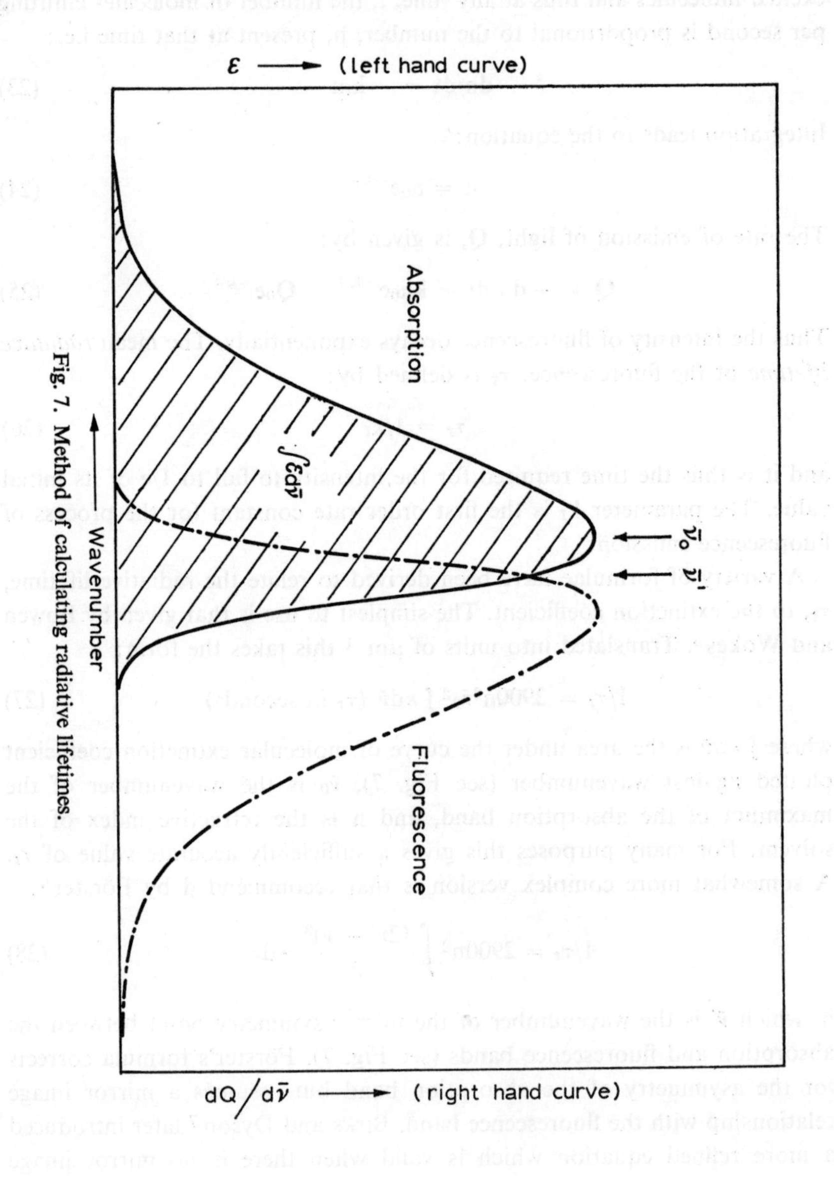

Fig. 7. Method of calculating radiative lifetimes.

emission spectrum as well as over the absorption spectrum. In units of μm^{-1} the Birks–Dyson equation is:

$$1/\tau_r = 2880 \frac{g_l}{g_u} \cdot \frac{n_f^3}{n_a} \cdot \frac{\int F d\bar{\nu}}{\int F \bar{\nu}^{-3} d\bar{\nu}} \cdot \int \frac{\varepsilon d\bar{\nu}}{\bar{\nu}} \tag{29}$$

where g_l and g_u are the multiplicities (see Section 1 C 1) of the ground and excited states, so that g_l/g_u is unity for the singlet–singlet transition giving rise to fluorescence. n_f and n_a are the mean refractive indices of the solvent over the fluorescence and absorption bands. F is the intensity of fluorescence at wavenumber $\bar{\nu}$. With compounds for which the difference between the nuclear configurations in the ground and excited states is small, the lifetimes calculated by equation 29 were found to agree with the experimentally determined values.

The simple expression 27 can be used to derive an even simpler formula for rough-and-ready calculation. Thus, the effective half-band widths ($\Delta\bar{\nu}$) of many absorption bands in the visible and near ultra-violet region are, at room temperature, of the order 0.2–0.5 μm^{-1}, say 0.3 μm^{-1} on average. We can then put $\int \varepsilon d\bar{\nu}$ equal approximately to $0.3\varepsilon_{max}$, where ε_{max} is the maximum extinction coefficient of the absorption band. With n^2 approximately 2, and choosing 2.5 μm^{-1} as the middle of the spectral range of most interest, we find by substituting in equation 27 that:

$$1/\tau_r \sim 10^4 \varepsilon_{max} \tag{30}$$

Thus, a strongly absorbing compound such as a dyestuff with ε_{max} about 10^5 would have a radiative lifetime of the order 10^{-9} sec.

It should be noted that the lifetimes calculated in this way are the *radiative* lifetimes, that is the lifetimes which would be observed in the absence of all other processes by which the molecule can return to the ground state. The observed lifetimes are nearly always less than the calculated values because of the competing radiationless processes. The radiative lifetime of the molecule in its lowest excited singlet state is one of the main factors governing the fluorescence intensity observed in solution at room temperature. If the lifetime is exceptionally long there is a much greater chance of the radiationless processes competing successfully with the radiative process that gives rise to fluorescence, and the intensity of the latter will therefore be lower. The significance of the relative values of the radiative and actual lifetimes will be dealt with quantitatively in Chapter 2.

In applying equations 27–29 it must be remembered that the integration has to be carried out only over the first absorption band. It will be noticed in Fig. 7 that the high wavenumber end of the first absorption band has conveniently been made to fall to zero. This state of affairs is rarely observed in practice (although the first absorption bands of some dyestuffs approximate to this condition) because the low wavenumber tail of the second absorption band almost invariably overlaps the high wavenumber end of the first. If the first absorption band is strong and the overlap small, it is possible by inspection to make the necessary minor correction for this overlap. If however the first absorption band is relatively weak and is overlapped by a strong second absorption band, it is difficult, and sometimes impossible, to determine the exact extent of the weak first absorption band. If the overlap is not too great and the 0–0 transition can be located in both the absorption and fluorescence spectra, the procedure illustrated in Fig. 8 is sometimes helpful. It is assumed that the extinction coefficient of the second absorption band is zero at the wavenumber of the 0–0 transition of the first absorption band. The scale of the fluorescence spectrum is then adjusted so that the height of its 0–0 band is equal to that of the first ab-

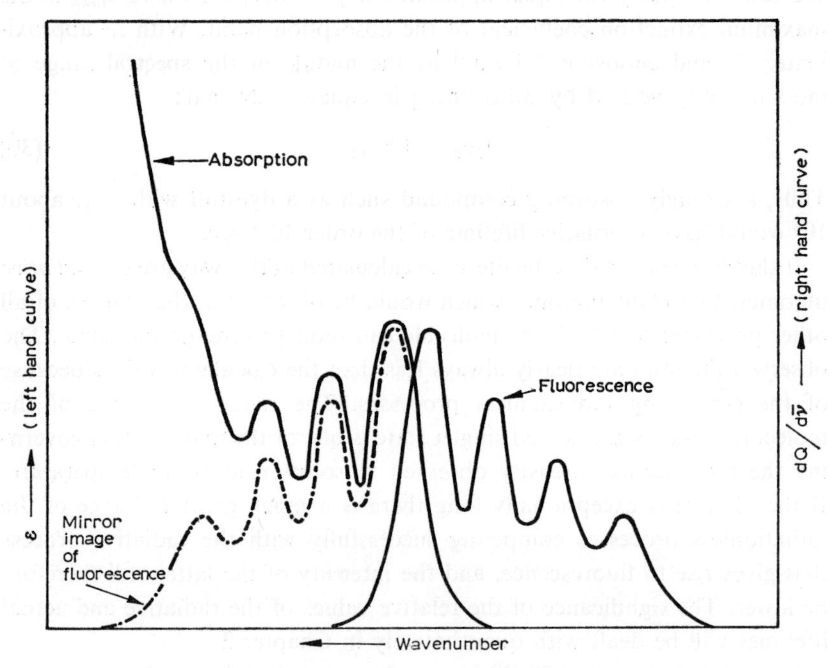

Fig. 8. Estimating the extent of the first absorption band.

sorption band. The fluorescence emission spectrum is then reflected about the mirror symmetry point and the curve so obtained is assumed to represent the absorption due to the first electronic transition alone. This is then integrated in the usual way. Unfortunately many compounds do not give good mirror symmetry and it is then difficult to locate the positions of the 0–0 bands. This procedure is therefore of limited application.

A concept often used in dealing with the probability of an electronic transition is the *oscillator strength*, f. This is proportional to the dipole strength of the transition, which is a measure of the difference between the degree of dipolar oscillation of the electron in the two orbitals involved in the transition. The value of f is also affected by the multiplicities of the two states (see Section 1 C 1) and the symmetries of their wavefunctions. A fully allowed transition has $f = 1$ but many important transitions have f values much smaller than this. The radiative lifetime of an excited state is related to the oscillator strength by the equation[8]:

$$\tau_r = \frac{1.5 \times 10^{-8}}{f\bar{\nu}^2} \tag{31}$$

From equation 27, with $n = 1$, we find that:

$$f = 4.3 \times 10^{-5} \int \varepsilon d\bar{\nu} \tag{32}$$

This means that, for a typical absorption band for which $\int \varepsilon d\bar{\nu}$ is approximately $0.3\varepsilon_{max}$, an oscillator strength of unity corresponds to $\varepsilon_{max} \sim 10^5$.

The radiative lifetime calculated by equations 27–29 refers to the *spontaneous* emission of light, and corresponds to the probability A_{nm} ($= 1/\tau_r$), that a molecule will undergo a radiative transition from an upper state n, to a lower state m, in the absence of radiation of frequency ν, corresponding to the energy difference between n and m. In general, the total probability of the transition is given by the sum of A_{nm} and a quantity $u_\nu B_{nm}$, in which u_ν is the density of radiation of frequency ν, and B_{nm} is a constant for the system concerned. The light emitted by the second process is known as *stimulated radiation*, and its phase is the same as that of the external stimulating light. The probability of stimulation, $u_\nu B_{nm}$, is the same as the probability, $u_\nu B_{mn}$, of the reverse transition, i.e. absorption. (In fact, according to the Einstein relations, $B_{mn} = B_{nm} = c^3 A_{nm}/8\pi h\nu^3$.) Thus in any system where the population in the ground state is greater than that in the excited state, the overall effect of exposure to light of frequency ν will be absorption of the latter. If by some means a system is produced in which the population in the excited state is greater than that in the ground state, exposure to light of frequency ν will result in additional radiation being emitted as a result of stimulation. This is the principle of the *laser*, the discussion of which is however beyond the scope of this book. Stimulated emission is most readily produced in systems emitting narrow spectral bands of luminescence, and its excitation by light absorption requires very high intensities of exciting light. We shall be concerned with systems in which, for various reasons, the stimulation is negligible, and in which therefore, the radiative decay of excited states is exponential with lifetime calculable by equations 27–29.

6. Types of Electronic Transition

The electron distributions and energies of a molecule in its ground and excited states are in principle given by the solutions of the Schrödinger equation and depend on electrostatic attractions between electrons and nuclei, electrostatic repulsions between electrons, internuclear vibrations, molecular rotation, as well as magnetic interactions resulting from electron spin and orbital motion, and nuclear spin. Precise solution of the equation is not possible except for the simplest of molecules, but approximate results can be obtained by treating the electron orbital motion separately from the electron spin and nuclear motions. The electrons are assumed to occupy *molecular orbitals*, each of which can accommodate two electrons, and a transition between two states of the molecule is considered as the movement of one electron from one orbital to another. We shall discuss qualitatively the electron distribution in several of the commonest types of orbital in the ground states of organic molecules, and the types of transitions that electrons occupying these orbitals can undergo on light absorption. The kind of electronic charge distribution associated with each type of orbital to be discussed can be deduced by reference to Figs. 9 and 10. It should be noted however that these figures are diagrammatic and are not drawn to scale. The envelopes represent the volumes of space where the probability of finding the electron is greatest. They are drawn with positive (unshaded) and negative (shaded) parts to indicate the sign of the amplitude of the wavefunction in this volume. The electronic charge density is represented by the square of the amplitude and is, of course, always positive.

The orbital of the electron pair forming a single bond is symmetrical around the bond axis with its electron charge density concentrated between the two atoms (see Fig. 9). This kind of orbital is known as a σ (bonding) orbital. Light absorption raises one of the pair of electrons into a σ^* orbital in which there is a node (i.e. zero charge density) in a plane at right angles to the bond. The two parts of the orbital wavefunction have opposite signs indicated by the shading in Fig. 9 and the electron in the new orbital introduces a repulsive force between the two atoms. The orbital is therefore called σ^* (anti-bonding). In designating optical transitions we shall adopt the rule of always writing the excited state first, and indicating the direction of the transition by an arrow. Thus the transition just described, brought about by light absorption, will be written $\sigma^* \leftarrow \sigma$. The reverse transition corresponding to light emission would be written $\sigma^* \rightarrow \sigma$. We shall not concern ourselves further with the σ^*–σ transitions because they require considerable energy, and only in saturated compounds absorbing at wavelengths shorter than about 200 nm do they correspond to the lowest energy transition. It may be mentioned in passing that in saturated molecules containing atoms having lone pairs of electrons, σ^* excited states may be produced by excitation of one of the lone pair (non-bonding) electrons.

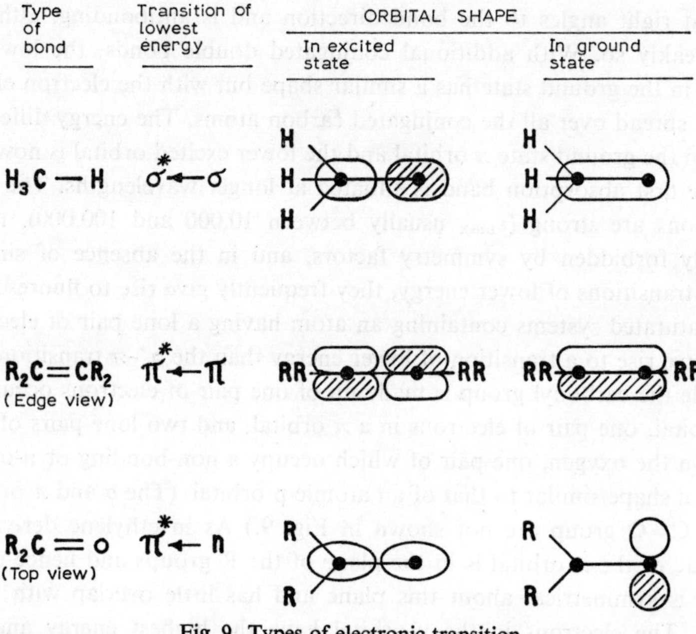

Fig. 9. Types of electronic transition.

Such $\sigma^* \leftarrow$ n transitions are probably the lowest energy transitions in aliphatic alcohols, ethers, amines, alkyl halides, etc.[9]. The absorption is diffuse and is situated at about 200 nm in alkyl ethers and alcohols, and at longer wavelengths in halogen compounds. Fluorescence is not generally observed from σ^* excited states.

Fluorescence is nearly always associated with the π electron system of an unsaturated molecule. If the unsaturated system does not include an atom such as oxygen, nitrogen or sulphur having a lone pair of electrons, the electronic transition of lowest energy corresponds to the promotion of one of the electrons occupying a π orbital in the ground state to an upper π^* orbital. In ethylene and its derivatives, for example, one of the pairs of electrons forming the double bond occupies a σ orbital of shape similar to that of a single bond. (The orbital of this pair of electrons is not shown in the diagram for $R_2C = CR_2$ in Fig. 9.) The other pair occupies a π orbital of higher energy which has a node in the plane of the R groups. The orbital is antisymmetric about the node, i.e. the two parts of the orbital wavefunction have opposite sign as indicated by the shading in Fig. 9, and the electron charge density is concentrated in the space between the carbon atoms so that the orbital is bonding. The upper π^* orbital has another nodal plane at right angles to the bond direction and is antibonding, although only weakly so. With additional conjugated double bonds, the lowest π orbital in the ground state has a similar shape but with the electron charge density spread over all the conjugated carbon atoms. The energy difference between the ground state π orbital and the lower excited orbital is now less, and the first absorption band is situated at longer wavelengths. The π^*–π transitions are strong (ε_{max} usually between 10,000 and 100,000), unless partially forbidden by symmetry factors, and in the absence of singlet–singlet transitions of lower energy, they frequently give rise to fluorescence.

Unsaturated systems containing an atom having a lone pair of electrons often give rise to a transition of lower energy than the π^*–π transition. For example the carbonyl group is made up of one pair of electrons occupying a σ orbital, one pair of electrons in a π orbital, and two lone pairs of electrons on the oxygen, one pair of which occupy a non-bonding or n-orbital having a shape similar to that of an atomic p orbital. (The σ and π orbitals of the C=O group are not shown in Fig. 9.) As in ethylene derivatives the node of the π orbital is in the plane of the R groups and hence the n-orbital is symmetrical about this plane and has little overlap with the π orbital. The electrons in the n-orbital have the highest energy and can undergo a transition to an upper π^* orbital, i.e. a $\pi^* \leftarrow$ n transition, which is generally the singlet–singlet transition of lowest energy. Such transitions have low probability ($\varepsilon_{max} < 2000$) and the π^*–n excited states have comparatively long radiative lifetime. They are thus much more subject to radiationless de-activation processes than π^*–π excited states and the fluorescence from them is very weak. Classification of transitions as π^*–π and π^*–n was first used by Kasha[10].

When more complex conjugated systems are considered, e.g. those involving an aromatic ring system and a nitrogen atom, it is necessary to subclassify the "lone pair" orbitals[11]. The reason for this may be illustrated by reference to the lone pair orbital of nitrogen in the three compounds pyridine, pyrrole and aniline (see Fig. 10). The orbitals associated with the N atom in pyridine consist of three hybrid trigonal sp^2 orbitals and a fourth p-type orbital. Two of the trigonal orbitals form the single bonds to the ring and the p-type orbital forms part of the π-orbital system of the ring. The third hybrid orbital corresponding to the lone pair of electrons has most of its electron charge density directed away from the ring. It is symmetric with the plane of the ring (see edge view in Fig. 10) and thus cannot conjugate with the π-orbital electrons of the latter since these are anti-symmetric to the plane of the ring. The lone pair orbital is in this respect similar to that of the carbonyl group already discussed and is classified as a non-bonding or n-orbital.

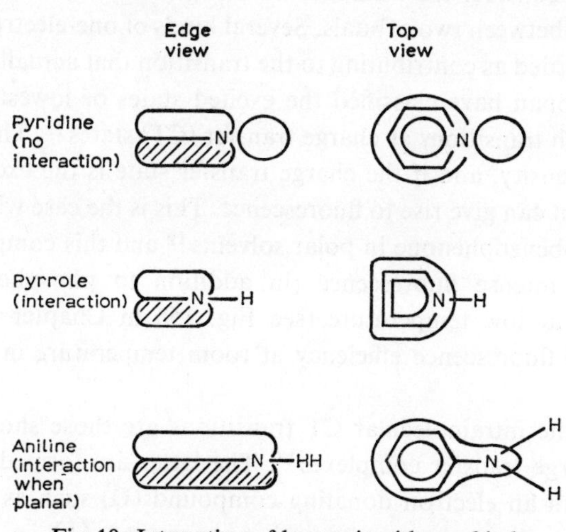

Fig. 10. Interaction of lone pair with π-orbital.

In pyrrole the nitrogen atom is singly bonded to the ring and the hydrogen atom. The lone pair electrons therefore occupy the remaining p orbital. This is antisymmetric to the plane of the ring and is conjugated with the π electrons of the carbon atoms in the latter. Only π^*–π transitions are therefore to be expected. In the co-planar form of aniline, the situation is similar to that of pyrrole. If the amino group is twisted at 90° to the plane, conjugation with the π electrons is not possible. A σ^*–n transition might occur at higher energies but still no π^*–n transition is to be expected. At intermediate angles of twist a transition from the lone pair orbital to the π electron system of the ring can occur. This results in a state having considerable charge transfer character. Kasha[11] has classified such intramolecular charge transfer transitions as a_π–l. Such transitions do not in general behave like true π^*–n transitions—in particular their radiative lifetime is closer to that of π^*–π.

Summarising the data in Fig. 10 we may expect compounds with the pyridine-like structure to have a weak π^*–n as the transition of lowest energy, and hence to be non-fluorescent. Pyrrole with a π^*–π transition of lowest energy should be fluorescent. Compounds such as aniline with lone pairs in l-orbitals will have comparatively intense long wavelength transitions and hence should in general show fluorescence.

When both an electron donor group (e.g. O^-, NH_2, OH) and an electron acceptor group (e.g. CO, NO_2) are connected with an electron system it is not possible to consider the transition to the excited state as the movement of one electron between two orbitals. Several kinds of one-electron transition have to be regarded as contributing to the transition that actually occurs and Porter and Suppan have classified the excited states of lowest energy that result from such transitions as charge transfer (CT) states[12]. The transitions are of high intensity, and if the charge transfer state is the excited state of lowest energy, it can give rise to fluorescence. This is the case with 4,4′-tetramethyldiaminobenzophenone in polar solvents[12] and this compound shows comparatively intense fluorescence (in addition to phosphorescence) in rigid solution at low temperature (see Fig. 158 in Chapter 5), although admittedly the fluorescence efficiency at room temperature in ethanol[13] is only 10^{-3}.

Related to the intramolecular CT transitions are those shown by intermolecular charge transfer complexes[14]. The latter are formed in solutions containing both an electron donating compound (D) such as an aromatic hydrocarbon, and an electron accepting compound (A) i.e. a substance containing strongly electron-attracting groups such as CN, CO or NO_2. Such a complex has an absorption band at a wavelength longer than either of the parent compounds and the absorption corresponds to the transition to a charge transfer excited state to which the configuration D^+A^- makes a major contribution. Such charge transfer complexes are frequently fluorescent in the solid state[15] and sometimes weakly so in solution[16]. Tanaka and co-workers[442] have concluded that the *phosphorescence* of various CT complexes at 77°K is due to transitions from the charge transfer *triplet* states.

It will now be clear that one of the main criteria for deciding whether or not a molecule is likely to show fluorescence is the type of its excited singlet state of lowest energy. Kasha[10] has described a series of experimental tests for characterising π^*–n transitions, and these were later extended by Sidman[17]. Porter[9] has described similar tests for transitions to charge transfer states. Three of the most useful tests are described in the following paragraphs.

Solvent Effect. Comparison of the wavenumbers of the absorption bands in non-polar and polar solvents shows that in the latter the π^*–π transitions are situated at slightly lower wavenumbers than in the former. The "red" shift in changing from a hydrocarbon solvent to ethanol is about 0.05 μm^{-1}. The CT transitions show a much larger red shift (about 0.2 μm^{-1}). In contrast, π^*–n transitions undergo a spectral shift in the opposite direction (i.e., a "blue" shift of about 0.08 μm^{-1}). To interpret these shifts in terms of the electron distributions in the ground and excited states it has to be remembered that the energy represented by the wavenumber of an absorption band corresponds to the difference between the energy of the equilibrium configuration of the ground state and the energy of the state reached immediately after light absorption, i.e., the "Franck–Condon" state (state *b* in Fig. 4). Thus the effect of change of solvent on the position of the absorption maximum must be interpreted in terms of the *net* effect of the change in the energies of the states *a* and *b* in Fig. 4. Now, in general the electron charge distribution of a π^*–π excited state is more extended than that of the ground state, and the excited state is therefore more *polarisible*. Change to a polar solvent increases the solvent interaction with both states, but the corresponding decrease in energy of the excited state is slightly greater than that of the ground state, and hence the absorption band shifts slightly towards the red. This is illustrated by the absorption spectra of benzophenone and its 4,4′-tetramethyldiamino derivative shown in

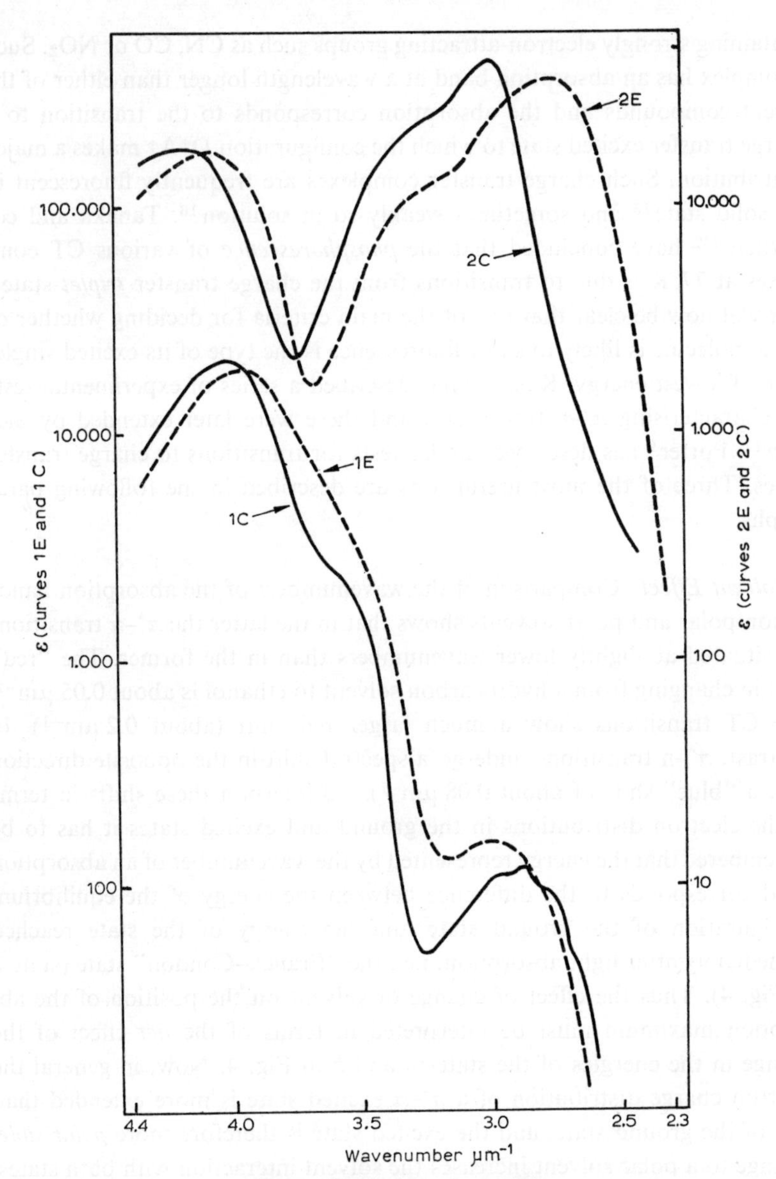

Fig. 11. Illustrating solvent shifts.
Benzophenone (1) and Michler's ketone (2) in cyclohexane (C) and ethanol (E).

Fig. 11. The second absorption bands of these compounds ($\sim 4.0\ \mu m^{-1}$) correspond to π^*–π transitions, and show a slight red shift on changing from cyclohexane to ethanol. The low wavenumber absorption band of Michler's ketone corresponds to a charge transfer transition. The CT state has a much greater permanent dipole moment than the ground state. Change to a polar solvent therefore results in a considerably larger reduction in the energy of the CT state than in that of the ground state, and the absorption band shows a large red shift (compare curves 2C and 2E in Fig. 11). The situation is reversed in the low wavenumber absorption band of unsubstituted benzophenone. This band is weak and corresponds to a π^*–n transition. In polar solvents the ground state is hydrogen bonded at the n-electron site and promotion of the non-bonding electron into the π-electron system reduces the hydrogen bonding forces in the excited state. Thus change to a polar solvent reduces the energy of the ground state to a greater degree than that of the excited state and the absorption band shifts to higher wavenumbers (compare curves 1C and 1E in Fig. 11).

Transition Probability. This is high for π^*–π and CT transitions ($\varepsilon_{max} \sim \sim 10^4$) and much lower for π^*–n transitions ($\varepsilon_{max} < 2000$). One cause of the low intensity in the latter is the relatively poor orbital overlap for π^*–n promotion.

Transition Energy. The π^*–n transition occurs at lower energy than the π^*–π transition observed in the hydrocarbon analogue of the compound concerned. For example formaldehyde and pyridine may be compared with ethylene and benzene respectively. In compounds having both donor and acceptor groups linked with the π-electron system, the CT absorption occurs at much longer wavelength than in the related molecules having no donor or acceptor substituents.

So far we have not concerned ourselves with the multiplicity of the various electronic states, i.e. we have not taken into account the resultant of the spins of the electrons in the molecule. This we consider in the following section.

C. PHOSPHORESCENCE AND DELAYED FLUORESCENCE

1. The Triplet State

Nearly all molecules having an even number of electrons, i.e. nearly all molecules other than free radicals, have a ground state in which their electrons occupy orbitals in pairs. Two electrons occupying the same orbital must, by the Pauli exclusion principle, have their spins opposed and hence the resultant spin quantum number for such molecules in the ground state is zero. However, when one of the electrons is promoted to an upper orbital, its spin may be oriented in the same or in the opposite direction to that of the electron remaining in the original orbital. If the spins are oriented in the same direction ("parallel") the resultant spin quantum number is $\frac{1}{2}+\frac{1}{2}=1$, and this vector can take up the three values $+1$, 0, or -1, when the system is situated in a magnetic field, whether this field be externally applied or produced by the orbital motions of the electrons in the system itself. The three values of the spin quantum number then correspond to three components of slightly different energy and the molecule is then said to be in a *triplet state*. The degree of splitting of the three components of the triplet in the absence of an external field depends on the coupling between the spin and orbital motions and this in turn depends on the masses of the atoms in the molecule. With the lighter atoms the coupling is small and the triplet behaves as though it were a single level, in the absence of external fields. In general the *multiplicity* of the state is equal to $2n+1$, where n is the number of unpaired electrons. Thus the ground states of most molecules with an even number of electrons have a multiplicity of 1, i.e. they are *singlet* states. There are however some molecules containing an even number of electrons for which the states of lowest energy have two unpaired electrons occupying different orbitals. The most important of these is molecular oxygen, of which therefore the ground state is a *triplet*.

The fact that the electron excited to an upper orbital may have its spin aligned in the same or in the opposite direction to that of the electron in the original orbital means that each of the kinds of transition so far discussed can in principle give rise to two kinds of excited state, the singlet and the triplet. We can therefore re-draw our diagram in Fig. 1 to include these triplet levels, one triplet level corresponding to each of the excited singlet levels shown in the original diagram. This has been done in Fig. 12. There is of course no triplet level corresponding to the singlet ground state because pairs of electrons in the same orbital must have their spins opposed. The orientations of the spin of the excited electron relative to that of the remaining electron are indicated in Fig. 12 in the rectangles adjacent to each level. Hund's rule for atoms appears to apply to molecules also and the triplet levels thus lie somewhat lower on the energy scale than the corresponding singlet levels. Although in atoms the triplet consists of three closely spaced energy levels, in molecules these three states cannot generally be distinguished and on our schematic energy diagram we therefore draw the triplet levels and their associated vibrational levels in exactly the same form as the singlet levels. For the sake of simplicity only one absorption transition to each of the upper singlet states is shown in Fig. 12 and both internal conversion and loss of vibrational energy are shown as a single wavy line.

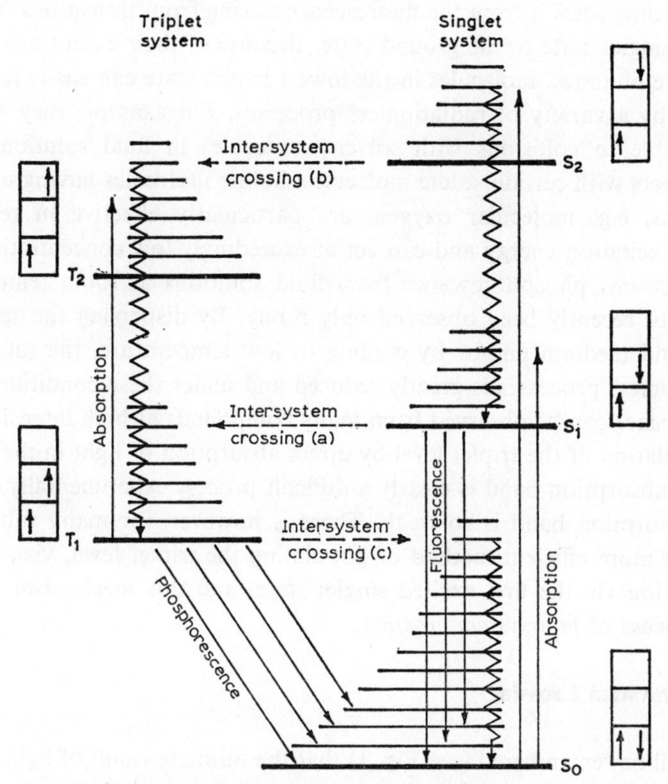

Fig. 12. Triplet and singlet energy levels and intersystem crossing.

Radiative transitions between states of different multiplicity e.g. between singlets and triplets, are theoretically forbidden. In practice, because of spin–orbit coupling, these transitions do take place, although with extremely low probability compared with singlet–singlet, or triplet–triplet transitions. Low transition probability means that the absorption band will be weak and the radiative lifetime for the reverse transition will be long. Thus while the radiative lifetime of the lowest excited singlet state for a strongly allowed transition, e.g. a π^*–π transition, is of the order 10^{-9}–10^{-8} sec, and even for the less probable singlet–singlet transitions, e.g. π^*–n transitions, rarely exceeds 10^{-6} sec, the radiative lifetime of the transition from the lowest triplet level to the ground state is rarely less than 10^{-4} sec, and for many molecules it is considerably greater than one sec. In this book we have reserved the term *phosphorescence* for the long-lived luminescence emitted in such radiative transitions direct from the triplet state to the ground state. This distinguishes it from the fluorescence arising from transitions from the lowest singlet state to the ground state. Because of their exceptionally long radiative lifetimes, molecules in the lowest triplet state can easily lose their energy by a variety of radiationless processes. For example they are very susceptible to collisions with solvent molecules in fluid solution, or to encounters with certain solute molecules. Solute molecules having unpaired electrons, e.g. molecular oxygen, are particularly effective in removing triplet excitation energy and can act at exceedingly low concentration. For these reasons phosphorescence from fluid solutions at room temperature had until recently been observed only rarely. By dispersing the substance in a rigid medium and/or by cooling to low temperature, the rate of the radiationless processes is greatly reduced and under these conditions phosphorescence can be observed from many compounds at high intensity.

Population of the triplet level by direct absorption of light in the triplet–singlet absorption band is clearly a difficult process experimentally because this absorption band is so weak. There is however, for many substances, a much more efficient method of populating the triplet level, viz., by light absorption via the first excited singlet state, and this mechanism involves the process of *intersystem crossing*.

2. Intersystem Crossing

It will be remembered (see Fig. 1) that the ultimate result of light absorption carrying the molecule to any of the upper excited singlet states is, in the absence of any intervening photochemical reactions, the accumulation of molecules near the lowest vibrational level of the first excited singlet state. If the latter is a π^* state the rate constant, k_f, for fluorescence emission has a magnitude of about 10^8 to 10^9 sec^{-1}, and the processes competing with fluorescence emission often have rate constants with similar values. One such process is intersystem crossing (see Fig. 12). The lowest vibrational level of the lowest triplet state is generally situated some way below that of the lowest excited singlet state but its upper vibrational levels reach to the bottom of the singlet level and intersystem crossing occurs by the molecule crossing over to one of these upper vibrational levels of the lowest triplet state (intersystem crossing (*a*) in Fig. 12). From here the molecule rapidly loses its excess vibrational energy and falls to near the lowest vibrational level of the triplet state. The process can also be visualised by means of a potential energy diagram similar to that drawn previously (Fig. 3) in connection with the description of the absorption and fluorescence processes. This diagram is reproduced in Fig. 13 without showing all the

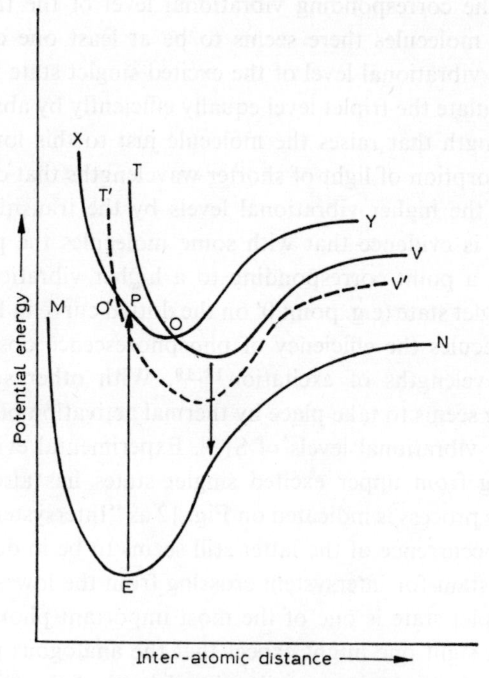

Fig. 13. Intersystem crossing.

vibrational levels, but in addition there is included the potential energy curve (TV) for the lowest triplet state. The lowest vibrational level of this curve is situated below that of the excited singlet state and generally the curves for the two excited states will cross at some point 0. Once again it should be noted that the diagram, as drawn, refers only to a di-atomic molecule and for a polyatomic molecule the potential energy would have to be represented by a multi-dimensional surface. The principle however is the same except that the crossing points between the excited singlet and triplet potential surfaces would generally be quite numerous instead of just the single point as indicated in Fig. 13. In Fig. 13 the population of the triplet state is then represented by the following series of processes. Light absorption raises the molecule from the ground state to an upper vibrational level of the excited singlet state (vertical line EP). The molecule in the excited singlet state rapidly loses its excess of vibrational energy, passing down the curve XY until it reaches the point 0. At this point the positions and momenta of the atomic nuclei are identical with those corresponding to a molecule in the lowest triplet state and the molecule can at this point cross over to the corresponding vibrational level of the triplet state. For many complex molecules there seems to be at least one cross-over point near the lowest vibrational level of the excited singlet state since it is found possible to populate the triplet level equally efficiently by absorption of light of long wavelength that raises the molecule just to this lowest vibrational level, as by absorption of light of shorter wavelengths that carries the molecule to one of the higher vibrational levels by the transition such as EP. However there is evidence that with some molecules the potential energy curves cross at a point corresponding to a higher vibrational level of the first excited singlet state (e.g. point 0′ on the dotted curve in Fig. 13), because with such molecules the efficiency of phosphorescence observed is greater for shorter wavelengths of excitation[18,19]. With other substances intersystem crossing seems to take place by thermal activation of molecules from lower to upper vibrational levels of S_1[20]. Experimental evidence for intersystem crossing from upper excited singlet states has also been put forward[21,22]. This process is indicated on Fig. 12 as "Intersystem Crossing (*b*)". However, the occurrence of the latter still seems to be in doubt[20].

The rate constant for intersystem crossing from the lowest excited singlet state to the triplet state is one of the most important photochemical parameters. At first sight one might expect that the analogous process of intersystem crossing from the lowest vibrational level of the triplet state to an upper vibrational level of the ground state (intersystem crossing (c) in Fig. 12) would take place with equal facility. In fact it is usually some 10^6–10^9 times slower than intersystem crossing from the excited singlet to the triplet. This enormous difference in rate for what appears to be a similar process is at first sight very puzzling. It is however a fortunate circumstance from the practical point of view because if it were not so, the production of long-lived photoluminescence would be much more difficult than it is. If radiationless $T_1 \rightarrow S_0$ transitions occurred with rate constants of 10^8 sec^{-1}, the molecule would have no time to emit $T_1 \rightarrow S_0$ phosphorescence by the spin-forbidden radiative process. On the other hand, if the $S_1 \rightarrow T_1$ intersystem crossing process were as slow as that of the $T_1 \rightarrow S_0$ radiationless process, then it could not compete effectively with fluorescence, and population of the triplet state by light absorption would be negligible except at very high light intensities. Phosphorescence would then be a rare phenomenon and photochemical behaviour would be very different.

The difference between the rates of the $S_1 \rightarrow T_1$ and the $T_1 \rightarrow S_0$ radiationless processes has been interpreted qualitatively in terms of the probability of intersection of the potential energy surfaces corresponding to the three states (see Fig. 13). Thus the potential surface of T_1 is likely to be closer to that of S_1 than to that of S_0. Hence it is expected to intersect that of S_1 in many places and the probability of $S_1 \rightarrow T_1$ intersystem crossing is expected to be high. The energy separation between the triplet level and the ground state is generally larger so that intersection of potential surfaces, and hence intersystem crossing, is expected to be less probable. Symmetry factors may also play a part. Even so, the fact that the two intersystem crossing rates differ by a factor so large as 10^6–10^9 is surprising, particularly with molecules such as anthracene for which the difference in energy between the T_1 and S_0 states is not much greater than that between the S_1 and T_1 states. With molecules such as anthracene it is possible that intersystem crossing from S_1 takes place to an *upper* triplet state, e.g. T_2. Such intersystem crossing has been postulated by Lim and co-workers [20]. Following the earlier work of Bowen and co-workers [23, 24], they measured the temperature dependence of the fluorescence efficiency and relative triplet formation efficiency of some 9- and 9,10-substituted anthracenes. They found the fluorescence efficiency to increase and the triplet formation to decrease as the temperature was lowered. The activation energy was the same for both processes and at 77°K triplet formation was very small. They interpreted the results in terms of intersystem crossing from S_1 to an upper triplet level, the activation energy being attributed to the higher energy of the triplet state relative to that of the excited singlet state, or to changes in the relative positions of the potential surfaces of the two states.

A third type of intersystem crossing which is of great importance in connection with delayed fluorescence, is the reverse of the $S_1 \rightarrow T_1$ process, i.e. the reverse of intersystem crossing (*a*) in Fig. 12. It can occur in molecules for which the triplet level is situated close to the upper singlet so that at room temperature an appreciable proportion of the triplet molecules gain sufficient vibrational quanta by thermal activation to raise them to a vibrational level of the triplet state equal in energy to the lowest vibrational levels of the excited singlet state. They can then cross to the excited singlet level by this reverse intersystem crossing process. For triplet molecules which have been thermally activated in this way, the probability of the intersystem crossing $S_1 \leftarrow T_1$ seems to be about the same as for the process (*a*), i.e. the rate constant is of the order of magnitude 10^8 sec^{-1}.

3. Definition of Phosphorescence and Delayed Fluorescence

Each of the radiative transitions $S_1 \rightarrow S_0$ and $T_1 \rightarrow S_0$ gives rise to a characteristic emission spectrum that serves to identify the excited state concerned and gives a basis for classifying the luminescence without reference to the processes by which the excited state was produced. We shall employ the general term *fluorescence* for all luminescence arising from the $S_1 \rightarrow S_0$ transition and reserve the name *phosphorescence* for that arising from the $T_1 \rightarrow S_0$ transition. There are several ways in which the S_1 state can be populated and these give rise to different kinds of fluorescence. The simplest kind is that already discussed in Section B. It is produced by those molecules that are excited from the ground state by light absorption, remain in the S_1 state for an average period equal to the lifetime of this state (typically 10^{-8} sec) and then return to the ground state by the radiative transition $S_1 \rightarrow S_0$. This emission we shall call *prompt fluorescence*. There are a variety of processes by which a molecule can pass to the excited singlet state from the *triplet* state. These processes also result in an emission characteristic of the $S_1 \rightarrow S_0$ transition but with a lifetime much longer than that characteristic of the S_1 state. We shall call such emissions *delayed fluorescence* and further sub-classify them into "E-type" and "P-type" delayed fluorescence, according to the mechanism by which the process $S_1 \leftarrow T_1$ occurs. Finally there are other forms of long-lived emission for which the long lifetime does not depend directly on the long lifetime of the triplet state. These emissions are observed when irradiation in rigid medium leads directly or indirectly to the decomposition of the solute into fragments that recombine slowly with the formation of an excited singlet or triplet molecule. We classify these emissions as *recombination delayed fluorescence* and *recombination phosphorescence*.

The relationships between the spectra of absorption, prompt fluorescence, delayed fluorescence and phosphorescence are summarised in Fig. 14. The spectra of prompt and delayed fluorescence usually appear immediately to the long-wavelength side of the first absorption band and are frequently an approximate mirror image of it. Phosphorescence (and recombination phosphorescence) appears at longer wavelengths because the triplet state has a lower energy than the first excited singlet state (see Fig. 12).

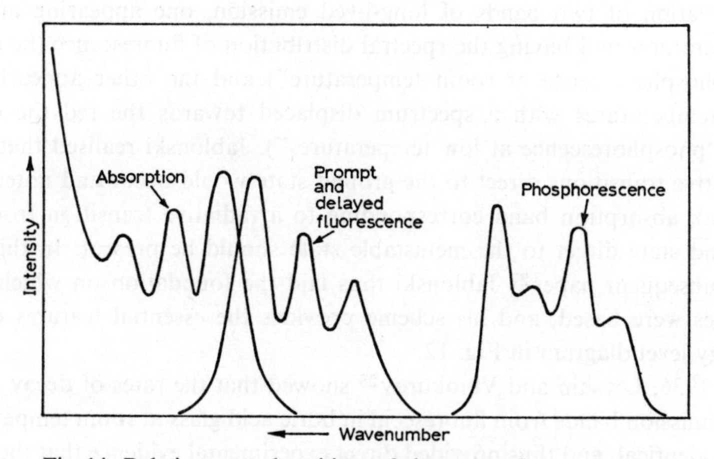

Fig. 14. Relative spectral positions of absorption, fluorescence and phosphorescence spectra.
The intensity of delayed fluorescence rarely exceeds a few percent of that of prompt fluorescence.

The various kinds of delayed fluorescence have been the subject of considerable controversy during recent years, and some aspects of the subject are still not fully understood. To get the field into proper perspective we shall present in the following sections an historical account of both delayed fluorescence and phosphorescence. The kinetics of these processes will be considered in more detail in Chapter 2.

4. E-Type Delayed Fluorescence

It had been known for many years that some organic compounds in rigid solution could give rise to a long-lived emission, or "after-glow" when exposed to ultra-violet light. In 1929 Perrin[25] suggested that the excited molecules could pass over to a metastable state of lower energy from which spontaneous emission was impossible, but that by thermal activation the molecules could return to the fluorescent state and ultimately emit fluorescence. The emission of this fluorescence would thus be "delayed" by the previous passage into the metastable state.

In 1933 Jablonski[26] proposed a scheme that was essentially an extension of that of Perrin, but included the possibility of a radiative transition of low probability direct from the metastable state to the ground state (i.e. phosphorescence as defined in Section 3 above). He thus explained the observation of two bands of long-lived emission, one appearing at high temperatures and having the spectral distribution of fluorescence (he called it "phosphorescence at room temperature") and the other appearing at low temperatures with a spectrum displaced towards the red (he called this "phosphorescence at low temperatures"). Jablonski realised that non-radiative transitions direct to the ground state would occur and noted that a weak absorption band corresponding to a radiative transition from the ground state direct to the metastable state should be present. In this and his subsequent paper[27] Jablonski thus laid the foundation on which later studies were based, and his scheme provides the essential features of the energy level diagram in Fig. 12.

In 1936, Levshin and Vinokurov[28] showed that the rates of decay of the two emission bands from fluorescein in boric acid glass at room temperature were identical, and thus provided direct experimental evidence that the same metastable state was returning to the ground state by two independent processes, each giving rise to a different emission. The fluorescein–boric acid system was investigated in more detail by Lewis, Lipkin and Magel[29]. They called the two emission bands "alpha phosphorescence" (corresponding to our E-type delayed fluorescence) and "beta phosphorescence" (corresponding to our phosphorescence). They also measured the absorption spectrum of the metastable state. They found that the decay of both emission bands was exponential and by making certain assumptions they calculated values for the decay constant of the alpha process as a function of temperature, from which they derived an activation energy for the alpha process.

This activation energy was found to be equal to the spectroscopic energy difference calculated from the frequency difference between the two emission bands and Lewis and co-workers therefore concluded that the alpha process did indeed proceed by thermal activation of the metastable state. (As we shall see later, the rate of thermal activation can be determined *precisely* without making any assumptions, by measuring the ratio of the *intensities* of the emission bands as a function of temperature.) In a later publication Lewis and Kasha[30] identified the metastable state with the triplet state, and this has been amply confirmed by subsequent work.

In fluid solution the rates of $T_1 \rightarrow S_0$ intersystem crossing and/or impurity quenching of the triplet state are much greater than the rate of the $T_1 \rightarrow S_0$ radiative process so that the phosphorescence efficiency is small, and until recently phosphorescence from fluid solutions had been observed only rarely. The first quantitative measurements were reported in 1930 by Mlle. Boudin[31]. She used a visual phosphoroscope and observed a long-lived photoluminescence from solutions of eosin in glycerol at room temperature, having a lifetime of 1.1 msec and an intensity about 1/400th of that of the prompt fluorescence. A few years later Kautsky[32] observed long-lived emission from de-oxygenated solutions of a variety of dyestuffs. The subject then received little attention until the advent of the technique of flash photolysis which was used by Bäckström and Sandros[33] to investigate the quenching of the triplet state of biacetyl, a substance long known for its phosphorescence.

The question of long-lived luminescence in fluid solution was taken up again by Parker and Hatchard[19] using a high sensitivity spectrophosphorimeter. They measured the spectrum of the long-lived emission from de-oxygenated solutions of eosin in glycerol and ethanol and found that it contained two bands—a visible band having a spectral distribution identical with that of the prompt fluorescence (this no doubt corresponds to the long-lived luminescence observed by Boudin[31]) and a second band in the extreme red. The quantitative results of this work are discussed in detail in Chapter 2. It suffices to say here that the results left no doubt that the visible delayed fluorescence was produced by thermal activation of the triplet state, i.e. by processes

$$T_1 \xrightarrow[\text{activation}]{\text{thermal}} S_1 \xrightarrow[\text{transition}]{\text{radiative}} S_0 \quad (33)$$

and that the extreme red band was the result of the direct $T_1 \rightarrow S_0$ radiative transition (i.e. it was phosphorescence by our definition). The two bands were thus completely analogous to those discussed by Jablonski[26,27] and to the "alpha and beta phosphorescence" observed by Lewis, Lipkin and Magel[29] with fluorescein in rigid medium. Because this type of delayed fluorescence was first clearly demonstrated by Boudin with eosin, Parker and Hatchard[34] proposed that it should be called *eosin-type* or *E-type* delayed fluorescence. Similar emissions were subsequently observed from fluid solutions of other dyestuffs[35,36]; they are to be expected with fluid solutions of all fluorescent compounds that have an appreciable triplet formation efficiency, and a triplet level lying close to the first excited singlet level.

5. P-Type Delayed Fluorescence

Interpretation of the early observations of P-type delayed fluorescence was confused by two factors. Firstly, the materials used in some of the experiments were impure, and secondly the long lifetime of the emissions was incorrectly attributed to the formation of long-lived excited dimers.

In 1950 Dikun[37] reported that the photoluminescence of phenanthrene vapour contained two components, one having a short lifetime characteristic of prompt fluorescence and the other having a much longer lifetime of about 10^{-3} sec. In 1958 Williams[38] reported the observation of long-lived emissions from the vapours of anthracene, perylene, pyrene and phenanthrene. With phenanthrene he found that the ratio of the intensity of the long-lived emission to that of the total emission increased as the vapour pressure of phenanthrene increased, and he therefore suggested that the long-lived emissions from all four compounds were due to long-lived excited dimers formed by reaction of the respective excited singlet monomers with a second molecule in the ground state:

$$^1A^* + {}^1A \rightarrow {}^1A_2^* \tag{34}$$

He supposed that the excited dimer dissociated by thermal activation to regenerate the singlet excited monomer, and that the emission from the latter thus gave rise to delayed fluorescence. He found that the spectra of the long-lived emissions from anthracene and perylene were indeed identical with those of the prompt fluorescence but he did not report on the spectrum of the long-lived emission from his phenanthrene. Later, Stevens and co-workers[39] repeated the work of Dikun and showed that the long-lived emission from phenanthrene vapour had a spectrum characteristic of anthracene fluorescence. They suggested that the anthracene was present as a trace impurity in the phenanthrene and that its delayed emission was sensitised by collision with the "long-lived excited dimer" of phenanthrene already proposed by Williams.

Several years earlier, Förster and Kasper[40] had published their celebrated paper on the prompt fluorescence of concentrated pyrene solutions, in which they had demonstrated that singlet excited pyrene does indeed react with a second molecule of pyrene in the ground state to form an excited dimer, and that the radiative transition of the latter to the ground state is responsible for the broad structureless band that appears in the emission spectrum of concentrated pyrene solutions, to the long-wavelength side of the monomer emission (this and related work is discussed as a special topic in Chapter 4). They did not determine the lifetime of the excited dimer, but in 1960, Stevens and Hutton[41] observed a long-lived emission from concentrated pyrene solutions and reported that it had a spectral distribution identical with that of the excited dimer emission of Förster and Kasper. They did not observe any long-lived emission from the pyrene monomer, although Parker and Hatchard[34] later showed that this was present also. Stevens and Hutton assumed that the pyrene dimer itself was long-lived, i.e. they adopted the mechanism of Williams. They proposed the term "excimer" for such "long-lived" excited dimers and subsequently explained the sensitised delayed fluorescence from the mixed vapours of other compounds by "excited dimer sensitisation"[39,42].

At this stage the experimental evidence for long-lived excited dimers ("excimers") seemed to be strong. However, in 1961 Parker and Hatchard[43] brought forward evidence that the greater part of the dimer emission from pyrene solutions had a lifetime less than 5×10^{-5} sec. Later, Birks and Munro[44] obtained a value of 4×10^{-8} sec by direct measurement with a fluorometer. It was then difficult to see how the long-lived component observed by Stevens and Hutton could be explained by the Williams mechanism. The problem was resolved by the work of Parker and Hatchard with solutions of anthracene and phenanthrene[45,46] and with pyrene[34]. They observed that the intensity of delayed fluorescence was proportional to the *square* of the rate of absorption of exciting light and showed that the delayed fluorescence was produced by the transfer of energy between two triplet molecules, whereby one of the molecules was ultimately raised to the excited singlet state from which it then emitted delayed fluorescence that decayed at a rate equal to twice the rate of decay of the triplets (see

Chapter 2). In brief, the Parker–Hatchard mechanism may be represented by:

$$^3A + ^3A \rightarrow (^1A_2^*) \rightarrow ^1A^* + ^1A \qquad (35)$$

in which the singlet excited dimer may or may not be formed as an intermediate. With pyrene solutions the intermediate formation of $^1A_2^*$ can be inferred from the presence of its characteristic band in the spectrum of long-lived luminescence from *dilute* solutions, in addition to the band characteristic of $^1A^*$. Parker and Hatchard suggested that the triplet–triplet interaction mechanism might explain the results of Williams[38] and Stevens and co-workers[39, 42] with aromatic hydrocarbons in the vapour phase, and this was subsequently found to be the case by Stevens, Walker and Hutton[47]. Thus the postulation of *long-lived* excited dimers ("excimers") was not necessary to explain any of the observed phenomena, whether in the gas phase or in solution. Nevertheless, the term "excimer" has been adopted by many workers for *short-lived* excited dimers of the Förster–Kasper type.

Because the mechanism of this kind of delayed fluorescence was first demonstrated for pyrene solutions, Parker and Hatchard[34] proposed the name "pyrene-type" or *P-type delayed fluorescence* to distinguish it from the E-type delayed fluorescence discussed in the previous section. The basic kinetics of P-type delayed fluorescence are dealt with in Chapter 2 and the formation of excited dimers is treated as a special topic in Chapter 4.

6. Recombination Photoluminescence

Lewis and Lipkin[48] showed that three kinds of primary photochemical reaction can occur in rigid media, viz.:

(a) photo-dissociation, e.g.:

$$Ph_3C\text{–}CPh_3 \xrightarrow{h\nu} 2Ph_3C^\bullet \qquad (36)$$

(b) photo-oxidation, e.g.:

$$Ph_2NH \xrightarrow{h\nu} Ph_2\overset{+}{N}H + e \qquad (37)$$

(c) photo-ionisation, e.g.:

$$Ph_2\overset{+}{N}H \xrightarrow{h\nu} Ph_2N^\bullet + H^+ \qquad (38)$$

Lewis and Kasha[30] suggested that these reactions could occur by two quite distinct mechanisms, viz. direct reaction after photon absorption by the ground state:

$$^1(AB) \xrightarrow{h\nu} A + B \tag{39}$$

or reaction as a result of photon absorption by the triplet state:

$$^1(AB) \xrightarrow{h\nu} {}^1(AB)^* \rightarrow {}^3(AB) \xrightarrow{h\nu'} A + B \tag{40}$$

In rigid media at low temperature the concentration of triplet molecules can build up to an appreciable value and photon absorption by the triplet is a quite likely process. The overall reaction 40 requires the successive absorption of two separate photons and the rate of such *bi-photonic* reactions is proportional to the square of the intensity of the exciting light at low light intensities.

One of the first investigations of the light emission resulting from recombination was that of Debye and Edwards[49]. They observed an emission of extremely long lifetime (> 100 seconds) from irradiated solutions of easily oxidisable compounds such as phenol and toluidine in rigid glasses at 77°K. They interpreted the non-exponential decay in terms of photoionisation (Lewis and Lipkin's "photo-oxidation" – equation 37 above) followed by diffusion of the trapped electrons back to the ionised molecules and their recombination to form an unspecified excited state:

$$A \xrightarrow{h\nu} A^+ + e \tag{41}$$

$$A^+ + e \rightarrow A \text{ (excited)} \tag{42}$$

$$A \text{ (excited)} \rightarrow A + h\nu' \tag{43}$$

Linschitz, Berry and Schweitzer[50] found that the absorption spectra of solutions of lithium in mixed amine glasses at low temperature showed an intense peak at 600 nm with a weaker background absorption extending into the infra-red. Illumination with light caused the 600 nm band to decrease and the long wavelength background to increase in intensity. They attributed the 600 nm band to strongly solvated electrons and the longer wavelength absorption to weakly solvated electrons. They then irradiated solutions of easily oxidisable organic compounds and were able to identify in the spectra of the irradiated glasses the bands due to solvated electrons as well as those due to the radicals or radical ions. Recombination at liquid nitrogen temperature was very slow, but as the irradiated solution was allowed to warm up, luminescence was emitted and the absorption bands of both radical and solvated electron decreased in intensity. This provided strong evidence that the luminescence was indeed due to the recombination of ion and electron (viz. equations 42 and 43). The spectrum of the luminescence was the same as that of the phosphorescence (i.e. the emission was recombination phosphorescence) and no luminescence corresponding to the $S_1 \rightarrow S_0$ transition was observed. This was not due to a high rate of intersystem crossing because photo-excitation produced intense prompt fluorescence. In this respect the results of Linschitz and co-workers are similar to those of Albrecht and co-workers [51–53] who found that when they exposed a photo-ionised rigid solution of tetramethylparaphenylenediamine to infrared radiation the solution emitted both recombination phosphorescence and recombination delayed fluorescence. The ratio of the intensities of the two emission bands was much greater than that observed for ordinary phosphorescence and prompt fluorescence of the same solution, indicating that the recombination process favoured the direct population of the triplet state, viz.:

$$A^+ + e \rightarrow {}^3A \tag{44}$$

The alternative process, i.e. population of the excited singlet state:

$$A^+ + e \rightarrow {}^1A^* \tag{45}$$

has been postulated by Lim and co-workers to explain the emission of delayed fluorescence lasting for several seconds when solutions of acriflavine and related dyestuffs are irradiated in ether–pentane–ethanol glasses at 77°K [54–56]. They also observed a transient absorption which they attributed to the positive radical ion formed by photo-ionisation. Except in the early stages, the delayed fluorescence decayed exponentially at a rate equal to that of the radical ion. Because the integrated intensity of the delayed fluorescence, and the initial concentration of radical ion were directly proportional to the intensity of the exciting light, they concluded that the excitation was a one-photon process. The efficiency of delayed fluorescence was greater with short wavelength excitation. To explain the results they employed a model similar to that suggested earlier by Albrecht and co-workers for the luminescence of tetramethylparaphenylenediamine, although Cadogan and Albrecht [57] have since re-interpreted the latter in terms of a bi-photonic process.

A delayed fluorescence dependent on the wavelength of excitation has also been observed by Stevens and Walker [58] with solutions of perylene in liquid paraffin at 77°K. They found that the excitation spectrum of the delayed fluorescence coincided approximately with the triplet–triplet absorption of perylene, and interpreted the results in terms of a bi-photonic mechanism involving photo-ionisation of the triplet state and recombination to give both excited singlet and triplet states:

$$^1P \xrightarrow{h\nu} {}^1P^* \rightarrow {}^3P \quad (46)$$

$$^3P \xrightarrow{h\nu} P^+ + e \quad (47)$$

$$P^+ + e \rightarrow \begin{cases} {}^1P^* \\ {}^3P \end{cases} \quad (48)$$

However, the validity of these observations has been questioned[59] because of the possibility of artifacts or trivial effects arising from phosphorescence of the containers, or impurities in the solution (see Section 3 N 8).

Porter and co-workers[60, 61] have made a study of bi-photonic photochemical processes resulting from the absorption of light by triplet states in rigid media at 77°K. They observed two kinds of process with solutions of aromatic compounds in aliphatic hydrocarbon glasses, viz.: ionisation of the solute, and sensitised dissociation of the solvent to give free radicals and hydrogen atoms. The latter then abstract hydrogen atoms from either the solvent or the solute, and so produce solute radicals. Exposure of many of the solutions to infra-red light, or gentle warming, resulted in the emission of both fluorescence and phosphorescence, caused by the recombination of ions and electrons. The phosphorescence/fluorescence ratio was higher than that observed in normal optical excitation, and in this respect the results are similar to those of Albrecht and co-workers which have since been reinterpreted in terms of a bi-photonic process[57].

D. POLARISATION OF PHOTOLUMINESCENCE

1. Polarised Light

The oscillations of the electric and magnetic fields forming a light wave take place in directions at right angles to one another and to the direction of travel. In considering the phenomenon of polarisation we shall fix our attention on the direction of the electric vector. "Natural" or unpolarised

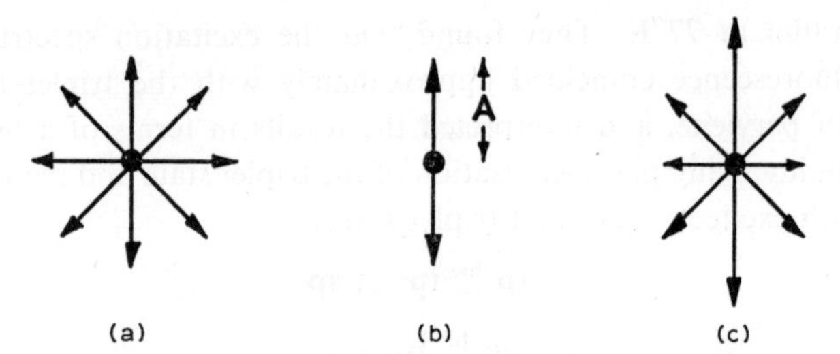

Fig. 15. Directions of electric vector in polarised and unpolarised light.
In unpolarised light (a), or partly polarised light (c), the vibrations take place at all angles perpendicular to the direction of travel: in polarised light (b) they are restricted to one angle.

light contains vibrations in all directions perpendicular to the direction of propagation. If a beam of unpolarised light is passed through a suitable crystal, such as tourmaline, or a polariser such as a nicol prism (see Section 3 P 1), the emergent beam has the oscillations of its electric vector confined to one plane and is said to be *plane polarised* or linearly polarised. The difference between natural and plane polarised light is shown in Fig. 15 (a) and (b) which represent sections across the direction of travel.

The process of absorption of polarised light by a molecule is shown diagrammatically in Fig. 16. The electric vector sets the electrons of the molecule into oscillation and if the frequency of the light has the appropriate value (equation 7) an electron is excited to a state of higher energy, the orbital of which has a nodal plane at right angles to the direction of the induced oscillations. If the molecule is oriented at right angles to the position shown in Fig. 16, i.e., with its transition moment at right angles to the electric vector, the probability of absorption is zero. At intermediate angles the probability of absorption will be finite but less than when the electric

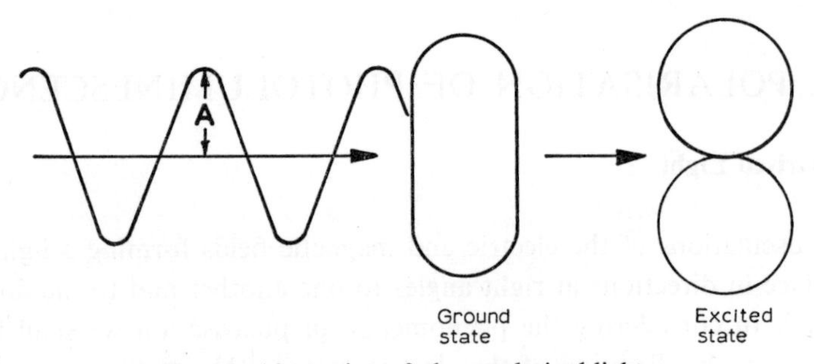

Fig. 16. Absorption of plane polarised light.

vector is parallel to the transition axis of the molecule. The dependence on orientation can be simply derived. The amplitude of the electric vector (A in Figs. 15 and 16) may be resolved into two vectors at right angles to one another. Thus the effective amplitude of the vertically polarised exciting light in a direction inclined at an angle θ, is $A \cos \theta$, and since the intensity of the beam is proportional to A^2, the *intensity* at an angle θ is $A^2 \cos^2 \theta$, and the probability of absorption in this direction is $\cos^2 \theta$. Thus if we pass a beam of unpolarised light through a sufficiently thick layer of similarly oriented molecules (as in a crystal of tourmaline) the components of all the vectors parallel to the transition axis will be absorbed, but the components at right angles to this direction will be transmitted and will combine to give a beam of plane polarised light.

2. Degree of Polarisation

If the slice of tourmaline crystal referred to in the previous paragraph is thin enough, the electric vector components parallel to the transition axis will not be completely absorbed and the emerging light will be only partly polarised (see Fig. 15 (c)). Partly polarised light may be produced in a variety of ways, e.g., by reflection from a glass surface, and light emerging from optical instruments, such as monochromators, is usually partly polarised. *The degree of polarisation*, or simply the *polarisation* of the beam, p, is defined as:

$$p = \frac{I_{\parallel} - I_{\perp}}{I_{\parallel} + I_{\perp}} \tag{49}$$

where $I_{\parallel}$ and $I_{\perp}$ are the intensities of the components of the beam resolved in directions parallel and perpendicular to the direction of partial polarisation. They correspond to the maximum and minimum intensities that are observed when the beam is passed through a 100% efficient polarising element (such as a nicol prism—see Section 3 P 1) and the element is rotated.

3. Polarisation of Luminescence from Oriented Molecules

If the excited state produced by light absorption is the lowest excited state of the molecule, it can return to the ground state with emission of its excess energy as fluorescence. The process is then precisely the reverse of the absorption process described in Section 1 D 1 and Fig. 16, and the plane of polarisation of the emitted light then contains the transition axis of the emitting molecule. The *direction* of the emitted ray need not be at right angles to the transition axis, but the latter is the most probable direction of emission.

The situation can be better appreciated by considering a hypothetical specimen in which all the fluorescent molecules are aligned in a rigid medium. (An approach to this is a stretched polymeric fibre containing elongated molecules that partly align themselves along the fibre.) In Fig. 17 the specimen is situated at the centre 0 of the rectangular co-ordinates XYZ, with OX and OZ in the plane of the paper and OY at right angles to this plane. The molecules are aligned with their axes parallel to OZ and are illuminated by light polarised in the plane XOZ. The fluorescence is emitted with maximum intensity in all directions in the plane XOY. For directions inclined to this plane at an angle α, the intensity is reduced by the factor $\cos^2 \alpha$. Hence the fluorescence along OZ is zero, and in other directions its intensity is represented by the length OP, where P traces out the locus represented by the dotted curve. The figure is symmetrical about the axis OZ, so that the three dimensional figure giving all possible directions of OP is obtained by rotating

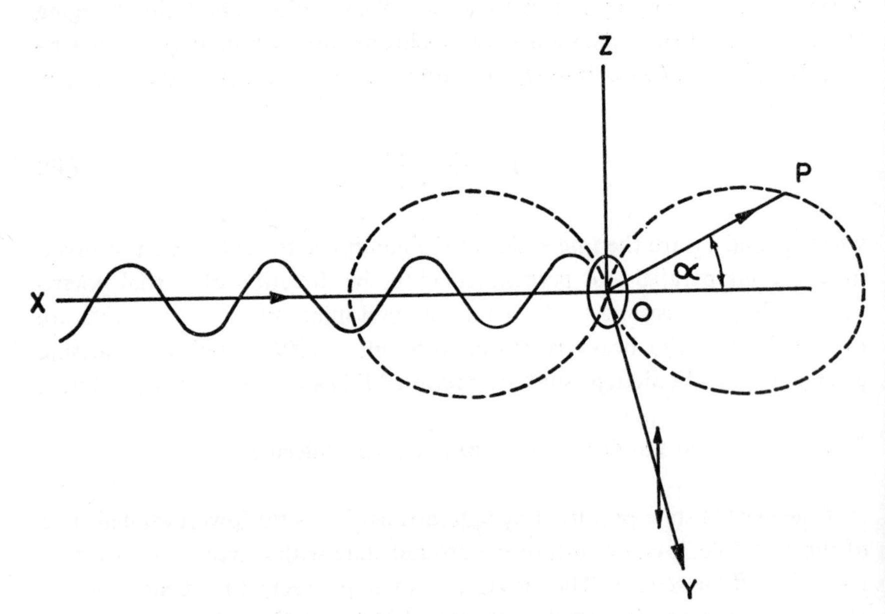

Fig. 17. Emission of polarised light from oriented molecules.

the dotted curve about the OZ axis. Whatever the direction of emission OP, the light is plane polarised in the plane ZOP.

In subsequent discussion of polarised luminescence we shall consider only light emitted in a direction at right angles to the beam of exciting radiation. It will be clear from Fig. 17 that if this direction is also at right angles to the transition axis of the oriented molecules (i.e. direction OY) the light will be completely vertically polarised ($p = 1$) and of maximum intensity. If the direction of observation is along the axis of the oriented molecules (i.e. along OZ) the intensity of fluorescence will be zero. It should also be noted that the emission along OY will still have a polarisation of unity even if the specimen is illuminated with unpolarised exciting light, since the horizontal component of the latter (i.e. the component in the direction XOY) will not be absorbed.

If the molecular axes are not parallel to OZ, but are inclined to it at an angle γ, the absorption of exciting light will be reduced by the factor $\cos^2 \gamma$. The fluorescence emitted will still be plane polarised, but in general the plane of polarisation will not be vertical (i.e. it will not be the plane ZOY). For example, if the molecular axes are parallel to OP in Fig. 17 (i.e. with $\gamma = (\pi/2 - \alpha)$) the plane of polarisation will be inclined at an angle γ to the vertical.

4. Polarisation of Luminescence from Randomly Oriented Molecules

Measurements are generally carried out on solutions (fluid or rigid) in which the fluorescent solute is uniformly distributed with the molecules oriented at random. We shall now consider the effect of this random orientation on the degree of polarisation of the fluorescence emitted in the direction at right angles to that of the beam of exciting light. For simplicity we shall consider first a solute for which the transition moment for absorption coincides with that for emission. (To a first approximation this corresponds to the case of the excitation $S_1 \leftarrow S_0$ and emission $S_1 \rightarrow S_0$ (but see Section 1 D 5) under conditions where depolarising factors are absent (i.e. in a rigid dilute solution—see Sections 1 D 5 and 1 D 6 below)). The polarisation in the absence of depolarising factors is known as the *principal polarisation* and is designated p_0.

In Fig. 18 consider molecules for which the transition moment is directed along OS and makes an angle θ with the axis OZ and an azimuthal an gle ϕ

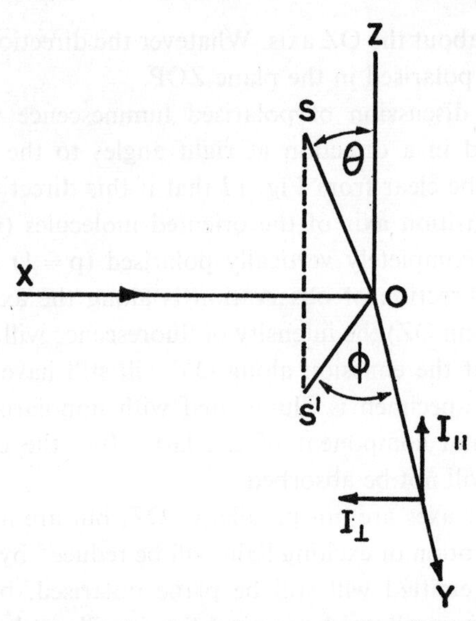

Fig. 18. Absorption and emission from randomly oriented molecules. Point S′ is the projection of point S on the plane XOY.

with the direction of observation OY (in Fig. 18 S′ is the projection of S on the plane XOY). Now the probability that such a molecule will be excited by light polarised with its electric vector directed along OZ will vary with the angle θ and is equal to $\cos^2 \theta$. Furthermore, for a *given number* of such molecules excited, the intensities of the vertical and horizontal components of the light emitted along OY will be given by:

$$I_{\parallel} \propto \cos^2 \theta \tag{50}$$

$$I_{\perp} \propto \sin^2 \theta . \sin^2 \phi \tag{51}$$

If the effects of variable probability of excitation (i.e., variable θ) together with equations 50 and 51 are integrated over all possible values of θ and ϕ, it is found that:

$$p_0 = \frac{I_{\parallel} - I_{\perp}}{I_{\parallel} + I_{\perp}} = \frac{1}{2} \tag{52}$$

This is for vertically polarised exciting light. If the exciting light is unpolarised, p_0 has the value $\frac{1}{3}$.

5. Polarisation Spectra

The value of p_0 given by equation 52 is the maximum that can be observed from a solution of randomly oriented molecules. It applies only when the transition moment for absorption coincides with that for emission. In general this will not be the case, e.g. when the molecules are raised to an upper excited state by light absorption and fall to the lowest excited singlet state before emitting fluorescence, or to the lowest triplet state before emitting phosphorescence. If the angle between the absorption and emission oscillators is β, it was shown by Perrin[62] and Jablonski[63] that the principal polarisation is given by:

$$p_0 = (3\cos^2\beta - 1)/(\cos^2\beta + 3) \qquad (53)$$

for vertically polarised exciting light. For unpolarised exciting light the relation is:

$$p_0 = (3\cos^2\beta - 1)/(7 - \cos^2\beta) \qquad (54)$$

For the condition $\beta = 0$, these equations give the maximum values $\frac{1}{2}$ and $\frac{1}{3}$, referred to in the previous section. For $\beta = \pi/2$, i.e. when the absorption and emission oscillators are mutually at right angles, the principal polarisations are $-\frac{1}{3}$ and $-\frac{1}{7}$ respectively for the two kinds of exciting light. In general, as β varies, p_0 will vary between the limits:

$+\frac{1}{2}$ and $-\frac{1}{3}$ for vertically polarised exciting light;

$+\frac{1}{3}$ and $-\frac{1}{7}$ for unpolarised exciting light.

The fact that the polarisation can be negative, i.e. that the fluorescence can be partially polarised in a horizontal direction can be visualised by reference to Fig. 18. Molecules for which θ is small are preferentially excited, and if β is large, the orientation of the emission oscillators of these molecules will be inclined to OZ at an angle close to $\pi/2$, so that the vertically polarised component of emission will be small, while the average value of the horizontal component will be large.

Now consider the hypothetical substance having the absorption spectrum shown by curve 1 in Fig. 19, in which the first absorption band, *a*, has its transition moment precisely aligned with the transition moment for emission, and the second absorption band, *b*, has its transition moment at right angles to the transition moment for emission. If the polarisation of the fluorescence (with vertically polarised exciting light) is plotted as a function of the

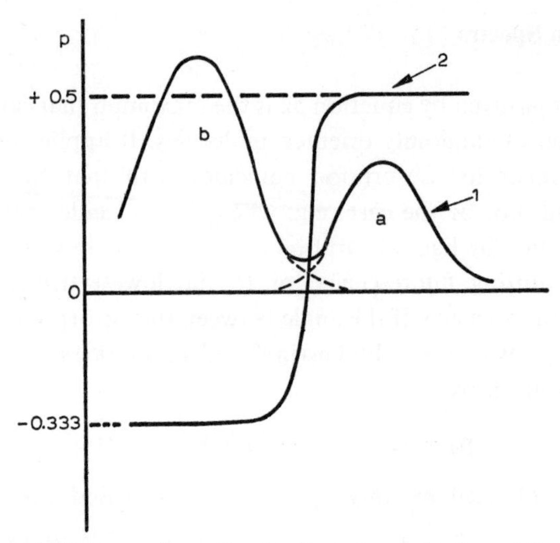

Fig. 19. Relationship between absorption bands and polarisation of fluorescence. Hypothetical ideal case. (1) absorption spectrum, (2) polarisation of the fluorescence excitation spectrum.

wavelength of the exciting light, the polarisation of the fluorescence excitation spectrum (curve 2) will be obtained. For this ideal case it will show a region of constant polarisation, with $p_0 = 0.5$, over most of the first absorption band, a region of rapidly decreasing polarisation where the two absorption bands overlap, and a second region of constant polarisation with $p_0 = -0.333$ over most of the second absorption band. In principle therefore, it is possible to use polarisation spectra to show the presence of two different overlapping electronic transitions. By the application of equation 53 it is also possible to determine the relative orientations (β) between the transition moments of the various absorption bands, and the transition moment of the emission band from the lowest excited state (either singlet or triplet depending on whether fluorescence or phosphorescence is observed).

In practice fluorescence polarisation spectra are generally more complex than Fig. 19. Many of the complexities can be explained by assuming the presence of several overlapping transitions, but even with an apparently "pure" $S_1 \leftrightarrow S_0$ transition, the full p_0 value of 0.5 is not always observed, and indeed p_0 may vary from one part of the vibrational structure of the fluorescence emission band to another, i.e. the polarisation of the fluorescence *emission* spectrum is not constant at constant excitation wavelength (see for example reference 64). The fact that the maximum theoretical values of the principal polarisation are not always observed has been attributed to molecular torsional vibrations in the excited state[65].

6. Rotational Depolarisation of Fluorescence

So far we have assumed that the solution is rigid, so that there is no change of orientation of the molecules between excitation and emission. If the solution has a low viscosity the molecules of solute will rotate rapidly, and if the lifetime of the excited molecules is long enough, they will have time to assume a random distribution of orientation before emitting luminescence. The emitted light will then be completely unpolarised. The degree of depolarisation will clearly depend on the average period of rotation (the rotational relaxation time, ρ) and the lifetime of the emission, τ. The relationships between the observed polarisation p, and the principal polarisation p_0 (i.e. the value observed in the absence of depolarising factors), and the lifetimes ρ and τ are as follows[25]:
for vertically polarised exciting light:

$$\left(\frac{1}{p} - \frac{1}{3}\right) = \left(\frac{1}{p_0} - \frac{1}{3}\right)\left(1 + \frac{3\tau}{\rho}\right) \tag{55}$$

and for unpolarised exciting light:

$$\left(\frac{1}{p} + \frac{1}{3}\right) = \left(\frac{1}{p_0} + \frac{1}{3}\right)\left(1 + \frac{3\tau}{\rho}\right) \tag{56}$$

in which, for a spherical molecule, ρ is related to the viscosity of the medium, η, by:

$$\rho = \frac{3V\eta}{RT} \tag{57}$$

where V is the molar volume. Thus if the polarisation of a solution is measured as a function of temperature under conditions where the lifetime, τ, is constant, the plot of $1/p$ against T/η will be linear. The value of p_0 can be derived from the intercept and the value of τ/V from the slope of the graph. Thus if τ is known, the effective molar volume V, may be calculated.

7. Depolarisation Caused by Energy Transfer

Although the process of energy transfer by dipole–dipole interaction between neighbouring molecules (see Section 2 B 6) is most efficient when the donor and acceptor molecules are parallel, it will also occur when they are not exactly parallel and the overal effect then results in a concentration-dependent depolarisation. According to Weber[66] the concentration-dependence may be expressed in the form:

$$\left(\frac{1}{p} \pm \frac{1}{3}\right) = \left(\frac{1}{p_0} \pm \frac{1}{3}\right)\left(1 + \frac{4\pi N R_c^6 \times 10^3}{15(2a)^3} c\right) \qquad (58)$$

in which the negative signs refer to excitation with vertically polarised light and the positive signs to excitation with unpolarised light; N is Avogadro's number, 2a the molecular diameter, c the concentration in moles litre^{-1}, and R_c is the critical distance between parallel dipoles at which the probability of emission is equal to the probability of energy transfer. This equation is equivalent to the empirical relationship found by Sveshnikoff and Feofilov[67], viz.:

$$\frac{1}{p} = \frac{1}{p_0} + Ac\tau \qquad (59)$$

in which τ is the lifetime of the excited state. Note that equation 59 is similar in form to the relationship between p and τ for rotational depolarisation (equations 55 and 56).

E. LIGHT SCATTERING

The possibility of interference by scattered exciting light must always be considered in the measurement of fluorescence spectra, and particularly at low concentration. Examples of its effect are discussed in later chapters. We shall outline here the mechanisms of light scattering and the characteristic properties of the scattered light.

1. Kinds of Light Scattering

Some light is scattered by liquids without change in wavelength and some is scattered with its wavelength changed by amounts depending on the nature of the scattering medium. Both processes are clearly distinguished from photoluminescence by the fact that they do not involve light absorption and take place within the period of vibration of the exciting light, i.e. within about 10^{-15} sec.

The intensity and spatial distribution of the light scattered by a suspension of "particles" depends on the size and shape of the particles, and it is convenient to discuss the phenomenon under three headings according to the size of the particles relative to the wavelength of light, viz.:

Rayleigh scattering—size $\ll \lambda$
Tyndall scattering—colloidal particles $< \lambda$ ($\sim$ 0.05–0.2 μm)
Large particles > 2 μm

All three processes produce scattered light of unchanged wavelength. Light scattering with change of wavelength—the Raman Effect—is treated in the later paragraphs.

2. Rayleigh Scattering

In a completely homogeneous transparent medium the induced vibrations set up in the medium by the passage of the light waves would give rise to secondary waves that would cancel one another exactly in all directions except that of propagation of the exciting light. Such a hypothetical medium would not give rise to light scattering. A liquid or gas made up of discrete molecules contains minute regions of fluctuating refractive index. As a result, the secondary waves generated by the induced vibrations do not exactly cancel one another and a small proportion of the exciting light is emitted in other directions than that of propagation of the incident beam. By considering the scattering centres to be spheres of radius r, much less than the wavelength of the light, and having equal polarisability in all directions, Rayleigh deduced that the intensity of scattered light is proportional to r^6/λ^4. The spatial distribution and polarisation of the light can be seen by reference to Fig. 20, which shows the spherical "particle" at 0 exposed to a beam of light polarised in the plane of the paper, i.e. with its electric vector parallel to ZZ′. The particle is set into forced vibration, the direction of vibration being that of the electric vector of the incident beam. The particle then re-emits light with its *plane* of polarisation such that it contains the vibration axis of the particle. Geometrically the situation is precisely the same as that described in Section 1 D 3 in connection with the fluorescence emission from a group of molecules all oriented with their

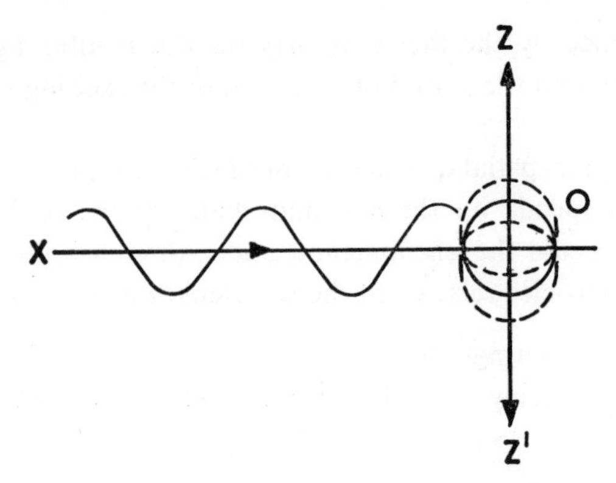

Fig. 20. Light scattering by spherical particles.

transition axes parallel to the direction of the electric vector of the exciting light, i.e. the emitted light is plane polarised and has an intensity proportional to $\cos^2 \alpha$ (see Fig. 17). Remembering that unpolarised exciting light can be treated as though it were a mixture of two polarised beams with their planes of polarisation at right angles, a little thought will show that the intensity of scattered light with unpolarised excitation will be proportional to $(1 + \cos^2 \alpha)$. This is the expression deduced by Rayleigh.

If we now consider only directions at right angles to that of the incident beam, the scattered light will be completely plane polarised whether the exciting light is polarised or not. With polarised exciting light, the intensity of scattered light will be a maximum along OY (in Fig. 17) and zero along OZ. In practice, Rayleigh scattering is found to be partly depolarised. Depolarisation factors ($I_{\perp}/I_{\parallel}$) are least (0.005) for the monatomic gases, and can be as high as 0.1 for polyatomic gases[68]. Values reported for many liquids at room temperature are greater than 0.1. The cause of depolarisation is discussed below.

The fact that Rayleigh scattering from solvents is strongly polarised can be put to practical use in high sensitivity spectrofluorimetry where scattered light can sometimes be objectionable. By inserting a polariser in the beam of fluorescence from a fluid solution so that only horizontally polarised light is passed to the analysing monochromator, the measured intensity of Rayleigh scattering is greatly reduced. The measured intensity of the fluorescence is reduced only by a factor of two if the fluorescence is largely depolarised by rotational relaxation (see Section 1 D 6).

3. Tyndall Scattering

This is produced by suspensions of colloidal dimensions, and as the size increases, the comparatively simple theory of Rayleigh becomes complicated by the occurrence of diffraction phenomena. However, the scattered light is usually strongly polarised. The *Tyndall Effect* depends on this polarisation. This effect is observed when a beam of vertically polarised light is passed through a fine suspension—the path of the beam is visible when viewed horizontally but is invisible from above or below.

The qualitative picture of the polarisation phenomenon may be obtained by assuming that the scattering particles are spherical and are set into forced vibrations by the light wave as discussed above for Rayleigh scattering. Depolarisation may be thought of as arising from the presence of non-spherical particles. In Fig. 21 consider a non-spherical particle with its direction of greatest polarisability AA′ at an angle to the direction of the electric vector of the incident wave. The forced oscillations of the particle now take place in some direction BB′ that is intermediate between the electric vector of the exciting light and the axis AA′.

The scattered light is emitted with its plane of polarisation such that it includes the line BB′ and hence in general, light observed in a direction at right angles to the incident beam will be only partly polarised. The situation is analogous to the emission of fluorescence from randomly oriented molecules discussed in Section 1 D 4.

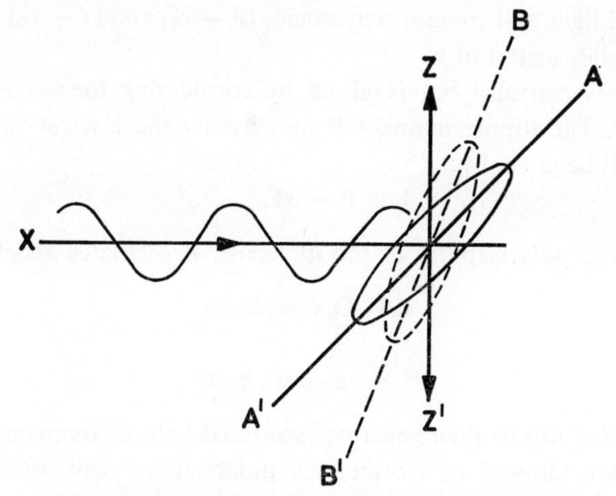

Fig. 21. Light scattering by non-spherical particles.

4. Scattering from Large Particles

As the size of the particles approaches that of the wavelength of light, the intensity of the scattered light is no longer symmetrically distributed, but becomes greater in the direction of the incident beam. For opaque particles larger than the wavelength, the light scattering is largely by reflection, and depolarisation is considerable.

5. The Raman Effect[69, 70]

When a beam of light of frequency ν passes through a transparent medium, the electrons of the molecules are set into forced vibration and emit a small proportion of scattered light of unchanged frequency, i.e. Rayleigh scattering, as already described in Section 1 E 2. If the nuclei of the atoms in the molecules were fixed in position, this would be the only scattering produced. Because the nuclei are capable of vibration and rotation, they can, during the scattering process, abstract part of the energy of the beam and convert it to vibrational energy. If the frequency of a particular vibration is ν_0, the amount of energy so abstracted will be $h\nu_0$, and hence the scattered photon will carry energy $h(\nu - \nu_0)$ and the scattered light will thus contain frequencies differing from ν. Further, if the irradiated molecule is already in an excited vibrational state, it can give up a quantum of vibrational energy during the scattering process. Thus for each vibration frequency ν_0, the scattered light will contain frequencies $(\nu - \nu_0)$ and $(\nu + \nu_0)$ in addition to the Rayleigh scatter at ν.

The process can also be visualised by considering the wave properties of the light. The dipole moment **P** induced by the electric vector of the wave, **E**, will be given by:

$$\mathbf{P} = \alpha \mathbf{E} \tag{60}$$

where α is the polarisability of the molecule. **E** oscillates according to:

$$\mathbf{E} = \mathbf{E}_0 \cos(2\pi\nu t) \tag{61}$$

and hence:

$$\mathbf{P} = \alpha \mathbf{E}_0 \cos(2\pi\nu t) \tag{62}$$

This oscillating dipole then generates scattered light of frequency ν. If now the nuclei are allowed to vibrate, the polarisability will oscillate with a frequency ν_0, i.e.,

$$\alpha = \alpha_0 + \alpha_1 \cos (2\pi\nu_0 t) \quad (63)$$

The oscillating polarisability will thus modulate the scattered light and the overall effect will be equivalent to the superposition of three oscillating dipole moments having frequencies $(\nu - \nu_0)$, ν and $(\nu + \nu_0)$. These will radiate to give the two Raman lines and the Rayleigh line. Quantitative evaluation shows that the intensity of emission at $(\nu + \nu_0)$ will be less than that at $(\nu - \nu_0)$ and for high values of ν_0, the intensity at $(\nu + \nu_0)$ will be very small.

6. Polarisation of Raman Emission

Raman emission is always polarised to some extent and certain vibrations give rise to a high degree of polarisation. Qualitatively it can be seen that for a symmetrical molecule executing a totally symmetrical vibration (e.g. a "breathing" vibration) the Raman emission, like the Rayleigh scattering, should be completely polarised in the OZ direction when viewed along OY (see Fig. 17). In Raman spectroscopy, the state of polarisation of the emission is normally expressed by the depolarisation ratio δ, which, in the terminology used in Section 1 D, is given by:

$$\delta = I_{\perp}/I_{\parallel} \quad (64)$$

Raman bands corresponding to vibrations that are *not* totally symmetrical, *always* have depolarisation ratios of $\frac{3}{4}$ for vertically polarised exciting light, or $\frac{6}{7}$ for unpolarised exciting light. A definite statement cannot be made about totally symmetric vibrations of molecules with unequal axes. When the axes are nearly equal, δ is nearly zero, i.e. the emission is almost completely polarised. With greater inequality of the axes, the depolarisation ratio of the Raman emission corresponding to a totally symmetrical vibration will be greater and may have a value almost as great as the limiting value of $\frac{3}{4}$ (or $\frac{6}{7}$ for unpolarised exciting light).

The degree of polarisation of the main Raman bands of some common solvents is of interest because these bands can interfere in high sensitivity spectrofluorimetry. If they have low δ values, their intensity can be reduced by insertion of a polariser to pass horizontally polarised light.

7. Raman Emission of Solvents

Eisenbrand and Picher [71] discussed the interference of Raman emission with the measurement of weak fluorescence. Using photographic recording they observed the main Raman band of water at 0.342 μm^{-1}, and the two bands of benzene at 0.993 and 0.306 μm^{-1}. In a later paper Eisenbrand [72] measured the integrated intensity of the Raman band of water excited by the 365 nm mercury line, relative to the integrated intensity of the fluorescence from a dilute solution of sodium hydroxypyrene-trisulphonate. He found that the Raman emission was equal to the fluorescence emission from about 10^{-9} g/ml of the fluorescent compound. From the extinction coefficient of the latter at 365 nm he calculated that 0.0023% of the exciting light was emitted in the Raman band.

Parker [73] used a photoelectric spectrofluorimeter to measure the Raman bands of five solvents commonly used in spectrofluorimetry. His spectra observed at two wavelengths of exciting light are shown in Fig. 22, and the complete set of data are given in Table 4. Some of the solvents contained impurities that fluoresced with certain wavelengths of exciting light, but the Raman bands were readily identified by the fact that when the wavenumber of the exciting light was varied, they appeared at wavenumbers differing by a constant amount from the wavenumber of the exciting light (compare the wavenumber shifts for each band in columns 2–5 of Table 4). Only the most prominent—or the two most prominent—bands are recorded in Table 4.

TABLE 4

RAMAN BANDS OBSERVED IN SELECTED SOLVENTS

(From Parker [73])

Solvent	Wavenumber shifts (μm^{-1}) with exciting light at				Mean wavenumber shift
	248 nm	313 nm	365 nm	436 nm	
Water	0.339	0.339	0.340	0.335	0.338
Ethanol	0.292	0.292	0.293	0.290	0.292
	0.143	0.143	0.141	0.134	0.140
Cyclohexane	0.287	0.287	0.291	0.285	0.288
	0.137	0.138	0.135	0.132	0.136
Carbon tetrachloride	—	0.073	0.079	0.071	0.074
Chloroform	—	0.073	0.073	0.065	0.070
	—	0.301	0.301	0.304	0.302

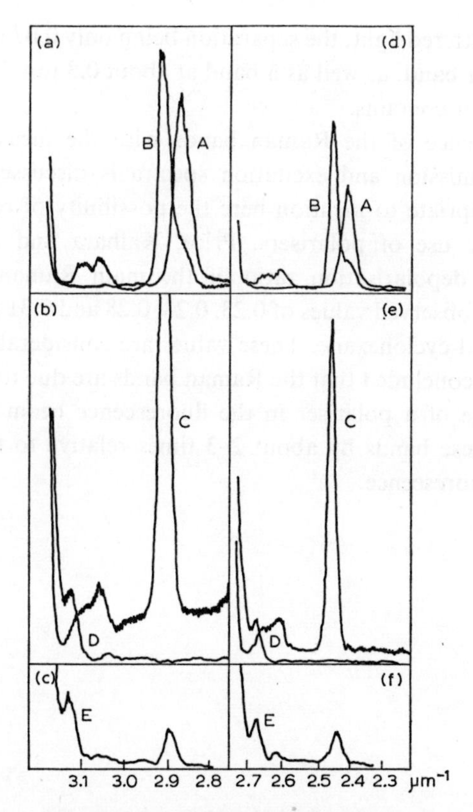

Fig. 22. Raman spectra of solvents.
Excited by (a), (b) and (c), 313 nm light; (d), (e) and (f), 365 nm light: curve A, water; curve B, ethanol; curve C, cyclohexane; curve D, carbon tetrachloride; curve E, chloroform (from Parker [73]).

All solvents containing hydrogen atoms show a band in the region of 0.3 μm^{-1}, that of ethanol having a subsidiary inflection on the low-wavenumber side. These bands are due to CH or OH vibrations. In water the wavenumber shift is considerably less if deuterium is substituted for hydrogen. Cyclohexane shows the narrowest and most intense band, in almost exactly the same position as the ethanol band. Both ethanol and cyclohexane also show a subsidiary double peak corresponding to a smaller frequency shift. The position of the more intense of these two peaks is indicated in Table 4. Carbon tetrachloride shows no significant bands at large wavenumber shifts owing to the absence of hydrogen atoms. The main band is close to the scattered light, the separation being only 0.07 μm^{-1}. Chloroform shows a similar band, as well as a band at about 0.3 μm^{-1} due to the hydrogen atom that it contains.

The interference of the Raman bands with the measurement of weak fluorescence emission and excitation spectra is discussed in Section 5 C, but it is appropriate to mention here the possibility of reducing this interference by the use of polarisers. Price, Kaihara and Howerton [74] have measured the depolarisation ratio of the main Raman band from four solvents. They observed values of 0.23, 0.23, 0.28 and 0.31 for water, ethanol, chloroform and cyclohexane. These values are considerably less than $\frac{6}{7}$ and they therefore concluded that the Raman bands are due to totally symmetric vibrations. Use of a polariser in the fluorescence beam would reduce the intensity of these bands by about 2–3 times relative to that of completely depolarised fluorescence.

Chapter 2

Kinetics of Photoluminescence

A. TRANSITION RATES

We give in Fig. 23 a simplified scheme showing singlet and triplet energy levels and summarising the kinds of transition so far discussed. For simplicity the vibrational levels associated with each electronic state have been omitted. Radiative transitions are shown as full arrows and radiationless transitions as pecked arrows. The letter against each arrow indicates the

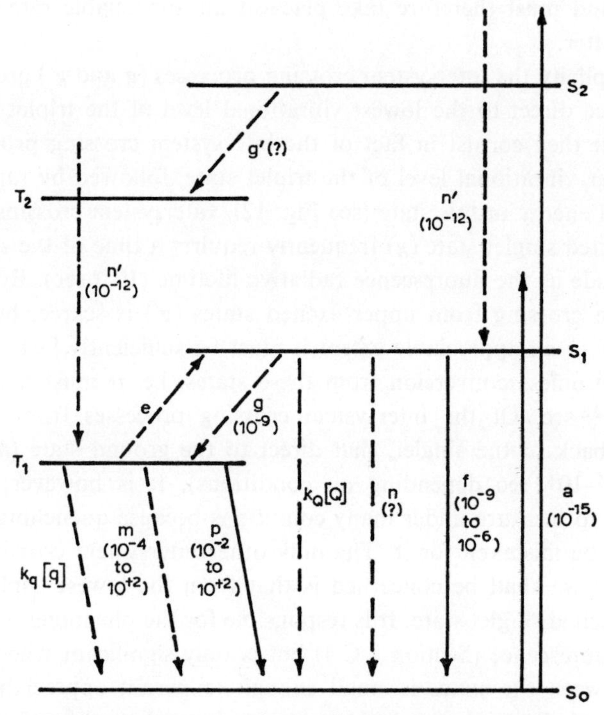

Fig. 23. Reciprocal rates of transitions (typical values).

kind of transition as described below, and the power of ten the approximate mean time in seconds for the occurrence of the process.

Of the radiative transitions, light absorption (*a*) takes place within the period of vibration of the light wave, i.e. within about 10^{-15} sec. Fluorescence (*f*) from the lowest excited singlet state has a radiative lifetime between 10^{-9} sec (for allowed transitions, e.g. $\pi^* \rightarrow \pi$) to 10^{-6} sec (for the less probable π^*–n transitions). Radiative lifetimes of triplet states (transition *p*) lie mainly within the range 10^{-2}–10^{+2} sec.

Of the radiationless transitions, internal conversion (*n'*) from upper excited states in both the singlet and triplet manifolds is rapid for most molecules, i.e. $\sim 10^{-12}$ sec, and the associated loss of vibrational energy is about equally fast. Internal conversion (*n*) from the lowest excited singlet to the ground state is now regarded by some workers as negligible for most molecules. However there is not yet sufficient precise data available to make this assumption, and we therefore include this process in Fig. 23. To make its presence felt it must compete with fluorescence and intersystem crossing, and must therefore take place at an appreciable rate compared with the latter.

For simplicity the intersystem crossing processes (*g* and *g'*) are shown as taking place direct to the lowest vibrational level of the triplet state concerned, but they consist in fact of the intersystem crossing process itself, to an upper vibrational level of the triplet state, followed by rapid loss of vibrational energy in this state (see Fig. 12). Intersystem crossing from the lowest excited singlet state (*g*) frequently requires a time of the same order of magnitude as the fluorescence radiative lifetime (10^{-9} sec). Evidence for intersystem crossing from upper excited states (*g'*) is scarce, but when it does occur to an appreciable extent it must be sufficiently fast to compete with radiationless conversion from these states, i.e. it must take place in about 10^{-12} sec. Of the intersystem crossing processes from the triplet manifold back to the singlet, that direct to the ground state (*m*) is often slow (10^{-4}–10^{+2} sec depending on conditions). It is however a difficult parameter to measure under many conditions because quenching processes can easily be mistaken for it. The only other intersystem crossing process with which we shall be concerned is that from the lowest triplet state to the first excited singlet state. It is responsible for the phenomenon of E-type delayed fluorescence (Section 1 C 4) but is only significant when the separation between the states is small enough to permit appreciable thermal activation of the triplet molecule to a level equal to that of the bottom of the excited singlet state. In Fig. 23, the symbol "*e*" refers to the rate of the overall process of thermal activation followed by intersystem crossing.

All the processes so far considered in Fig. 23, with the exception of light absorption, are first order reactions having rate constants equal to the reciprocals of the lifetimes quoted. The lowest excited singlet state and the lowest triplet state have sufficiently long lifetimes to permit chemical reaction or luminescence quenching by encounter with molecules of solutes. These are of course second order processes, but since the concentration of quenching solute is often much greater than that of the excited molecules, the processes can frequently be represented as first order reactions with rate constants $k_Q[Q]$ or $k_q[q]$, in which k_Q and k_q are the bimolecular rate constants and [Q] and [q] the concentrations of quenching molecules. Photochemical reactions direct from upper excited states must compete with the rapid internal conversion and are probably restricted to molecular dissociations which can take place within one or a very few molecular vibrations.

B. FLUORESCENCE

1. Fluorescence Efficiency and Lifetime

Consider a system similar to that in Fig. 5 in which the liquid is illuminated with a beam of exciting light of constant intensity. Consider also that the optical density for the thickness of solution considered is small, so that the rate of light absorption, I_a, is substantially constant throughout the volume of solution, and is thus given by:

$$I_a = 2300\varepsilon c I_0 \tag{65}$$

where I_a is measured in einstein litre^{-1} sec^{-1} and I_0 is measured in einstein cm^{-2} sec^{-1} (see equations 17–21). After a period of illumination that is long compared with the fluorescence lifetime, a steady state is set up, in which the rate of production of excited singlet molecules (in S_1) is just balanced by their rate of disappearance via the various processes shown in Fig. 23. In the absence of photochemical reaction or intersystem crossing from upper excited singlet states, the rate of production of excited molecules in S_1 is equal to the rate of light absorption, I_a, measured in einstein litre^{-1} sec^{-1}. If $[S_1]$ is the concentration of excited singlet molecules when the steady state has been achieved, we have:

$$I_a = (k_f + k_n + k_g + \Sigma\, k_Q[Q])[S_1] \qquad (66)$$

where the k's are the first order rate constants of the relevant processes in Fig. 23 and $\Sigma\, k_Q[Q][S_1]$ is the sum of the rates of disappearance of S_1 by quenching by a variety of quenching molecules Q. We can regard the internal conversion represented by k_n as simply another form of quenching reaction, i.e. quenching by the solvent molecules themselves. Alternatively, if we wish to investigate quenching by one particular solute, and maintain all other solute concentrations constant, we can combine the $k_Q[Q]$ terms for the latter in a composite first order rate constant k_n. Equation 66 then simplifies to:

$$I_a = (k_f + k_n + k_g + k_Q[Q])[S_1] \qquad (67)$$

Now the total rate of emission of fluorescence, **Q**, is equal to $k_f[S_1]$, and thus (see equation 6):

$$k_f[S_1] = \mathbf{Q} = I_a\phi_f = (k_f + k_n + k_g + k_Q[Q])[S_1]\phi_f \qquad (68)$$

in which ϕ_f is the fluorescence efficiency.
Hence:

$$\phi_f = k_f/(k_f + k_n + k_g + k_Q[Q]) \qquad (69)$$

When the exciting light is shut off, the fluorescence decays exponentially with a lifetime, τ, given by:

$$1/\tau = k_f + k_n + k_g + k_Q[Q] \qquad (70)$$

Since the radiative lifetime, τ_r, is $1/k_f$,

$$\phi_f = \tau/\tau_r \qquad (71)$$

We shall find that the kinetics of phosphorescence and delayed fluorescence can also be dealt with by setting up steady state equations like equation 67, although the steady state concentration of interest will then be that of the triplet state.

2. The Stern–Volmer Equation and Fluorescence Quenching

From equation 69 it follows that the fluorescence efficiency (ϕ_f^0) of solutions containing zero concentration of quencher is:

$$\phi_f^0 = k_f/(k_f + k_n + k_g) \qquad (72)$$

and hence:

$$\phi_f^0/\phi_f = 1 + k_Q[Q]/(k_f + k_n + k_g) = 1 + \mathbf{K}[Q] \tag{73}$$

or, if τ_0 is the lifetime in the absence of quencher, we have:

$$1/\tau_0 = k_f + k_n + k_g \tag{74}$$

and:

$$\phi_f^0/\phi_f - 1 = k_Q\tau_0[Q] = \mathbf{K}[Q] \tag{75}$$

Thus if we measure the fluorescence of a series of solutions with an instrument giving a response (R) proportional to the intensity of light received, a plot of $(R^0/R - 1)$ against concentration of quencher will give a straight line of slope **K**. The relationships 73 and 75 were first used by Stern and Volmer [75] and **K** is generally known as the Stern–Volmer quenching constant.

The fluorescent solute itself may act as quencher, i.e. encounters between excited and ground state molecules may result in quenching of the former. Special effects resulting from such *self-quenching* are discussed in Section 4 D.

Fluorescence quenching sometimes causes an increase in the amount of triplet formed (see Section 4 A 4). Insufficient information has been accumulated to show whether the reciprocity of fluorescence quenching and triplet formation is a general phenomenon.

It is important to note that the observation of a Stern–Volmer quenching relationship does not necessarily prove that fluorescence quenching is taking place by a collisional mechanism. A similar relationship will be observed if the added solute forms a non-fluorescent compound with the fluorescent molecule. Thus in the equilibrium:

$$B + Q \rightleftarrows BQ \tag{76}$$

let b be the total concentration of fluorescer added to the solution, of which a fraction α is converted to the non-fluorescent complex, BQ, by the addition of a large excess, [Q], of the substance Q. If Q does not itself quench the excited state of B, and if neither Q nor BQ act as inner filters (see Section 3 H), the observed fluorescence intensity (F) will always be proportional to the concentration of uncomplexed B present, i.e.:

$$F = k(1 - \alpha)b \tag{77}$$

and

$$F_0 = kb \tag{78}$$

thus

$$\frac{F_0 - F}{F} = \frac{\alpha}{1 - \alpha} = K_c[Q] \tag{79}$$

where K_c is the equilibrium constant for the formation of the complex. Thus the system will apparently follow the Stern–Volmer quenching law, but the quenching constant will now be the equilibrium constant of complex formation.

3. Diffusion-Controlled Processes

Bimolecular reactions in solution proceed via "encounters" between reacting molecules, each encounter consisting of perhaps a hundred actual collisions, because the surrounding solvent molecules act as a kind of "cage", preventing the separation of the reactants after each collision. If the activation energy (ΔE) for the reaction is high, the observed rate will be given by the usual kinetic expression, i.e. the rate will be equal to the product of the *collision rate*, a "steric factor" and the Boltzmann factor, $\exp(-\Delta E/\mathbf{R}T)$. If however the activation energy is very low, reaction will take place on each *encounter*, and the rate of reaction will then be controlled solely by the rate at which reactants can diffuse together. The diffusion-controlled bimolecular rate constant (k_c) for the ideal case of large spherical molecules depends only on the temperature and the viscosity of the solvent, and is given by[76]:

$$k_c = 8\mathbf{R}T/3000\eta \tag{80}$$

where η is the viscosity of the solvent in poises.

Values of k_c for some common solvents, calculated from equation 80, are given in Table 5. Values for ethanol as a function of temperature, calculated from viscosity data[77] are plotted in Fig. 24. It will be noted that the plot of $\log_{10}(k_c)$ against $1/T$ is nearly linear. This means that the diffusion-controlled rate behaves as though it had an activation energy, equal to about 3.7 kcal mole^{-1} in the case of ethanol.

Although in this book we shall frequently refer to encounter rates and diffusion-controlled rate constants calculated by equation 80, the latter has to be applied with care. It is derived[78, 79] from the Smoluchowski[80] equation (equation 81) relating rate constant for a bimolecular diffusion-controlled reaction with the encounter distance (σ_{AB}) and the diffusion coefficients ($\mathbf{D}_A$ and $\mathbf{D}_B$) of reactants:

$$k_c = 4\pi\sigma_{AB}N(\mathbf{D}_A + \mathbf{D}_B) \times 10^3 \tag{81}$$

TABLE 5

DIFFUSION-CONTROLLED RATE CONSTANTS

($k_c = 8\,RT/3000\eta$)

Solvent	Temperature (°C)	η (poise)	k_c (litre mole^{-1} sec^{-1})
Water	+20	0.010	6.5×10^9
	+80	0.0036	2.2×10^{10}
Ethanol	−98	0.44	8.8×10^7
	+20	0.012	5.4×10^9
	+60	0.0059	1.2×10^{10}
n-Hexane	+20	0.0032	2.0×10^{10}
Cyclohexane	+20	0.0095	6.9×10^9
Benzene	+20	0.0065	1.0×10^{10}
	+60	0.0039	1.9×10^{10}
Glycerol	−42	6.7×10^4	7.6×10^2
	0	121	5.0×10^5
	+20	15	4.3×10^6
	+30	6.3	1.1×10^7

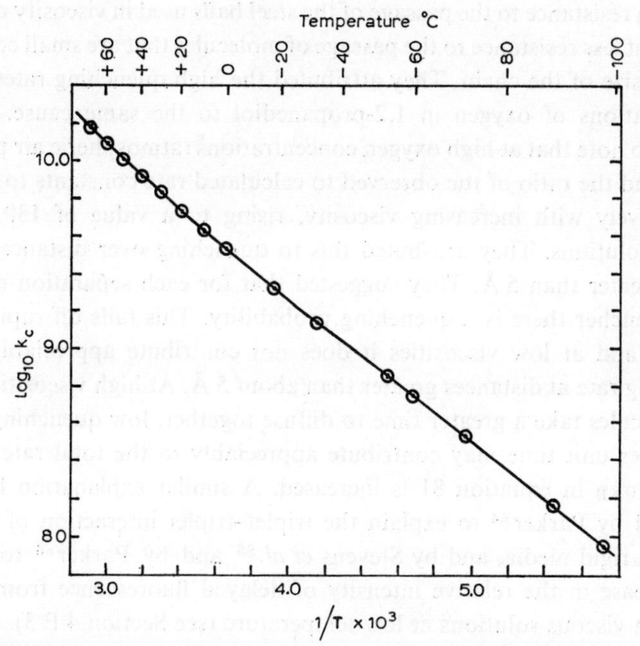

Fig. 24. Diffusion-controlled rate constant in ethanol (calculated by equation 80).

and the Stokes–Einstein[81,82] equation:

$$D_A = kT/6\pi r_A \eta \tag{82}$$

with the assumption that:

$$\sigma_{AB} = r_A + r_B = 2r_A \tag{83}$$

It was pointed out by Osborne and Porter[83] that if it is assumed that the coefficient of sliding friction is zero, so that the solute can "slip" in contact with the solvent, the following modified expression for equation 80 is obtained:

$$k_c' = 8RT/2000\eta \tag{84}$$

Osborne and Porter measured the rate constants for quenching of triplet naphthalene by α-iodonaphthalene in various solvents. In 1,2-propanediol, glycerol and 50/20 liquid paraffin/n-hexane, the observed values were close to those calculated by equation 84. In liquid paraffin alone, however, the observed rate was more than four times the calculated one. They attributed this to the fact that liquid paraffin is composed of long chain molecules which offer high resistance to the passage of the steel balls used in viscosity measurements, but less resistance to the passage of molecules that are small compared with the size of the chain. They attributed the high quenching rates by low concentrations of oxygen in 1,2-propanediol to the same cause. It is of interest to note that at high oxygen concentrations (atmospheric air pressure) they found the ratio of the observed to calculated rate constants to increase progressively with increasing viscosity, rising to a value of 130 in very viscous solutions. They attributed this to quenching over distances appreciably greater than 5 Å. They suggested that for each separation of triplet from quencher there is a quenching probability. This falls off rapidly with distance and at low viscosities it does not contribute appreciably to the quenching rate at distances greater than about 5 Å. At high viscosities where the molecules take a greater time to diffuse together, low quenching probabilities per unit time may contribute appreciably to the total rate, i.e. the effective σ_{AB} in equation 81 is increased. A similar explanation has been advanced by Parker[84] to explain the triplet–triplet interaction of phenanthrene in rigid media, and by Stevens *et al.*[85] and by Parker[86] to explain the decrease in the relative intensity of delayed fluorescence from excited dimers in viscous solutions at low temperature (see Section 4 E 3).

A further consideration must be borne in mind when using diffusion-controlled rate constants. If a system is initially populated with a *random* distribution of reactive molecules, some pairs of molecules will be situated close together and will react rapidly. Hence, the measured initial rate "constant" will be high, and it will decrease with time as such close "pairs" are consumed. Ultimately a steady state will be established in which the rate of reaction is balanced by the rate of diffusion of reactive species towards one another. The diffusion-controlled rate constant (equation 80) corresponds to the situation in this steady state. In fluid solutions the establishment of the steady state may require times as long as 10^{-7} sec. If a fast first order reaction (e.g. fluorescence emission with $k_f \sim 10^8$ sec^{-1}) is competing with the fast bimolecular reaction it is not strictly correct to use the calculated diffusion-controlled rate constant for the latter. A mathematical treatment of transient rate kinetics has been given by Noyes[87] including its application to fluorescence quenching.

In spite of these complications equation 80 is very useful for approximate calculations in connection with the photochemistry and photoluminescence of solutions, because many processes of interest are diffusion-controlled. Equation 80 may be used to calculate an approximate upper limit for the rate of a bimolecular reaction and hence limiting values of other parameters can be deduced. One example will suffice to illustrate this. It is of considerable practical interest, e.g. in analytical chemistry, to know the order of magnitude of solute concentrations that are likely to produce appreciable quenching of fluorescence. This can be derived from the Stern–Volmer equation. Thus for 50% quenching ($\phi_f^0/\phi_f = 2$), and in a diffusion-controlled quenching process, k_Q in equation 75 is put equal to k_c. We then have:

$$k_c \tau_0 [Q] \sim 1 \tag{85}$$

With a fluorescent compound showing a strong first absorption band ($\tau_r \sim 10^{-8}$ sec) and strong fluorescence ($\tau_0 \sim \tau_r$), we find in water at 20°C:

$$[Q]_{50\%} \sim 0.01\text{M} \tag{86}$$

For 10% quenching ($\phi_f^0/\phi_f = 1.11$):

$$k_c \tau_0 [Q] \sim 0.11 \tag{87}$$

and

$$[Q]_{10\%} \sim 0.001\text{M} \tag{88}$$

Thus for a strongly allowed transition, collisional quenching by solutes can be reduced to negligible proportions simply by diluting the solution until the concentration of all solutes is reduced to less than 10^{-3}M. It does not of course follow that all solutes at concentrations greater than 10^{-3}M will produce quenching. Many will not do so even at much higher concentrations. The solute may reduce the intensity of fluorescence at lower concentrations for other reasons, e.g. it may react with the fluorescent substance in the ground state, or it may introduce inner filter effects (see Section 3 H). These effects are not true quenching as defined in this book, i.e. they are not processes which reduce the actual quantum efficiency of the luminescence process.

4. Fluorescence from Upper Excited States

It has been stated as a general rule (Section 1 B 1 and Fig. 1) that fluorescence always occurs from near the lowest vibrational level of the first excited singlet state, no matter to what level the molecule is originally excited by light absorption. This follows also from the lifetimes of the molecule in the

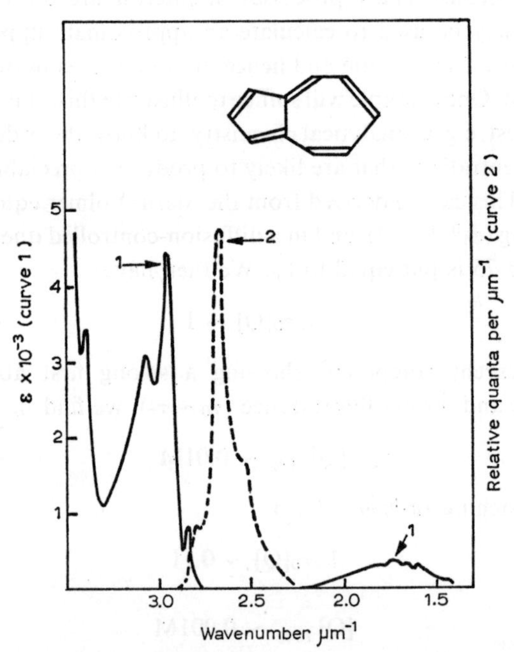

Fig. 25. Absorption and fluorescence emission spectra of azulene in ethanol at 20°C. Curve 1, absorption spectrum; curve 2, emission spectrum excited by light of wavenumber 3.19 μm^{-1} (313 nm).

various excited states as indicated in Fig. 23, compared with the radiative lifetimes calculated from equation 27. The only compounds with which fluorescence from an upper excited state has been observed and confirmed are azulene (see Fig. 25) and some of its derivatives[88,89]. Azulene has absorption bands in the ultra-violet region similar to those of the aromatic hydrocarbons, but in addition it has weak absorption bands in the red region of the spectrum. Excitation in the ultra-violet region, i.e. in the second absorption band, causes the emission of fluorescence in the region 2.5 to 2.8 μm^{-1}, having an approximate mirror image relationship with the second absorption band (see Fig. 25). Despite determined efforts[88,89], emission from the first absorption band has not yet been detected even at low temperature. One reason for this may be the relative insensitivity of photodetectors in the near infra-red region. A second reason may be that the long wavelength transition is weak, so that the radiative lifetime is comparatively long and hence quenching processes may overwhelm the radiative process. It has been suggested that the red absorption bands of azulene are due to another species, e.g. a tautomer, and not to azulene itself. Investigations have so far failed to prove this however. The fact that emission is observed at moderate intensities from the second excited state of azulene implies that internal conversion from this state to the lowest excited state must be a comparatively slow process. Possible reasons for this have been suggested, one of which is that the large separation between the two states reduces the possibility of crossing of the potential energy surfaces.

Extinction coefficients for transitions to upper excited levels are often high. The radiative lifetimes are therefore short and in the absence of actual dissociation of the excited molecules, there must be a finite probability of emission. The expected order of magnitude of the quantum efficiency of such an emission may be calculated by using the approximate value for $k_{n'}$ given in Fig. 23. Consider for example the absorption spectrum of benzophenone (Fig. 11). The radiative lifetime calculated from the integrated absorption band having a maximum at 4.0 μm^{-1} is between 10^{-9} and 10^{-8} sec. If we assume that the actual lifetime in this state is about 10^{-12} sec, we should expect an emission in the region of 3.5 μm^{-1} with ϕ_f between 10^{-4} and 10^{-3}. This should easily be detectable with a good fluorescence spectrometer. Measurements[90] with benzophenone in cyclohexane excited by 250 nm radiation (optical density at 250 nm was 0.2) gave a weak emission in the region 3.6–2.9 μm^{-1} having a quantum efficiency of not more than 3×10^{-5}. This implies that the lifetime in the upper state is less than

3×10^{-13} sec. This calculation is based on the assumption that all molecules excited to upper vibrational levels of S_2 cascade down to the lowest vibrational level (Processes A in Fig. 26) at rates much greater than the rate of internal conversion from the latter (Process B in Fig. 26). In fact there is probably not a great deal of difference between these rates, and it is then necessary to take into account the process of fluorescence from upper vibrational levels of S_2 (including resonance radiation from the level to which the molecule was originally excited by light absorption), as well as from the lowest vibrational level. In the case of benzophenone part of the fluorescence arising from upper vibrational levels would be distributed in the band 4.0–3.6 μm^{-1} and would not be accessible to observation in the experiment already described owing to the overlapping intense Raman

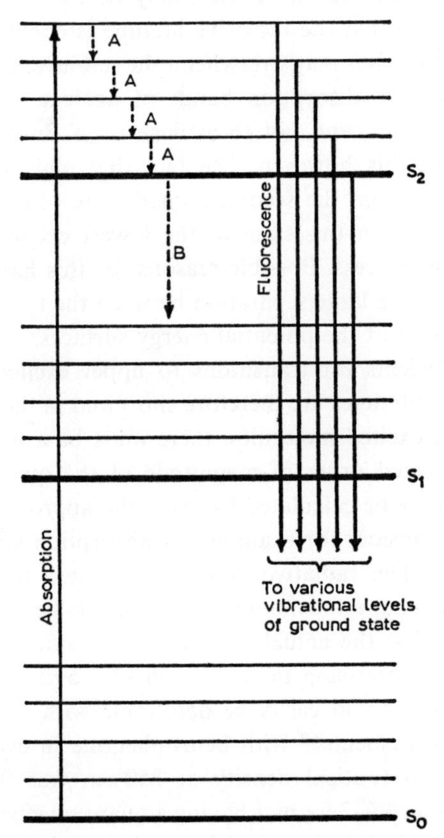

Fig. 26. Fluorescence from upper excited singlet states.

emission from the solvent. The overall fluorescence efficiency from the upper excited state could thus have been greater than the observed value of 3×10^{-5}.

Interpretation of the results of such experiments is further complicated by the possibility of interference by traces of fluorescent impurities or photodecomposition products. In fact the main difficulty in measuring the weak emission from upper singlet states is not the difficulty of detecting it with a high sensitivity instrument, but in proving that any emission that is detected is indeed due to the $S_2 \rightarrow S_0$ transition, and not to a small amount of impurity in the substance under investigation. We shall have much to say about the question of purity later, but it may be mentioned at this stage that claims have been made be several workers to have observed fluorescence emission from the upper excited state of biphenylene[91], claims which were subsequently proved to be ill-founded because the emission was found to be due to an impurity[92].

5. Temperature Dependence of Prompt Fluorescence

In the absence of quenchers the fluorescence efficiency (ϕ_f) depends on the relative rates of the radiative process (k_f) on the one hand, and the radiationless processes of intersystem crossing (k_g) and internal conversion (k_n) on the other (see equation 69). The rate of the radiative process is not expected to vary with temperature and hence the variation of ϕ_f reflects the variations of k_g and k_n. These should increase as the temperature increases because at higher temperatures a larger proportion of the molecules are raised to upper vibrational levels of S_1 and the probability of passing through potential surface intersections is increased. At low temperatures both rates should tend to a limiting value corresponding to the probability of intersystem crossing or internal conversion from the lowest vibrational level of S_1. Thus a substance having a fluorescence efficiency of close to unity at room temperature will not show much change on lowering the temperature although it may show a decrease at temperatures above room temperature. A weakly fluorescent substance may become strongly fluorescent at low temperature. The effect of temperature on the fluorescence efficiency of a variety of aromatic compounds has been investigated by Bowen and co-workers[23, 24]. They found that fluorescence normally diminishes with temperature although with some substances the change was small. It reached a limiting value at low temperatures in accordance with the above qualitative predictions. The fluorescence yields were related to the temperature by the following equation:

$$1/\phi_f - 1 = k_1 \exp(-E/\mathbf{RT}) + k_2 \tag{89}$$

With rigid molecules such as anthracene the value of E was little affected by the viscosity of the solvent, but with flexible molecules such as di-9-anthryl-ethane the value of E varied considerably and was in fact close to the value of ΔE characterising the variation of solvent fluidity with temperature. Bowen and Seaman suggested that excited molecules of this type can degrade their energy by relative thermal diffusional movements of the two large rings joined by the flexible chain (i.e. by "internal collisional quenching").

More recently Lim, Laposa and Yu[20] made an interesting study of the effect of temperature on the fluorescence efficiency, and relative triplet formation efficiency, of 9-, and 9,10- substituted anthracenes. They found that the fluorescence efficiency followed a law similar to equation 89 with $k_2 = 0$, and that the relative triplet formation efficiency, as measured by the intensity of triplet–triplet absorption (A), varied with temperature as follows:

$$1/A - p = K \exp(+E/\mathbf{RT}) \tag{90}$$

The value of E was found to be similar for both 89 and 90, suggesting that the temperature dependence of ϕ_f results from a temperature dependence of the rate of intersystem crossing $S_1 \rightarrow T_1$. They concluded that the internal conversion $S_1 \rightarrow S_0$ was negligible for these compounds and suggested that the overall process $S_1 \rightarrow T_1$ occurred by initial intersystem crossing $S_1 \rightarrow T_2$, in which T_2 is situated slightly above S_1.

The temperature effects so far considered refer to solutions without any quenching solute, i.e., solutions in which the degradation of singlet excited states occurs only by collision with "inert" solvent molecules. In the presence of added quencher, the Stern–Volmer equation (73) will apply, and there will then be an additional temperature effect introduced as a result of the variation in the rate of encounters between the fluorescer and the quencher, caused by the varying solvent viscosity. Again this will cause the fluorescence efficiency to decrease with increasing temperature.

6. Energy Transfer between Singlet States

The process of internal conversion involves the ultimate transfer of excitation energy to the surrounding medium where it appears as heat, and this process should strictly speaking be included under the heading of energy transfer. However, we shall be concerned here with processes of energy transfer in which the transferred energy appears as electronic excitation energy of the acceptor molecule. Two such long-range processes have to be considered in solutions. The simplest involves the emission of fluorescence light by the donor molecule and the re-absorption of this fluorescence by the acceptor molecule. For this process to take place the absorption spectrum of the acceptor must overlap the emission of the donor and the intervening medium must be transparent to the light concerned. This so-called "trivial" process is of no theoretical interest, but it can give rise to practical difficulties and lead to spurious results in the measurement of both prompt and delayed emission spectra as will be discussed in Chapter 3.

The experimental evidence for energy transfer of the second kind came first from measurements of the polarisation of fluorescence of dyestuff solutions, and indicated that excitation energy could be transferred between molecules as far apart as 50 Å, i.e. over distances considerably greater than the collision diameter. Thus when a dilute solution of a fluorescent dye in a highly viscous medium is irradiated with polarised light, the fluorescence emitted is strongly polarised (see Section 1 D 4). It was observed however[93, 94] that the polarisation of fluorescence decreases sharply with increasing concentration, and with fluorescein for example, the degree of polarisation drops to half its maximum value at a concentration in the region of 10^{-3}M[67]. This concentration-depolarisation can only be interpreted by the assumption that molecules other than those primarily excited become the radiators of fluorescence. It can be shown that operation of the trivial process of emission and re-absorption is not sufficient to account for the effect. Perrin[95, 96] suggested that a transfer of energy could take place by direct electrodynamic interaction between excited and unexcited molecules, which were regarded as high frequency oscillators. The theory of this type of transfer has been quantitatively developed by Förster[97, 98]. In general it may be represented by the equation:

$$^{1}D^{*} + {}^{1}A \rightarrow {}^{1}D + {}^{1}A^{*} \qquad (91)$$

The conditions favouring energy transfer are (i) a large overlap between the first absorption band of the acceptor molecule (A) and the emission band of the donor molecule (D), and (ii) high fluorescence yield of the donor molecule. Donor and acceptor may of course be molecules of the same substance provided that there is appreciable overlap of absorption and fluorescence spectra. The efficiency of intermolecular dipole–dipole transitions may be expressed in terms of a critical separation distance R_c at which the probability of transfer is equal to that of spontaneous decay. From Försters equation R_c is given by[99]:

$$R_c^6 \sim \frac{9000(\ln 10)\,\kappa^2\phi_f}{128\pi^6 n^4 N} \int_0^\infty \frac{f_D \varepsilon_A . d\bar{\nu}}{\bar{\nu}^4} \qquad (92)$$

in which N is Avogadro's number, n is the refractive index of the medium, κ^2 is an orientation factor (approximately 2/3), ϕ_f is the fluorescence efficiency, f_D is the fluorescence emission spectrum of the donor normalised to unity, ε_A is the absorption spectrum of the acceptor and $\bar{\nu}$ is the wavenumber in cm^{-1}. Förster calculated a critical transfer distance of 50 Å units for fluorescein, and 80 Å units for chlorophyll. Quinine bisulphate with a considerably broader absorption spectrum, lower ε_{max}, and smaller overlap, gave a critical transfer distance some 10 times smaller. A mean molecular separation of 100 Å units corresponds to a concentration of about 2×10^{-3}M. Thus fluorescence self-quenching by energy transfer requires concentrations nearly as great as those required for collisional quenching at unit encounter efficiency in highly fluid solvents (see equation 86), but with the added requirement of overlap between absorption and fluorescence spectra. It is safe to assume therefore that if a fluid solution has been diluted sufficiently to avoid fluorescence self-quenching at a diffusion-controlled rate, quenching by long-range energy transfer will also be negligible.

The efficiency of the non-radiative transfer process can however be considerably greater than this when the donor and acceptor are different molecular species having a favourable absorption/emission overlap and high donor fluorescence efficiency. There have been a number of experimental confirmations of the occurrence of energy transfer in such systems. Three donor–acceptor pairs were studied quantitatively by Bowen and Livingston[100], viz., chloranthracene–perylene, chloranthracene–rubrene, and cyanoanthracene–rubrene. They found that the rate constants for energy transfer between the different species exceeded the diffusion-controlled rate by factors between 15 and 30, in benzene and chloroform. In the more viscous liquid paraffin the difference was even greater. The results thus demonstrated that the energy exchange was not a diffusional process. It follows from these results that with quenchers having a very favourably situated absorption band and fluorescers of high efficiency, the concentration of quencher must be reduced to 10^{-4}M or less before the quenching effect of singlet energy transfer on the intensity of fluorescence can be neglected.

It may be noted that if the overlap between the emission spectrum of the donor and the absorption spectrum of the acceptor is negligible, the process 91 can still occur by actual *encounter* between $^1D^*$ and ^1A, provided of course that the singlet energy of the acceptor is less than that of the donor[101].

7. Intramolecular Energy Transfer

The process of long range singlet-to-singlet energy transfer described in the previous section results in the emission of *sensitised prompt fluorescence* of the acceptor when the solution is sufficiently concentrated that an appreciable proportion of acceptor molecules are within the critical transfer distance. We might expect therefore that, if the "donor" and "acceptor" molecules were permanently tied together by means of a saturated molecular chain, the energy transfer would be observed even in dilute solutions. Two such systems were investigated by Weber and Teale[102], viz., 1-dimethylaminonaphthalene-5-(N-benzyl)-sulphonamide (I) and 1-dimethylaminonaphthalene-5-(N-phenyl)-sulphonamide (II):

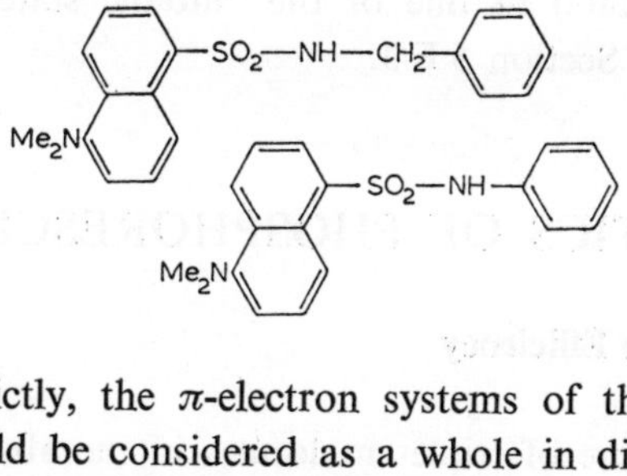

SO_2—NH—CH_2— I

Me_2N

SO_2—NH— II

Me_2N

Now although, strictly, the π-electron systems of the naphthalene and benzene rings should be considered as a whole in discussing the energy levels of these two compounds, the systems are not conjugated, and to a first approximation, contribute separately to the absorption spectrum.

Nevertheless both compounds emit a fluorescence band similar to that of 1-dimethylaminonaphthalene-5-sulphonamide (III):

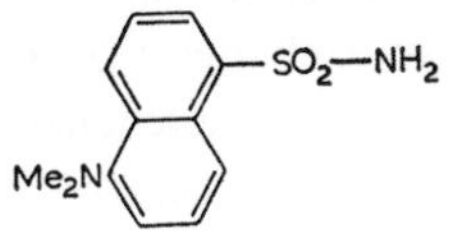

Me_2N — SO_2—NH_2 III

Neither compound was found to emit fluorescence in the region expected of a benzene derivative. The fluorescence efficiency of compound (I) was constant from long wavelengths down to wavelengths shorter than the region corresponding to benzenoid absorption, and Weber and Teale concluded that with this compound the energy transfer to the naphthalene ring system was able to compete with radiationless de-activation of the benzene ring. With compound (II) the fluorescence yield fell by about 30% at 260 nm, i.e. in the region of benzenoid absorption, and it was concluded that in this compound radiationless deactivation of the benzene ring was much more rapid than transfer to the naphthalene ring system.

A somewhat analogous process takes place in the organic compounds of some rare-earth ions, e.g. europium trisdibenzoylmethide. These exhibit broad absorption spectra characteristic of the organic ligand. Part of the excitation energy is transferred to one of the *4f* electrons of the central rare earth atom which is raised to a higher unfilled *4f* orbital. With many rare earths this excitation energy is then degraded by non-radiative processes but with europium, terbium, dysprosium and samarium, line emissions are observed corresponding to one or more $4f_n \rightarrow 4f_m$ transitions. Although this phenomenon is usually termed *intramolecular energy transfer* it may also be regarded as a form of intersystem crossing from a π^* state (in fact the lowest triplet state) to one of the "atomic states". This process is discussed further in Section 5 E 3.

C. KINETICS OF PHOSPHORESCENCE

1. Triplet Formation Efficiency

Because the lifetime of triplet molecules in the absence of very strong quenching is some orders of magnitude greater than that of molecules in the singlet state, the processes involving the triplet state can be dealt with separately by steady state theory, and the kinetic treatment is thus greatly simplified. To set up the kinetic equation we first define the triplet formation efficiency, ϕ_t, as the number of triplet molecules formed per quantum of exciting light absorbed. In the absence of photochemical reaction or inter-system crossing from upper excited singlet states, the rate of population of the lowest excited singlet state, S_1, is equal to the rate of light absorption, I_a, and the rate of formation of triplet molecules by crossing from S_1 is equal to $I_a\phi_t$. An expression for ϕ_t can be derived in the same manner as the expression for ϕ_f (equation 69), by considering the steady-state concentration of *singlet* molecules. We then find that:

$$\phi_t = k_g/(k_f + k_n + k_g + k_Q[Q]) \qquad (93)$$

and since quenching of the singlet state of strongly absorbing substances requires quencher concentrations of at least 10^{-3}M, we can ignore the factor $k_Q[Q]$ in the absence of deliberately added quenchers.

2. Phosphorescence Efficiency and Lifetime

At low light intensities the rates of the various processes by which the triplet is produced and consumed (see Fig. 23) are then represented by the following equations:

$$S_0 \overset{h\nu}{\rightarrow} S_1 \rightarrow T_1 \qquad \text{rate} = I_a\phi_t \qquad (94)$$

$$T_1 \rightarrow S_0 + h\nu' \qquad k_p[T] \qquad (95)$$

$$T_1 \rightarrow S_0 + \text{heat} \qquad k_m[T] \qquad (96)$$

$$T_1 + q \rightarrow S_0 + q \qquad k_q[T][q] \qquad (97)$$

$$T_1 \xrightarrow[\text{activation}]{\text{thermal}} S_1 \qquad k_e[T] \qquad (98)$$

in which q represents the concentration of all quenching impurities in the solution, $k_q[q]$ a composite pseudo first order rate constant for impurity quenching, and $k_e[T]$ represents the rate of thermal activation of triplet molecules back to the excited singlet level. We then have in the steady state:

$$I_a\phi_t = (k_p + k_m + k_q[q] + k_e)[T] \qquad (99)$$

Since the rate of emission of phosphorescence is $k_p[T]$, the phosphorescence efficiency ϕ_p, is given by:

$$\phi_p = k_p[T]/I_a = k_p\phi_t/(k_p + k_m + k_q[q] + k_e) \quad (100)$$

in which k_p is related to the *radiative* lifetime of the triplet molecules τ_R by:

$$k_p = 1/\tau_R \quad (101)$$

The *actual* lifetime of the triplet, τ, is given by:

$$1/\tau = k_p + k_m + k_q[q] + k_e \quad (102)$$

and equation 100 can thus be expressed in the alternative form:

$$\phi_p/\phi_t = \tau/\tau_R \quad (103)$$

At low temperatures k_e becomes negligible, and in rigid media diffusion is very small so that $k_q[q]$ can also be neglected. At low temperatures in rigid media therefore:

$$\phi_p/\phi_t = k_p/(k_p + k_m) \quad (104)$$

Now τ_R or k_p are fundamental characteristics of the substance and are independent of temperature. To determine their values we need to determine not only ϕ_p and τ, which can be obtained by direct measurement, but also ϕ_t, which until recently was difficult to determine. The triplet formation efficiency, ϕ_t, is dependent on the processes undergone by the singlet state (see equation 93) and in the absence of added singlet quenchers is given by:

$$\phi_t = k_g/(k_f + k_n + k_g) \quad (105)$$

To avoid the difficulty of determining ϕ_t directly, many workers have assumed that, because of its short lifetime, internal conversion of the first excited singlet state (n in Fig. 23) is negligible. We then have:

$$\phi_t \sim k_g/(k_f + k_g) \quad (106)$$

$$\phi_f \sim k_f/(k_f + k_g) \quad (107)$$

and thus,

$$\phi_t \sim (1 - \phi_f) \quad (108)$$

The limited experimental data so far available have indicated that for some compounds equation 108 is at least approximately valid[103, 104] although other workers have obtained different results[105]. The rate of internal conversion from the singlet state, k_n, is expected to decrease with temperature and hence equation 108 is expected to be more precisely obeyed at low temperatures. Under these conditions, the effect of tempera-

TABLE 6

PHOSPHORESCENCE EFFICIENCIES AND LIFETIMES OF 7×10^{-5} M EOSIN DI-ANION

°C	τ (sec)	ϕ_p	ϕ_p/τ
In glycerol			
− 196	0.0107	0.0643	6.0
− 70	0.0055	0.0328	6.0
− 20	0.0030	0.0160	5.3
+ 25	0.0026	0.0124	4.8
+ 70	0.0009	0.0047	5.2
In ethanol			
− 196	0.0089	0.0237	2.7
− 70	0.0039	0.0099	2.5
− 20	0.0026	0.0070	2.7
+ 25	0.0014	0.0039	2.8
+ 70	0.0007	0.0015	2.1

ture on ϕ_t will then depend only on the variation of the rate of intersystem crossing k_g with temperature. Although the absolute values of ϕ_t are difficult to determine, relative values as a function of temperature can be derived from equation 103 by measuring ϕ_p and τ. Some results obtained by Parker and Hatchard[19] for eosin in glycerol are shown in Table 6, from which it will be observed that ϕ_p/τ varies very little with temperature over the range + 70 to − 196°C. It is therefore a reasonable assumption that over this temperature range for eosin equation 108 is obeyed. Since ϕ_f for eosin in glycerol is 0.45, substituting $\phi_p/\tau = 6.0$ and $\phi_t = 0.55$ in equation 103, we find that the radiative lifetime of eosin is about 100 msec, i.e. some ten times greater than the observed lifetime at liquid nitrogen temperatures.

Methods used to determine ϕ_t are dealt with in Chapter 4.

3. Temperature Dependence of Phosphorescence

As discussed in Section 2 B 5 the efficiency of prompt fluorescence generally decreases as the temperature is raised. The effect is moderate, i.e. ϕ_f rarely changes by more than a factor of 10 between 77°K and room temperature. In contrast the efficiency of $T_1 \rightarrow S_0$ phosphorescence may decrease by several orders of magnitude over this temperature range and is critically dependent on the viscosity of the medium. The difference in behaviour is due to the long radiative lifetime of triplet states and the correspondingly greater susceptibility to impurity quenching and to $T_1 \rightarrow S_0$ intersystem crossing (see equation 100). With some compounds there is in addition the superimposed effect of variation in ϕ_t (see Section 2 B 5). This operates to oppose the quenching effect and sometimes leads to very small phosphorescence efficiencies at 77°K. However, with many compounds, for which ϕ_t varies comparatively little with temperature, the magnitude of the change in ϕ_p in passing from rigid medium at 77°K to fluid solution at room temperature may be estimated as follows. It is found experimentally that the lifetimes of many triplet molecules in deoxygenated fluid solution at room temperature lie within the range 1–10 msec, irrespective of their radiative lifetimes, provided that the triplet does not undergo chemical reaction with the solvent. If the triplet lifetime at 77°K is known, the degree of quenching at room temperature can be calculated from equation 103. Thus triplet eosin has a comparatively short radiative lifetime ($\sim$ 100 msec). At 77°K its observed lifetime is about 10 msec and its phosphorescence is

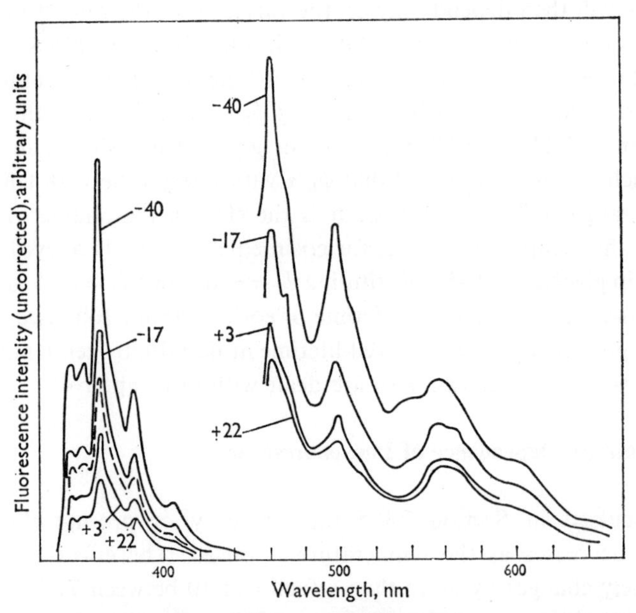

Fig. 27. Phosphorescence (right hand curves) and delayed fluorescence (left hand curves) of phenanthrene in ethanol.

Rate of light absorption was about 10^{-5} einstein litre^{-1} sec^{-1} at 313 nm. Temperatures are indicated against each curve. Sensitivities were 8,500 and 108 times that used to measure the prompt fluorescence (dotted curve) (from Parker[441]).

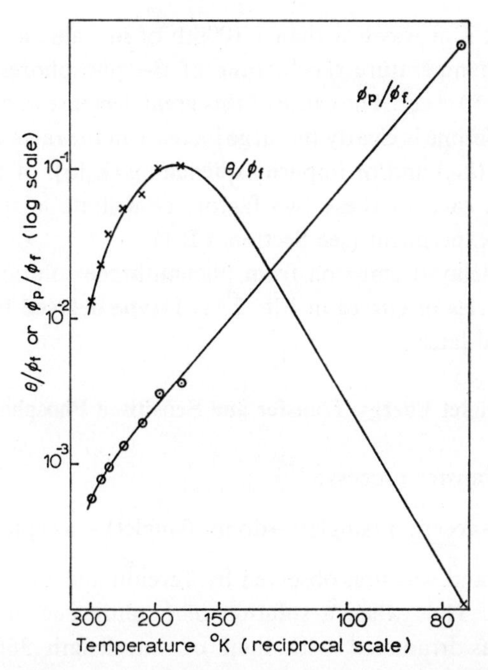

Fig. 28. Variation of efficiencies of phosphorescence and delayed fluorescence of phenanthrene with temperature.

thus quenched by a factor of about 10 under these conditions. At room temperature in glycerol ($\tau_p = 2.6$ msec) its phosphorescence is quenched by an additional factor of only about 4, and it can be readily observed. The triplet lifetime was measured as early as 1930 by Boudin [31] using a visual phosphoroscope. On the other hand the radiative lifetimes of many triplets, e.g. the triplets of the aromatic hydrocarbons, are of the order 1 to 10 sec, and hence it is to be expected that their phosphorescence efficiency in fluid solution at room temperature will be reduced by a factor of 1000 or more. Phosphorescence from an aromatic hydrocarbon in fluid solution at room temperature was first observed by Parker and Hatchard [35] in 1962. Spectra obtained with phenanthrene at several temperatures are shown in Fig. 27 (right hand curves) and values of $\log_{10}(\phi_p/\phi_f)$ plotted against 1/T covering a wider range of temperatures are shown in Fig. 28. At 77°K the phosphorescence intensity was about the same as that of the prompt fluorescence. At -98°C the phosphorescence intensity was already some 200 times less than at 77°K. As the temperature rose the phosphorescence intensity rapidly fell and at + 22°C it was less than 1/1000th of its value at 77°K. Over the same range of temperature the lifetime of the phosphorescence fell from 3.6 sec to about 10^{-3} sec. The cause of this great decrease in phosphorescence efficiency and lifetime is clearly the large increase in the rates of the radiationless conversion (k_m) and/or impurity quenching ($k_q[q]$) of the triplet state. To what extent each of these two factors contribute is still a matter for discussion and experiment (see Section 4 B 1).

The intense delayed emission from phenanthrene solutions indicated by the left hand series of curves in Fig. 27 is P-type delayed fluorescence and will be discussed later.

4. Triplet-to-Singlet Energy Transfer and Sensitised Phosphorescence

The energy transfer process:

donor (triplet) + acceptor (singlet) → donor (singlet) + acceptor (triplet) (109)

is spin-allowed and was first observed by Terenin and Ermolaev[106] in rigid media at 77°K. Thus when a solution of naphthalene in ethanol at this temperature was irradiated with light of wavelength 366 nm, no lumi-

TABLE 7

SYSTEMS INVESTIGATED BY TERENIN AND ERMOLAEV
(Adapted from Terenin and Ermolaev[106])

Compound	Height of triplet level (μm^{-1})	Phosphorescence lifetime (sec)	Long wavelength limit of absorption spectrum (μm^{-1})
Donors			
Acetophenone	2.585	2.3×10^{-3}	2.75
Benzoin	2.565	2.4×10^{-3}	2.80
Benzaldehyde	2.52	1.5×10^{-3}	2.675
Diphenylamine	2.52	1.85	3.10
o-Hydroxy-benzaldehyde	2.46	—	2.60
Carbazole	2.448	7.25	2.95
Benzophenone	2.42	4.7×10^{-3}	2.60
Acceptors			
Diphenyl	2.28	4.4	3.40
Naphthalene	2.13	2.30	3.10
α-Methylnaphthalene	2.12	—	3.05
α-Chlornaphthalene	2.07	0.30	—

nescence was observed because the naphthalene does not absorb at this wavelength. However, if benzaldehyde was added to a solution containing the same concentration of naphthalene a bright phosphorescence emission characteristic of the $T_1 \rightarrow S_0$ transition of naphthalene appeared. Clearly the exciting light was absorbed by the benzaldehyde and the excitation energy passed on to the naphthalene to produce the *sensitised phosphorescence*. It could not have been passed by singlet-to-singlet transfer because the singlet state of benzaldehyde is situated at a lower energy than the singlet state of naphthalene. Terenin and Ermolaev showed that sensitisation would occur provided that the triplet level of the donor was situated above that of the acceptor. For example all the compounds in the top half of Table 7 were found to be capable of acting as donors to those in the bottom half of the table. They proposed a scheme involving the following essential processes:

$$D \xrightarrow{h\nu} D^* \rightarrow {}^3D \qquad (110)$$

$${}^3D \rightarrow {}^1D + h\nu' \qquad (111)$$

$${}^3D \rightarrow {}^1D \qquad (112)$$

$${}^3D + {}^1A \rightarrow {}^1D + {}^3A \qquad (109)$$

$${}^3A \rightarrow A + h\nu'' \qquad (113)$$

Assuming that the probability of energy transfer (equation 109) is proportional to the concentration of acceptor, the quenching of the donor triplet may be written:

$$1/\tau - 1/\tau_0 = k_{tr}c_A \qquad (114)$$

where τ and τ_0 are the lifetimes of donor phosphorescence with and without acceptor present at a concentration c_A and k_{tr} is a second order rate constant. They found values of $k_{tr}\tau_0$ between 0.7 and 2.3 litre mole^{-1} in ethanol–ether solution at 77°K. Thus quite high concentrations of acceptor were required to produce appreciable donor triplet quenching and acceptor sensitisation.

Triplet-to-singlet energy transfer in fluid solution was first investigated by Bäckström and Sandros. They found[33] that substances capable of quenching the phosphorescence of biacetyl could be divided into two classes. The first class comprised substances having a loosely bound hydrogen atom, such as primary and secondary amines, phenols, aldehydes and alcohols. With these compounds they assumed that the quenching action resulted from photochemical reaction with triplet biacetyl in which a hydrogen atom was abstracted from the quencher molecule. Quenching by the second class of compounds was due to triplet-to-singlet energy transfer (equation 109). They suggested that the probability of transfer increases with the extent of overlap of the phosphorescence spectrum of the biacetyl and the $T_1 \leftarrow S_0$ absorption spectrum of the quencher as well as with the oscillator strengths of the individual transitions. Bäckström and Sandros[107] also observed *sensitised phosphorescence* in fluid solution, using as acceptors the phosphorescent diketones, biacetyl, benzil and anisil. As sensitiser they used benzophenone whose triplet level lies 0.4 μm^{-1} above that of biacetyl. This difference was sufficiently large to justify the assumption that the transfer of triplet energy was diffusion-controlled. Thus by measuring the intensity of sensitised phosphorescence of biacetyl as a function of the concentration of biacetyl, they were able to calculate values for the triplet lifetime of benzophenone at 20°C in solutions of benzene (1.9×10^{-6} sec) and isopropanol (5.7×10^{-8} sec).

Porter and Wilkinson[108] investigated triplet-to-singlet energy transfer by the method of flash absorption spectroscopy. They were able to observe directly both the quenching of the donor triplet and the formation of the acceptor triplet and they were able to exclude the process:

$$^1D^* + {}^1A \rightarrow {}^1D + {}^3A \tag{115}$$

by the fact that the fluorescence yield of the donor was unchanged by the presence of the acceptor. Process 115 is improbable on grounds of spin conservation. They were in agreement with many of the conclusions of Bäckström and Sandros but found no indication that the quenching efficiency decreases as the separation of the acceptor triplet below that of the donor increases. They concluded that energy transfer occurs between molecules during an encounter at normal collisional separation. Under these conditions overlap of the orbitals of the two molecules occurs, the electrons are then indistinguishable and the acceptor may emerge in an electronically excited state.

Porter and Wilkinson[108] found that for the donor–acceptor pairs that they investigated, the triplet-to-singlet energy transfer took place on every encounter provided that the triplet level of the acceptor was situated well below that of the donor. Sandros and Bäckström[109], and Parker, Hatchard and Joyce[110], later showed that the process can also take place when the triplet levels of donor and acceptor are situated close together, although the encounter efficiency is then lower. We can visualise the process by a diagram such as that shown in Fig. 29. When the triplet level of the acceptor lies well below that of the donor (diagram A) energy transfer results in the formation of a vibrationally excited triplet of the acceptor with energy equal to that held originally by the donor. The acceptor then rapidly loses its

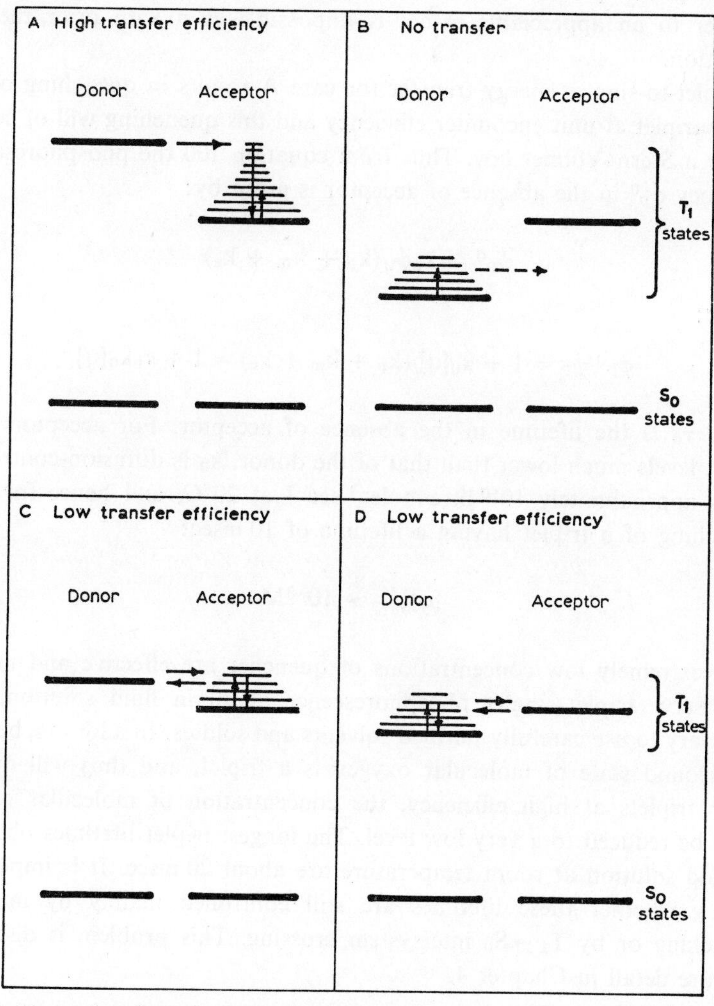

Fig. 29. Illustrating dependence of triplet-to-singlet transfer efficiency on the relative positions of the triplet levels.

excess of vibrational energy after which it does not have sufficient energy to permit transfer back to the donor. On the other hand if the triplet levels lie close together as in diagrams C and D (whether the acceptor is just above or just below the donor) transfer is possible in both directions—it can proceed from the lower to the higher level by prior thermal activation. The overall probability of transfer per encounter is thus reduced. If however the acceptor triplet lies well above that of the donor (diagram B), energy transfer to an appreciable extent is impossible even with prior thermal activation.

Triplet-to-singlet energy transfer for case A results in quenching of the donor triplet at unit encounter efficiency and this quenching will of course follow a Stern–Volmer law. Thus from equation 100 the phosphorescence efficiency $\phi_p{}^0$ in the absence of acceptor is given by:

$$\phi_p{}^0 = k_p\phi_t/(k_p + k_m + k_e) \tag{116}$$

Hence:

$$\phi_p{}^0/\phi_p = 1 + k_q[q]/(k_p + k_m + k_e) = 1 + \tau_0 k_q[q] \tag{117}$$

where τ_0 is the lifetime in the absence of acceptor. For acceptors with triplet levels much lower than that of the donor, k_q is diffusion-controlled, i.e. is approximately 10^{10} litre mole^{-1} sec^{-1} at 20°C, and hence for 50% quenching of a triplet having a lifetime of 10 msec:

$$[q]_{50\%} \sim 10^{-8}\text{M} \tag{118}$$

Thus extremely low concentrations of quencher are effective and indeed, to observe triplet–singlet phosphorescence at all in fluid solution, it is necessary to use carefully purified solvents and solutes. In addition, because the ground state of molecular oxygen is a triplet, and thus will quench other triplets at high efficiency, the concentration of molecular oxygen must be reduced to a very low level. The longest triplet lifetimes observed in fluid solution at room temperature are about 20 msec. It is impossible to say whether these lifetimes are still controlled mainly by impurity quenching or by $T_1 \rightarrow S_0$ intersystem crossing. This problem is discussed in more detail in Chapter 4.

D. KINETICS OF DELAYED FLUORESCENCE

1. E-Type Delayed Fluorescence

The emission of E-type delayed fluorescence is the result of thermal activation of molecules from the lowest triplet state to the first excited singlet state followed by radiative transition from there to the ground state (see Section 1 C 4). It is the kind of delayed fluorescence considered by Perrin, and by Jablonski, and investigated by Lewis and co-workers in rigid medium, and by Parker and Hatchard in fluid solution. The work of the latter[19] with solutions of eosin provide the most complete quantitative measurements, and we shall now consider these in more detail. The kinetics of the emission of E-type delayed fluorescence can be readily deduced by reference to the reaction scheme given by equations 94 to 98 (Section 2 C 2). Thus, the efficiency of phosphorescence (see equation 100) is given by:

$$\phi_p = k_p\phi_t/(k_p + k_m + k_q[q] + k_e) \qquad (119)$$

Similarly if we assume that all molecules thermally activated to the excited singlet state (equation 98) fluoresce with the usual fluorescence efficiency, ϕ_f, the efficiency of delayed fluorescence is given by an expression analogous to 119, viz.:

$$\phi_e = k_e\phi_t\phi_f/(k_p + k_m + k_q[q] + k_e) \qquad (120)$$

and hence

$$\phi_e/\phi_p = k_e\phi_f/k_p \qquad (121)$$

The ratio of the intensities of the bands of E-type delayed fluorescence (high wavenumber bands in Fig. 30) and phosphorescence (low wavenumber bands in Fig. 30) should thus be independent of ϕ_t and of all triplet quenching processes.

Now k_e represents a thermal activation to an upper vibrational level of the triplet state followed by intersystem crossing to the first excited singlet state and we can therefore write:

$$k_e = \mathbf{A} \exp(-\Delta E/\mathbf{RT}) \qquad (122)$$

where **A** is a frequency factor and ΔE the activation energy required. Hence from equations 101, 121 and 122,

$$\phi_e/\phi_p = \tau_R\phi_f\mathbf{A} \exp(-\Delta E/\mathbf{RT}) \qquad (123)$$

Plots of ln (ϕ_e/ϕ_p) determined from the spectra in Fig. 30, against 1/T were linear (see Fig. 31) and the value of ΔE calculated from the slopes agreed within experimental error with the energy difference between the maxima of the two emission bands, thus confirming the proposed mechanism.

It should be noted that the rate of emission of E-type delayed fluorescence is equal to $k_e[T]$, i.e. like phosphorescence, it is directly proportional to the concentration of triplet molecules present. When the exciting light is shut off, it thus decays exponentially with a lifetime equal to that of the

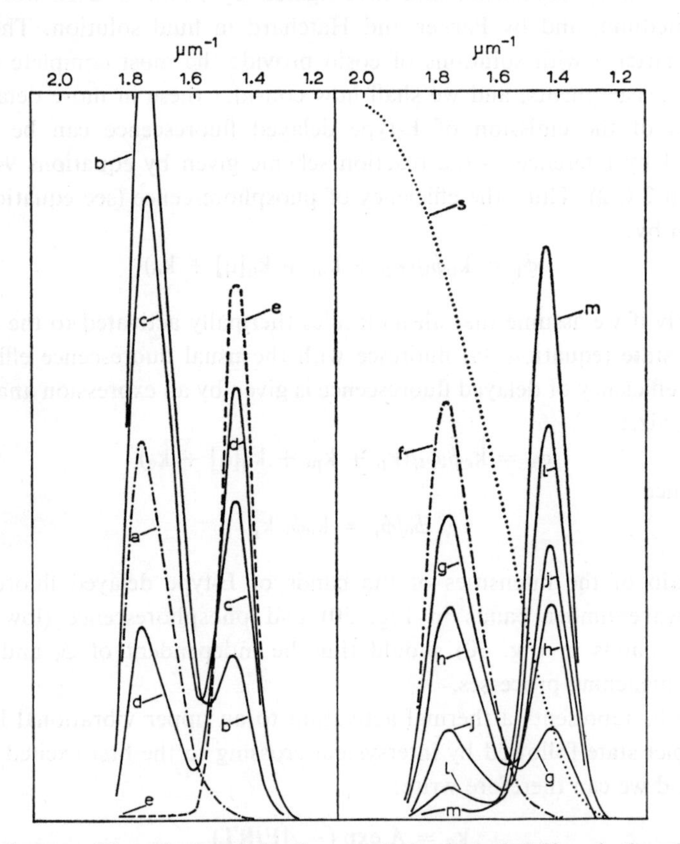

Fig. 30. Spectra of long-lived emission from eosin in glycerol (left) and ethanol (right). Prompt fluorescence (a) + 30°C, and (f) + 22°C. Long-lived emission at (b) + 69°C, (c) + 48°C, (d) + 18°C, (e) − 40°C, (g) + 71°C, (h) + 43°C, (j) + 22°C, (l) − 7°C, (m) − 58°C. Intensity scales for (b)–(e) are 200 times less than for (a); intensity scales for (g)–(m) are 1000 times less than for (f). Curve S is spectral sensitivity of 9558 photomultiplier with quartz monochromator ($S_{\bar{\nu}}$) (from Parker and Hatchard[19]).

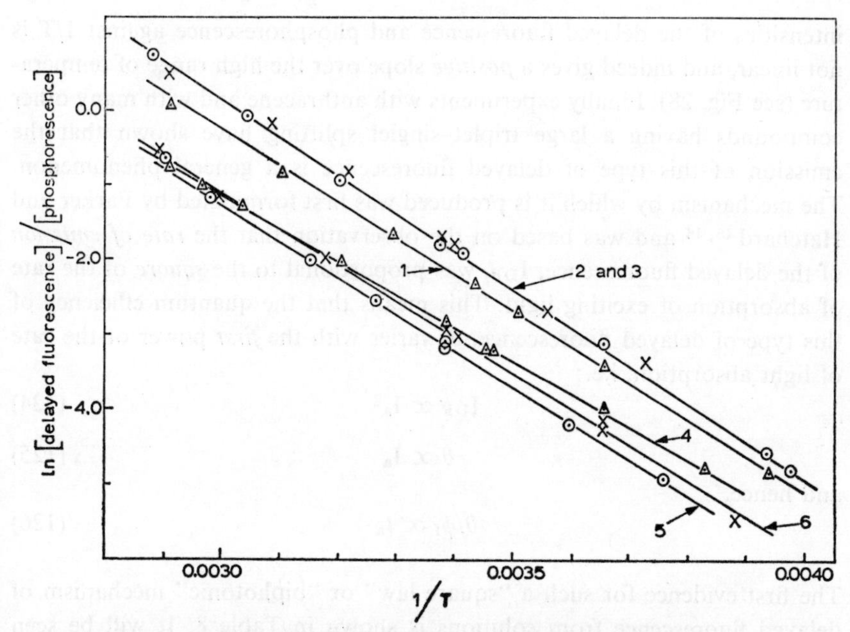

Fig. 31. E-Type delayed fluorescence as a function of temperature.
Curves 1, 2, 3: eosin in glycerol at concentrations 7×10^{-5} M, 1.5×10^{-5} M and 0.08×10^{-5} M; curves 4, 5, 6: eosin in ethanol at concentrations 7×10^{-5} M, 1.5×10^{-5} M and 0.3×10^{-5} M (from Parker and Hatchard[19]).

phosphorescence. In this it differs from the kind of delayed fluorescence we shall describe in the next section.

The application of measurements of E-type delayed fluorescence to the determination of approximate rates of intersystem crossing processes, and triplet energies, is described in Chapter 4.

2. Triplet-to-Triplet Energy Transfer and P-Type Delayed Fluorescence

The emission of delayed fluorescence (left hand curves in Fig. 27) as well as phosphorescence (right hand curves in Fig. 27) from solutions of phenanthrene suggests at first sight that this is another case of delayed fluorescence similar to that observed with eosin. However, the following facts show clearly that the emission cannot be E-type delayed fluorescence. First, the separation of the triplet and first excited singlet levels of phenanthrene is so large (19 kcal) that thermal activation at room temperature must be very small. Second, a plot of the logarithm of the ratio of the intensities of the delayed fluorescence and phosphorescence against 1/**T** is not linear, and indeed gives a *positive* slope over the high range of temperature (see Fig. 28). Finally experiments with anthracene and with many other compounds having a large triplet–singlet splitting have shown that the emission of this type of delayed fluorescence is a general phenomenon. The mechanism by which it is produced was first formulated by Parker and Hatchard[45,46] and was based on the observation that the *rate of emission* of the delayed fluorescence, I_{DF}, was proportional to the *square* of the rate of absorption of exciting light. This means that the quantum efficiency of this type of delayed fluorescence, θ, varies with the *first* power of the rate of light absorption, i.e.:

$$I_{DF} \propto I_a^2 \tag{124}$$

$$\theta \propto I_a \tag{125}$$

and hence,

$$\theta/\phi_f \propto I_a \tag{126}$$

The first evidence for such a "square law" or "biphotonic" mechanism of delayed fluorescence from solutions is shown in Table 8. It will be seen that the ratio θ/ϕ_f (which with E-type delayed fluorescence is independent of rate of light absorption) varies directly with the rate of light absorption. Similar evidence for such square law delayed fluorescence has since been obtained with many other compounds in fluid solution[34–36,86,111–115] and with several compounds in the vapour phase[47]. The proposed mechanism[46] was based on the assumption that encounter between two triplet molecules gives rise to a species X which subsequently dissociates into excited and ground state singlet molecules:

$$S_0 \xrightarrow{h\nu} S_1 \rightarrow T \qquad \text{rate} = I_a\phi_t \tag{127}$$

$$T \rightarrow S_0 \qquad k_h[T] \tag{128}$$

$$T + T \rightarrow X \qquad k_x[T]^2 \tag{129}$$

$$X \rightarrow \text{products} \qquad k_r[X] \tag{130}$$

$$X \rightarrow S_1 + S_0 \qquad k_s[X] \tag{131}$$

$$S_1 \rightarrow S_0 + h\nu' \qquad \text{fast} \tag{132}$$

in which k_h represents the sum of all first order processes by which the triplet is consumed (see equations 95 to 98). At low rates of light absorption (I_a)

the rate of triplet self-quenching (equation 129) is small compared with the rate of first order consumption (equation 128). Hence in the steady state:

$$I_a\phi_t = k_h[T] = [T]/\tau \qquad (133)$$

where τ is the lifetime of the triplet. If it is then assumed that the dissociation of the unknown intermediate X is rapid, so that equation 129 is the rate-controlling process for the emission of delayed fluorescence, then, since two triplets are required for the production of each excited singlet molecule, which then emits with an efficiency ϕ_f, we have:

$$I_{DF} = \tfrac{1}{2}k_x[T]^2\phi_f k_s/(k_r + k_s) \qquad (134)$$

and hence from equation 133,

$$\theta/\phi_f = \tfrac{1}{2}k_x I_a(\phi_t\tau)^2(k_s/(k_r + k_s)) \qquad (135)$$

where θ is the efficiency of delayed fluorescence.

It will be observed that the overall effect of reactions 129 and 131 is the transfer of triplet energy from one triplet molecule to a second *triplet*

TABLE 8

EFFECT OF LIGHT INTENSITY ON THE DELAYED FLUORESCENCE OF ANTHRACENE SOLUTIONS

(The highest intensity of exciting light was approximately 1.4×10^{-8} einstein cm^{-2} sec^{-1} at 2.73 μm^{-1} (366 nm))

Relative intensity of exciting light	5×10^{-4} M		5×10^{-5} M	
I_0	$10^4\theta/\phi_f$	$\theta/(\phi_f I_0)$	$10^4\theta/\phi_f$	$\theta/(\phi_f I_0)$
1.00	93	0.009	92	0.009
0.29	36	0.012	30	0.010
0.112	17.5	0.016	10.4	0.009
0.081	12.9	0.016	—	—
0.032	4.5	0.014	2.3	0.007
0.010	1.4	0.014	—	—

Note

θ was the efficiency of delayed fluorescence.

ϕ_f was the efficiency of prompt fluorescence.

The values of θ/ϕ_f for the two solutions are not directly comparable because the exciting light was strongly absorbed in the more concentrated solution and the luminescence had to be observed by frontal illumination. The precision of these measurements was not as good as those of later experiments but was sufficient to indicate the dependence of θ on the rate of light absorption.

These results are taken from the original paper of Parker and Hatchard[46].

molecule, i.e. it results in the concentration of part of the energy from two separately absorbed quanta on to one molecular species. We can thus refer to the processes represented by equations 129 and 131 as triplet-to-triplet energy transfer. We shall see in Chapter 4 that with some compounds, of which pyrene is the best example, the species X consists of an excited dimer S_2^* and has been identified as such by the fact that it emits characteristic fluorescence before it has time to dissociate completely, i.e.:

$$\begin{array}{ccc} T + T \rightarrow & S_2^* \longrightarrow & S_1 + S_0 \\ & \downarrow h\nu'' & \downarrow h\nu' \\ & S_0 + S_0 & S_0 \end{array} \qquad (136)$$

With other compounds, emission from the excited dimer is not observed and it was originally assumed that in these cases the dimer dissociated too rapidly for its emission to be appreciable. There is now, however, considerable evidence[84–86] that triplet-to-triplet energy transfer can also take place over distances greater than the encounter distance. This second process may be truly represented as pure energy transfer, viz.:

$$T + T \rightarrow S_1 + S_0 \qquad (137)$$

It is therefore desirable to express equation 135 in terms of parameters more relevant to the concept of triplet-to-triplet energy transfer as represented by equation 137. This we do as follows. First it should be noted that measurements by flash absorption spectroscopy[108] have shown that the rate constant for bimolecular triplet–triplet quenching is usually close to the diffusion controlled value. It is therefore convenient to express the product $k_x(k_s/(k_r + k_s))$ in equation 135 as the product $p_c k_c$ in which k_c is the diffusion-controlled bimolecular rate constant and p_c is the probability that an encounter between two triplets will give rise ultimately to an excited and ground state singlet molecule. Equation 135 then reduces to the following form:

$$\theta/\phi_f = \tfrac{1}{2} p_c k_c I_a (\phi_t \tau)^2 \qquad (138)$$

When the exciting light is shut off the triplet concentration decays exponentially with a lifetime, $\tau = 1/k_h$. Since the rate of emission of phosphorescence ($I_p = k_p[T]$) is always proportional to the triplet concentration, it decays at the same rate as the triplet, viz.:

$$I_p = I_p^0 \exp(-t/\tau) \qquad (139)$$

Now the rate of emission of delayed fluorescence (see equation 134) is at all times proportional to $[T]^2$, and hence when the exciting light is shut off:

$$I_{DF} = I_{DF}^0 [\exp(-t/\tau)]^2 = I_{DF}^0 \exp(-2t/\tau) \qquad (140)$$

Thus the intensity of delayed fluorescence should also decay exponentially, but with a lifetime equal to *one half* of that of the phosphorescence. Results obtained with phenanthrene in ethanol[46], with acenaphthene, benzanthracene, fluoranthene and pyrene in liquid paraffin[115] and with dyestuffs in ethanol at low temperatures[36] have confirmed this prediction. Since in fluid solution phosphorescence is usually much weaker than delayed fluorescence, the measurement of the lifetime of delayed fluorescence gives a convenient method for measuring triplet lifetimes in fluid solution. Values of $\tau_{DF}(=\tau/2)$, θ/ϕ_f and I_a, all measured with a single de-oxygenated solution, together with independently determined values of ϕ_t can be inserted in equation 138 to provide a value for p_c. This and other applications are discussed in Chapter 4.

3. P-Type Delayed Fluorescence in Rigid Media

As the temperature of a solution is reduced its viscosity rises and hence the value of k_c in equation 138 decreases. At the same time the lifetimes of most triplet molecules increase and it is difficult to predict from equation 138 how the efficiency of P-type delayed fluorescence will be affected at intermediate temperatures. At very low temperatures in very rigid solvents, k_c is very small, and in fact it is found that rates of light absorption sufficient to give quite high efficiencies of delayed fluorescence in fluid media, give negligible efficiencies in rigid media. However, if the rate of light absorption is increased greatly by using high solute concentrations, delayed fluorescence can still be observed from some compounds in rigid medium. Somewhat conflicting results have been obtained by different workers. This is probably due to the fact that the physical properties of a frozen glass depend critically on its composition, and it is not always possible to be sure that the solute is retained completely in solution. Furthermore great care must be taken to distinguish P-type delayed fluorescence in rigid medium from the recombination delayed fluorescence discussed in Section 1 C 6 because the latter can be produced by a biphotonic mechanism involving the successive absorption of two quanta, and hence can also have an intensity proportional to the square of the intensity of the exciting light.

The first observation of delayed fluorescence in rigid medium, attributed to triplet-to-triplet energy transfer was that of Czarnecki[116] using solutions of naphthalene in rigid methylmethacrylate polymer. Muel[117] found that the intensity of the delayed fluorescence from solutions of benz(a)pyrene at 77°K was proportional to the square of the intensity of the exciting light and decayed exponentially with a lifetime equal to one half of that of the triplet. He proposed that the delayed fluorescence was produced via a process of triplet-to-triplet energy transfer analogous to equation 137.

Azumi and McGlynn[118, 119] observed delayed fluorescence from several aromatic hydrocarbons in ether–pentane–ethanol glass at 77°K and found that although the phosphorescence decayed exponentially (or nearly so) the delayed fluorescence could be represented by the sum of two first order processes, the faster of which had a lifetime considerably shorter than one half that of the phosphorescence. They assumed that diffusional transport in the glassy medium was negligible and interpreted the delayed fluorescence in terms of direct resonance transfer between triplets—i.e. triplet-to-triplet energy transfer over distances greater than the encounter distance. They did not however explain what was the rate controlling process responsible for the long lifetime of the delayed fluorescence.

Dupuy[120] found no delayed fluorescence from chrysene, naphthalene or phenanthrene in certain *crystalline* hydrocarbon matrices, but observed that the presence of a small quantity of benzene was sufficient to cause the emission of delayed fluorescence. He suggested that this was due to the formation of dimers as a result of concentration of the fluorescent hydrocarbon in the "islets" of benzene formed by rejection of the benzene from the crystallising matrix. This mechanism, like that of Azumi and McGlynn, assumed that diffusional transport in the glassy medium was negligible, and thus accounted for the absence of oxygen quenching. It did not however explain what is the rate controlling process governing the lifetime of the delayed fluorescence. Heterogeneous matrices were also investigated by McGlynn and co-workers[121]. With pyrene in pure isopentane glass they observed only phosphorescence, but if small amounts of water were introduced, the slightly foggy glasses showed both phosphorescence and delayed fluorescence. At intermediate concentrations the delayed fluorescence showed bands due both to monomer and dimer, the intensity of the former being proportional to the square of the exciting light intensity and the latter to the first power of the exciting light intensity. They at first proposed that the dimer emission was due to the formation of a *long-lived* excited dimer but later attributed it to crystallites of pyrene. It is clear that the situation existing in such heterogeneous glasses is complex and more work is required to substantiate the findings, in particular to decide whether some of the other effects may be due to micro crystals of solute.

Detailed kinetic data have been obtained by Parker[84] for phenanthrene in clear rigid glasses at 77°K. He found that concentrated solutions in clear ether–pentane–ethanol glass gave a delayed fluorescence spectrum identical with that of the prompt fluorescence and with an intensity proportional to the square of the intensity of the exciting light. Finely crystalline suspensions of phenanthrene in isopentane at 77°K also gave delayed fluorescence. The spectrum of this was identical with that of the prompt fluorescence of the *crystals*, but was different from the spectrum obtained in the clear glass. Unlike that from the clear glass the intensity was proportional to the first power of the intensity of the exciting light. Parker therefore concluded that the delayed fluorescence from the glass was not due to fine crystals but to phenanthrene held in true solution. The possibility that the delayed fluorescence was due to recombination following photo-ionisation of the triplet state, was ruled out by the results obtained with varying wavelengths of exciting light. For *complete* light absorption the rate of emission of delayed fluorescence (I_{DF}) was found to obey the relation:

$$I_{DF} \propto I_0^2 \varepsilon_S \qquad (141)$$

where ε_S is the extinction coefficient of the ground state. A mechanism involving direct absorption by the triplet state would require

$$I_{DF} \propto I_0^2 \varepsilon_T \qquad (142)$$

The observed results would then require $\varepsilon_S \propto \varepsilon_T$ at various wavelengths, which seems unlikely.

Although the phosphorescence decayed exponentially with a lifetime of over 3 sec, the decay of the delayed fluorescence was non-exponential and the initial rate of decay was many times faster than the rate of decay of phosphorescence (see Fig. 32). The intensity of delayed fluorescence was proportional to the square of the intensity of the phosphorescence (and also proportional to the square of the intensity of the prompt fluorescence). This in itself does not prove that it was produced by triplet–triplet annihilation, but merely shows that two absorbed quanta are required for the production of one quantum of delayed fluorescence. That triplet molecules are involved was shown by observing the rate of *growth* of the delayed fluorescence (see Fig. 33). Thus when the exciting light was turned on after a short period of darkness (about 1 sec) the delayed fluorescence intensity

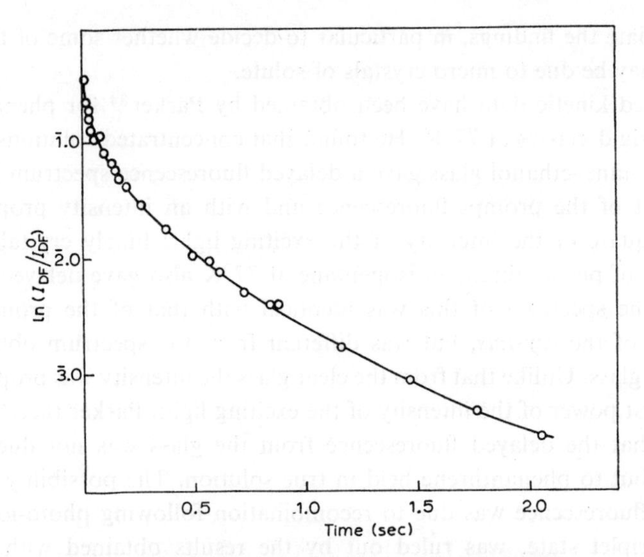

Fig. 32. Decay of delayed fluorescence in rigid medium.
Phenanthrene in ether–pentane–ethanol glass at 77°K; rate of light absorption was 6×10^{-5} einstein litre^{-1} sec^{-1} at 313 nm (from Parker[84]).

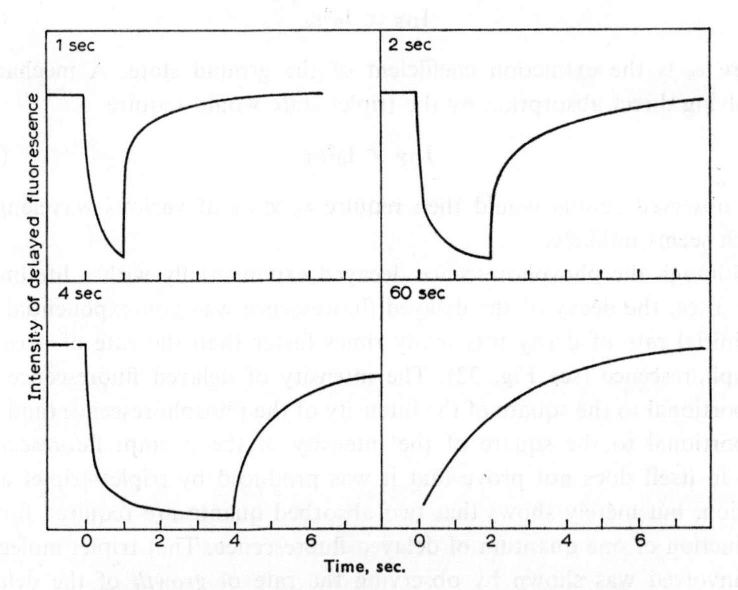

Fig. 33. Growth of delayed fluorescence in rigid medium.
Phenanthrene in ether–pentane–ethanol glass at 77°K; duration of preceding period of darkness as indicated (from Parker[84]).

rose rapidly to a value close to the maximum, and then more slowly until the maximum was reached. The extent of the initial rapid rise was smaller the longer the preceding period of darkness, and after a period of darkness long enough for the triplet concentration to have fallen to a negligible value, the delayed fluorescence rose approximately exponentially at a rate only a little less than the rate of growth of the triplet concentration as indicated by the phosphorescence. These results are consistent with the view that the delayed fluorescence is produced by mutual annihilation of neighbouring triplet molecules[84]. That either diffusion or rotation of the triplets is the rate controlling process is suggested by the results in Table 9 which show that as the rigidity of the glass increases the efficiency of triplet–triplet annihilation, as measured by $\theta/(\phi_t I_a)$, decreases.

Parker interpreted the results in terms of long range triplet–triplet annihilation involving one of two alternative mechanisms. The first assumed slow diffusive transport and the establishment of a non-random distribution of triplet molecules as a result of their interaction. The alternative mechanism assumed that diffusive transport was negligible and attributed the delayed fluorescence to the interaction between those pairs of triplets initially formed at sites within the interaction distance, the delay in emission corresponding to the time taken for rotation of the triplets into a favourable configuration for interaction. Quantitative treatment of the results indicated that the first mechanism would require a triplet–triplet interaction distance of at least

TABLE 9

EFFECT OF SOLVENT ON THE EFFICIENCY OF THE DELAYED FLUORESCENCE OF PHENANTHRENE AT 77°K

Solvent	Rigidity	Rate of light absorption $10^5 I_a$ (einstein litre^{-1} sec^{-1})	θ/ϕ_t	$\theta/(\phi_t I_a)$	Triplet lifetime (sec)
E.P.A.	low	6.1	0.0041[a]	67	3.6 ± 0.1
	↓	5.8	0.0038[d]	66	
Ethanol	↓	7.6	0.0011[d]	14	3.7 ± 0.2
Liquid paraffin	↓	6.7	0.00047[a]	7	3.4 ± 0.2
PMM	high	7.3	0.00012[a]	1.6	2.9
		7.8	0.00013[d]	1.7	

[a] aerated, [d] de-aerated.
E.P.A. = ethyl ether 2 vol, isopentane 2 vol, ethanol 1 vol.
PMM = polymethylmethacrylate.

30 Å units. Assessment of the second mechanism indicated that an interaction distance greater than 15 Å units was required to explain the results.

It is possible that both diffusive and rotational movements of molecules in highly viscous media occur simultaneously with about equal difficulty and the system would then be described by a combination of the two mechanisms. Non-steady state conditions, whether they arise from slow diffusive transport, or hindered rotation, are likely to be frequently encountered in situations where photoluminescence is produced via a bimolecular process in a glassy medium. Transient rate kinetics[84,87] must then be used to give a quantitative description of the system (see also Section 2 B 3).

4. Annihilation-Controlled Delayed Fluorescence

The overall rate of disappearance of a triplet species in a fluid solution containing no other triplet species is equal to the sum of the rates of the four first order processes given in equations 95–98, together with a combination of second order processes. The latter include the formation of an excited dimer (equation 129), the direct formation of excited monomer (equation 137), and possibly the formation of other products (see Section 4 B 3). We shall represent the sum of the rates of the first order processes by a composite first order rate constant k_h, and the sum of the rates of the second order processes by a composite second order rate constant k_a. The rate of disappearance of the triplet molecules of a compound A is then given by:

$$d[^3A]/dt = -k_h[^3A] - k_a[^3A]^2 \qquad (143)$$

The square-law dependence of the intensity of P-type delayed fluorescence holds only with certain limits, i.e. only if the rate of light absorption is limited to values for which the second term on the right hand side of equation 143 is negligible compared with the first. It is of interest to express the limits in terms of molecular parameters, and to enquire what values of these parameters favour high efficiency of delayed fluorescence. We shall define a factor p_a as the probability that an annihilation of two triplet molecules will ultimately lead to the formation of an excited singlet molecule. The rate of formation of the latter is then equal to $\frac{1}{2}p_a k_a[^3A]^2$. For simplicity we shall also assume that triplet–triplet annihilation is diffusion-controlled, i.e.:

$$k_a = k_c = 8\mathbf{RT}/3000\eta \qquad (144)$$

Then for square-law delayed fluorescence:

$$k_c[^3A]^2 \ll k_h[^3A] \tag{145}$$

and the efficiency of delayed fluorescence is given by an equation analogous to 138, viz.:

$$\theta = \tfrac{1}{2}\phi_f p_a k_c I_a(\phi_t \tau)^2 \tag{146}$$

A high value of θ will be favoured by low solvent viscosity (high k_c), high probability factor p_a and long triplet lifetime. If two compounds have the same triplet lifetime, the one having the higher value of $\phi_f \phi_t^2$ will show the higher θ. The maximum value will be obtained with a compound for which $\phi_f = \frac{1}{3}$ and $\phi_t = \frac{2}{3}$.

To determine the maximum possible efficiency of "square law" delayed fluorescence we have to set an arbitrary limit to the proportion of second order triplet decay that we can tolerate. We shall choose as this arbitrary limit the condition in which the rate of disappearance of triplets by annihilation has risen to $\frac{1}{5}$th of that for first order decay. Thus at the arbitrary limit:

$$k_c[^3A]^2 = \tfrac{1}{5}k_h[^3A] \tag{147}$$

and the steady state equation for this condition is:

$$I_a\phi_t = k_h[^3A] + k_c[^3A]^2 = 1.2k_h[^3A] \tag{148}$$

i.e. from equation 147:

$$I_a\phi_t = 0.24k_h^2/k_c = 0.24/(\tau^2 k_c) \tag{149}$$

Thus in ethanol solution at room temperature ($k_c = 0.6 \times 10^{10}$ litre mole^{-1} sec^{-1}) with a compound having a triplet lifetime of 5 msec, the limit will be reached when $I_a\phi_t$ is about 2×10^{-6} einstein litre^{-1} sec^{-1}. The efficiency of delayed fluorescence at the arbitrary limit is no longer given precisely by equation 146. It may be derived as follows:

$$\theta = \tfrac{1}{2}\phi_f p_a k_c[^3A]^2/I_a \tag{150}$$

and by substitution from equations 147 and 149:

$$\theta = \tfrac{1}{12}p_a\phi_f\phi_t \tag{151}$$

Hence, provided that I_a is increased sufficiently to allow the conditions corresponding to the limit to be attained, the value of θ is *independent of all triplet reaction rates*. It attains its maximum value of 0.02 for a compound having $p_a = 1.0$ and $\phi_f = \phi_t = 0.5$.

As the rate of light absorption increases, the rate of triplet–triplet annihilation ($k_a[^3A]^2$) will ultimately become large compared with $k_h[^3A]$. θ will then increase more slowly than predicted by equation 146 and the triplet decay curve will contain an increasing proportion of second order component. Finally at very high rates of light absorption, $k_a[^3A]^2$ will become very large and the triplet decay curve will become substantially second order. Delayed fluorescence under these conditions is known as *annihilation-controlled delayed fluorescence*.

In the above treatment (equation 151) we have ignored the fraction of the singlet excited molecules formed by triplet–triplet annihilation that return to the triplet state, because at low rates of light absorption they represent only a small proportion of the triplets originally formed. However at very high rates of light absorption, for which the delayed fluorescence is annihilation controlled, this is no longer true. Under these conditions *all* triplets undergo annihilation and the efficiency of delayed fluorescence is then given by:

$$\theta = \phi_f(\tfrac{1}{2}p_a\phi_t + \tfrac{1}{4}(p_a\phi_t)^2 + \tfrac{1}{8}(p_a\phi_t)^3 + \ldots) \qquad (152)$$

If p_a is unity, this has a maximum value of 0.17 when $\phi_t = 0.59$ and $\phi_f = 0.41$. Higher efficiencies of *directly excited* P-type delayed fluorescence cannot be attained under any circumstances.

Although at high rates of light absorption the intensity of P-type delayed fluorescence is no longer proportional to the square of the *rate of light absorption*, it is still proportional to the square of the intensity of the $T_1 \rightarrow S_0$ emission, because the latter is always proportional to $[^3A]$ and the former to $[^3A]^2$. Thus at high values of I_a, the intensity of P-type delayed fluorescence increases at the expense of the intensity of phosphorescence. Ultimately, when the delayed fluorescence becomes annihilation controlled, the *intensity* of P-type delayed fluorescence becomes proportional to I_a, and the *intensity* of phosphorescence becomes proportional to $(I_a)^{\frac{1}{2}}$.

It is of considerable practical importance to recognise experimentally when the rate of triplet–triplet annihilation becomes sufficiently great to introduce deviations into the square law dependence of P-type delayed fluorescence. We shall therefore consider what is the effect of a given proportion of triplet–triplet annihilation on the shape of the decay curve of delayed fluorescence, and on the shape of the square law intensity plot.

The rate of triplet decay when the exciting light is shut off is given by equation 143 and at all times the rate of emission of delayed fluorescence is given by:

$$I_{DF} = \tfrac{1}{2}\phi_f p_a k_a[^3A]^2 \tag{153}$$

Differentiation of equation 153 and substitution from 143 gives:

$$dI_{DF}/dt = -2k_h I_{DF} - (k_a/p_a\phi_f)^{\frac{1}{2}}(2I_{DF})^{\frac{3}{2}} \tag{154}$$

Integration gives:

$$\ln\left[\frac{I_{DF}}{I_{DF}^0}\right] - 2\ln\left[\frac{k_h + (k_a/p_a\phi_f)^{\frac{1}{2}}(2I_{DF})^{\frac{1}{2}}}{k_h + (k_a/p_a\phi_f)^{\frac{1}{2}}(2I_{DF}^0)^{\frac{1}{2}}}\right] = -2k_h t \tag{155}$$

where I_{DF}^0 is the intensity when $t = 0$, i.e. when the exciting light is on. Now let the ratio of second order to first order triplet decay at zero time be R, i.e.:

$$k_a[^3A^0]^2 = Rk_h[^3A^0] \tag{156}$$

Then,

$$[^3A^0] = Rk_h/k_a \tag{157}$$

and from 153:

$$I_{DF}^0 = \tfrac{1}{2}p_a\phi_f R^2 k_h^2/k_a \tag{158}$$

or

$$k_a^{\frac{1}{2}} = Rk_h(p_a\phi_f/2I_{DF}^0)^{\frac{1}{2}} \tag{159}$$

Substituting in equation 155:

$$\ln\left[\frac{I_{DF}}{I_{DF}^0}\right] - 2\ln\left[\frac{1 + R(I_{DF}/I_{DF}^0)^{\frac{1}{2}}}{1 + R}\right] = -2k_h t \tag{160}$$

Clearly when R is zero, I_{DF} decays exponentially with a lifetime, τ_{DF}, equal to $\frac{1}{2}k_h$, as previously described, and under these conditions a plot of equation 160 against time in units of τ_{DF} will have a slope of -1 as indicated by the straight line in Fig. 34. The deviation from exponential decay for three other values of R is indicated by the other curves in Fig. 34. It will be observed that when $R = 0.2$, i.e. the limiting conditions chosen for the maximum tolerable amount of second order decay in the previous calculations, the plot over two half-lives is very nearly linear with slope about 10% less than for linear decay. Even when R is as great as unity, the curvature of the plot taken over one lifetime is comparatively small, but the derived "lifetime" from this part of the curve would be only about 60% of the exponential lifetime. When $R = 10$, the decay curve approximates to that

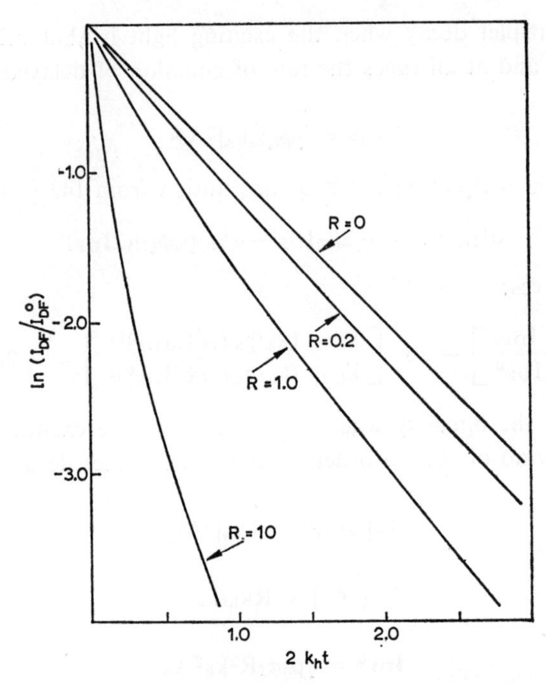

Fig. 34. Decay curves of P-type delayed fluorescence for various ratios (R) of the initial rates of second order and first order decay.

of annihilation-controlled delayed fluorescence. By the methods normally used for measuring decay curves of delayed fluorescence it is difficult to obtain accurate plots over more than about two lifetimes, and in view of the difficulty of detecting the small curvature introduced by quite large proportions of second order decay, it is better to make measurements with

TABLE 10

RELATIONSHIP BETWEEN R AND RELEVANT MOLECULAR PARAMETERS FOR VARIOUS AMOUNTS OF DEVIATION FROM SQUARE LAW DELAYED FLUORESCENCE AS SHOWN IN Fig. 35

R	$k_a\tau^2(I_a\phi_t)$	$(I_a\phi_t)$ for $k_a = 0.6 \times 10^{10}$ litre mole^{-1} sec^{-1}	
		$\tau = 1$ msec	$\tau = 10$ msec
0.01	0.0101	1.7×10^{-6}	1.7×10^{-8}
0.05	0.0525	8.8×10^{-6}	8.8×10^{-8}
0.1	0.11	1.8×10^{-5}	1.8×10^{-7}
0.2	0.24	4.0×10^{-5}	4.0×10^{-7}

at least two intensities of exciting light differing by a factor of at least 3. If the derived lifetimes do not differ by more than 5%, it can be safely concluded that the value obtained at the lower intensity is close to the true value. It should be noted that the derivation of equation 160 and the curves in Fig. 34 do not take into account the fraction of singlet excited molecules formed by triplet–triplet annihilation that return to the triplet state (see, for example, equation 152). When $p_a\phi_t$ is small, the error introduced by this omission is small at all values of R, but when $p_a\phi_t$ is large, the error is significant at values of R greater than unity.

With low rates of triplet formation ($I_a\phi_t$), the *efficiency* of delayed fluorescence (θ) is (by re-arrangement of equation 138 and substitution of p_ak_a for p_ck_c) given by:

$$\frac{\theta}{\frac{1}{2}p_a\phi_f\phi_t} = \frac{k_a}{k_h^2}(I_a\phi_t) \tag{161}$$

i.e. a plot of $\theta/(\frac{1}{2}p_a\phi_f\phi_t)$ against $k_a(I_a\phi_t)/k_h^2$ will be linear and have unit slope. The shape of this plot for values of R greater than zero can be calculated as follows. The steady state equation during illumination is:

$$I_a\phi_t = k_h[^3A^0] + k_a[^3A^0]^2 = k_h[^3A^0](1 + R) \tag{162}$$

and hence from equation 157:

$$(k_a/k_h^2)(I_a\phi_t) = R(1 + R) \tag{163}$$

Also from equations 153, 157 and 163:

$$\frac{\theta}{\frac{1}{2}p_a\phi_f\phi_t} = \frac{R}{1 + R} \tag{164}$$

Thus the required curve is obtained by plotting R/(1 + R) against R(1 + R). The curve for values of R covering the range 0.01 to 0.2 is shown in Fig. 35. It will be observed that for values of R up to about 0.05 the curve differs little from unit slope, i.e. the delayed fluorescence follows closely the square law. With R = 0.2, i.e. the limiting conditions chosen for the maximum tolerable amount of second order decay in the previous calculations, the efficiency of delayed fluorescence is already 30% lower than would be predicted by the square law.

Some values of R and $k_a\tau^2(I_a\phi_t)$, and corresponding values of $I_a\phi_t$ for ethanol at room temperature (assuming $k_c = k_a = 0.6 \times 10^{10}$ litre mole^{-1} sec^{-1}) with triplet lifetimes of 1 msec and 10 msec are shown in Table 10,

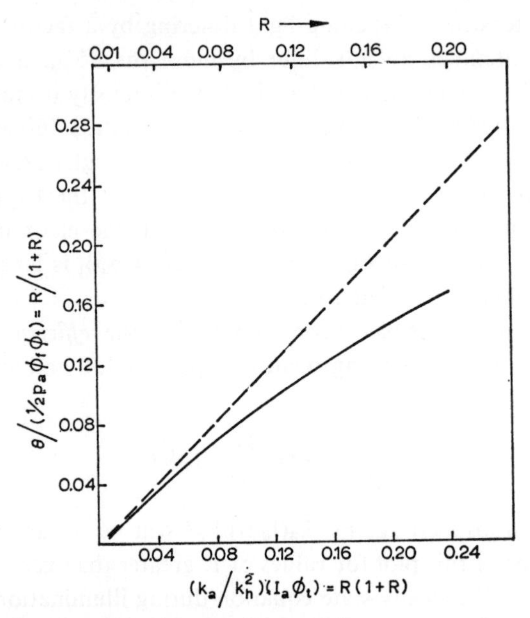

Fig. 35. Relationship between efficiency of delayed fluorescence and rate of light absorption for conditions in which second order triplet decay is significant.

from which maximum permissible $I_a\phi_t$ values for various degrees of deviation from square law dependence may be derived by reference to Fig. 35.

5. Sensitised Delayed Fluorescence

The fact that the process:

$$^3D + {}^1A \rightarrow {}^1D + {}^3A \tag{165}$$

can take place with high encounter efficiency in fluid solution (see Section 2 C 4) means that it is possible to populate the triplet level of an acceptor molecule (A) when A is present in the solution only at very low concentrations (see equation 118). It is to be expected therefore that in a solution containing both D and A, light absorbed by D will give rise to delayed fluorescence of A, and that only a very small concentration of A will be required. Such *sensitised delayed fluorescence* in solution was first observed with phenanthrene as donor and anthracene as acceptor[46] (see Fig. 36). In these original experiments with 10^{-3}M phenanthrene, emission from anthracene was observed when its concentration was as low as 10^{-8}M. Such a concentration was far too low to give rise to P-type delayed fluorescence by direct excitation, and in fact the prompt fluorescence spectrum of the solution showed only emission from the phenanthrene. Later measurements have shown that concentrations as low as 10^{-9}M can be detected, for example using phenanthrene as donor and pyrene or benz(a)pyrene as acceptor (see Section 5 E 4).

In the following discussion we shall for simplicity ignore the formation of the intermediate X of equation 129 and assume that triplet-to-triplet energy transfer results directly in the formation of excited and ground state singlet molecules, i.e., according to equation 137. To make our treatment completely general we shall express all second order rate constants in terms of the diffusion-controlled rate, k_c, multiplied by a probability factor, p (with appropriate subscript), as we did in equation 138. There are in general two mechanisms by which delayed fluorescence of acceptor can be produced.

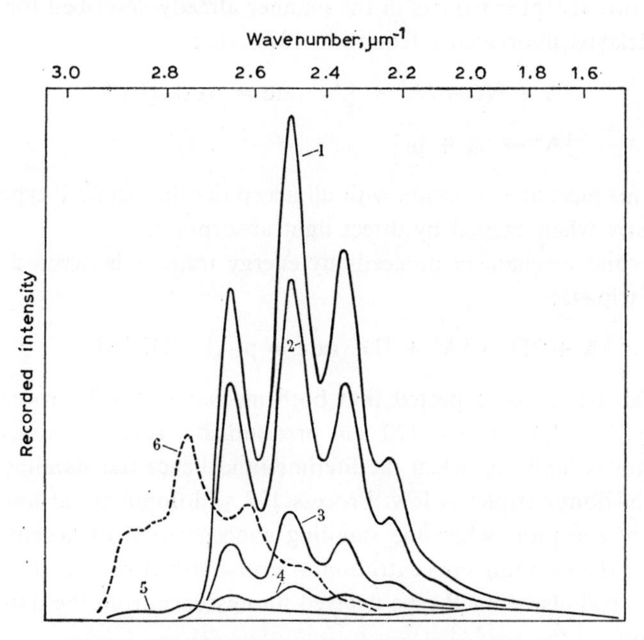

Fig. 36. Sensitised delayed fluorescence of anthracene.
All solutions contained 10^{-3} M phenanthrene in ethanol; concentrations of anthracene were (1) 10^{-6} M, (2) 5×10^{-7} M, (3) 10^{-7} M, (4) 10^{-8} M, (5) 10^{-9} M. Curve (6) is prompt fluorescence of solution (1) at about 260 times lower sensitivity. Intensity of exciting light was about 2.7×10^{-9} einstein cm^{-2} sec^{-1} at 3.19 µm^{-1} (313 nm) (from Parker and Hatchard[46]).

Both require for their initial stages the population of the triplet level of the donor, D, by light absorption, followed by triplet energy transfer to the acceptor, A:

$$^1D \xrightarrow{h\nu} {}^1D^* \rightarrow {}^3D \qquad \text{rate} = \phi_t{}^D I_a \qquad (166)$$

$$^3D + {}^1A \rightarrow {}^1D + {}^3A \qquad p_e k_c[^3D][^1A] \qquad (167)$$

$$^3D \rightarrow {}^1D \qquad k_h{}^D[^3D] \qquad (168)$$

$$^3A \rightarrow {}^1A \qquad k_h{}^A[^3A] \qquad (169)$$

$\phi_t{}^D$ is the triplet formation efficiency of the donor. $k_h{}^D$ and $k_h{}^A$ are the composite first order rate constants for consumption of the triplets: they include the pseudo first-order contributions due to quenching impurities as well as the constant, k_p, for radiative decay.

Following process 167, the first mechanism proceeds by energy transfer between two acceptor triplets in the manner already described for directly-excited delayed fluorescence (equation 137), viz.:

$$^3A + {}^3A \rightarrow {}^1A^* + {}^1A \quad \text{rate} = p_{AA}k_c[^3A]^2 \qquad (170)$$

$$^1A^* \rightarrow {}^1A + h\nu' \qquad (171)$$

Clearly this mechanism occurs with all acceptors that show P-type delayed fluorescence when excited by direct light absorption.

The second mechanism proceeds by energy transfer between donor and acceptor triplets:

$$^3A + {}^3D \rightarrow {}^1A^* + {}^1D \quad \text{rate} = p_{DA}k_c[^3D][^3A] \qquad (172)$$

In general it is to be expected that both mechanisms will proceed simultaneously, but that process 170 will predominate when the concentration of acceptor is high, i.e. when the lifetime (and hence the standing concentration) of donor triplet is low. Process 172 will dominate at low concentrations of acceptor, when the standing concentration of acceptor triplet is low and the standing concentration of donor triplet is high. The solution will in general also emit P-type delayed fluorescence from the donor in the usual way:

$$^3D + {}^3D \rightarrow {}^1D^* + {}^1D \quad \text{rate} = p_{DD}k_c[^3D]^2 \qquad (173)$$

$$^1D^* \rightarrow {}^1D + h\nu'' \qquad (174)$$

If a system is chosen in which the acceptor has a triplet energy of less than one half of that of its first excited singlet state, but in which the sum of the triplet energies of donor and acceptor is greater than the energy of the acceptor singlet, process 170 will be energetically impossible but process 172 can still occur. With such a system it is impossible to excite P-type delayed fluorescence with light absorbed by the acceptor, but sensitised delayed fluorescence can still be emitted via process 172. Such a system is anthracene (donor) with naphthacene (acceptor). The triplet energy of naphthacene[122] is 1.025 μm^{-1}. It is thus slightly less than one half of that of the first excited singlet level (2.11 μm^{-1}), but is sufficient in combination with the energy of the anthracene triplet (1.47 μm^{-1}) to make process 172 energetically possible.

Experimental results are in accord with these predictions[123]. Thus, directly excited P-type delayed fluorescence from naphthacene alone is extremely weak, but anthracene strongly sensitises the delayed fluorescence of naphthacene with naphthacene concentrations as low as 10^{-8}M (see Fig. 37). As the concentration of naphthacene is increased from this value

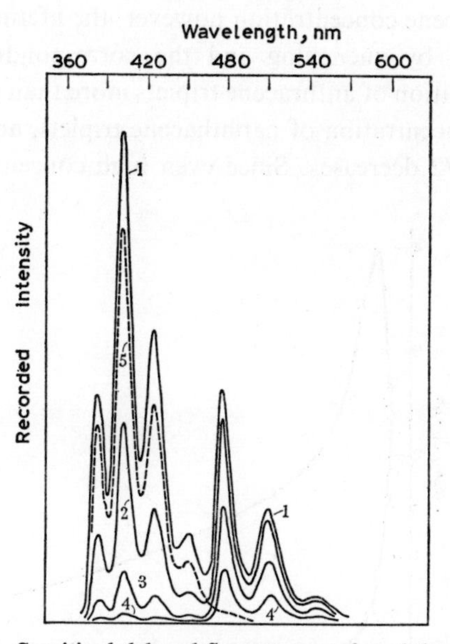

Fig. 37. Sensitised delayed fluorescence of naphthacene.
All solutions contained 5×10^{-5} M anthracene in ethanol; rate of light absorption was 0.4×10^{-5} einstein litre^{-1} sec^{-1} at 2.73 μm^{-1} (366 nm). Concentrations of naphthacene were (1) 4×10^{-8} M, (2) 8×10^{-8} M, (3) 1.5×10^{-7} M, (4) 3×10^{-7} M. Curve (5), prompt fluorescence of (4) at 1000 times lower sensitivity (from Parker[123]).

the intensity of the delayed fluorescence from the donor (anthracene) is quenched as predicted by equation 167. The lifetime ($\tau_{DF}{}^{D}$) of the delayed fluorescence of the anthracene is equal to one half that of triplet anthracene (τ^{D}) and it thus follows the Stern–Volmer quenching law, viz.:

$$(\tau_{DF}{}^{D})_0/(\tau_{DF}{}^{D}) - 1 = p_e k_c \tau_0{}^{D}[A] \tag{175}$$

where [A] is the concentration of acceptor (naphthacene) and $\tau_0{}^{D}$ is the lifetime of the triplet of anthracene when [A] is zero. The rate of quenching was found to be close to the diffusion-controlled value[123]. Since the intensity of delayed fluorescence of the donor is proportional to $(\tau^{D})^2$ (equation 138), the intensity of delayed fluorescence of the anthracene is quenched much more rapidly than its lifetime.

As the concentration of naphthacene increases the intensity of sensitised delayed fluorescence of the naphthacene at first increases (see Fig. 38), passes through a maximum and then decreases again. The initial increase corresponds to increased formation of naphthacene triplet and hence a greater degree of mixed triplet-quenching (equation 172). With further increase of naphthacene concentration however, the lifetime of the anthracene triplet is reduced by quenching and the corresponding decrease in the standing concentration of anthracene triplets more than compensates for the increase in the concentration of naphthacene triplets, and hence the overall rate of process 172 decreases. Since even high concentrations of naphtha-

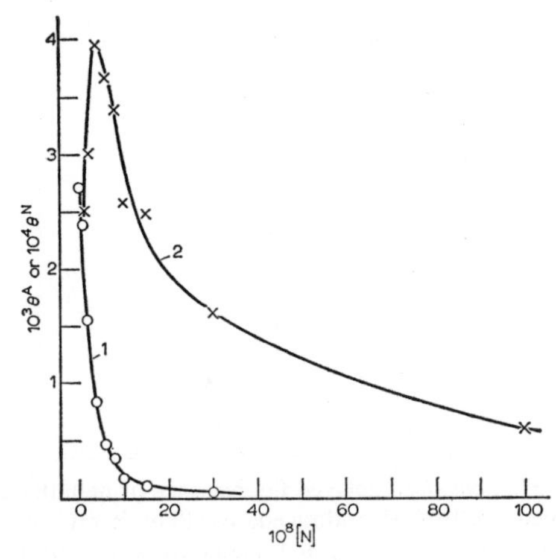

Fig. 38. Effect of naphthacene concentration on the efficiencies of delayed fluorescence of anthracene (curve 1) and naphthacene (curve 2) (from Parker[123]).

cene triplet are unable to undergo mutual energy transfer (equation 170), the overall intensity of sensitised delayed fluorescence decreases.

The results obtained with the system anthracene/naphthacene thus provide clear evidence for the occurrence of the mixed triplet quenching mechanism and it is reasonable to assume that the same mechanism occurs in other systems for which the occurrence of process 170 is also possible. The total efficiency of sensitised delayed fluorescence is then equal to the sum of the efficiencies of the sensitised delayed fluorescence produced by processes 170 and 172. We shall now derive expressions for each of these efficiencies in terms of the rate parameters indicated in equations 166 to 173. We assume that the rate of light absorption is low so that the rates of the triplet-to-triplet transfer reactions (170 and 172) are negligible compared with the composite rates of first order decay (equations 168 and 169). For the donor triplet, therefore, the steady state equation is:

$$I_a \phi_t^D = [^3D]/\tau^D \tag{176}$$

where the lifetime of the donor, τ^D, is given by:

$$1/\tau^D = k_h^D + p_e k_c[A] \tag{177}$$

For the acceptor triplet the steady state equation is:

$$p_e k_c[^3D][A] = k_h^A[^3A] = [^3A]/\tau_0^A \tag{178}$$

since the lifetime of the acceptor is assumed to be unaffected by the total concentration of acceptor [A]. Hence from equations 176 and 178

$$[^3A] = p_e k_c \tau_0^A [A] I_a \phi_t^D \tau^D \tag{179}$$

The rate of emission of delayed fluorescence of acceptor produced by self quenching (process 170) is:

$$I_{DF}' = \tfrac{1}{2}\phi_f p_{AA} k_c [^3A]^2 \tag{180}$$

and its efficiency is thus given by:

$$\theta'/\phi_f = [\tfrac{1}{2} p_{AA} p_e^2 k_c^3 I_a (\phi_t^D \tau_0^A)^2][\tau^D[A]]^2 \tag{181}$$

The rate of emission of delayed fluorescence produced by the mixed triplet interaction (process 172) is:

$$I_{DF}'' = \phi_f p_{DA} k_c [^3A][^3D] \tag{182}$$

and its efficiency is thus given by:

$$\theta''/\phi_f = [p_{DA}p_e k_c^2 I_a(\phi_t^D)^2 \tau_0^A][(\tau^D)^2[A]] \tag{183}$$

For a particular donor/acceptor pair the quantities in the first brackets in equations 181 and 183 are constant. It is of interest to plot θ'/ϕ_f and θ''/ϕ_f for typical conditions met in fluid solution at room temperature. For this we choose the following values of parameters:

k_c (for ethanol at 22°C) $= 0.57 \times 10^{10}$ litre mole^{-1} sec^{-1}

$p_e = p_{AA} = p_{DA} = 1$

$I_a = 10^{-5}$ einstein litre^{-1} sec^{-1}

$\phi_t^D = 0.1$

$\tau_0^D = \tau_0^A = 0.005$ sec

Plots of θ'/ϕ_f and θ''/ϕ_f for these conditions are shown in Fig. 39 together with θ/ϕ_f for the donor ($p_{DD} = 1.0$), which is given by equation 138.

Several features of these curves are worth noting. Firstly the delayed fluorescence of the donor is already appreciably quenched, and θ''/ϕ_f has

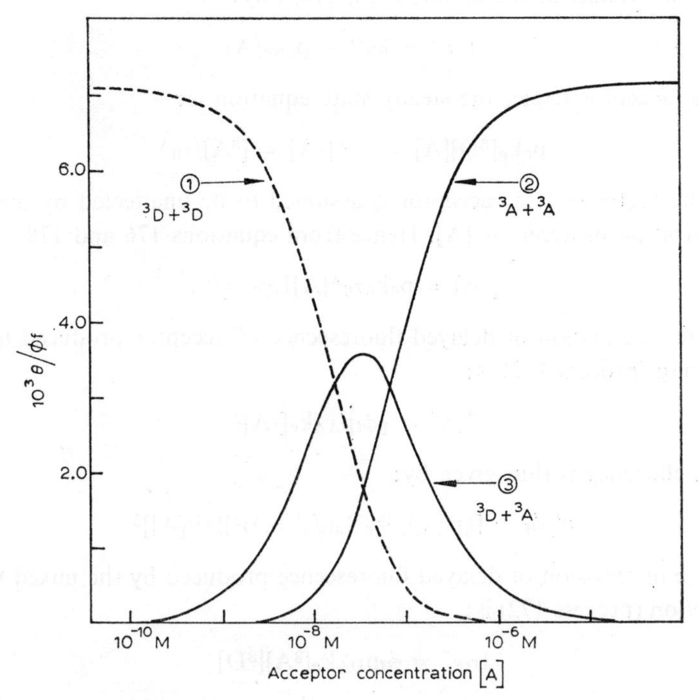

Fig. 39. Calculated efficiencies of delayed fluorescence for a typical donor–acceptor system.

already risen to a significant value with an acceptor concentration of 10^{-9}M. θ'/ϕ_f does not become significant until the acceptor concentration has risen to about 10^{-8}M, at which point the delayed fluorescence of the donor is already nearly 50% quenched. At higher concentrations the contribution of θ' to the total efficiency of delayed fluorescence $(\theta' + \theta'')$ increases rapidly. Ultimately at high acceptor concentrations, for which the donor is completely quenched, the value of θ'/ϕ_f of the acceptor rises to a value equal to that of the delayed fluorescence of the donor in the absence of acceptor. Under these conditions the donor triplets are converted quantitatively by process 167 into acceptor triplets. The relative shapes of the curves and their position on the [A] axis will of course depend on the unquenched lifetimes of donor and acceptor as indicated by equations 177 and 179. For a chosen donor/acceptor pair, the values of θ/ϕ_f for all three curves will increase in proportion to I_a and to $(\phi_t^D)^2$ (i.e. the sensitised delayed fluorescence will obey the square law), provided of course that the rate of light absorption is not sufficiently great to make the rates of the triplet-to-triplet transfer reactions significant compared with the total rates of first order triplet decay.

Comparison of equations 181 and 183 indicates that θ'/θ'' is proportional to $k_c\tau_0^A[A]$. Thus for a given acceptor concentration, long lifetime of acceptor triplet and high diffusion-controlled rates (i.e. solvents of low viscosity) favour the triplet self-quenching mechanism.

The lifetime of the delayed fluorescence of the donor will be equal to one half of that of the phosphorescence of the donor at all points on curve 1, i.e. it will decrease as the donor triplet is quenched. Similarly the lifetime of the delayed fluorescence of the acceptor produced by triplet self-quenching will be equal to one half of the phosphorescence of the acceptor for all points on curve 2, and will be constant. The decay of the delayed fluorescence of the acceptor produced by the mixed triplet interaction will be exponential, but its lifetime will vary from point to point along curve 3, and will be equal to $\tau_0^A\tau^D/(\tau_0^A + \tau^D)$. This has been verified[123] in the system anthracene/naphthacene discussed above. The observed total delayed fluorescence of the acceptor will thus in general be made up of *two* exponential decay processes—that corresponding to curve 2 and that corresponding to curve 3.

Finally, it is noted that with a high concentration of acceptor and a rate of light absorption such that decay of the acceptor is annihilation-controlled, the efficiency of sensitised delayed fluorescence can reach high values. Thus, if ϕ_t^D and ϕ_f^A are both unity, and $p_{AA} = 1$, the efficiency of annihilation-controlled delayed fluorescence will be 0.5. This may be compared with the value 0.17 for the maximum efficiency of directly excited P-type delayed fluorescence (see equation 152).

6. Anti-Stokes Delayed Fluorescence and Mutual Sensitisation

The process of triplet-to-singlet energy transfer (see Section 2 C 4) can occur provided that the triplet level of the acceptor lies close to or below that of the donor. If a system is chosen in which the first excited singlet level of the donor is lower than that of the acceptor (see Fig. 40) it is possible to populate the triplet level of the acceptor by light absorbed only by the donor (via process 1 in Fig. 40). Triplet-to-triplet energy transfer between donor and acceptor triplets (process 2 in Fig. 40), or between two acceptor triplets, will then give rise to excited singlets of the acceptor and the resulting sensitised delayed fluorescence will have a wavelength shorter than that of the exciting light. Such sensitised anti-Stokes delayed fluorescence was first observed for the system phenanthrene (donor)/naphthalene (acceptor)[124] (see Fig. 41), and has since been observed with a variety of aromatic hydrocarbons sensitised by dyestuffs[110, 125]. For example the delayed fluorescence of perylene sensitised to light of wavelength 546 nm by eosin at room

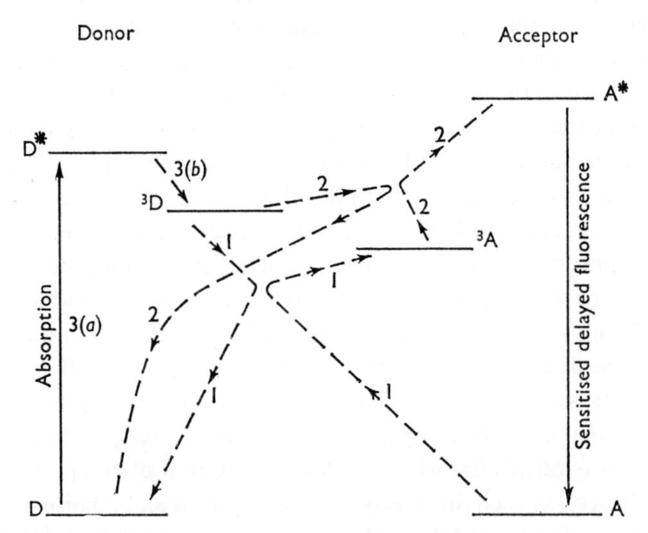

Fig. 40. Mechanism of sensitised anti-Stokes delayed fluorescence by the mixed triplet mechanism (from Parker, Hatchard and Joyce[125]).

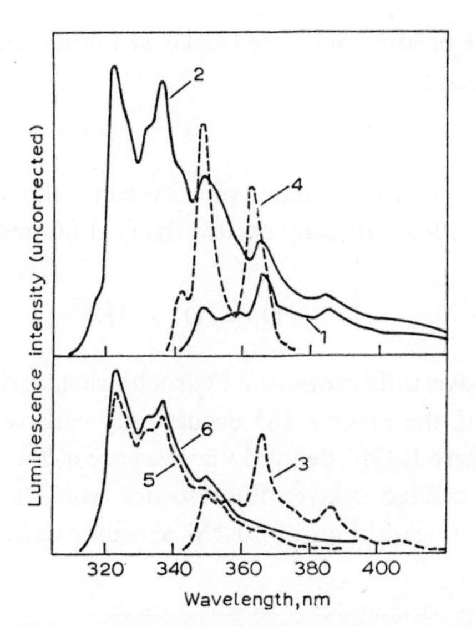

Fig. 41. Sensitised anti-Stokes delayed fluorescence of naphthalene. Curves 1 and 2, delayed fluorescence from 10^{-3} M phenanthrene, and from 10^{-3} M phenanthrene plus 3×10^{-3} M naphthalene in ethanol. Curve 3, prompt fluorescence from 1 and 2 at 260 times less sensitivity. Curve 4, spectral distribution of exciting light for curves 1, 2 and 3, approximately 0.7×10^{-6} einstein litre^{-1} sec^{-1} absorbed. Curves 5 and 6, prompt and delayed fluorescence from 3×10^{-3} M naphthalene in ethanol excited by 313 nm, approximately 2×10^{-6} einstein litre^{-1} sec^{-1} absorbed (curve 5 at 100 times lower sensitivity than curve 6). Temperature for all curves was -72 ± 3°C (from Parker and Hatchard[124]).

temperature is shown in Fig. 42. The fact that the quanta of delayed fluorescence emitted from such systems carry considerably greater energy than those of the exciting light provides convincing proof that part of the energy from two separately absorbed quanta has been concentrated in a single molecular species.

Kinetically, sensitised anti-Stokes delayed fluorescence follows the same rules as those derived in the previous section for the delayed fluorescence of the acceptor, i.e. as indicated by equations 181 and 183, and Fig. 39. However, the process of mixed triplet interaction (equation 172) can in general give rise also to some singlet excited *donor* molecules provided that the sum of the triplet energies is greater than the singlet energy for the donor, which is clearly the case in systems exhibiting sensitised anti-Stokes delayed fluorescence. In these systems, therefore, the addition of low concentrations of acceptor should produce sensitised delayed fluorescence of the *donor* by the following process:

$$^3D + {}^3A \rightarrow {}^1D^* + {}^1A \tag{184}$$

followed by equation 174. In many such systems, this donor sensitisation is obscured by the intense directly excited delayed fluorescence of the donor itself:

$$^3D + {}^3D \rightarrow {}^1D^* + {}^1D \tag{185}$$

However with the dyestuffs proflavine hydrochloride, acridine orange hydrochloride, and eosin, the process 185 occurs only with very low probability, and at $-75°C$ (where E-type delayed fluorescence of the donor is negligible) very little directly excited delayed fluorescence from the donor is observed (see Fig. 43 curve 1). Addition of 10^{-6}M acceptor causes almost complete

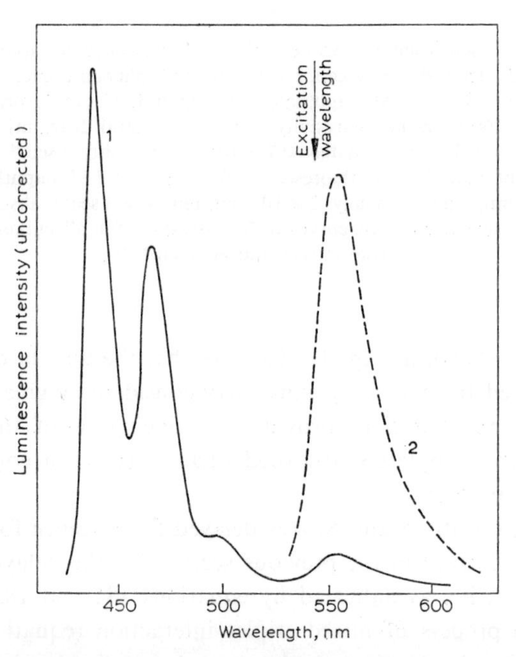

Fig. 42. Sensitised anti-Stokes delayed fluorescence of perylene.
5×10^{-6} M eosin disodium salt and 10^{-6} M perylene in ethanol at 22°C. Curve 1, delayed fluorescence excited by 546 nm (1.83 μm^{-1}) light (1.1×10^{-5} einstein litre^{-1} sec^{-1} absorbed). Curve 2, prompt fluorescence at 1400 times lower sensitivity (from Parker, Hatchard and Joyce[125]).

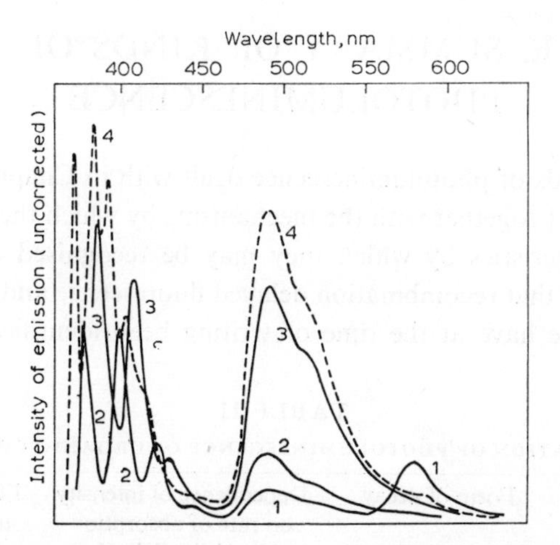

Fig. 43. Illustrating mutual sensitisation of delayed fluorescence.
Delayed fluorescence at $-75°C$ of 8×10^{-6} M proflavine hydrochloride in ethanol with (1) no addition, (2) 10^{-6} M anthracene, (3) 10^{-6} M 1,2-benzanthracene, (4) 10^{-6} M pyrene. Rate of light absorption was 6×10^{-6} einstein litre^{-1} sec^{-1} at 436 nm (from Parker, Hatchard and Joyce[110]).

quenching of the phosphorescence of the donor but at the same time there is a considerable increase in the delayed fluorescence from the *donor* (e.g. curves 2, 3 and 4 in Fig. 43). This has been attributed to the operation of process 184 above. Kinetically such sensitised donor delayed fluorescence should obey a law identical to that pertaining to the delayed fluorescence of acceptor produced by the mixed triplet quenching mechanism (equation 183 and curve 3 in Fig. 39). However, it does not appear to do so. At high acceptor concentrations the donor triplet is very strongly quenched and process 184 should be completely suppressed. Nevertheless with, for example, benzanthracene as acceptor at concentrations as high as 10^{-4}M, the sensitised delayed fluorescence of the donor was still observed at an intensity comparable with that produced with low concentrations of acceptor.

Parker and co-workers, who observed mutual sensitisation in the systems just discussed[110], have since considered[59] the possibility that the long-lived emission from the donors might have been due to the operation of the trivial effect (see Section 3 N 8). They re-investigated one system and concluded that the trivial effect could not have been responsible for more than a small proportion of the donor emission. Nevertheless these systems require further investigation in view of the difficulty of interpreting the results.

E. SUMMARY OF KINDS OF PHOTOLUMINESCENCE

The various kinds of photoluminescence dealt with in Chapters 1 and 2 are listed in Table 11 together with the mechanisms by which they are produced and the characteristics by which they may be recognised experimentally. It may be noted that recombination delayed fluorescence and recombination phosphorescence have at the time of writing been identified only in rigid

TABLE 11

CLASSIFICATION OF PHOTOLUMINESCENCE OF ORGANIC COMPOUNDS

Type	Form of decay	Dependence of intensity on rate of absorption of exciting light (I_a)	Effect of temperature on intensity
Prompt fluorescence	Exponential	$\propto I_a$	Moderate decrease
Phosphorescence	Same as that of T_1 and exponential at low I_a with lifetime τ_p	$\propto I_a$ at low values $\propto I_a^{\frac{1}{2}}$ at high values	Rapid decrease ($\propto \tau_p$)
Sensitised phosphorescence	Ditto	Ditto	Complex
E-type delayed fluorescence	Lifetime $= \tau_p$	Ditto	$\phi_e/\phi_p \propto \exp(-\Delta E/\mathbf{R}T)$
Sensitised E-type delayed fluorescence	Ditto	Ditto	Complex
P-type delayed fluorescence	Lifetime $= \frac{1}{2}\tau_p$ at low I_a—non-exponential at high I_a or in rigid medium	$\propto (I_a)^2$ at low values $\propto I_a$ at high values	θ/ϕ_p not $\propto \exp(-\Delta E/\mathbf{R}T)$
Sensitised P-type delayed fluorescence	(a) (With high [A]) lifetime $= \frac{1}{2}\tau_A$ at low I_a (b) (With low [A]) lifetime $= \tau_A\tau_D/(\tau_A + \tau_D)$ at low I_a	Ditto $\propto (I_a)^2$ at low I_a	Complex but greatly reduced in rigid media at low I_a
Recombination phosphorescence	Non-exponential and often slow	(a) (Dissociation of singlet) $\propto I_a$ (b) (Dissociation of triplet) $\propto (I_a)^2$ at low values	Often increases on warming *after* irradiation (thermoluminescence)
Recombination delayed fluorescence	Ditto	As for recombination phosphorescence	Ditto

media. Sensitised P-type delayed fluorescence has not so far been specifically identified in rigid media but it could no doubt be produced under these conditions with high concentrations of solute. Sensitised phosphorescence in fluid solution will clearly be emitted, though generally at low intensity, from all fluid solutions exhibiting sensitised P-type delayed fluorescence. It was first observed by Terenin and Ermolaev[106] in rigid media containing high concentrations of acceptor i.e. in the classic experiments that first demonstrated the occurrence of triplet-to-singlet energy transfer (see Section 2 C 4). Sensitised E-type delayed fluorescence also has not been specifically investigated. It must clearly be emitted (together with sensitised P-type delayed fluorescence where appropriate) from all sensitised solutions in which the acceptor has a triplet level lying close to its excited singlet level.

Table 11 includes only transitions characteristic of organic compounds. The introduction of rare earth atoms into a molecule produces a new set of energy levels due to electrons in the shielded *4f* shell and the number of possible types of radiative transition is increased. This subject was mentioned briefly in Section 2 B 7 and will be discussed further in Section 5 E 3.

Chapter 3

Apparatus and Experimental Methods

A. GENERAL COMMENTS

In this chapter we shall deal with the basic principles of design of photoluminescence spectrometers, and the experimental techniques involved in their application. The object is to provide sufficient information so that the chemist can judge the relative merits of commercial equipment for his particular purpose, or set up equipment for himself if he so desires. Some confusion has arisen in the past over the names given to the various kinds of instrument. As noted by Pringsheim[126], any photometer can be used to measure the intensity of fluorescence but photometers for this purpose have been advertised as either "fluorometers", or "fluorimeters", the former mainly in the U.S.A. and the latter in Europe. Since the term "fluorometer" has also been applied to instruments for measuring fluorescence lifetime, it might be argued that the term fluorimeter should be reserved for equipment to measure fluorescence intensity. However, Pringsheim has suggested the term "fluoro-photometer" for the latter, and an apparatus incorporating a dispersing unit would then be called a spectrofluorophotometer. On the other hand Bowman, Caulfield and Udenfriend[127] have coined the term "spectrophotofluorometer". In Europe the term "spectrofluorimeter" is often used for such an instrument. The situation is now complicated by the introduction of instruments designed specifically for the measurement of phosphorescence and on both sides of the Atlantic these have been referred to as spectrophosphorimeters. Whatever term is met in the literature, it seems that the prefix "spectro-" implies that at least one spectrometer is incorporated in the equipment. In this book we shall use the terms "fluorimeter" and "spectrofluorimeter" for instruments designed to measure the *intensity* of fluorescence. Instruments for measuring the lifetimes of fluorescence are generally of quite different design and for these we shall reserve the term "fluorometer". The design of fluorimeters and spectrofluorimeters is generally such that they provide a measure of the *total* photoluminescence from the specimen whatever its lifetime. Instruments designed to separate the long-lived from the short-lived luminescence are generally capable of measuring both the intensity and the lifetime of the *long-lived* luminescence and we shall use the same name for both functions, i.e. phosphorimeter or spectrophosphorimeter as the case may be.

Since spectrofluorimetry is an extension and refinement of fluorimetry, many of the basic principles are common to both types of equipment. We shall therefore deal first with these and subsequently discuss the additional factors introduced in spectrophosphorimeters and in fluorometers. Every fluorimeter or spectrofluorimeter contains three basic items: (1) a source of light with which to excite fluorescence in the specimen, (2) a specimen holder, and (3) a detector to observe or measure the fluorescence emitted. Instruments vary according to the degree to which the wavelength and intensity of the exciting light can be controlled, the degree to which specific wavelengths of emitted light can be selected, and the sensitivity and precision with which the selected light can be measured. In filter fluorimeters, selection of wavelength is made by inserting filters in the beams of exciting light and fluorescence light. In spectrofluorimetry, a monochromator is used to select the required wavelengths from one or both of the beams. The spectrofluorimeter can therefore be used to measure the spectral distribution of the fluorescence light emitted—the *fluorescence emission spectrum*—or to measure the variation of fluorescence intensity with wavelength of the exciting light—the *fluorescence excitation spectrum*.

A diagram of a typical general-purpose spectrofluorimeter is shown in Fig. 44. It consists of a source giving a continuum of visible and ultra-violet light, a monochromator to select the required wavelength for excitation, a sample container, and a second monochromator fitted with a photomultiplier to analyse the fluorescence light. With such an apparatus, one can select a narrow band of wavelengths of exciting light and measure the spectrum of the fluorescence emitted by the sample, or one can set the fluorescence monochromator on the wavelength of the fluorescence band of the substance of interest, and observe how the intensity of this fluorescence varies with the wavelength of the exciting light used. The principle is simple, but in practice it is difficult to achieve high sensitivity with this system because so much light is wasted. The light from the source is emitted in all directions, and only a limited proportion finds its way into the first monochromator. Of this light, only a narrow band of wavelengths is selected for passing to the sample. It is often arranged that less than 1% of this light

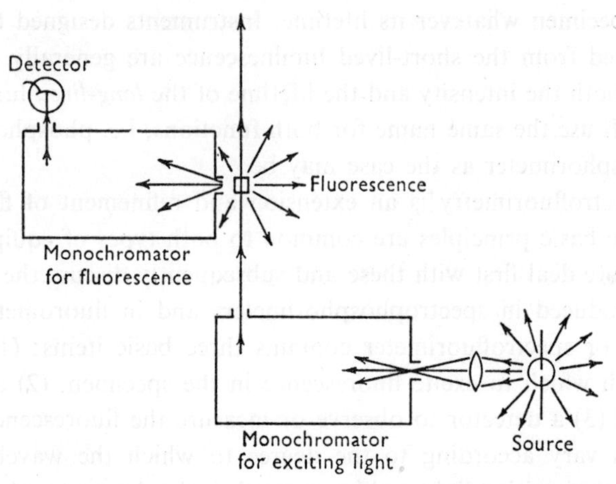

Fig. 44. Arrangement of general purpose spectrofluorimeter (from Parker and Rees[150]).

is absorbed, the remainder passing on and being of no further use. The resulting fluorescence light is again emitted in all directions, and again, of necessity only a comparatively small proportion is collected by the second monochromator. This light is finally dispersed again, and a narrow band of wavelengths is collected, so that the final intensity falling on the detector is a small fraction of the light originally emitted by the source. There are several ways in which the limitations imposed by these light losses can be minimised. Obviously it is desirable to use the most powerful lamp, the most sensitive detector and the largest monochromators available. Even with very large and expensive monochromators, the sensitivity will still be rather low if it is attempted to take measurements with very narrow slits on both monochromators simultaneously. Fortunately it is not usually necessary to do this and a considerable increase in sensitivity can be attained by using wide slits on one or other monochromator. For example Parker[73] recorded well resolved excitation spectra of anthracene at a concentration of less than one part in 100 million using an early instrument having comparatively low power monochromators, but with wide slit settings on the fluorescence monochromator. Similar sensitivity and resolution can be obtained in measuring the fluorescence emission spectrum by using wide slits on the excitation monochromator. Instead of using a monochromator with wide slits, it is sometimes possible to dispense with one monochromator altogether and to select either the exciting wavelengths or the fluorescence wavelengths by means of filters. When a mercury lamp and filters are used to give the exciting light, a much larger intensity can be obtained than is possible even with a large monochromator. As discussed later, the limitations imposed by the intensity of the exciting light are most severe at wavelengths shorter than 300 nm. This region is important because many simple organic compounds do not absorb at longer wavelengths and to obtain adequate sensitivity, the lower discrimination associated with the use of filters may have to be accepted. Such single-monochromator experiments have been used by many workers (for example see references 102, 128–131). We shall give examples of the application of one-monochromator and two-monochromator instruments in later chapters. Examples of the intensities of exciting light that have been obtained by the use of monochromators or filters will be given under the heading LIGHT SOURCES below.

B. MONOCHROMATORS

1. Working Principles

The principles of monochromator operation will be discussed in sufficient detail to enable the reader to judge the comparative merits of different commercial instruments for a particular purpose, and to obtain the best performance from the chosen instrument. The design of monochromators, particularly those giving high performance, requires consideration of many other factors and is outside the scope of this book. It may be mentioned however that Johns and Ralph[132] have described the design, construction and performance of large water-prism and grating monochromators capable of giving very high intensities for use in photochemical and photobiological studies.

All monochromator systems contain the following components (see Fig. 45): an entrance slit, W_1, of adjustable width; a collimator, M_1, which may be a mirror or a lens; a dispersing element, G, which may be a grating or a prism; a mirror, M_2, or a lens, to focus the dispersed light; and an exit slit, W_2, also of adjustable width. We shall assume for the sake of simplicity that the heights of W_1 and W_2 are equal and that the focal lengths of M_1 and M_2 are also equal. This is usually the case for monochromators used in spectrofluorimetry. Light from the source, S, is passed via the lens, L_1, to the entrance slit, W_1, of the monochromator. The diverging light from W_1 is rendered parallel by the collimator, M_1, and passed to the

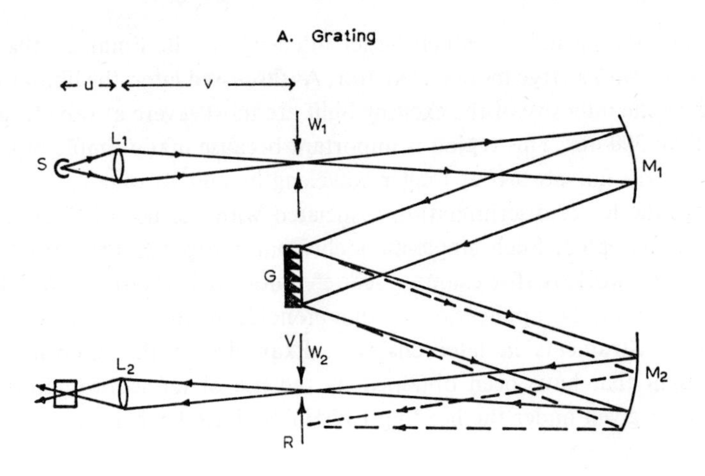

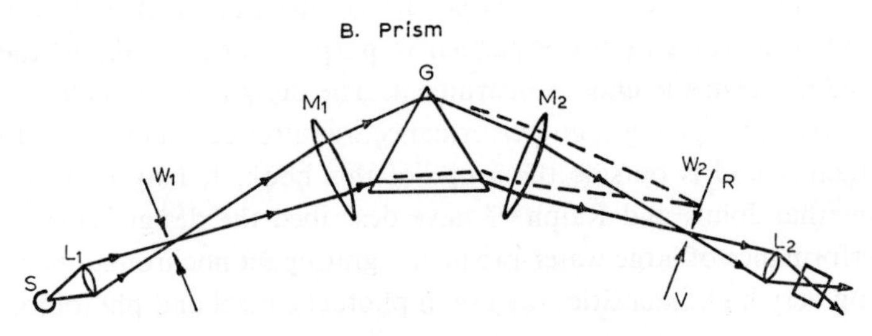

Fig. 45. Working principle of a monochromator.

dispersing element, G. Different wavelengths of light leave G at slightly different angles and the dispersed light is focussed by the mirror or lens, M_2, on the plane VR, so that each specific wavelength produces an image of the entrance slit at some point along VR. The images corresponding to the various wavelengths in the light from the source thus form a band of light varying in wavelength from one end of VR to the other, i.e. a spectrum of the source. The exit slit W_2, is situated at a suitable point along VR and thus selects a narrow band of wavelengths which are passed to the specimen (or the detector as the case may be), via another lens L_2, or other suitable optics. On rotating the dispersing element G, the spectrum VR moves laterally across the exit slit so that any particular wavelength region can be made to pass to the specimen or to the detector. The wavelength may be selected by manual rotation of G, or more frequently in spectrofluorimeters, G may be rotated at constant speed by a synchronous motor so that the complete spectrum can be scanned automatically and presented on a recorder chart.

2. Angular and Linear Dispersion

Let the angular dispersion, α, be defined as the difference in angle between two rays leaving G that differ in wavelength by 1 nm. The dispersing element G may be either a grating or a prism (see Fig. 45 A or B). For rays passing the exit slit, the dispersion of a grating is almost independent of wavelength, but that of a prism varies with wavelength. For example the dispersion of a quartz prism varies by about two orders of magnitude from the near infra-red (1000 nm) to the short-wavelength end of the quartz-ultra-violet region (200 nm). We shall find therefore that in comparing the performance of prism instruments with those employing gratings, care has to be taken to specify the wavelength region of interest.

If α is measured in radians, the distance, m, along VR occupied by the unit band of wavelengths is given by:

$$m = \alpha f \tag{186}$$

where f is the focal length of the mirror, M_2. The quantity m is the linear dispersion of the monochromator and is conveniently expressed in millimetres per nanometre. Manufacturers of grating instruments quote either the value of m, or more frequently its reciprocal. Prism instruments are usually provided with a graph showing 1/m (or a stated multiple of this) plotted as a function of wavelength.

3. Spectral Purity with a Continuous Source

Consider first an arrangement in which the entrance slit W_1 is uniformly illuminated by a source having constant wavelength distribution, i.e. a source that emits the same number of einsteins per second within each unit band of wavelengths throughout the spectrum. Suppose that the entrance slit is adjusted to a very narrow width, w_1. The image of W_1, produced at a point along VR by a specific wavelength λ will also have a width w_1. This image will be partially overlapped by the images produced by adjacent wavelengths within the range $\lambda \pm \Delta\lambda_1$, because the distance w_1 along VR corresponds to a wavelength interval given by:

$$\Delta\lambda_1 = w_1/m \tag{187}$$

Since, however, w_1 is assumed to be very small, the wavelength interval $\Delta\lambda_1$, will also be small, and we can consider the position on VR occupied by the image produced by light of wavelength λ to be illuminated by essentially monochromatic light. The spectrum along VR will thus be "pure" and of constant intensity.

Let the exit slit now be opened by a finite width w_2. The centre of the exit slit corresponds to a wavelength λ, and each extreme edge to a wavelength $\lambda \pm \Delta\lambda_2/2$ where

$$\Delta\lambda_2 = w_2/m \tag{188}$$

The situation is as depicted in column A of Fig. 46, in which section 1 represents the relative slit openings and section 2 shows the images of the narrow entrance slit that fall on the two extreme edges of the exit slit, i.e. the images of the entrance slit produced by light of wavelengths $\lambda \pm \Delta\lambda_2/2$. Under these conditions it will be seen that the light passing the exit slit will include wavelengths within the band $\lambda \pm \Delta\lambda_2/2$ and will have a constant intensity within this range, as indicated by section 3 under column A in Fig. 46. It is of interest to note that the widths of the exit and entrance slits can be interchanged without altering the wavelength distribution of the light issuing from the exit slit. The spatial distribution of the light will of course be different, i.e. it will now be concentrated over the area of the very narrow exit slit.

Now let the width of the entrance slit be increased to a comparatively large value. The images of the entrance slit corresponding to wavelengths $\lambda \pm \Delta\lambda_2/2$ are also shown in Fig. 46 under column B. It will be seen that only one half of these images pass through the exit slit, but in addition smaller fractions of images outside this wavelength range also pass the exit slit. A little thought will show that the overlapping images produce an intensity-wavelength distribution as shown in section 3 under column B in Fig. 46. The band width of the light passing the exit slit is now $(\Delta\lambda_2 + \Delta\lambda_1)$. Only wavelengths between $\lambda \pm (\Delta\lambda_2 - \Delta\lambda_1)/2$ will pass at full intensity, and outside these limits the intensity will fall off linearly.

If the entrance slit is opened to the same width as the exit slit, the situation

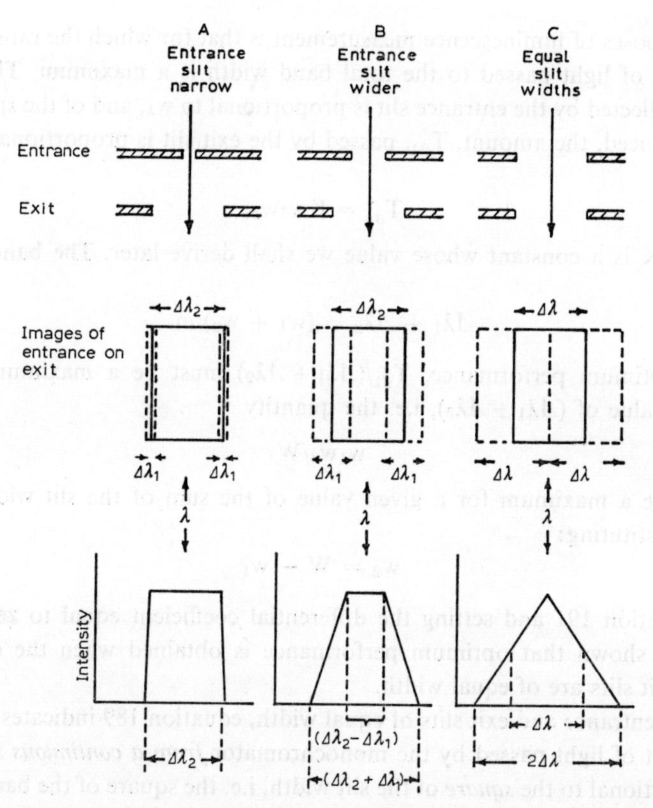

Fig. 46. Spectral distribution of light passing a monochromator from a continuous source. The upper section of the diagram represents the widths of the slits. The middle section represents the exit slits as full line rectangles of width equivalent to a band-width $\Delta\lambda_2$, and the images of the entrance slits as dotted line rectangles of width equivalent to a band width $\Delta\lambda_1$; the two images shown on each figure are those produced by light of wavelengths $\lambda \pm \Delta\lambda_2/2$. The lower section of the diagram represents the wavelength distributions of the light passing the exit slit for each of the three slit width combinations shown in the upper section. In the lower section the intensity scales are not comparable between A, B and C.

is that indicated under column C of Fig. 46. Only the central wavelength λ now passes at full intensity and the spectral distribution has a triangular shape. The band of wavelengths passed by the exit slit is now $2\Delta\lambda$ and at wavelengths $\lambda \pm \Delta\lambda/2$ the intensity is reduced to one half of the maximum value. For this reason $\Delta\lambda$ is sometimes referred to as the *half-band width*.

It is important to determine the optimum relative slit widths required to give the best performance for any chosen band width. The best performance for purposes of luminescence measurement is that for which the ratio of the amount of light passed to the total band width is a maximum. The total light collected by the entrance slit is proportional to w_1, and of the spectrum so produced, the amount, $T_{\Delta\lambda}$, passed by the exit slit is proportional to w_2. Hence:

$$T_{\Delta\lambda} = Kw_1w_2 \tag{189}$$

where K is a constant whose value we shall derive later. The band width is:

$$\Delta\lambda_1 + \Delta\lambda_2 = (w_1 + w_2)/m \tag{190}$$

For optimum performance, $T_{\Delta\lambda}/(\Delta\lambda_1 + \Delta\lambda_2)$ must be a maximum for a given value of $(\Delta\lambda_1 + \Delta\lambda_2)$, i.e. the quantity

$$w_1w_2/W \tag{191}$$

must be a maximum for a given value of the sum of the slit widths, W. By substituting:

$$w_2 = W - w_1 \tag{192}$$

in equation 191 and setting the differential coefficient equal to zero, it is simply shown that optimum performance is obtained when the entrance and exit slits are of equal width.

For entrance and exit slits of equal width, equation 189 indicates that the amount of light passed by the monochromator *from a continuous source* is proportional to the *square* of the slit width, i.e. the square of the bandwidth. Thus if twice the band width can be tolerated for a particular application, the intensity can be increased four times by doubling the slit width.

4. Spectral Purity with a Discontinuous Source

The light emission from mercury vapour lamps is concentrated mainly in relatively few intense lines of specific wavelengths that are superimposed on a much weaker continuous background. A monochromator can be used to isolate one of these intense lines and in this way a high intensity of substantially monochromatic light can be obtained for excitation of luminescence. Let us assume for the moment that the continuous background from the lamp is negligible, and consider what is the maximum slit width that can be used to isolate a particular line of wavelength λ_1 from its nearest neighbour at λ_2 using a monochromator of constant linear dispersion, m. First consider the case corresponding to column B in Fig. 46, in which the monochromator is set at a wavelength λ_1. Light of this wavelength produces a monochromatic image of the entrance slit in the plane of the exit slit. The width of this image is w_1 and it is centred on the rectangular opening of the exit slit of width w_2. Clearly all the light making up the image will pass the exit slit provided that w_1 is not greater than w_2.

The light from the second line of wavelength λ_2 will also produce an image of the entrance slit in the plane VR. This image will also have a width w_1 but its centre will be separated from that of λ_1 (see Fig. 47) by a distance equal to:

$$m(\lambda_2 - \lambda_1) \text{ millimetres} \tag{193}$$

Since each image itself occupies a distance w_1 the maximum opening of the entrance slit that can be used without the two images overlapping (see Fig. 47) is given by:

$$w_1 = m(\lambda_2 - \lambda_1) \tag{194}$$

Under these conditions the two wavelengths can just be separated if the exit slit is opened to the same width as the entrance slit. If a larger exit slit is employed, some of λ_2 will pass; with a smaller slit, not all of λ_1 will be collected. The optimum working condition is thus again equal slit widths. It should be noted however that unlike the case of the continuous source dealt with above, the flux of light passing the exit slit is proportional to the *first* power of the slit widths.

The fact that measured line intensities vary as the first power of the band

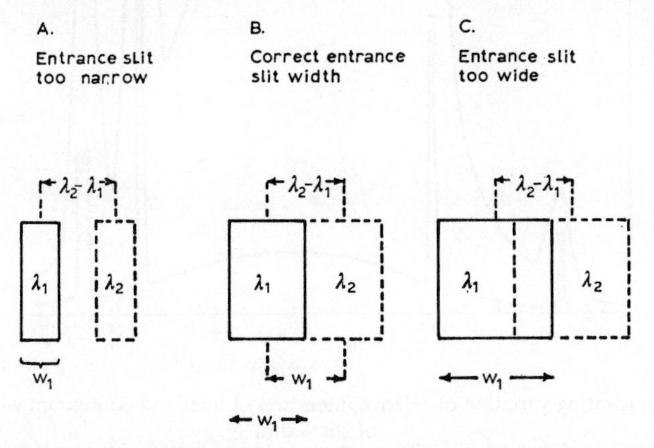

Fig. 47. Images of entrance slit produced by lines at wavelengths λ_1 and λ_2. For optimum collection of λ_1, the monochromator is set to λ_1, and both slits are opened so that $w_1 = w_2 = m(\lambda_2 - \lambda_1)$.

width, while the intensities of continuous spectra vary as the second power of the band width means that with a source emitting both lines and a continuum, the relative intensities of the lines and the continuum in the measured spectrum will vary with the band width used. This is illustrated in Fig. 48 which shows the spectra obtained from a dilute solution of quinine bisulphate excited by light from a mercury lamp containing the wavelengths 405, 436 and 546 nm. The first of these wavelengths was weakly absorbed by the quinine bisulphate and excited the broad band of fluorescence. Because of the low intensity of this emission, high instrumental sensitivity had to be

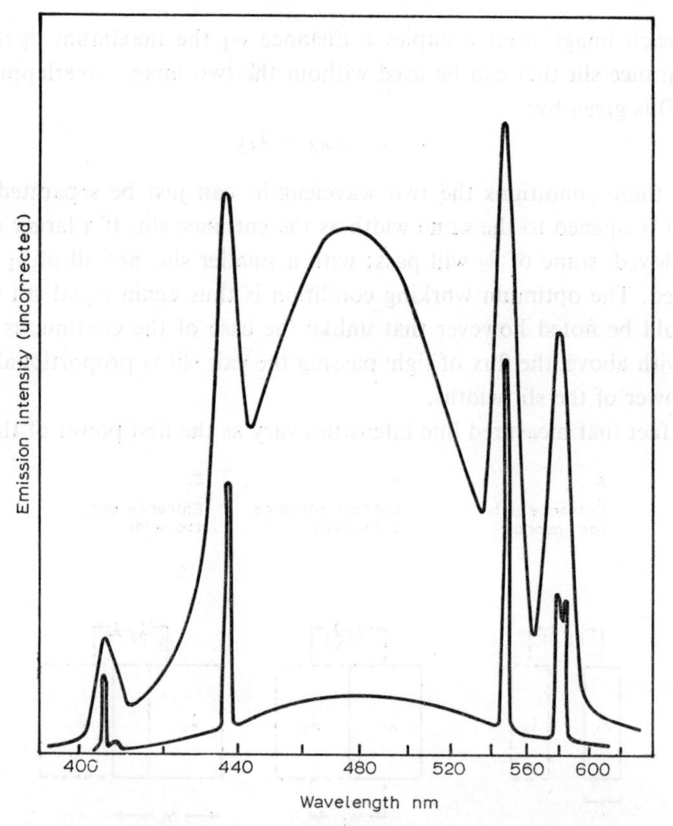

Fig. 48. Illustrating variation of relative intensities of lines and continuum with variation of slit width.

Emission and scatter from 100 μg per ml of quinine bisulphate in 0.1 N sulphuric acid irradiated by light from mercury lamp filtered through 2 cm of 0.8% w/v sodium nitrite in water. Slit widths were 0.5 mm for upper curve and 0.05 mm for lower curve.

used and the intensities of scattered exciting light were relatively large. It will be seen that when a very narrow slit width was used, the observed intensity of the band emission was very low in comparison with the lines. When a large slit width was used (with lower instrumental sensitivity) the observed intensity of the band emission was as great as that recorded for the line emission.

This method of emphasising line emission applies also to very narrow *band* emission and can be put to practical use. For example the Shpol'skii spectra to be discussed in Section 4 F 5 contain bands less than 0.1 nm wide and by the use of narrow slits, the Shpol'skii bands of one substance can be seen superimposed on the broad band emission from other substances present in greater quantity. The effect of band width is also of value when measuring emission spectra that are overlapped by Raman spectra from the solvent. The main Raman bands from solvents are narrow and can easily be recognised when narrow band widths are used. With large band widths the Raman bands may be mistaken for new fluorescence bands (see Section 5 C 1).

Because of the different dependence of the intensities of line and band emission on band width, the spectral distribution of sources giving rise to both line and continuous emission must be expressed in a special way. Thus, when we scan the spectrum of such a source with wide and equal spectrometer slits corresponding to a half-band width $\Delta\lambda$, we obtain a spectrum like the upper spectrum in Fig. 48 in which the continuum produces a smooth curve on which are superimposed approximately triangular peaks of half-band width $\Delta\lambda$. We therefore imagine the whole spectrum to be split up into narrow rectangular slices of width $\Delta\lambda$. Each mercury line will be represented by one of these rectangles of height equal to the peak intensity observed at the wavelength of the mercury line, less the height of the rectangle representing the continuum at this wavelength. The height of the latter is deduced from that of adjacent rectangles where there are no lines. A slightly different method of measuring mixed line and continuous spectra will be described in Section C dealing with light sources, where examples will be given.

5. Determination of Linear Dispersion

If dispersion data have not been provided with the instrument, they can be determined with reasonable precision by the method about to be described.

The monochromator should be fitted with a photometer, e.g. the photomultiplier–amplifier combination normally used for measuring fluorescence spectra. A magnesium oxide screen is set up in front of the entrance slit at an angle of about 60° to the optic axis of the monochromator as shown in Fig. 49A. The magnesium oxide screen, R, is illuminated by the light from a mercury lamp, DS, or other source giving a line spectrum. No focussing lens is needed since the intensity of illumination of the magnesium oxide screen need not be great. With a very narrow entrance slit, open the exit slit sufficiently to pass a band width, $\Delta\lambda_2$, of several nm. Consider now the response of the photometer when the monochromator is scanned across the wavelength, λ, of the chosen mercury line. In the upper portion of Fig. 50 the full line rectangle represents the exit slit opened to a width w_2 millimetres, corresponding to the half-band width $\Delta\lambda_2$ (see equation 188). If the entrance slit is opened to a smaller width w_1, corresponding to a half-band width $\Delta\lambda_1$,

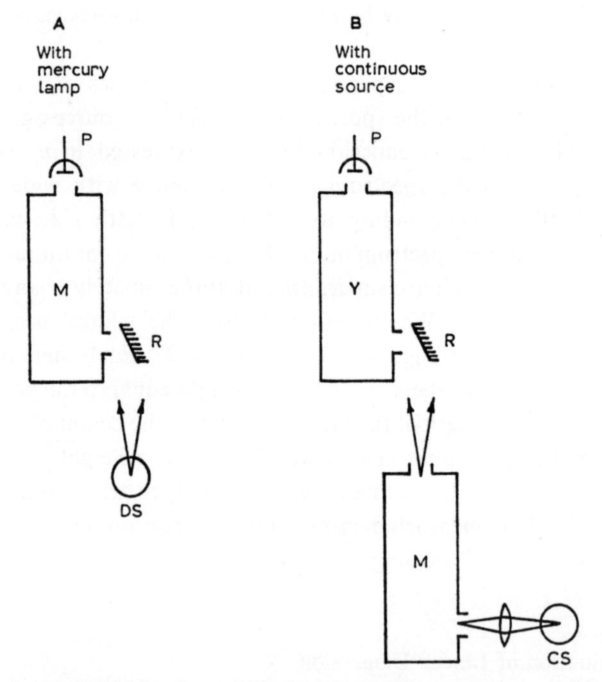

Fig. 49. Arrangements for determining the linear dispersion of a monochromator. M is the monochromator being examined. In method B, Y is the analysing monochromator. DS and CS are line and continuous sources, RR are magnesium oxide screens and PP are photomultiplier detectors.

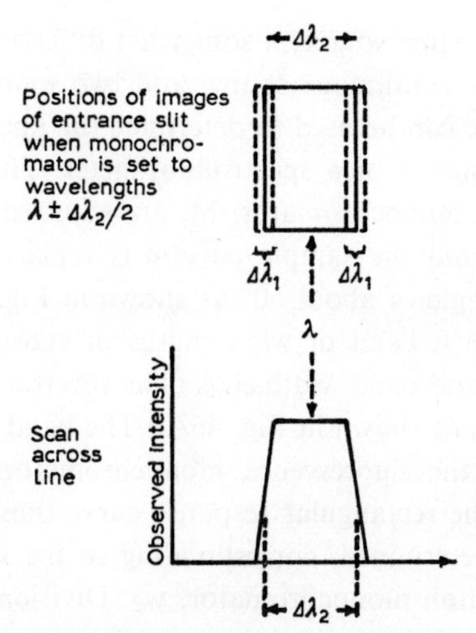

Fig. 50. Method of measuring linear dispersion using a line source.

the image of the entrance slit will fall completely within the rectangle of the exit slit for all wavelength settings between $\lambda \pm (\Delta\lambda_2 - \Delta\lambda_1)/2$. For settings $\lambda \pm \Delta\lambda_2/2$, the centre of the image of the entrance slit will fall exactly on either edge of the exit slit. These two positions of the image are shown by the narrow dotted rectangles in Fig. 50, each of width $\Delta\lambda_1$. At each of these two positions exactly one half of the light will pass through the exit slit. At wavelength settings $\lambda \pm (\Delta\lambda_2 + \Delta\lambda_1)/2$ the image of the entrance slit will have just passed out of the rectangle of the exit slit and the light flux passing the latter will have fallen to zero. Thus when the monochromator is scanned across the line, the observed intensity will vary as shown in the lower part of Fig. 50. The difference in wavelength between points half way up each sloping section of the curve will be equal to the half-band width $\Delta\lambda_2$. By making the width of the entrance slit very small compared with the exit slit, the sloping sections will be almost vertical and the required wavelength settings $((\lambda_2 - \lambda_1) = \Delta\lambda_2)$ can be determined precisely. The slit width in millimeters divided by the observed value of $\Delta\lambda_2$ gives the linear dispersion, m (see equation 188). There are some dozen suitable lines in the mercury spectrum between 250 and 600 nm and determination of m at each of these wavelengths is generally sufficient to draw the dispersion curve for a prism instrument. For a grating instrument, the linear dispersion should be nearly constant throughout the wavelength range.

If the linear dispersion has to be determined at more wavelengths than are available from the line source, a somewhat different procedure may be used. This requires a continuous source and two monochromators. Thus this second procedure can be used to determine the linear dispersion of the *excitation* monochromator of a spectrofluorimeter. The entrance and exit slits of the excitation monochromator, M, are adjusted as described in the previous paragraph and the sample cuvette is replaced by a magnesium oxide screen at an angle of about 60° as shown in Fig. 49B. The screen is thus illuminated with a band of wavelengths of substantially rectangular spectral distribution and band width $\Delta\lambda_2$. (The spectral distribution of light passing the exit slit is as shown in Fig. 46A.) The band width is then determined by scanning the fluorescence monochromator, Y, set with very narrow slits. From the rectangular response curve thus obtained, the half-band width $\Delta\lambda_2$ is determined, corresponding to the known width of the exit slit of the excitation monochromator, w_2. Division of w_2 by $\Delta\lambda_2$ then gives the value of m of the excitation monochromator for the wavelength concerned. In certain regions of the spectrum, the intensity of the continuous source (CS in Fig. 49B) may not be great enough to permit the use of a narrow entrance slit on monochromator M *and* narrow slits on monochromator Y. In these circumstances an alternative procedure that will provide a much greater intensity of illumination of the magnesium oxide screen is to set both slits on M to large and equal values. The situation is then that indicated in Fig. 46C. On scanning the analysing monochromator with narrow slits a triangular response curve will be obtained from which the half-band width can be determined by measurement of the wavelengths corresponding to intensities equal to one half of the maximum intensity.

6. Optimum Entrance Slit Illumination

Referring back to Fig. 45 the light from the source S, is fed to the monochromator by focussing an image of S on the entrance slit W_1 by means of the lens L_1. For optimum light collection the focal length of the lens L_1 should be such that the collimating mirror M_1 is completely filled with light, and the image of the source on the entrance slit just covers the area of the latter when it is opened to its maximum working width. The distances u and v will clearly depend on the relative sizes and shapes of the source and the fully opened entrance slit. Let us assume for simplicity that the ratio of width to height of source is the same as that of the fully opened entrance slit (this condition corresponds to the optimum utilisation of the source energy). If it is then arranged that the height of the image of the source is equal to that of the slit, the latter will be filled with light at all working slit widths. The relationship between u and v is then given by the lens magnification formula, viz.:

$$v/u = h/s \tag{195}$$

where h is the height of the entrance slit and s the height of the source. Let the rate of light emission from the source (in einsteins sec^{-1}) be equal to I_λ per unit solid angle within a unit band of wavelengths (i.e. within 1 nm). Let θ_1 be the solid angle subtended by the cone of rays reaching L_1 from S, and let F be the aperture of the monochromator defined by:

$$F = d/f \tag{196}$$

where f is the focal length of the collimating mirror M_1, and d is the effective height of the dispersing element. The projected area of the dispersing element, i.e. the cross section of the parallel beam of rays from M_1 dispersed by the prism or grating, is assumed to be square, and hence its area is d^2. (Since M_1 is normally circular, its diameter will be somewhat greater than d.) The solid angle, θ_2, subtended by the effective area of M_1 at the entrance slit, W_1, is thus given approximately by:

$$\theta_2 = d^2/f^2 = F^2 \tag{197}$$

From equation 195:

$$\theta_1/\theta_2 = v^2/u^2 = h^2/s^2 \tag{198}$$

and hence from equations 196 and 197:

$$\theta_1 = F^2h^2/s^2 \tag{199}$$

Clearly the amount of light collected by L_1 and brought to a focus on the plane of the entrance slit is equal to $I_\lambda\theta_1$ einsteins sec^{-1} nm^{-1}. If the area of the source is A_u, the area of the image of the source, A_v, on the plane of the entrance slit is given by:

$$A_v/A_u = h^2/s^2 \tag{200}$$

From equations 199 and 200 the *intensity* of illumination on the plane of the entrance slit is given by:

$$I_\lambda\theta_1/A_v = I_\lambda F^2/A_u \text{ einstein mm}^{-2}\text{ sec}^{-1}\text{ nm}^{-1} \tag{201}$$

(for low values of F). It is not possible to increase the intrinsic intensity from a source by focusing with a lens. What we have achieved with the lens, however, is to ensure that the entrance slit is filled with light no matter how great its size in relation to the size of the source.

If the entrance slit of the monochromator is now opened to a width w_1, the amount of light passing through the slit and collected by the effective area of M_1 is equal to:

$$\left(\frac{I_\lambda}{A_u}\right) F^2 h w_1 \text{ einstein sec}^{-1} \text{ nm}^{-1} \qquad (202)$$

Thus to collect the greatest possible amount of light at the entrance slit we require a source with the greatest possible value of I_λ/A_u, i.e. the greatest intrinsic brightness (einsteins mm^{-2} sec^{-1}). For a source of a given power, this implies the smallest area of source, and with a shape to match that of the fully opened entrance slit.

So far we have assumed that the source is of uniform brightness over its whole area, and under these conditions the best utilisation of source energy is achieved when the image of the source just fits the fully opened entrance slit. In practice however sources are not of uniform brightness, and by focussing the source in this way, the entrance slit will not be uniformly illuminated. To obtain greater uniformity of illumination it is usually desirable to move the source slightly nearer to the lens L_1 than the position corresponding to an exact focus, so that a "fuzzy" patch of light is produced on the entrance slit, having a somewhat greater area than the latter. Setting the lamp slightly off-focus in this way has the additional advantage that slight movement of the arc during operation of the lamp will produce less variation in intensity at the exit slit than if the image is exactly focussed. The best compromise should be obtained by trial and error.

7. Light Gathering Power

To obtain a measure of the performance of a monochromator we shall derive an expression for its light gathering power, that is, a measure of the amount of light within a given band width that the monochromator will pass from a source of given brightness. The light gathering power depends on the slit height, h, the focal length, f, the angular dispersion, α, and the working height, d, of the prism or grating. The height of the slits cannot be increased indefinitely for a given focal length, because light rays far removed from the optic axis are less efficiently collected so that the top and bottom of the exit slit are less intensely illuminated. Slits that are very high in relation to the focal length also give rise to distortion and loss of resolution. We shall assume therefore that the instrument has already been designed with an optimum slit height and that this is the same for both entrance and exit slit. This implies that the focal lengths of M_1 and M_2 are equal, i.e. that the monochromator does not produce any magnification. We shall further assume that light losses due to inefficiency of the optical parts of the instrument are zero, and that the entrance and exit slits are always set to equal width w, i.e. to the condition for obtaining maximum efficiency as described in Section 3 B 3. The expression to be derived for the light gathering power refers to conditions in which the illumination of the entrance slit by the light from the source is adjusted to an optimum as described in the previous section.

The light collected by the entrance slit is distributed along VR with a linear dispersion of m millimetres per nanometre, and since we have assumed no light losses within the monochromator, each section of VR of length m, will, by equation 202 receive:

$$\left(\frac{I_\lambda}{A_u}\right) F^2 hw \text{ einstein sec}^{-1} \tag{203}$$

Since the exit slit is also opened to a width w, the rate of passage of light through the exit slit, $T_{\Delta\lambda}$, is given by:

$$T_{\Delta\lambda} = \left(\frac{I_\lambda}{A_u}\right) \frac{F^2 hw^2}{m} \text{ einstein sec}^{-1} \tag{204}$$

This light will have the spectral distribution shown in Fig. 46 column C, with a *full* band width of $2\Delta\lambda$, where $2\Delta\lambda$ is given by equation 190, viz.:

$$2\Delta\lambda = 2w/m \tag{205}$$

Now the light gathering power is related to the amount of light passed when the monochromator slits are adjusted to give unit half-band width. For this condition we must set w equal to m in equation 204, and thus:

$$T_{1\,nm} = \left(\frac{I_\lambda}{A_u}\right) F^2 hm \tag{206}$$

The quantity in brackets is the brightness of the source and the quantity F^2hm is the light gathering power (LGP) of the monochromator. By substituting from equations 186 and 196 the light gathering power can be expressed in an alternative form, viz.:

$$\mathrm{LGP} = \mathrm{F^2hm} = \alpha \mathrm{hd^2/f} \tag{207}$$

either of which can be used for comparing monochromators, depending on which parameters are given. Thus, the LGP is proportional to slit height, linear dispersion and the square of the aperture. Alternatively for two monochromators employing the same dispersing element, e.g. two quartz prism monochromators (for which α is the same), the LGP is proportional to d^2/f and to the slit height. It is obvious that to obtain the greatest monochromator performance requires the largest possible prisms or gratings (i.e. large value of d) and the greatest possible ratio of slit height to focal length, consistent with absence of serious distortion and non-uniformity of exit slit illumination.

Using the LGP value, equation 204 for the light flux passed by the monochromator with given slit width settings can be expressed in the following form:

$$\mathrm{T}_{\Delta\lambda} = \left(\frac{\mathrm{I}_\lambda}{\mathrm{A_u}}\right)(\mathrm{LGP})\left(\frac{\mathrm{w}}{\mathrm{m}}\right)^2 = \left(\frac{\mathrm{I}_\lambda}{\mathrm{A_u}}\right)(\mathrm{LGP})(\Delta\lambda)^2 \tag{208}$$

This is equivalent to equation 189 for equal slits, in which

$$\mathrm{K} = \left(\frac{\mathrm{I}_\lambda}{\mathrm{A_u}}\right)(\mathrm{LGP})/\mathrm{m}^2 \tag{209}$$

The equation giving the intensity passed from a line source can be derived in a manner similar to that for equation 208. Thus if J is the rate of light emission in the line per unit solid angle from the source, the intensity of illumination on the plane of the entrance slit (see derivation of equation 201) is given by:

$$\left(\frac{\mathrm{J}}{\mathrm{A_u}}\right)\mathrm{F}^2 \text{ einstein mm}^{-2}\text{ sec}^{-1} \tag{210}$$

The amount collected by the entrance slit is thus

$$\mathrm{T}_\lambda = \left(\frac{\mathrm{J}}{\mathrm{A_u}}\right)\mathrm{F^2hw_1} \tag{211}$$

and provided that the exit slit is opened to a width not less than w_1, this light all passes the latter. Equation 211 may be written in the alternative form:

$$\mathrm{T}_\lambda = \left(\frac{\mathrm{J}}{\mathrm{A_u}}\right)(\mathrm{LGP})\ (\Delta\lambda) \tag{212}$$

Thus the flux passing the monochromator is proportional to the LGP of the monochromator and the *first* power of the band width to which the slits are set. The light is of course monochromatic.

8. Comparison of LGP Values

The characteristics of a selection of prism and grating monochromators are shown in Table 12. With the exception of that having a water prism, these instruments are similar to some commercially available. Since the linear dispersion of grating monochromators is substantially independent of wavelength, their LGP values expressed in terms of wavelength apply at all wavelengths. The dispersion of prism materials varies considerably with wavelength and LGP values are shown in Table 12 for selected wavelengths only. Values of LGP as a function of wavelength for some of the monochromators are plotted in Fig. 51.

Inspection of the data in Table 12 and Fig. 51 provides some interesting comparisons. Thus the grating monochromator no. 1 costs about the same as the prism monochromator no. 2, but has a greater LGP at all wavelengths longer than 200 nm. In the middle of the visible region at 500 nm the LGP value of the grating is 25 times better than that of the quartz prism. The performance of the second instrument is improved by a factor of three in the visible if a glass prism is used, but in terms of LGP alone, it is still

TABLE 12

COMPARISON OF MONOCHROMATORS

No.	Instrument type	h (mm)	d (mm)	F	m ($mm\ nm^{-1}$)	(LGP) F^2hm ($mm^2\ nm^{-1}$)
1	Grating	25	100	0.20	0.30	0.30
2	Quartz prism	18	45	0.14	0.033 (500)	0.012
					0.22 (280)	0.077
2a	Glass prism	18	45	0.14	0.10 (500)	0.035
3	Grating	11	50	0.29	0.14	0.13
4	Quartz prism	15	40	0.06	0.067 (500)	0.0036
					0.44 (280)	0.024
5	Quartz prism	50	115	0.18	0.066 (500)	0.11
					0.44 (280)	0.71
5a	Water prism	50	115	0.18	0.28 (280)	0.45

Note

The (LGP) values ignore light losses and assume uniform exit slit illumination. Actual values may be considerably lower, depending on the wavelength.

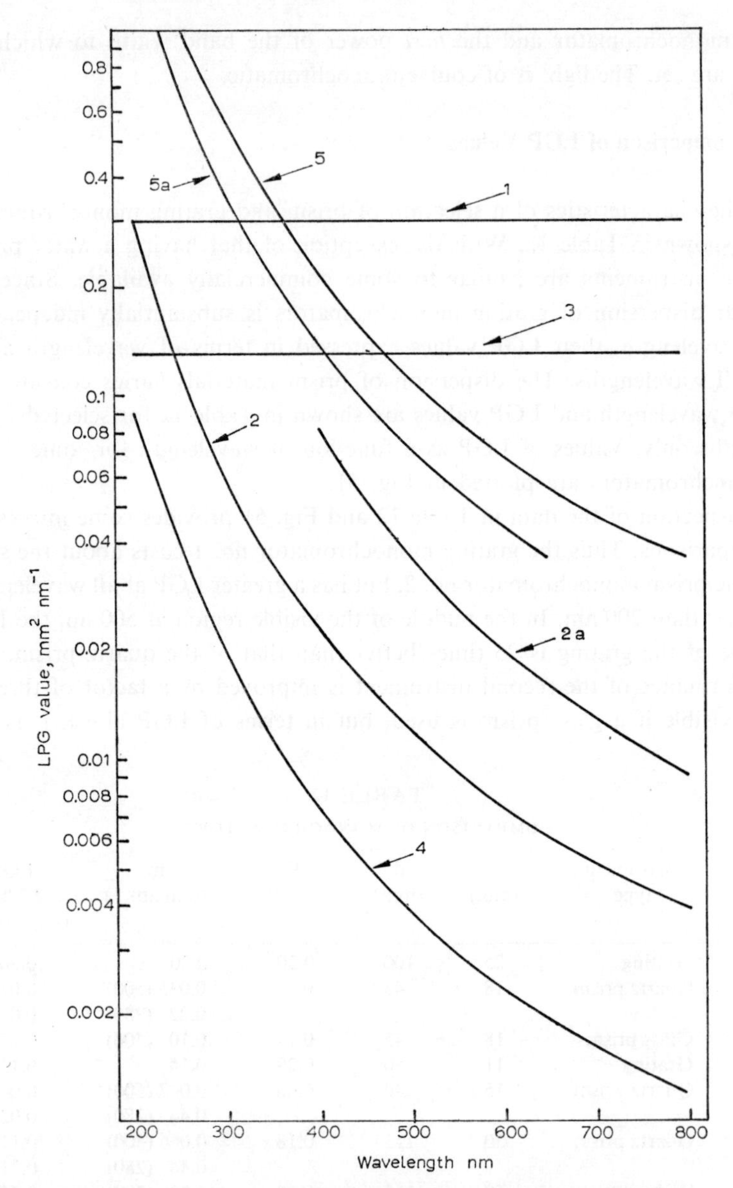

Fig. 51. LGP values of monochromators (wavelength basis).
The LGP values take no account of light losses. Practical values are considerably lower.

considerably less efficient than the grating instrument. To obtain an LGP value comparable with that of the grating instrument over a wide range of wavelengths requires the use of a very large and expensive quartz prism monochromator (no. 5), and even with this, the LGP is less than that of the grating at wavelengths longer than 360 nm. To avoid the use of costly large quartz prisms, hollow quartz or silica prisms filled with water have been used. These are of course much cheaper, but unfortunately the dispersion of water is less than that of quartz over the wavelength range of interest, as indicated by the comparison between curves 5 and 5a in Fig. 51. Other liquids of higher dispersion can be used for selected wavelength regions, but in view of the fact that large gratings can now be purchased at moderate cost, it is doubtful whether the saving in cost by the use of a water prism is worth-while in comparison with the total cost of constructing a monochromator.

Prism monochromator no. 4 corresponds approximately to those used in several manually operated ultra-violet absorption spectrophotometers. Its cost is about the same as that of the grating monochromator no. 3 and is about one-third of that of nos. 1 and 2. Again it will be seen that on the basis of LGP alone, the grating instrument is considerably better than the prism.

It can, however, be misleading to compare monochromators solely on the basis of their LGP values. Other factors to be taken into account are (a) The efficiency of the various components of the monochromator (these have been assumed to be 100% efficient in deriving the LGP value), (b) The interference produced by the presence of other spectral orders in grating monochromators, (c) Spectral impurities introduced by scattered light, (d) The size and divergence of the beam issuing from the exit slit in relation to the purpose for which the monochromator is required. Each of these additional factors will now be briefly discussed.

9. Grating Inefficiencies

A grating reflects the incident light into several spectral orders and thus the light in any one order will be less than that from a prism instrument of equal LGP. Monochromator gratings are usually "blazed", i.e. the rulings on the grating surface are cut at an angle so as to throw as much light as possible into the first order spectrum. The grating blaze is designed to be most efficient in a particular wavelength region, and in this region the efficiency may be as high as 50%. In other wavelength regions the efficiency may fall to below 20%. Hence in studying the LGP values in Table 12 and Fig. 51, this additional variable factor operating in favour of prism instruments must be borne in mind.

The fact that some of the light is thrown into different spectral orders implies that the beam issuing from the exit of a grating monochromator will contain additional wavelengths far removed from that to which the monochromator wavelength drum is set. For example an instrument set at 500 nm for the first order spectrum, will pass light from the second order of wavelength 250 nm. If the monochromator is being used to isolate monochromatic light for the excitation of a substance absorbing at 500 nm, other substances not absorbing at 500 nm, but absorbing at 250 nm will also be excited to a small degree. If the other substances happen to be present in large concentration, a large error will be introduced. Similarly if a grating monochromator is being used to analyse fluorescence emission from a specimen excited by light of wavelength 250 nm, a spurious signal will be observed at a wavelength setting of 500 nm due to scattered exciting light passed by the monochromator in the second order spectrum.

Most light sources give relatively little light of wavelength shorter than 200 nm and hence for monochromator settings between 200 and 400 nm the light issuing from the exit slit will be free of second order interference. The insertion of a filter absorbing all wavelengths shorter than about 380 to 400 nm (e.g. 1 cm of a 1% w/v solution of sodium nitrite in water) will clear the spectrum of second order interference for settings between 400 and 800 nm. It should be noted however that the insertion of filters between the fluorescing specimen and the monochromator used to measure the fluorescence emission spectrum can give rise to interference in high sensitivity work unless care is taken to ensure that the filter itself is non-fluorescent. Otherwise fluorescence of the filter excited by exciting light scattered by the specimen may be sufficient to swampt the weak fluorescence from a low concentration of the fluorescing solute. The choice of filters is dealt with in a later section.

10. Scattered Light

Apart from second order spectral interference, the light issuing from the exit slit should in theory contain no wavelengths outside the band width to which the slits are set. In practice however the exit beam contains a small proportion of light of all wavelengths that enter the entrance slit. This unwanted light arises mainly from scattering in a random manner from the grating surface and from imperfections or dust particles on the focussing mirror M_2 (Fig. 45). Scattered light is generally larger in grating instruments than in prism instruments, and its magnitude is in addition governed by the quality of the optical components and the excellence of the design of the monochromator. It can be particularly troublesome when measuring a weak emission band at one wavelength in the presence of a very strong emission band at another wavelength. Scattered light in the excitation monochromator can also give rise to interference in the same way as does light from the second order spectrum already mentioned. Examples of the interference produced by scattered light in high sensitivity work are discussed in Chapter 5. It can often be removed by the use of appropriate filters, but when very pure exciting light is required, e.g. when exciting a small concentration of one compound in the presence of a high concentration of another compound having an absorption band at a slightly shorter wavelength, the wavelength discrimination obtained by the use of filters may not be sufficient, and a double monochromator may then have to be used.

11. Geometry of Exit Beam and Specimen Illumination

The light gathering power as defined by equation 207 provides a measure of monochromator performance in terms of total flux of light passing the exit slit, but it gives no indication of the geometry of the exit beam. A monochromator with a large slit and large aperture will have a high LGP, but the light will be distributed over the large area of the exit slit and the intensity per unit area may be comparatively low. If the beam is too large to pass directly into the sample cuvette, it may of course be focussed to give a smaller image by means of the exit slit lens L_2 in Fig. 45. The distance between lens and specimen may then however be too short to allow the interposition of beam splitters etc., or the aperture may be too large to interpose a polariser, or to allow the beam to enter a sample compartment having a restricted aperture, e.g. a Dewar vessel. Thus the size and aperture of the beam as well as the LGP of the monochromator must be taken into account when designing a spectrofluorimeter for a particular purpose.

So far much of our discussion has been concerned with the choice and arrangement of monochromator to isolate the exciting light. The irradiation of the specimen and the collection of the resulting luminescence is of course equally important. The various modes of specimen irradiation, and the general design of sample compartment will be discussed later. We must consider here the requirements for efficient collection of the luminescence by the second monochromator. The considerations regarding the size of light source and its focussing on the entrance slit of the excitation monochromator all apply equally to the emission monochromator, except that the illuminated area of the specimen is now the source. Therefore, to gather as much luminescence as possible, the area of illumination on the specimen should be small and it should be focussed on the entrance slit of the emission monochromator so as to fill both the latter, and the collimator, with light. Clearly, the shorter the focal length of the exit slit lens of the excitation monochromator, the smaller the area of specimen illuminated and hence in principle the larger the proportion of luminescence that can be directed into the emission monochromator. It may not be possible to press this principle to its limit, however, because under a very high intensity of illumination the specimen may undergo photochemical decomposition too rapidly for measurements to be made.

12. Resolving Power

There is a theoretical maximum resolving power associated with any prism or grating system. In spectrofluorimetry it is rarely possible, or necessary, to use an instrument at resolutions near the limit, even when the instrument itself is capable of achieving the limit, because of the narrow slits required and the low light fluxes passed. However, it should be noted that some instruments are not designed for working at very high resolution. For example a grating monochromator of large LGP may be designed specifically for isolating high intensities of light of comparatively broad band width, and indeed may be supplied only with a series of interchangeable fixed slits of comparatively large width. Other instruments fitted with variable slits may still have comparatively low resolving power owing to imperfection of design, resulting for example in curvature of the image falling on the exit slit, so that when the slit is closed beyond a certain width, the band width does not decrease any further. For many fluorimetric applications in which the fluorescence bands are very broad, these limitations in resolving power are of no consequence, but in measuring Shpol'skii spectra for example, the bands can be exceedingly narrow, and for this purpose a precision prism instrument may be found to give superior results to a grating instrument although the latter may have a much higher LGP value.

13. Choice between Wavelength and Wavenumber

It was pointed out in Chapter 1 that because light absorption and emission are quantised processes in which the size of the quantum is equal to the difference in energy between two molecular states, it is more meaningful to plot spectra in terms of wavenumber rather than wavelength. The reader will nevertheless have noticed that the discussion in this chapter has been in terms of wavelength. This is because most monochromators are calibrated in wavelength, and furthermore with grating instruments the linear dispersion expressed in wavelength is substantially independent of wavelength setting. This means that if the spectrum is scanned by driving the wavelength drum of a grating monochromator at a constant speed, the spectrum will automatically be recorded on a strip chart recorder with a linear wavelength scale, and this is a great convenience. It should be remembered however that if a large spectral region is scanned, an undue amount of chart space is taken up with the long wavelength region of the spectrum and the short wavelength region, in which there is generally more information per nanometre, is bunched together. If the drum of a prism monochromator is driven at constant speed, the spectrum obtained is linear in neither wavelength nor wavenumber. It is however more nearly linear in the latter and for some purposes a more useful spectrum is obtained than would have been obtained had the scan been made with a grating monochromator. If an instrument is to be designed that records fully corrected spectra using an X–Y recorder (see Section 3 K 4) this consideration does not arise because both types of instrument can be made to record a spectrum linear in either wavelength or wavenumber.

Whatever type of instrument is used it is worth making oneself familiar with both scales, and their interconversion. For ease of conversion the units recommended are the nanometre (nm = millimicron) for wavelength and the reciprocal micron (μm^{-1}) for wavenumber. Thus the spectral region of most interest in photoluminescence is 200–1000 nm, or 5.0–1.0 μm^{-1}. Because both types of instrument have been used for recording the spectra reproduced in this book, both types of plot have been used, to avoid the labour and inaccuracies involved in re-calculation and replotting. In those spectra where frequent reference to the horizontal scale is necessary, both scales have been given.

Some care is required in converting *bandwidths* from one scale to the other. The equation for conversion may be derived as follows. Let λ be the wavelength expressed in nm and $\bar{\nu}$ the wavenumber expressed in μm^{-1}. Then:

$$\bar{\nu} = 1000/\lambda \tag{213}$$

Differentiating we find that:

$$\Delta\bar{\nu} = -\frac{1000}{\lambda^2}\cdot\Delta\lambda \tag{214}$$

or,

$$\Delta\lambda = -\frac{1000}{\bar{\nu}^2}\cdot\Delta\bar{\nu} \tag{215}$$

Thus the band width in reciprocal micrometres (μm^{-1}) corresponding to a band width of 1 nm varies rapidly with wavelength, as indicated by the values in Table 13. It is 25 times greater at 200 nm than at 1000 nm.

Because in photochemistry wavenumber is of more significance than wavelength it is also more significant to express band widths in wavenumbers rather than wavelengths. It would therefore have been more appropriate from the theoretical point of view to have expressed the LGP values of monochromators in terms of wavenumber. Such values would give a measure of the light flux through a monochromator (with slits set to pass a band width of 1 μm^{-1}) from a continuous source emitting unit intensity in one μm^{-1} band width per unit solid angle. LGP values expressed on a wavenumber basis may be simply calculated from those expressed on a

TABLE 13

RELATIONSHIP BETWEEN BAND WIDTHS EXPRESSED IN WAVELENGTH AND WAVENUMBER

λ (nm)	$\bar{\nu}$ (μm^{-1})	$\Delta\bar{\nu}$ corresponding to 1 nm
200	5.0	0.025
250	4.0	0.016
300	3.33	0.0111
400	2.5	0.00625
500	2.0	0.004
600	1.67	0.00278
800	1.25	0.00156
1000	1.00	0.001

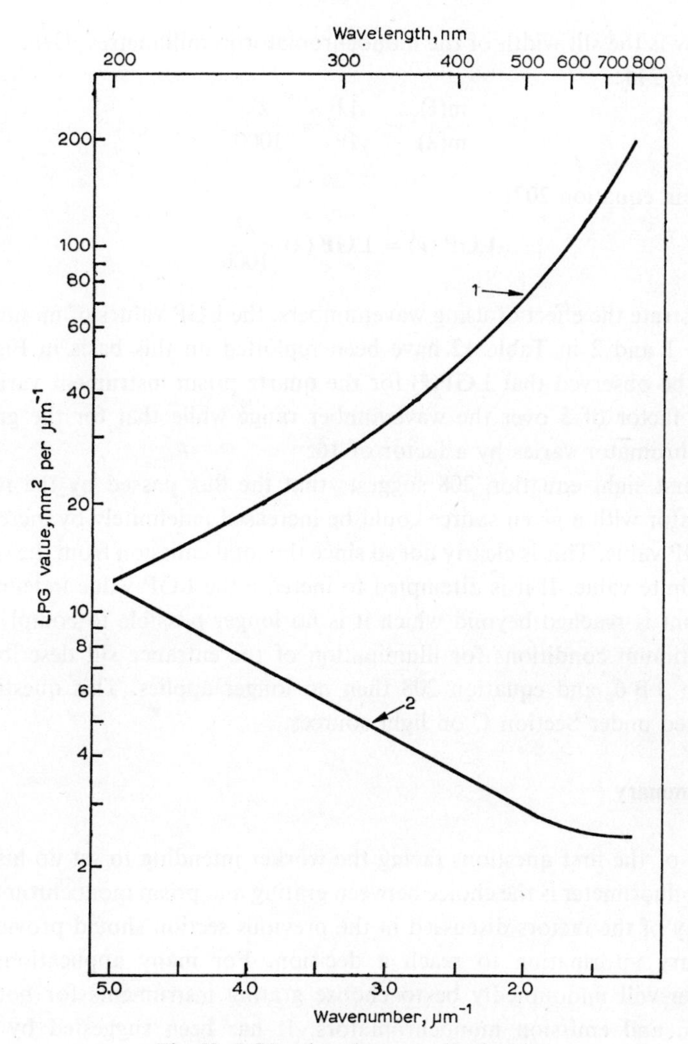

Fig. 52. LGP values (wavenumber basis).

wavelength basis as follows. Denote the linear dispersion expressed in millimetres per manometre by m (λ), and expressed in millimetres per reciprocal micrometre as m ($\bar{\nu}$). By equation 187:

$$m(\lambda) = w/\Delta\lambda \tag{216}$$

and

$$m(\bar{\nu}) = w/\Delta\bar{\nu} \tag{217}$$

where w is the slit width of the monochromator in millimetres. Hence from equation 214:

$$\frac{m(\bar{\nu})}{m(\lambda)} = \frac{\Delta\lambda}{\Delta\bar{\nu}} = \frac{\lambda^2}{1000} \tag{218}$$

and from equation 207:

$$\text{LGP}(\bar{\nu}) = \text{LGP}(\lambda)\,\frac{\lambda^2}{1000} \tag{219}$$

To illustrate the effect of using wavenumbers, the LGP values of monochromators 1 and 2 in Table 12 have been replotted on this basis in Fig. 52. It will be observed that LGP($\bar{\nu}$) for the quartz prism instrument varies by only a factor of 5 over the wavenumber range while that for the grating monochromator varies by a factor of 16.

At first sight equation 208 suggests that the flux passed by the monochromator with a given source could be increased indefinitely by increasing the LGP value. This is clearly not so since the total emission from the source has a finite value. If it is attempted to increase the LGP value indefinitely, the point is reached beyond which it is no longer possible to comply with the optimum conditions for illumination of the entrance slit described in Section 3 B 6, and equation 208 then no longer applies. This question is discussed under Section C on light sources.

14. Summary

One of the first questions facing the worker intending to set up his own spectrofluorimeter is the choice between grating and prism monochromators. A study of the factors discussed in the previous section should provide the necessary information to reach a decision. For many applications this decision will undoubtedly be to choose grating instruments for both excitation and emission monochromators. It has been suggested by some workers that a prism instrument should be chosen for the excitation monochromator, since this monochromator generally has to work at shorter wavelengths where the dispersion of quartz is greater, and purer exciting light can thus be obtained. To gain the advantages of both types of instrument the ideal arrangement is to employ double monochromators, each consisting of a combination of a grating and a prism. The grating then provides the high resolution while the prism rejects unwanted spectral orders as well as scattered light of unwanted wavelengths far removed from the wavelength required. Such instruments are of course much more expensive and for many purposes unnecessary.

To summarise, the maximum overall light gathering power of the spectrofluorimeter will be achieved by observing the following rules:

(i) Use monochromators having the largest possible LGP (see equation 207) consistent with the size of source used and the maximum working slit widths required.

(ii) Use a source having the highest possible intrinsic brightness (einstein mm^{-2} sec^{-1} nm^{-1}) in the wavelength region of interest, and with size and shape such that condition (iii) below can be satisfied (see also Section 3 C 9).

(iii) Choose the source condensing lens so that both the entrance slit at its full working width, and the collimator, are just filled with light.

(iv) Focus the monochromatic light from the excitation monochromator to the smallest possible area on the specimen (but see Section 3 C 9). This area should preferably have a shape to match that of the entrance slit of the second monochromator.

(v) Choose the condensing lens for the second monochromator so that the entrance slit at its full working width and the collimator of the second monochromator are just filled with fluorescence light from the specimen.

In practice other considerations prevent these rules being fully obeyed and a compromise has to be reached as already discussed. This compromise will vary according to the purpose for which the instrument is designed.

15. Precautions

The optical components of monochromators are quite easily damaged and the following precautions are recommended. The monochromator is best kept in a room maintained at constant temperature—rapid changes in temperature can cause condensation of moisture on mirror, prism or grating surfaces. As an extra precaution, indicator silica gel driers may be kept in the instrument case and changed as soon as they are exhausted. Chemicals giving off corrosive vapours should be excluded from the laboratory where the instrument is housed. The monochromator case should be kept closed as far as possible, to avoid dust settling on the optical surfaces. When making adjustments, care should be taken to avoid breathing over the optical surfaces. After prolonged use the mirror surfaces may become dull, and the performance of the instrument will be seriously reduced. It is then advisable to return the mirrors to the manufacturer for re-aluminising. It is hardly necessary to say that the surfaces of the prisms, mirrors or gratings should never be touched with the bare fingers.

Maintenance and use of the slits requires care. They should be cleaned by carefully rubbing the jaws with slit-sticks, or with a sliver of match-stick, to remove dust and fine grit. This is particularly important if the slits are frequently adjusted to narrow widths. In some monochromators it is possible to clamp the slit jaws together by incautious adjustment. Great care should be taken to avoid doing this when adjusting the slits to narrow widths, because if the slit jaws are slightly damaged it will no longer be possible to obtain the highest resolution of which the monochromator is capable.

C. LIGHT SOURCES

1. Measurement of Spectral Distribution

With the exception of tungsten filament lamps, which give a true continuous spectrum, and low pressure mercury lamps, which give a substantially pure line spectrum, all sources to be discussed give a mixed spectrum containing both line and continuous emissions. This presents a special problem in representing quantitatively the relative intensities of the lines and the background continuum, since the flux from a monochromatic line is expressed in einstein sec^{-1} while that from a continuum is expressed as einstein sec^{-1} per unit band width. This problem has already been mentioned in the previous section and it will now be discussed in more detail so that the reader can extract the maximum of information from the spectral distribution of the sources to be discussed.

It was shown in Section 3 B 4 (Fig. 48) that when a mixed spectrum is scanned by a spectrometer with equal entrance and exit slits, each monochromatic line emission gives rise to a triangular response of half-band width equal to the half-band width setting of the monochromator slits, superimposed on the continuous spectrum. The relative value of the light flux passed from a line and from the adjacent continuum is obtained by comparing equations 208 and 212 from which it is seen that the line flux is proportional to $\Delta\lambda$, while the flux from the continuum is proportional to $(\Delta\lambda)^2$. For the purpose of recording mixed emission spectra of lamps it is more useful to scan the spectrum with a very narrow entrance slit and a comparatively broad exit slit. Under these conditions each *monochromatic* line will produce a substantially rectangular response of band width equal to that corresponding to the width of the exit slit, as explained in Section 3 B 5 and Fig. 50. The continuum will of course still produce a continuous curve. The relationship between the intensity of the line and that of the *adjacent* continuum can be calculated from such a recording in the following way. It may be simply shown by methods analogous to those used to derive equations 208 and 212, that for *unequal* monochromator slits of widths corresponding to band widths $\Delta\lambda_1$ and $\Delta\lambda_2$ (where $\Delta\lambda_1 < \Delta\lambda_2$), the light fluxes passed from a continuum of intensity I_λ, and from a monochromatic line of intensity J, are given by:

$$T_{\Delta\lambda} = (I_\lambda/A_u)(LGP)(\Delta\lambda_1)(\Delta\lambda_2) \tag{220}$$

and

$$T_\lambda = (J/A_u)(LGP)(\Delta\lambda_1) \tag{221}$$

Thus in the region of any particular line at wavelength λ,

$$\frac{\text{Recorder deflection due to continuum}}{\text{Recorder deflection due to line}} = \frac{T_{\Delta\lambda}}{T_\lambda} = \frac{I_\lambda(\Delta\lambda_2)}{J} \tag{222}$$

But $\Delta\lambda_2$ is the width of the rectangular response produced by the line on the spectral scan, and thus if the response curve of the continuum is divided into rectangles of width equal to $\Delta\lambda_2$ (see Fig. 53), comparison of the areas of these rectangles within a chosen waveband between λ_1 and λ_2, with that of the rectangle due to the line, gives a direct measure of the relative numbers of quanta emitted by the continuum in the band between λ_1 and λ_2, and by the line. Thus in Fig. 53, the proportion of *monochromatic* light of wavelength λ emitted in the waveband between λ_1 and λ_2 is given by the ratio of the areas $a/(a + \Sigma b)$. Alternatively the proportion of light of wavelengths $\lambda \pm \Delta\lambda_2/2$ emitted in the waveband between λ_1 and λ_2 is given by the ratio of the areas $(a + b_4)/(a + \Sigma b)$. In this discussion we have used wavelengths but of course the method is equally valid for wavenumber plots. This method is particularly useful for measuring the spectral purity of the light given by a filter system, when for example it has been attempted to isolate a monochromatic mercury line by means of a narrow band pass filter.

Area comparisons of the kind just described give reasonably accurate results over restricted wavelength or wavenumber intervals, without further correction, provided that the spectral sensitivity of the spectrometer does not vary too rapidly. If however precise comparisons between widely separated spectral regions are required, the recorded spectra must be corrected. The recorded continuum is corrected in exactly the same way as that used to correct a recorded luminescence spectrum (see Section 3 K 3) by dividing the observed intensity at each wavelength or wavenumber, by the value of the spectral sensitivity expressed in appropriate units, i.e. by the value of S_λ or $S_{\bar{\nu}}$, at that wavelength or wavenumber. The recorded intensities of the lines are independent of the band width settings of the spectrometer but are proportional to the product PL (see Section 3 K 3), where

$$PL = S_\lambda / B_\lambda = S_{\bar{\nu}} / B_{\bar{\nu}} \tag{223}$$

in which L is the transmission factor of the monochromator and P the photomultiplier response per quantum, and B_λ or $B_{\bar{\nu}}$ is the relative band-width of the monochromator, which in general varies with wavelength or wavenumber. Thus with a grating instrument for which the linear dispersion expressed in wavelengths is constant, the same correction curve may be used for the intensities of both the lines and the continuum, provided that the

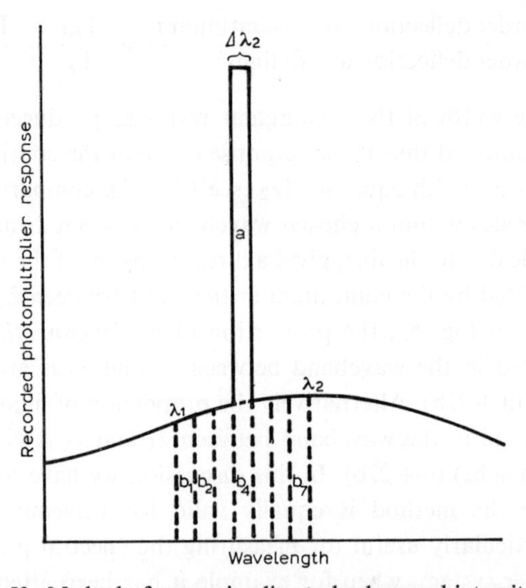

Fig. 53. Method of measuring line to background intensity.
The spectrum is scanned with a very narrow entrance slit, and broad exit slit corresponding to a band width $\Delta\lambda_2$. The line produces the rectangular response of area "a" and the background produces a smooth curve. Comparison of "a" with the area under the curve between λ_1 and λ_2 gives a direct measure of the ratio of line emission to background emission within these two wavelengths.

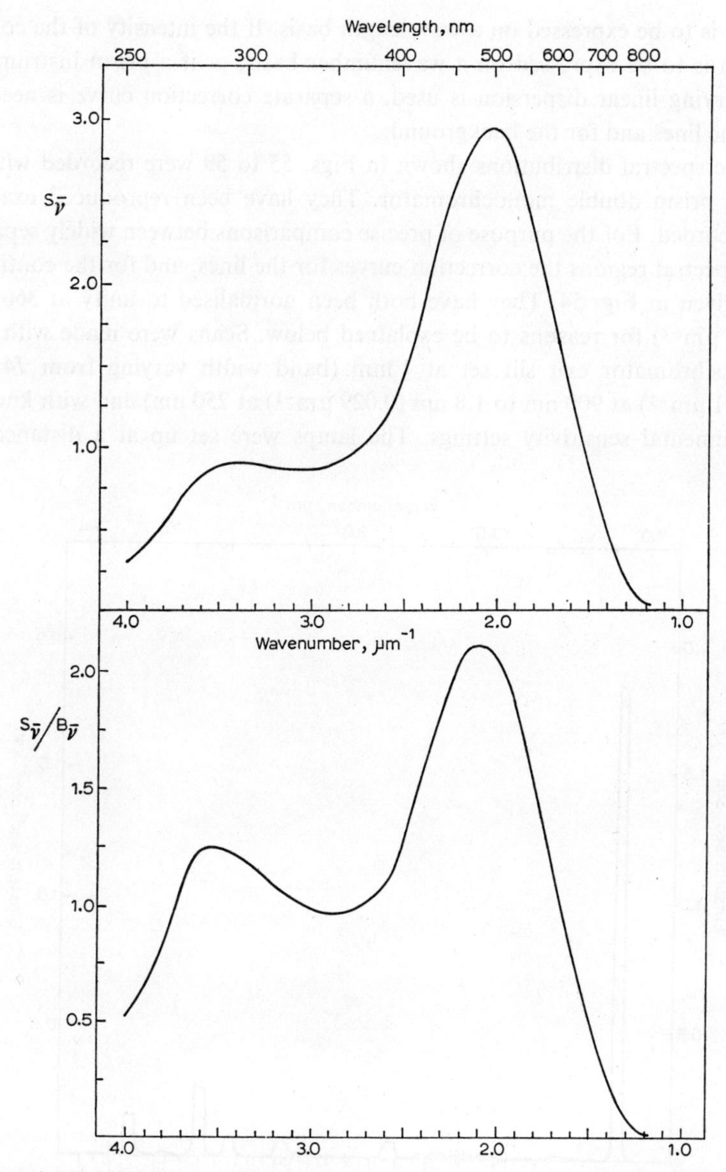

Fig. 54. Spectral sensitivity curves of double silica prism monochromator fitted with 9558Q photomultiplier.

latter is to be expressed on a wavelength basis. If the intensity of the continuum is to be expressed on a wavenumber basis, or if a prism instrument of varying linear dispersion is used, a separate correction curve is needed for the lines and for the background.

The spectral distributions shown in Figs. 55 to 59 were recorded with a silica prism double monochromator. They have been reproduced exactly as recorded. For the purpose of precise comparisons between widely separated spectral regions the correction curves for the lines, and for the continua are given in Fig. 54. They have both been normalised to unity at 366 nm (2.73 μm^{-1}) for reasons to be explained below. Scans were made with the monochromator exit slit set at 1 mm (band width varying from 74 nm (0.091 μm^{-1}) at 900 nm to 1.8 nm (0.029 μm^{-1}) at 250 nm) and with known instrumental sensitivity settings. The lamps were set up at a distance of

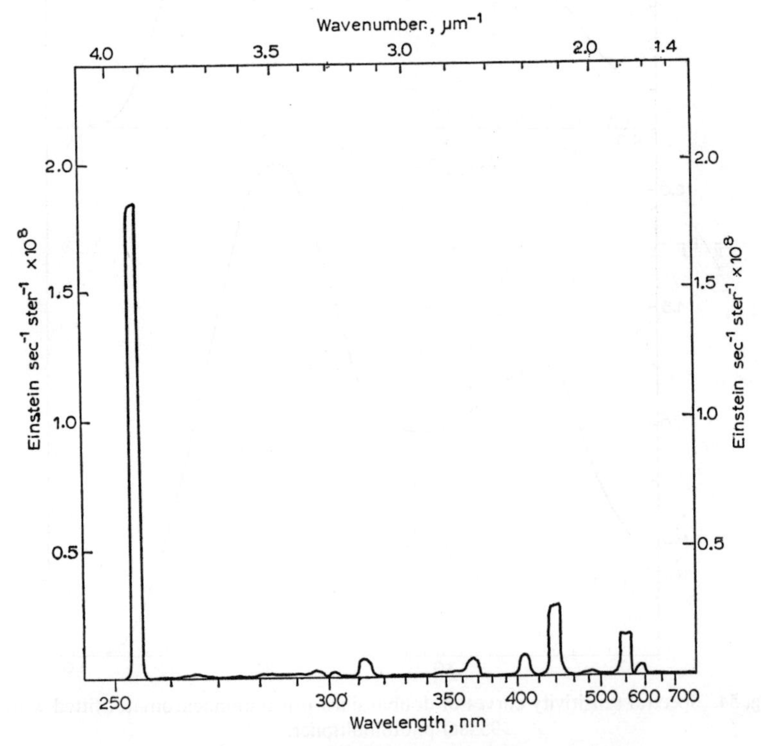

Fig. 55. Spectrum of 6 W low pressure mercury lamp.
Emission observed from tube length of 40 mm; total tube length was 340 mm.
(See text and Fig. 54).

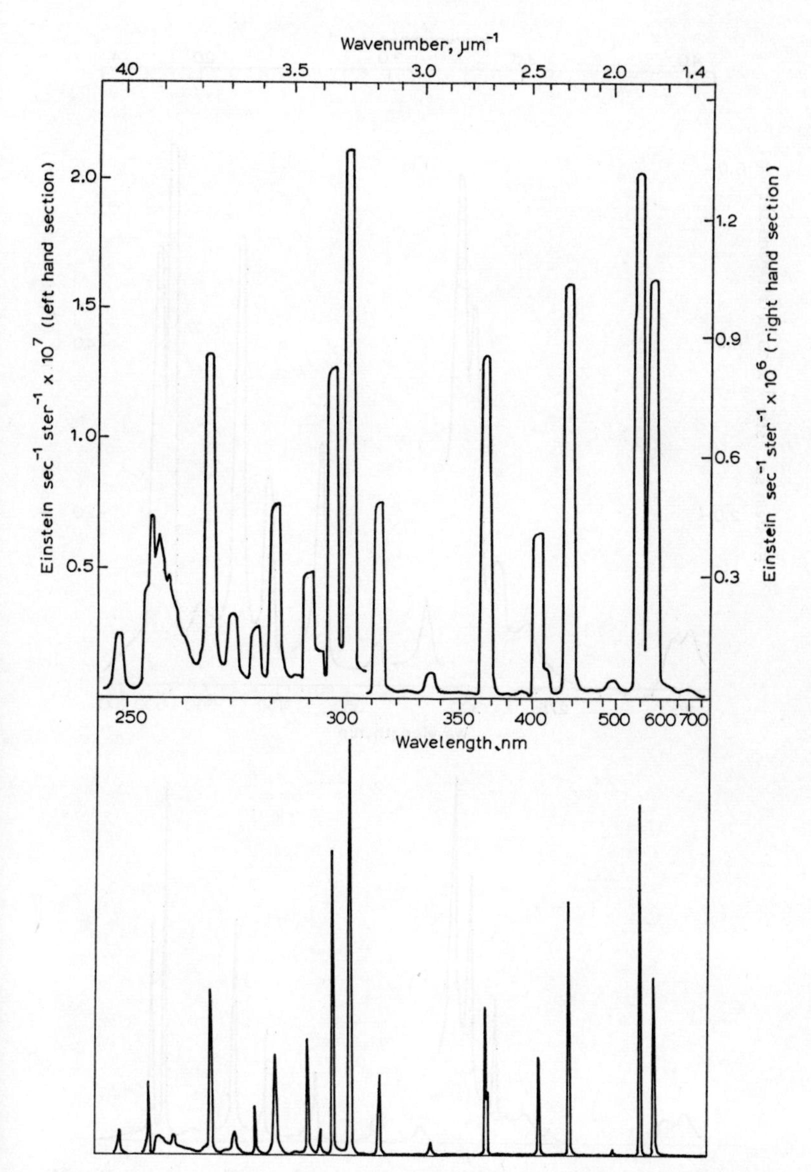

Fig. 56. Spectrum of 125 W medium pressure mercury lamp. (See text and Fig. 54).

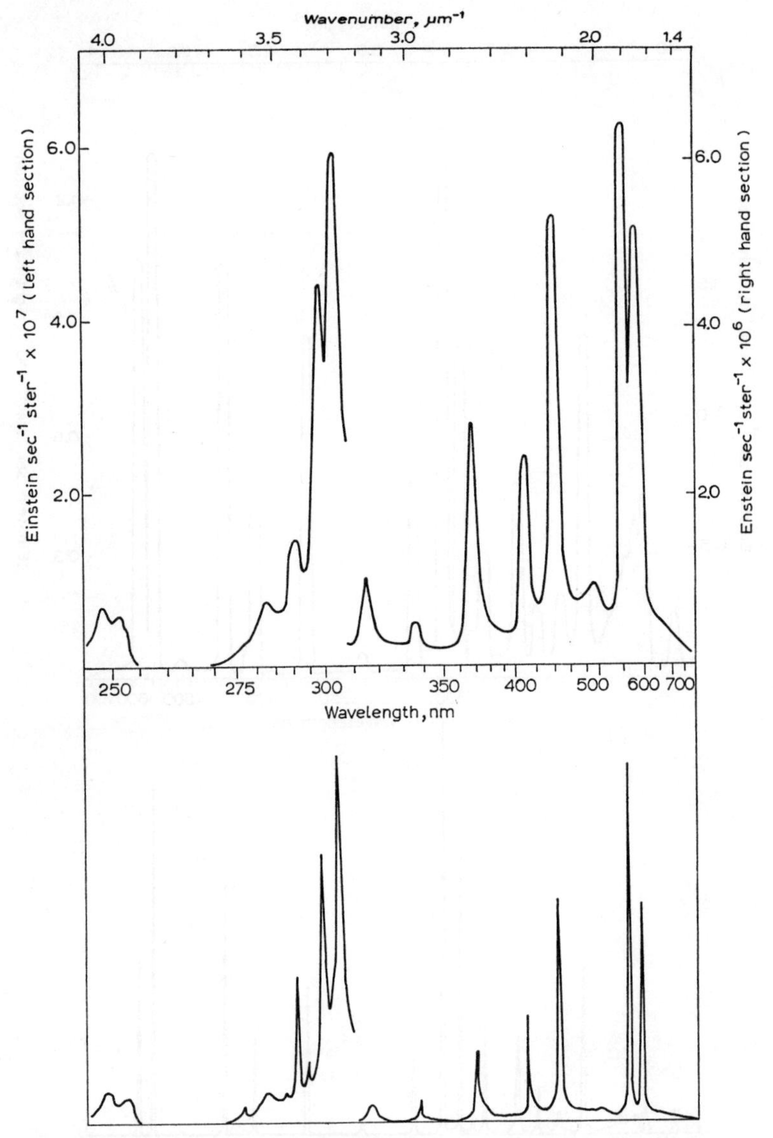

Fig. 57. Spectrum of extra-high pressure mercury lamp. (See text and Fig. 54).

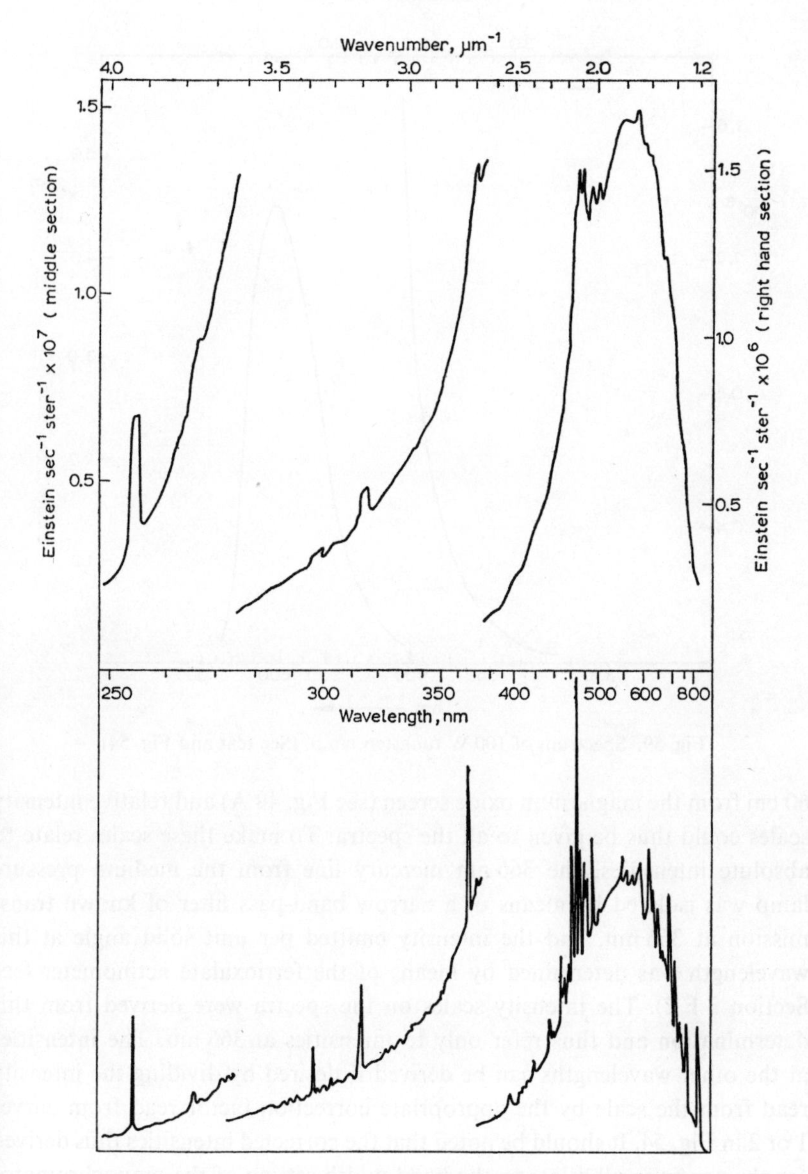

Fig. 58. Spectrum of 350 W xenon lamp. (See text and Fig. 54).

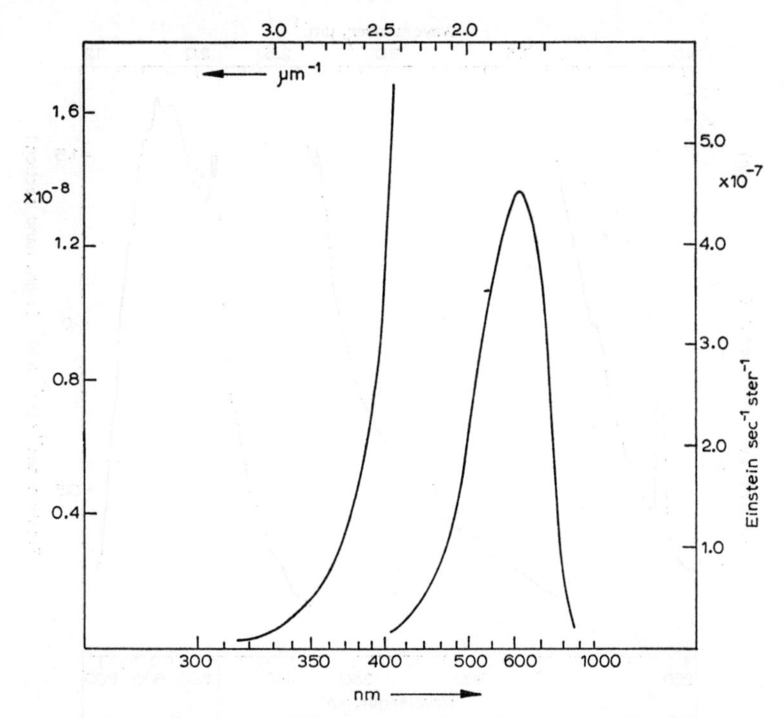

Fig. 59. Spectrum of 100 W tungsten lamp. (See text and Fig. 54).

60 cm from the magnesium oxide screen (see Fig. 49 A) and relative intensity scales could thus be given to all the spectra. To make these scales relate to absolute intensities, the 366 nm mercury line from the medium pressure lamp was isolated by means of a narrow band-pass filter of known transmission at 366 nm, and the intensity emitted per unit solid angle at this wavelength was determined by means of the ferrioxalate actinometer (see Section 3 F 2). The intensity scales on the spectra were derived from this determination and thus refer only to intensities at 366 nm. The intensities at the other wavelengths can be derived if desired by dividing the intensity read from the scale by the appropriate correction factor read from curves 1 or 2 in Fig. 54. It should be noted that the corrected intensities thus derived for the continua all relate to the band width setting of the monochromator at 366 nm (2.73 μm^{-1}) viz. 7 nm or 0.052 μm^{-1}.

The spectra of the mercury lamps cover the range 250–700 nm. Two vertical scales are generally given on each figure. These refer to the left-hand and right-hand sections of the spectra which were scanned at different sensitivities (with an overlap at 310 nm for the medium and high pressure mercury lamps). The low pressure mercury lamp was scanned throughout at the same sensitivity and the xenon lamp was scanned at three sensitivities. The left-hand scale on the latter refers to the *middle* section of the curve. The left hand section of the spectrum was scanned at a sensitivity nine times greater than this. In addition to the scans made with the monochromator exit slit set at 1 nm, qualitative scans were also made with a narrow exit slit so that the lines were recorded as sharp narrow peaks and recorded intensities of the continua were greatly reduced. Closely spaced lines were thus resolved in the qualitative scans. (They are not however all seen to be resolved in the Figures owing to the difficulties of reproduction.)

Examinations of Figs. 55 to 57 will reveal two peculiarities. Firstly, with monochromatic lines (e.g. the lines from the low pressure mercury lamp, and many of the lines from the medium pressure lamp), the sides of the "rectangle" corresponding to the line emission are substantially vertical, as required by the theory already discussed. The tops of the "rectangles" however are sloping instead of horizontal. This is due to the variable sensitivity across the face of the photomultiplier. As the line is scanned, the narrow image of the entrance slit traverses the area of the wide exit slit as described in Fig. 50, and thus impinges on different regions of the photomultiplier surface during the scan. The second peculiarity is the shape of the line scans with the extra high pressure lamp. These no longer have vertical sides and this is due to the fact that the "lines" from these lamps are no longer monochromatic, but are *pressure-broadened* as will be explained later.

The data in Figs. 55–59 relate to the *total* emission from all parts of the lamps concerned, including in most cases that from the incandescent electrodes. The light flux obtained from a monochromator depends on the brightness of the source (I_λ/A_u or J/A_u) as described in Section 3 B 7, and hence to compare the sources for the purpose of spectrofluorimetry, their size and shape, as well as their total emission, must be taken into account. The size of the source is an indefinite quantity because brightness varies from point to point. Rough estimates of the dimensions of the brightest regions of the sources are given in Table 14 together with the approximate proportion of the total visible emission that they represent. Because these estimates were made visually, they are only very approximate, but are useful for purposes of rough comparison between sources. It must be remembered however that the wavelength distribution may also vary in

TABLE 14

DIMENSIONS OF SOURCES

Lamp	Region of most intense illumination	
	% of total emission	Dimensions (mm)
6 W low pressure Hg	12[a]	40[a] × 6
125 W medium pressure Hg	60	20 × 1.2
1000 W high pressure Hg (run at 700 W)	60	6.6 × 1.2
200 W extra-high pressure Hg	80	2.5 × 1.6
25 W Cd	75	26 × 3.3
25 W Zn	~ 100	22 × 4.9
350 W xenon	~ 100	5.9 × 3.6
100 W tungsten projection	~ 100	17 × 2.1

[a] The low pressure lamp was in the form of a U tube of total length about 340 mm of which the emission from 40 mm was observed. The lamp could be run at ten times the quoted power with corresponding increase in intensity of the spectrum.

different areas of the source, and the source sizes in the ultra-violet region may differ appreciably from the visual estimates.

The method of using the data in Figs. 55–59 is illustrated by two examples. First, suppose it is required to derive the flux from the continuum of the xenon lamp at 280 nm. Reference to Fig. 58 indicates an uncorrected intensity

TABLE 15

FLUX FROM 200 W EXTRA-HIGH PRESSURE MERCURY LAMP THROUGH GRATING MONOCHROMATOR

λ (nm)	Chart reading (einstein sec^{-1} ster^{-1})	S_ν/B_ν	J (einstein sec^{-1} ster^{-1})	T_λ (einstein sec^{-1})	
				Calculated	Observed
436	5.3×10^{-6}	1.90	2.8×10^{-6}	7.3×10^{-7}	0.31×10^{-7}
405	2.5	1.37	1.8	4.7	0.22
366	2.9	1.00	2.9	7.5	0.43
313	1.1	1.08	1.0	2.6	0.31

Effective area of source (see Table 14) was $2.5 \times 1.6 \div 0.8$ mm^2 i.e. 5.0 mm^2. LGP value of monochromator was 0.13 and half-band width was 10 nm.
Substituting in equation 212:

$$T_\lambda = \left(\frac{J}{5.0}\right)(0.13)(10) = 0.26J$$

of 0.19×10^{-7} einstein sec^{-1} ster^{-1} at 280 nm. The value of $S_{\bar{\nu}}$ (from Fig. 54) is 0.86. Hence the lamp emission is 0.22×10^{-7} einstein sec^{-1} ster^{-1} for a band width of 0.052 μm^{-1}. Conversion of this band width to wavelength (equation 215) gives a lamp emission of 0.54×10^{-8} einstein sec^{-1} ster^{-1} nm^{-1}. This is the value of I_{λ} for insertion in equation 208 if required. Derivation of line intensities is illustrated by the data in Table 15 for the 200 w extra high pressure mercury lamp. The values in column 2 were read directly from Fig. 57 and those in column 3 from Fig. 54. The calculated values of T_{λ} refer to a grating monochromator of LGP value 0.13 working at a half-band width of 10 nm. The calculated values are between 8 and 24 times greater than the light fluxes actually observed through this monochromator. The discrepancy is accounted for by the following losses in the optics: (a) 7 pairs of transmitting surfaces having total transmission factor of ~ 0.5, (b) mirror losses, (c) grating inefficiency (reflectivity decreasing at longer wavelengths), and (d) incomplete collection of the image of the source at the entrance slit. The comparison provides a good example of how far practical light fluxes fall short of the theoretical values.

2. Incandescent Sources

Light emitted from a body solely as a result of its high temperature is known as *incandescence* or thermal radiation, and is thus distinguished from that emitted as a result of excitation by other means—i.e. the various types of luminescence (see Section 1 A 1). The ideal *black body* absorbs all radiant energy, of whatever wavelength, that falls on it, and the spectral distribution of light emitted by such a body is related to its temperature by Planck's law. This is usually expressed in the form:

$$dE/d\lambda = \frac{2\pi h\mathbf{c}^2 A\lambda^{-5}}{[\exp(h\mathbf{c}/\mathbf{k}\lambda T) - 1]} = \frac{3.7403 A\lambda^{-5} \times 10^{-5}}{[\exp(1.4384/\lambda T) - 1]} \qquad (224)$$

in which $dE/d\lambda$ is the energy emitted (ergs sec^{-1}) within unit band width (1 cm) at the wavelength λ (expressed in cm) by an ideal black body of surface area A at a temperature T degrees Kelvin. The physical constants are Planck's constant, h, velocity of light, **c**, and Boltzmann's constant, **k** (1.3805×10^{-16} erg per degree per molecule). Expressed in the more convenient units of watts (1 watt $= 10^7$ erg sec^{-1}) and micrometres, Planck's equation is:

$$dE/d\lambda = \frac{3.7403A\lambda^{-5} \times 10^4}{[\exp(14384/\lambda T) - 1]} \text{ watt per } \mu\text{m interval} \qquad (225)$$

In photoluminescence work it is often more convenient to use einsteins instead of energy, and wavenumber instead of wavelength. Expressed in this way, Planck's equation becomes:

$$dQ/d\bar{\nu} = \frac{0.3127A\bar{\nu}^2}{[\exp(14384\bar{\nu}/T) - 1]} \text{ einstein sec}^{-1} \text{ per } \mu\text{m}^{-1} \text{ interval} \qquad (226)$$

$\bar{\nu}$ is now expressed in μm^{-1}. Planck's equation applies strictly to the emission into a space at absolute zero, but for wavelengths in the visible and ultraviolet regions from incandescent sources, this is substantially the same as emission into a space at room temperature.

The spectral distribution from a black body takes the form shown in Fig. 60 and has a maximum at a wavelength that decreases as the temperature increases. Temperatures of the order 6000°K (e.g. the surface of the sun) are required to move the maximum (expressed on an energy–wavelength basis) into the middle of the visible region. The most convenient form of incandescent source is a tungsten filament lamp. These operate at temperatures in the region of 3000°K. The curves in Fig. 60 refer to 2800°K and correspond to a slightly "under-run" tungsten lamp. This approximates to the temperature of standard lamps used for spectral calibration. It will

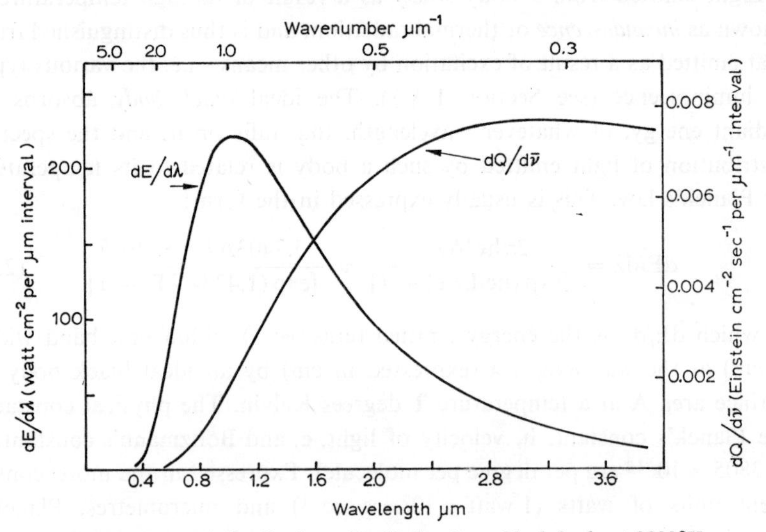

Fig. 60. Spectral distribution of radiation from black body at 2800°K.

be observed that the maximum $dE/d\lambda$ is situated in the near infra-red at just over 1 μm and the curve falls very steeply towards shorter wavelengths. The proportion of visible light emitted is rather low, and the proportion of ultra-violet ($\lambda < 400$ nm) is very low. The spectral distribution is even less favourable towards the shorter wavelengths when it is expressed on an einstein–wavenumber basis (see Fig. 60). The proportion of short wavelength emission can be increased quite considerably by over-running the lamp at higher temperature, but the life of the lamp is then shortened. The life of the lamp filament is extended by the presence of iodine vapour in the envelope and such lamps, having an envelope of quartz, are used to provide increased emission in the near ultra-violet region.

The tungsten filament is not, in fact, a true black body. Its *total* emission is less than that shown in Fig. 60. However, the *shape* of the emission spectrum in the visible region can be approximated to that of a black body emitter at a certain temperature. Temperatures measured in this way are known as *colour temperatures*. The colour temperature of a tungsten filament is somewhat higher than its true temperature: for example the emission at 2800°K corresponds to a colour temperature of about 2880°K, i.e. the *proportion* of short wavelength emission is a little greater than that shown in Fig. 60. The actual emission is less at all wavelengths.

The uncorrected observed emission from a 100 W strip filament lamp (filament size 17 × 2.1 mm) is shown in Fig. 59 and comparison with Fig. 58 shows that it can compete with the xenon arc lamp (source size 5.9 × 3.6 mm) only in the longer wavelengths of the visible spectrum, and even here its intrinsic brightness is considerably less. For excitation in this region it has the advantages of a completely continuous spectrum and, when run from a constant voltage transformer, a very high stability. For long wavelength excitation, particularly if high intensities are not required, the tungsten lamp is thus well worth consideration. For excitation in the violet end of the visible spectrum it is rarely used and in the ultra-violet region it must for most purposes be replaced by a discharge lamp.

Probably the most important application of the tungsten lamp in photoluminescence work is its use as a standard for determining the spectral sensitivity curve of a spectrometer in the visible region of the spectrum. Standard lamps can be purchased with instructions to operate them at a known colour temperature, from which the spectral distribution of the visible light emitted can be calculated by equations 225 or 226. This method of calibration is described in Section 3 K 3.

3. Xenon Arc Lamps

The determination of a complete fluorescence excitation spectrum requires a lamp giving a high intensity continuous spectrum throughout the visible and ultra-violet region, preferably with a constant quantum output at all frequencies to minimise correction factors. Unfortunately the sources at present available fall far short of this ideal. The high-power tungsten-filament lamp can be used for excitation in the visible region, but its intensity falls off rapidly towards the violet, and in the near ultra-violet region is too low for most purposes. The hydrogen lamp provides a reasonably constant energy distribution in the ultra-violet region, but its intensity is generally too low for measuring fluorescence excitation spectra. It can however be used for calibration of spectrofluorimeters in this region (see Section 3 K 3). For most work on excitation spectra, a compact source xenon arc lamp available commercially in sizes from 250 watts to 2.5 kW is used. Although desirable from the point of view of high intensity, the larger sizes of lamp, together with their associated control equipment, are expensive, and also the dissipation of the heat is an inconvenience. Lamp powers of 250 to 500 watts are most generally used. The source is small and of high intrinsic brightness, so that a comparatively large proportion of the light can be focussed on the entrance slit of the excitation monochromator. The xenon arc lamp has two disadvantages; the intensity falls off in the higher wave-numbers of the ultra-violet region (see Fig. 58) and the visible spectrum consists of a continuum on which are superimposed a large number of lines (many lamps also contain traces of mercury, which also produces a weak line spectrum) (see Fig. 58). When working at low resolution these lines are not troublesome, but at high resolution they can be.

In common with all gas discharge or arc lamps, xenon lamps have a "negative resistance", i.e. the voltage across the lamp terminals *increases* as the current *decreases*. An arc lamp cannot therefore be operated directly from a low impedance source or it will extinguish or destroy itself. It is necessary to incorporate in the circuit a "ballast" in the form of a resistance, or for A.C. lamps, an inductance or choke coil. Part of the voltage from the mains is dropped across this "ballast" and an increase in current through the circuit automatically causes a reduction in the voltage across the lamp terminals. A typical circuit for operating a xenon lamp is shown diagrammatically in Fig. 61. Xenon lamps require special starting gear which is normally supplied by the manufacturers, who will also provide a suitable

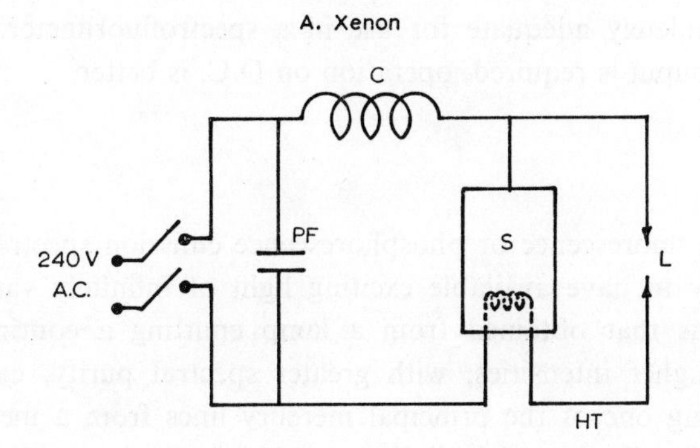

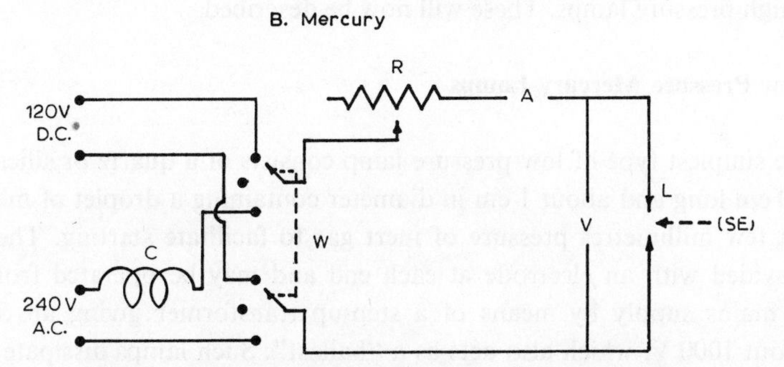

Fig. 61. Circuits for operating xenon and mercury lamps.
L, lamp; SE, starter electrode (3-electrode lamps); V, voltmeter; A, ammeter; S, starter unit; R, rheostat; W, 2-pole-3-position switch; C, choke coils; PF, power factor capacitor. The HT lead between starter unit and lamp, and the lead through the lamp housing to the starter electrode should have insulation capable of withstanding at least 50 kV, to avoid flash-over to earth. The AC supply and choke coil in the lower figure may be omitted with lamps having a third (starter) electrode. A Tesla coil is then required to supply HT to the latter.

circuit diagram for operation from A.C. or D.C. supplies. The operating voltage is about 20 V and a 500 W lamp thus draws a current of about 25 amperes. For operation from 240 V A.C. mains a large choke coil is wired in series with the lamp (see Fig. 61) to act as the ballast and drop most of the mains voltage. To avoid drawing heavy wattless current from the mains a power factor capacitor should also be included. Operated in this way the light output from xenon lamps is reasonably stable and if a beam-splitter monitor is incorporated in the equipment (see Section 3 G 3) its stability is completely adequate for use in a spectrofluorimeter. If a highly stable light ouput is required, operation on D.C. is better.

4. Mercury Lamps

When measuring fluorescence or phosphorescence emission spectra it is often not necessary to have available exciting light of infinitely variable wavelength, such as that obtained from a lamp emitting a continuous spectrum. Much higher intensities, with greater spectral purity, can be obtained by isolating one of the principal mercury lines from a mercury lamp. The spectral distribution of the light emitted by mercury arcs depends on the pressure at which the arc operates, and mercury lamps are divided into three types: low pressure or "resonance" lamps, medium pressure lamps and high pressure lamps. These will now be described.

5. Low Pressure Mercury Lamps

The simplest type of low pressure lamp consists of a quartz or silica tube 30–50 cm long and about 1 cm in diameter containing a droplet of mercury and a few millimetres pressure of inert gas to facilitate starting. The tube is provided with an electrode at each end and may be operated from the A.C. mains supply by means of a step-up transformer giving an output of about 1000 V, which also acts as a "ballast". Such lamps dissipate some tens of watts and operate at comparatively low temperatures, so that the vapour pressure of the mercury is not much greater than its room temperature value ($\sim 10^{-3}$ mm). Under these conditions nearly all the radiation is concentrated in the lines at 253.7 and 185.0 nm, corresponding to the transitions $^3P_1 \rightarrow {}^1S_0$ and $^1P_1 \rightarrow {}^1S_0$ (see Fig. 62). The latter is a strongly allowed transition. The former, which would be spin-forbidden in lighter elements, is in fact quite strongly allowed because of intercombination between multiplicities, arising from the fact that for the heavier elements the singlet states have a degree of triplet character and the triplet states have some singlet character.

If the lamp is constructed of pure quartz or synthetic silica, the line at 185.0 nm is very intense, but is absorbed by impure quartz. It is also absorbed by air and produces ozone which is objectionable because it absorbs at 253.7 nm, and also is highly toxic. Since in spectrofluorimetry the only likely use for the low pressure lamp is to provide 253.7 nm radiation the use of impure quartz to prevent ozone formation is an advantage. The total output of 253.7 nm is high (see Fig. 55) but is distributed over the whole length of the tube and it is therefore impossible to pass more than a small proportion through a monochromator. The best method of using these lamps for luminescence excitation is therefore to isolate the 253.7 nm line by means of filters. The lamps may be purchased in a coiled form which can be arranged to surround a cylindrical cuvette so as to obtain the greatest light flux through the specimen. In many cases such an arrangement is inconvenient for luminescence excitation, and since in addition low pressure lamps provide only one wavelength at a useful intensity, they are used

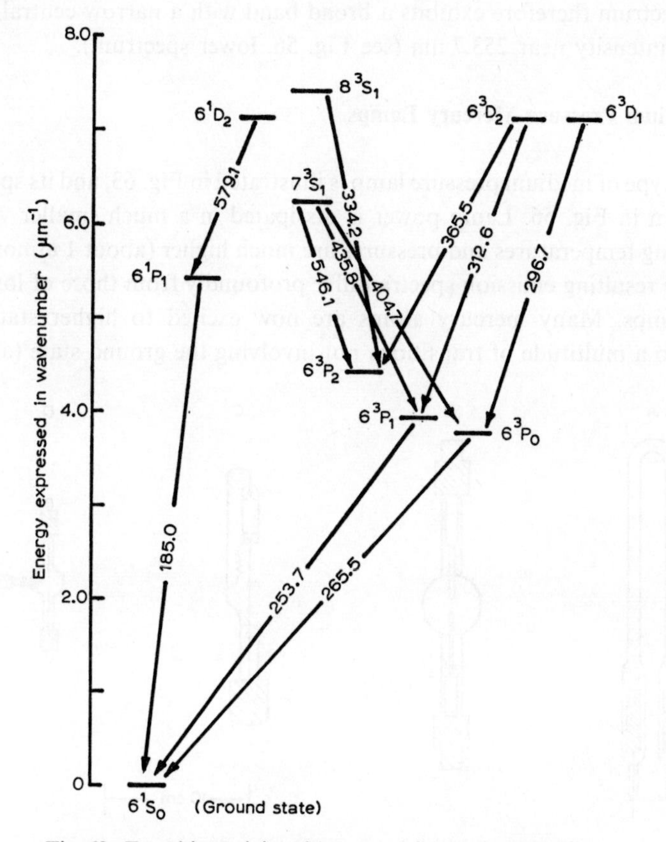

Fig. 62. Transitions giving rise to some intense mercury lines.
Term symbols for excited states are given alongside the appropriate energy levels. Wavelengths of lines, in nm, are indicated for each transition.

comparatively rarely in spectrofluorimetry. For many photochemical investigations however the light is sufficiently monochromatic at 253.7 nm without filtration and is often used in this way.

The mercury lines at 253.7 and 185.0 nm correspond to transitions to the ground state and they may thus be readily observed in *absorption* also (they are therefore often referred to as *resonance* lines). It is for this reason that they can only be obtained from low pressure mercury lamps. At higher mercury pressures such as those produced by the higher operating temperature of medium pressure lamps, the line becomes broadened and the hot emitting region is surrounded by a sufficiently high concentration of cooler mercury vapour to absorb almost completely the central region of the line. The spectrum therefore exhibits a broad band with a narrow central region of low intensity near 253.7 nm (see Fig. 56, lower spectrum).

6. Medium Pressure Mercury Lamps

One type of medium pressure lamp is illustrated in Fig. 63, and its spectrum is shown in Fig. 56. Lamp power is dissipated in a much smaller volume, operating temperatures and pressures are much higher (about 1 atmosphere) and the resulting emission spectra differ profoundly from those of low pressure lamps. Many mercury atoms are now excited to higher states and undergo a multitude of transitions not involving the ground state (a few of

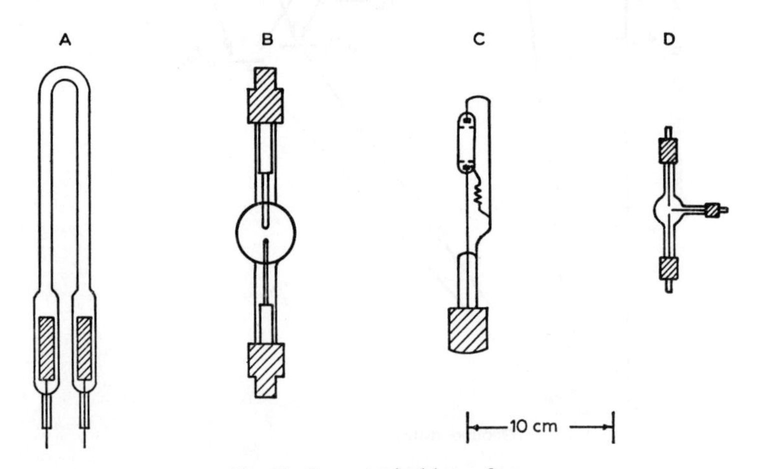

Fig. 63. Some typical lamp forms.
A, low pressure mercury; B, 500 W high pressure xenon; C, 100 W medium pressure mercury; D, 200 W extra-high pressure mercury.

these transitions are indicated in Fig. 62). The light emitted in these transitions is not re-absorbed because the population of atoms in the excited states is still low, and a variety of wavelengths of nearly monochromatic light are thus available. The lines are somewhat pressure-broadened as a result of collisions with other atoms and the central region of the resonance line at 253.7 nm is missing due to self-absorption as explained in the previous section. There is also an appreciable continuous emission in some spectral regions. Two other points about the spectra are worth noting. The first relates to the transition $6^3P_0 \rightarrow 6^1S_0$. This transition is doubly forbidden (i.e. it is a triplet–singlet transition, and it is a transition between two states both having zero total electronic angular momentum), and is not observed in absorption. The $^3P_0$ state is however populated by collisional deactivation of the $^3P_1$ state, and if no activation of the metastable state occurs, the atom remains there until it emits at 265.5 nm, and returns to the ground state. Because this is not a resonance line, it is not re-absorbed, and it is emitted at a useful intensity from medium pressure lamps (see Fig. 56). The second point concerns the region between 255 and 275 nm. In the spectra from medium pressure lamps this contains some useful lines but in the spectra of high pressure lamps this region is almost devoid of emission.

Several of the intense lines are situated close together and these are normally isolated together for the purpose of excitation. (They are not resolved in the lower spectrum of Fig. 56 owing to the small scale used.) Examples are the 577.0–579.1 nm yellow pair, the 365.0–366.3 group of three, the 313 nm group and the 297–303 group. For many purposes such groups of lines are sufficiently "monochromatic", but in some experiments, where precise rates of light absorption have to be determined with a solute having a steep absorption curve in the region of the group of lines, some difficulty is experienced (see for example Section 4 A 5). The difficulty is accentuated with high pressure lamps because in addition, the lines are considerably broadened and the intensity of the continuum is much greater.

Many commercial medium pressure lamps having powers of 100–200 W are designed to operate from 240 V A.C. mains with a choke coil in series. Some can be succesfully operated from a D.C. source using a resistor as ballast, in a manner similar to that shown in Fig. 61 for high pressure lamps. A.C. operation is more convenient and because the light output from A.C. operated medium pressure lamps is reasonably stable, this method is usually employed. It should be remembered however that all arc lamps operated from A.C. give light that is almost 100% modulated at twice the frequency of the mains. For measurement of prompt fluorescence this is of no consequence, but for quantitative measurements of long-lived luminescence with a spectrophosphorimeter employing light-choppers, a modulated light source is sometimes objectionable, and operation from a smoothed D.C. supply, or from a rectified 3-phase A.C. supply is desirable.

7. High Pressure Mercury Lamps

These operate at still higher temperatures and at pressures of tens or hundreds of atmospheres. Temperature and pressure broadening of the spectral lines is therefore greater and a higher intensity of continuum is also present (see Fig. 57). As was mentioned earlier, spectral emission in the region 255–275 nm is almost absent from these lamps. The most useful types are the high pressure compact source lamps. They consist of a small quartz bulb with two heavy electrodes, having a comparatively small arc gap, and are similar in design to the xenon lamps already described. The arc is thus compressed into a small volume and very high intrinsic brightness is obtained. Some types have a third starting electrode sealed through the side of the bulb. With their small size of source and extreme brightness these lamps are particularly valuable for use in conjunction with a monochromator to isolate the broad mercury "lines" at wavelengths above 297 nm. They also give a useful "band" emission in the region of 250 nm. The emission spectrum of the high pressure mercury lamp (see Table 14) is similar in form to that of the extra high pressure lamp shown in Fig. 57, but the lines show somewhat less pressure broadening. The light output per watt of input power is about the same.

Both the two- and three-electrode types are supplied for operation from A.C. with a choke in series, but operated in this way the arc is generally not stable. It moves from one point on the electrode to another, and if focussed on the entrance slit of the monochromator, the output from the latter can undergo wide fluctuations. Much greater stability can be achieved by operation from a D.C. supply using a ballast resistor. For operation of the two-electrode types the manufacturers generally supply special equipment to provide a high-voltage pulse through the electrodes to start the lamp. An alternative arrangement is shown in Fig. 61. A small choke is incorporated in the circuit. The lamp is started on A.C. in the usual way and after a few seconds it is switched to the D.C. supply. This method has been used for some years by the author to operate a 1 kW lamp of this type. To obtain high stability and long life the ballast resistor is adjusted so that the lamp is under-run at about 700 W. The light output is thereby halved, but very long lamp life and high stability is obtained. Some D.C. lamps are however not designed for A.C. operation and the manufacturers should be consulted before A.C. is used to start the lamp. The three-electrode models can be started by connecting the 120 V D.C. supply and applying the output from a Tesla coil to the third electrode. The A.C. supply and choke in Fig. 61 can then be dispensed with.

8. Lamps Containing Elements other than Mercury

Sodium lamps are well-known as sources for spectrometer calibration, for refractive index determination and similar purposes. Their output is generally too low for spectrofluorimetry. A variety of lamps giving atomic spectra has been described by Elenbaas and Riemens[133]. Typical outputs of the most intense lines emitted by 25 W cadmium and zinc lamps are given in Table 16. The sizes of these sources were somewhat greater than that of the 125 W medium pressure mercury lamp (see Table 14), and since the line intensities were considerably lower than those of the mercury lamp (compare Table 16 with Fig. 56) they are generally used only for purposes for which none of the mercury lines are suitable. Medium or high pressure lamps containing mercury in addition to cadmium and/or zinc are more useful for spectrofluorimetry. A compact source high pressure mercury–cadmium lamp has been described by Nelson[134]. These lamps operate from D.C. supplies and the manufacturers should be consulted for details of the control gear required. A typical spectral distribution of lines from such a lamp is given in Table 17.

9. Minimum Usable Source Size

From equation 206 it will be observed that the light flux passing the monochromator set to unit half-band width, is equal to the product of the brightness of the source (I_λ/A_u) and the LGP value of the monochromator. This indicates that for sources of the same output, the light flux through the monochromator will be greater with the source of smaller area, A_u. However there is clearly a limit to the flux that can be obtained since the flux through the monochromator cannot exceed the total emission from the source towards the collecting lens, i.e. it cannot exceed $2\pi I_\lambda$. In practice it is difficult and expensive to devise a lens system that will collect light over a solid

TABLE 16

OUTPUTS OF MOST INTENSE LINES OR GROUPS OF LINES FROM 25 W CADMIUM AND ZINC LAMPS

Cadmium		Zinc	
Wavelength (nm)	Einstein sec^{-1} $ster^{-1}$	Wavelength (nm)	Einstein sec^{-1} $ster^{-1}$
643.8	0.8×10^{-8}		
		636.2	0.6×10^{-8}
508.6	2.1		
		481.0	1.9
480.0	2.0		
		472.2, 468.0	2.5
467.8	1.2		
361.5, 361.3, 361.1	0.9		
346.8, 346.6	0.7		
340.2	0.3		
		334.6, 334.5	0.5
		330.3	0.6
		328.2	0.2
326.1, 325.3	1.5		
313.3	0.05		
		307.6, 307.2	0.6
298.1	0.11		
288.1	0.09		
283.7	0.05		
		280.1	0.05
		277.1	0.05
277.5, 276.4, 275.7	0.04		
228.8	2.5		
		213.9	0.7

TABLE 17

RELATIVE INTENSITIES OF LINES FROM $2\frac{1}{2}$ kW HIGH PRESSURE MERCURY–CADMIUM LAMP

Element	Wavelength (nm)	Relative quantum intensity
Cd	276	1
Cd–Hg	284/285	2
Cd–Hg	288/289	3
Cd–Hg	297/298	7
Hg	302	5
Cd	308	3
Cd–Hg	313	8
Cd	325/326	7
Hg	334	3
Cd	340	12
Cd	347	20
Cd	361	25
Hg	365/366	20
Hg	405/408	8
Hg	436	12
Cd	468	30
Cd	480	60
Cd	509	80
Hg	546	20
Hg	577/579	20
Cd	644	100

angle approaching 2π steradians and we shall assume that the maximum practicable aperture of the collecting lens (L_1 in Fig. 45) is 3 steradians. On this assumption the minimum source *height* that can be effectively used when the monochromator is operating with *narrow* slits is given by equation 199 viz.:

$$s_{min} = Fh/\sqrt{3} \qquad (227)$$

If the slit width is increased, the point will be reached where the breadth of the image of the source no longer fills the entrance slit, and the breadth of the source, b, now becomes the limiting factor. The minimum source breadth for a given maximum slit opening w_{max} may also be calculated from equation 199, since,

$$b/s = w/h \qquad (228)$$

Hence

$$b_{min} = w_{max}F/\sqrt{3} \qquad (229)$$

The minimum usable heights and widths of source calculated from the data for the monochromators referred to in Table 12 are shown in Table 18.

It will be observed that all sources described in Table 14 have widths greater than or equal to those shown in column 4 of Table 18. Some have heights less than the minimum required. For example the 200 W extra high pressure lamp could not be focussed to fill the full height of the slit of monochromator No. 5 and still satisfy the conditions for optimum illumination. It would therefore be more efficient to use the 1000 W high pressure lamp with this monochromator.

TABLE 18

MINIMUM USEFUL HEIGHT AND WIDTH OF SOURCE

Monochromator	s_{min} (mm)	w_{max} (mm)	b_{min} (mm)
1	2.9	10	1.2
2	1.5	1.75	0.14
3	1.8	3	0.50
4	0.52	2	0.07
5	5.2	6	0.62

10. Examples of Light Fluxes obtained

Given lamp data such as that in Figs. 55–59 or Table 16, and the LGP value of a monochromator, it is possible to calculate the theoretical maximum light flux for a given band width with any combination of lamp and monochromator by applying equations 208 or 212. In applying these equations the LGP value of the monochromator must be multiplied by its transmission. This may be as low as 0.1 for a grating instrument at an unfavourable wavelength. The light fluxes actually obtained are generally quite considerably less than the calculated values (see Table 15) because the source is rarely of uniform brightness and it is desirable to de-focus its image on the entrance slit for the reasons given in Section 3 C 6. Some further examples of light fluxes into fluorimeter cuvettes that have been obtained, using monochromators or filters, are given in Tables 19 and 20. Two points are worth noting. First, the intensities of monochromatic light obtained by the use of a mercury lamp with monochromator No. 2 (see

TABLE 19

LIGHT FLUXES THROUGH QUARTZ PRISM MONOCHROMATOR NO. 2
(For LGP values see Fig. 51)

Wavelength (nm)	Full band width i.e. extreme wavelengths passed (approx.) (nm)	Light flux (einstein sec^{-1})	
		700 W H.P. mercury lamp	375 W Xe lamp
436	473–406	14×10^{-9}	4×10^{-9}
405	435–381	8	2
366	390–350	15	1
313	329–302	5	0.3
248	243–253	0.8	0.02

Table 19) were considerably greater than those obtained from a xenon lamp, although the half-band widths were very large and the light isolated from the xenon lamp was far from monochromatic. Second, for a given lamp power, a greater intensity of mercury line was obtained by the use of

TABLE 20

TYPICAL VALUES OF LIGHT FLUX FROM MERCURY LINES ISOLATED BY MONOCHROMATORS OR FILTERS

Wavelength (nm)	Light flux (einstein sec^{-1})		
	A	B	C
577/9	–	1.4×10^{-8}	0.7×10^{-8}
546	–	1.9	1.5
436	–	3.1	3.1
405	–	2.2	0.8
366	4.8×10^{-8}	4.3	3.2
334	1.7	1.3	–
313	2.7	3.1	–
297/302	2.1	2.0	–
248 (band)	–	0.2	–

(A) Large quartz prism monochromator (No. 5 in Fig. 51) with 700 W high pressure mercury lamp. Half-band width 10 nm at 300 nm.
(B) Grating monochromator (No. 3 in Fig. 51) with 200 W high pressure mercury lamp. Half-band width 10 nm.
(C) 125 W medium pressure mercury lamp with filters.

Note

A 100 W low pressure lamp with 2 cm of each limb exposed gave a flux of $\sim 3 \times 10^{-8}$ einstein sec^{-1} at 254 nm over an aperture of area 4 cm^2 situated at a distance of 2 cm from the lamp.

filters than by the use of a monochromator of large LGP value (see Table 20). However, the filtered light contained a fair proportion of unwanted wavelengths, viz.: 2.5, 7.3, 12.7, 14.1 and 8.6% respectively for the wavelengths referred to in column 4 of Table 20. The use of filters is discussed in more detail in Section 3 D.

11. Precautions to be Observed in Using Ultra-violet Lamps

Short-wavelength ultra-violet light, even at low intensity, is harmful to the eyes, and the greatest care should be taken to shield the operator of a spectrofluorimeter both from the direct light from the lamp, and from the light issuing from the exit slit of the monochromator. It should be remembered that when working in a darkened room, the eyes accept a much greater proportion of ambient light and small "leaks" in lamp housings etc., that would be harmless in a well lighted room, can give rise to unpleasant eye-strain if exposure in the dark is prolonged. When viewing fluorescent specimens illuminated by ultra-violet light, protective goggles should always be worn. The choice of filter glasses for the goggles is somewhat critical because many substances fluoresce at the extreme violet limit of the visible spectrum and a filter glass that appears almost colourless will absorb violet light almost completely. The ideal is a filter glass with sharp cut-off at about 380 nm. It is a wise precaution to check the transmission characteristics of goggle filters with an absorption spectrophotometer before use. It should rarely be necessary to look directly at an ultra-violet lamp. Goggle filters used for this purpose should also be checked for absence of infra-red transmission—dark green glasses are sometimes found to have surprisingly high transmission in the near infra-red. When viewing intense sources the skin must also be protected or unpleasant burns may be caused.

Care should be taken when handling high pressure lamps because they contain gas under pressure at *room* temperature. When hot there is always a distinct possibility of explosion. They should never be operated without a protective cover, and after switching off, the cover should not be removed until the lamp has cooled. The quartz bulbs of medium and high pressure lamps should be handled as little as possible. Finger marks, grease spots and other contaminants "burn in" to the quartz during operation at high temperature and decrease the transmission. It is recommended that after installation and before operation, the quartz should be cleaned by wiping with tissue moistened first with ethanol, then with water and finally again with ethanol.

Light of wavelengths shorter than 200 nm is absorbed sufficiently by the oxygen of the surrounding air to produce quite high concentrations of ozone. The envelopes of many ultra-violet sources are fairly transparent in the region of the 185 nm mercury line and the resulting ozone formation is objectionable for two reasons. First, it is highly toxic, and second it absorbs strongly in the region below 300 nm. If the lamp is in an air-cooled housing it is desirable to extract the ozone through a hood, or at the least to operate the lamp in a room from which the air is continuously extracted. The ozone formation can be prevented by continuously flushing the housing with a stream of nitrogen gas. An alternative is to operate the lamp in a completely enclosed housing so that a steady-state concentration of ozone is built up, and fluctuations in its absorption are minimised. If the output of light of wavelength less than 200 nm is large, however, this arrangement can lead

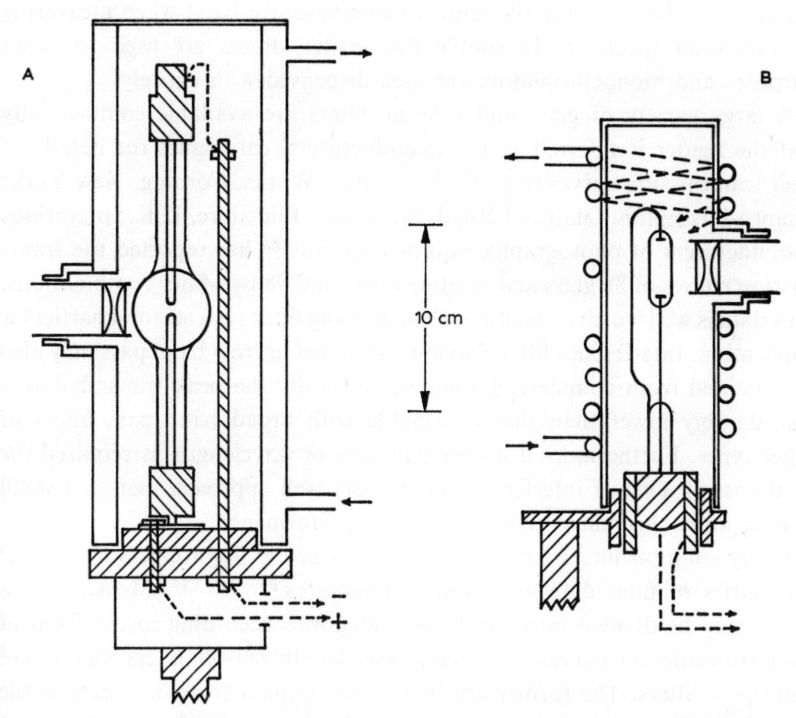

Fig. 64. Two kinds of water-cooled lamp housing.

to a serious loss of intensity at wavelengths in the region 240–280 nm. Enclosed housings must be water-cooled and two designs of such housings are shown in Fig. 64. By combining a small short-focus lens in the housing in the manner shown, the entrance slit optics are kept small and escape of light is completely prevented. The housing of type A in Fig. 64 gives more efficient cooling, and is adequate for a 1 kW lamp. It is however more difficult to construct than type B.

D. FILTERS

1. General Comments

Filters may be used as an inexpensive substitute for a monochromator of large LGP value, or they may be used in conjunction with a monochromator to increase its efficiency. They may be used for isolating a band of wavelengths of exciting light when measuring a luminescence emission spectrum, or for isolating the required luminescence band when measuring an excitation spectrum. In simple fluorimeters filters are used for both purposes and monochromators are then dispensed with entirely.

A large variety of glass and gelatine filters are available commercially and the reader is referred to the manufacturers' catalogues for details of their transmission curves (e.g., Corning Glass Works, Corning, New York; Chance-Pilkington, Glascoed Road, St. Asaph, Flintshire, U.K.; or various manufacturers of photographic equipment). Sill[135] has reported the transmission curves of 88 glass and gelatine filters and 98 two-filter combinations. This data is well worth consulting when devising filter systems for a particular application. Interference filters having a specified narrow band pass may also be obtained from commercial sources. Generally the peak transmission is considerably lower than that obtainable with broad band pass filters of other types. Furthermore if a wide coverage of wavelengths is required the total cost of a set of interference filters may well approach that of a small grating monochromator giving equivalent performance.

Many solution filters have been described in the literature. Some useful summaries of filter data are given in references[99, 136, 137]. Most solution filters may be divided into two broad categories according to the form of their transmission curves, viz., short-wavelength cut-off filters and broad band pass filters. The former are by far the simplest to devise because the electronic absorption spectra of many substances extend throughout the

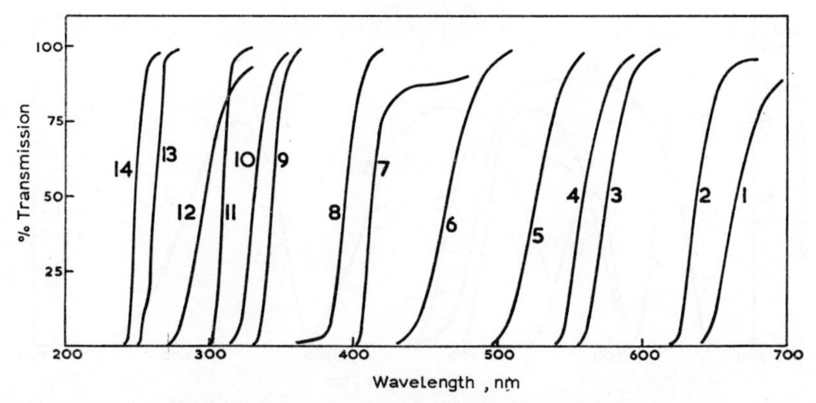

Fig. 65. Transmission curves of short-wavelength cut-off filters.
(1) 0.02% w/v methyl violet in water, 1 cm (also transmits below 460 nm). (2) 0.2% w/v rhodamine B in water, 1 cm. (3) 50% w/v $Na_2Cr_2O_7 . 2H_2O$ in water, 1 cm. (4) 10% w/v $K_2Cr_2O_7$ in water, 1 cm. (5) 0.5% w/v $K_2Cr_2O_7$ in water, 1 cm. (6) 0.1% w/v K_2CrO_4 in water, 1 cm. (7) 75 g of $NaNO_2$ in 100 ml of water, 1 cm. (8) 1% w/v $NaNO_2$ in water, 1 cm. (9) 2 M KNO_3 in water, 2 cm. (10) 0.2 M KNO_3 in water, 2 cm. (11) 0.5% w/v potassium hydrogen phthalate in water, 1 cm. (12) 1.5% w/v $CuSO_4 . 5H_2O$ in water, 1 cm. (13) 0.17% w/v KI in water, 1 cm. (14) 4 N acetic acid in water, 1 cm.

short-wavelength region up to the long wavelength limit characteristic of the substance concerned. Thus concentrated solutions of many substances provide very effective short-wavelength cut-off filters. If the transmission of the filter in a particular short-wavelength region is still too great, it is a simple matter to combine it with a second substance having a shorter wavelength cut-off, that absorbs strongly in the region of low absorption by the first filter. The transmission curves of some useful liquid cut-off filters are shown in Fig. 65. These are given mainly for guidance and it is desirable for the user himself to measure the transmission of the chosen filter, particularly if very low transmission is required at all wavelengths shorter than the cut-off point. Some variation in the precise cut-off wavelengths may be achieved by varying the concentration and/or the optical depth.

The choice of substances suitable as band-pass filters is more limited because these require a compound that has a large wavelength separation between the first and second absorption bands, with low extinction coefficients in the intermediate region. A variety of dyestuffs are suitable for wavelengths in the visible region, and solutions of transition metal salts are also widely used. Two very useful glass filters have been included in Fig. 66 because they provide a convenient means of isolating the near and middle

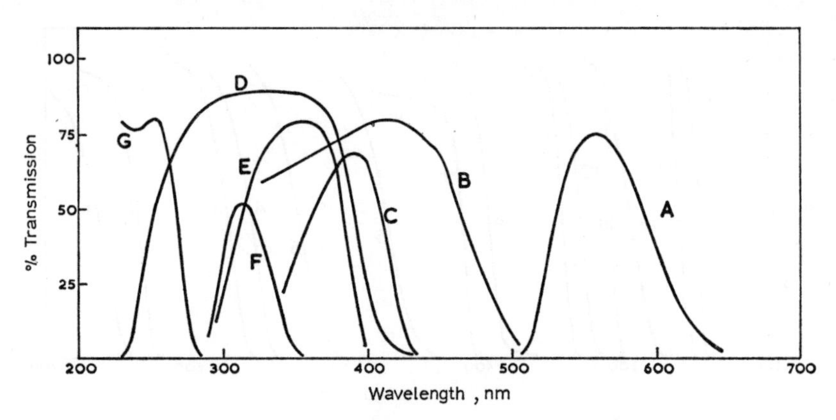

Fig. 66. Transmission curves of band-pass filters.
(A) 5% w/v $CuCl_2.2H_2O$ in 8 N HCl, 1 cm. (B) 2.5% w/v $CuSO_4.5H_2O$ in 5 N ammonium hydroxide, 1 cm. (C) 0.75% w/v iodine in carbon tetrachloride, 1 cm (also transmits beyond 650 nm). (D) Chance–Pilkington OX7 glass, 2 mm (also transmits beyond 660 nm). (E) Chance–Pilkington OX9A glass, 3 mm (also transmits beyond 680 nm). (F) 15% w/v $KCr(SO_4)_2.12H_2O$ in 1.0 N H_2SO_4, 2 cm (also has a window of low transmission in the visible region). (G) chlorine at one atmosphere in 4 cm silica cell (also transmits beyond 380 nm). Note: Materials for ultra-violet filters must be free from ferric iron. Chrome alum is best prepared by reducing twice-recrystallised $K_2Cr_2O_7$ with ethanol followed by recrystallisation.

ultra-violet region, while absorbing almost the whole of the visible spectrum. One gaseous filter (chlorine) has been included. This is simply prepared by passing dry chlorine gas into a silica optical cell through a narrow side arm. The delivery tube is then withdrawn and the side arm sealed off. The presence of a little air in the chlorine is of no consequence, and if the cell is carefully cleaned and dried before filling, such a filter will retain its transmission characteristics for long periods. A 4 cm depth cuts off all wavelengths between about 285 and 385 nm.

To obtain a sufficiently narrow band pass, it is customary to combine a broad band-pass filter having the required long-wavelength cut-off with a suitable short-wavelength cut-off filter. Examples of such filter combinations are described below.

2. Filters for Isolating Exciting Light

Filter combinations for isolation of one or more of the principal lines from the spectrum of a mercury vapour lamp are given in Table 21. Most of these give relatively good spectral purity although those combinations employing OX9A or OX7 glasses transmit also in the extreme red or near infra-red. This region may be absorbed by the use in addition of 1 or 2 cm of 10% w/v copper sulphate in 0.1 N sulphuric acid, but this solution absorbs strongly below 320 nm. Greater spectral purity with, in some cases, reduced transmission, may be obtained by means of the filters recommended by Calvert and Pitts[99]. This excellent compilation provides detailed information of spectral purity and photo-stability of liquid filters for the isolation of mercury lines. It should be noted, however, that when high intensities of exciting light are required it is sometimes an advantage to use a broad band-pass filter to separate a combination of two or more mercury lines. This applies particularly to wavelengths shorter than 300 nm where emission from lamps is weaker than at longer wavelengths. The combination of chlorine gas and OX7 glass is useful for isolating both the 254 nm band and the 265 nm line from a medium pressure lamp. If a high pressure lamp is used this filter combination will give essentially the band at 250 nm only since there is little lamp emission in the region 260–275 nm (see Fig. 57). This filter is also very useful in conjunction with a grating monochromator when short-wavelength excitation with very small long-wavelength impurity content is required. Such an application is illustrated in Fig. 67. This refers to the examination by frontal illumination of a specimen of lubricating oil adsorbed on filter paper. Here the amount of exciting light scattered by the

TABLE 21

FILTERS FOR ISOLATING MERCURY LINES
(See Figs. 65 and 66)

Wavelength(s)	Filter combination
250 + 265	G + D
302 + 313 + little 334	F
313 + little 334	F + 11
334	F (1 cm) + 10
366	E + 9
405 + 436	B + 8
405	C + 8
436	B + 7
546 + 578	A (+ 5)
546	A + H (+ 5)
578	A + 3

Note
H = 2 mm of Chance–Pilkington didymium glass—transmits ~ 80% at 546 nm and ~ 1% at 578 nm.

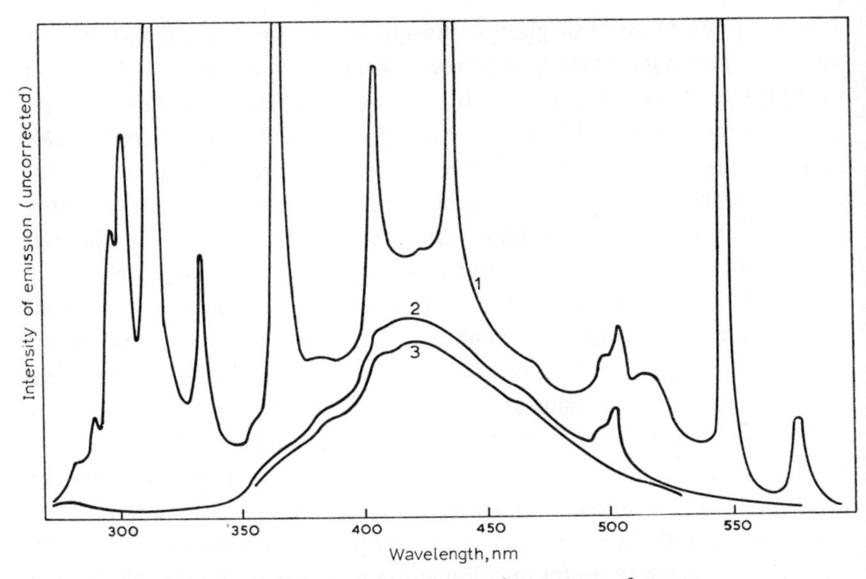

Fig. 67. Use of filters to improve monochromator performance.
Mineral oil on filter paper by frontal illumination with ~250 nm radiation from high pressure mercury lamp, (1) isolated by grating monochromator, (2) isolated by grating monochromator with chlorine-OX7 filters, (3) as (2) but with 1 cm of 1.5% $CuSO_4.5H_2O$ in the beam of luminescence.

specimen is very large and the small proportion of stray light of all wavelengths passed by the grating monochromator is sufficient to interfere with the measurement of the emission spectrum of the specimen (curve 1). The use of the chlorine/OX7 filter in conjunction with the monochromator removes this interference (curve 2). In very high sensitivity work the unwanted scattered light from a grating monochromator can be objectionable even with solutions and right angle illumination. An example of this is given in Fig. 150 (Chapter 5).

It is sometimes necessary to use a filter to remove second order spectra from the exciting light isolated by a grating monochromator. For example, when set to pass exciting light of wavelength 500 nm, the monochromator will also pass a small proportion of light of wavelength 250 nm. This is simply removed by the choice of a suitable cut-off filter—in this example a plate glass would cut out the 250 nm component.

In some applications an extremely high filter rejection efficiency is required. For example, in determining benz(a)pyrene in frozen crystalline n-octane (see Section 5 E 8) a large amount of exciting light is scattered into the analysing monochromator and the small transmission of 3 mm of OX9A filter at 405 nm is sufficient to introduce a large "blank" if this filter is used to isolate the 366 nm line for excitation. The 405 nm transmission is reduced to an acceptably low level by the use of three 3 mm pieces of OX9A glass—the transmission of the combined filter at 366 nm is still high (47%).

3. Isolation of Fluorescence or Phosphorescence

One of the main functions of a filter used to isolate a fluorescence band is to absorb the exciting light scattered by the specimen. The filter must therefore be non-fluorescent, otherwise a large spurious fluorescence signal will be observed. For this reason many glass or gelatine filters are unsuitable and an inorganic liquid filter in a non-fluorescent cuvette is preferred. Frequently it is sufficient to choose a short-wavelength cut-off filter to absorb the scattered exciting light and one of the filters illustrated in Fig. 65 will then be adequate. Such filters are often used when measuring the excitation spectrum of a solution containing a single fluorescent component, or for measuring fluorescence polarisation spectra. Such a filter then replaces the analysing monochromator which, when used for this purpose, would be operated with wide slits to obtain the necessary sensitivity. The choice of filters for separating the fluorescence emission band of two components is much more critical because comparatively narrow band-pass filters are then required. Measurement of the separate excitation spectra of components of a mixture of fluorescent substances is best made with a monochromator to separate the different fluorescence bands.

When making measurements at high sensitivity it is sometimes necessary to use filters to increase the efficiency of the monochromator used to analyse the emission spectrum. For example, the second order spectrum passed by a grating monochromator can introduce an objectionable signal due to scattered light. Thus in Fig. 67 curve (2), the peak at ~ 500 nm is caused by scattered exciting light of wavelength ~ 250 nm passing through the analysing monochromator in the second order spectrum. It is removed (see curve 3 in Fig. 67) by inserting a copper sulphate filter (No. 12 in Fig. 65) that absorbs at 250 nm but transmits freely above 300 nm.

If a weak luminescence spectrum in one spectral region has to be measured in the presence of a very intense luminescence in another spectral region, it is often desirable to use a filter to prevent the intense luminescence from entering the analysing monochromator. For example, solutions of phenanthrene in rigid solvents at 77°K give rise to both P-type delayed fluorescence in the region ~ 350 nm and intense phosphorescence in the green region[84]. At low rates of light absorption the delayed fluorescence may be 1000 times less intense than the phosphorescence. At high instrumental sensitivity the small amount of green phosphorescence scattered in the monochromator gives rise to a large signal at all wavelength settings of the latter and thus interferes with the observation of the weak delayed fluorescence. In the example quoted the interference was simply removed by inserting a sheet of OX9A glass between the phenanthrene specimen and the analysing monochromator. This absorbed the green phosphorescence but transmitted the ultra-violet delayed fluorescence.

4. Choice of Filters for Filter Fluorimetry

The filter used to isolate the exciting light (primary filter) and that used in the beam of fluorescence (secondary filter) are always chosen to be "complementary", i.e. each filter absorbs all light transmitted by the other. For measurements at *moderate* sensitivity a pair of such complementary filters is adequate and generally simple to devise. For example, if the fluorescence is excited by the mercury line at 366 nm and the fluorescence band appears in the blue region of the spectrum, a suitable primary filter would be a combination of No. 9 (Fig. 65) with E (Fig. 66). This combination does not transmit beyond 400 nm and hence filter no. 7 (Fig. 65) would be suitable for a secondary filter. Similarly if the mercury line at 405 nm were used for excitation, the combination 8C as primary with no. 7 as secondary would be suitable.

For work at high sensitivity with solvents containing hydrogen atoms (e.g. water, ethanol, cyclohexane) the filter combinations just described would give rise to unacceptably high "blank" values due to the main Raman bands from the solvents (see Sections 1 E 5 and 5 C 1). Thus the main Raman band of water excited by light of wavelength 366 nm appears at 416 nm and to reject this the secondary filter must cut off at longer wavelength—for example filter no. 6 would be suitable. Similarly, if the mercury line at 405 nm is used for excitation, the main Raman band of water appears at 469 nm and neither filter no. 7 nor filter no. 6 would be suitable. An example of the use of 405 nm for excitation in a high sensitivity method is given in Chapter 5 where the choice of filters is described.

E. DETECTION AND MEASUREMENT OF LIGHT BY PHYSICAL METHODS

1. Types of Photo-Detector

There are three types of *physical* instruments that are generally used to detect or measure light in the visible and ultra-violet regions of the spectrum: the thermopile, the various types of photo-cell, and the photo-electron multiplier tube (photomultiplier). For the measurement of fluorescence and phosphorescence, particularly at low intensity, the photomultiplier is used almost exclusively because of its extremely high sensitivity. For the measurement of higher intensities, e.g. in the beam of radiation from an excitation monochromator, the *vacuum photocell* is sometimes more suitable, while the thermopile is the primary standard for absolute radiation measurements. We shall therefore describe the construction and method of operation of all three types of detector.

Some other types of photocell may be mentioned in passing. *Photo-conductive cells* consist of suitable crystalline semi-conductor material that has a high resistance in the dark. On exposure to light, electrons are excited into the conduction band and the conductivity increases. Photo-conductive cells can be made to operate at wavelengths well into the infra-red region. *Photovoltaic* cells have the advantage that they convert the visible light energy into a voltage sufficient to operate a galvanometer or microammeter, and they do not therefore need an external electrical supply. They are however, relatively insensitive and their response is linear only over a narrow intensity range. The *gas-filled photocell*, like the vacuum photocell to be described below, relies for its action on the photo-emission of electrons from a sensitive cathode exposed to light. On travelling towards the anode under the influence of the applied voltage, the electrons collide with the atoms of the argon gas filling and cause ionisation. The secondary electrons so produced can cause further ionisation and the primary photo-current may thus be multiplied by a factor of 10 to 20. Gas-filled photocells are less stable than vacuum photocells and have a shorter life.

Apart from the three types of photo-detector mentioned above, there are two further devices that are of considerable value in connection with photoluminescence measurements. The first is the fluorescence quantum counter, which permits direct comparison of quantum intensities of beams of light of different wavelengths. The second is the chemical actinometer, which provides a simple and reliable method for the determination of light flux in absolute units. The first is described in Section 3 E 5 below; the second is dealt with in Section 3 F.

2. Thermopiles

The thermopile absorbs all radiation incident upon it and converts the energy into heat. This raises the temperature of a series of thermocouples relative to the cold junctions which are shielded from the radiation, and the resulting voltage is proportional to the intensity of the incident radiation. The principle of construction is shown diagrammatically in Fig. 68 A. The light to be measured falls on the surfaces of a series of small and very thin strips of metal coated with lamp black. Each strip has a pair of wires of dissimilar metals forming the thermocouple soldered to the back. The exposed elements are connected alternately to an identical set of elements which are shielded from the light. To obtain the maximum sensitivity the elements are made as thin as possible so as to have a low heat content and hence give the greatest rise in temperature. The connecting wires are very thin to minimise heat conduction, and the cold elements are made identical to the exposed elements so that zero "drift" due to external temperature changes is minimised. Thermopiles with their elements in air are very susceptible to slight draughts. The zero drift is greatly reduced if the thermopile surface is covered with a quartz plate. Much greater stability and increased sensitivity is achieved with a vacuum thermopile. In practice the thin connecting wires act as suspensions for the receiving elements. The thermopile is thus easily damaged by mechanical shock and must be treated with care.

Absolute calibration of the thermopile voltage in terms of incident radiant flux is carried out by one of two methods. The first is by reference to the total radiation from a true black body at a known temperature, as given by Stefan's Law (derived by integration of the Planck relation, equation 224):

$$\text{Rate of emission of total radiation} = 5.670 \times 10^{-5} T^4 \text{ erg cm}^{-2} \text{ sec}^{-1} \qquad (230)$$

In the second method a specially designed thermopile is used. This is so designed that the sensitive elements can be heated by the passage of an electric current through a conductor attached to the elements. The thermovoltage produced in this way is then directly related to the known rate of dissipation of electrical energy in the conductor. Absolute thermopile cali-

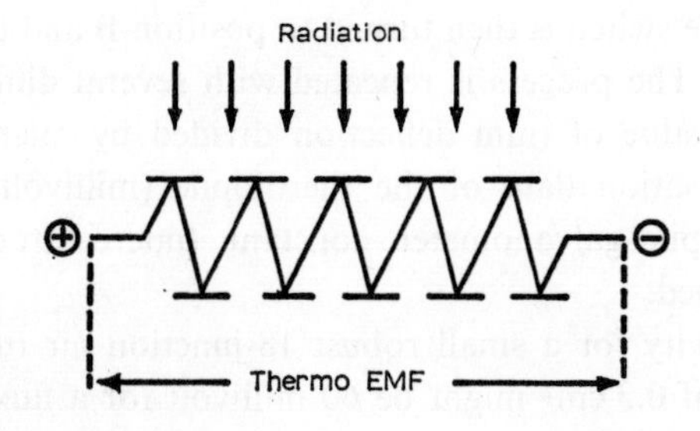

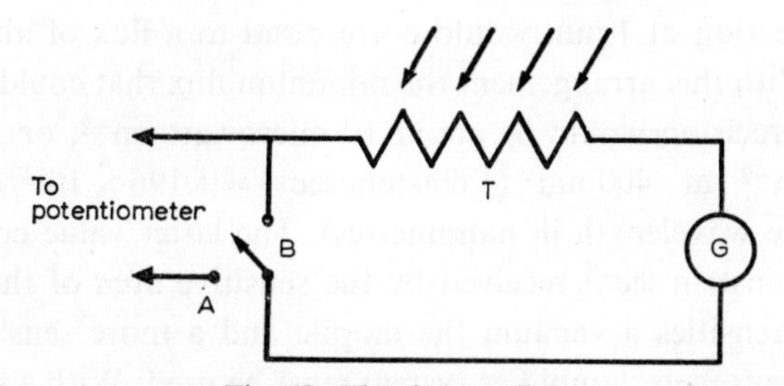

Fig. 68. The thermopile.
A, diagrammatic representation. B, calibration of thermopile–galvanometer circuit.

bration is a highly specialised operation, and is not normally undertaken by the chemist. The usual practice is to purchase a thermopile and have it calibrated by the National Physical Laboratory, or to purchase a calibrated standard lamp that gives a stated total radiation flux at a specified distance, and to use this to standardise the thermopile. Perhaps the best course is to obtain both standard lamp and standardised thermopile, so that the operator can check the precision of his own standardisation procedure.

The calibration data is provided in terms of the millivolts generated when the thermopile is exposed to a uniform flux of 1 watt per square centimetre. The thermopile is a low impedance low voltage source, and a low resistance galvanometer of high voltage sensitivity must be used as a detector. The e.m.f. may be determined directly with a precision potentiometer, but it is sometimes more convenient, though slightly less precise, to take direct readings of galvanometer deflection. To do this, the thermopile–galvanometer circuit must first be calibrated in terms of the open circuit e.m.f. delivered by the thermopile. This may be achieved with the simple circuit shown in Fig. 68 B. The thermopile T is exposed to a constant flux from a lamp and the e.m.f. measured directly by the potentiometer with the switch in position A. The switch is then turned to position B and the galvanometer deflection noted. The process is repeated with several different fluxes, and from the mean value of (mm deflection divided by microvolts), and the known standardisation data of the thermopile (millivolt $watt^{-1}$ cm^2) the required thermopile–galvanometer constant (microwatt cm^{-2} per mm deflection) is obtained.

Typical sensitivity for a small robust 18-junction air thermopile having a sensitive area of 0.3 cm^2 might be 60 millivolt for a flux of 1 watt cm^{-2}. With a thermopile resistance of 25 ohm, a galvanometer having a resistance of 25 ohm and a voltage sensitivity of 16 millimetres per microvolt at 1 metre, a deflection of 1 mm would correspond to a flux of about 2 microwatt cm^{-2}. With this arrangement the minimum flux that could be measured with useful precision would be about 10 microwatt cm^{-2}, or 3×10^{-11} einstein sec^{-1} cm^{-2} at 400 nm (1 einstein $sec^{-1} = 1.196 \times 10^{14}/\lambda$ microwatt, where λ is the wavelength in nanometres). The latter value corresponds to about 10^{-11} einstein sec^{-1} received by the sensitive area of the thermopile. For lower intensities a vacuum thermopile and a more sensitive galvanometer or galvanometer/amplifier system must be used. With a small vacuum thermopile of the highest sensitivity, operated under ideal conditions, the minimum detectable flux at 400 nm might be 10^{-14} einstein sec^{-1} cm^{-2}, or as little as 10^{-16} einstein sec^{-1} concentrated on the thermopile area. Thermopile measurements at low light intensities require tedious precautions and they can generally be circumvented by using the less sensitive thermopile to calibrate a vacuum photocell, using a high intensity beam of the required wavelength. The photocell can then be used to make absolute measurements at low intensities.

In spectrofluorimetry a thermopile is rarely used to measure *absolute* light intensities directly. Its main use is to extend the calibration data obtained with the ferrioxalate actinometer to wavelengths greater than 500 nm, for which the actinometer is not suitable. This is described in Section 3 F.

It should be remembered that a thermopile measures the total *energy*. It cannot therefore be used to determine the quantum intensity of a polychromatic beam unless the spectral distribution is also known. When it is used to compare the quantum intensity of two beams of different wavelengths, the appropriate wavelength correction must be made. Thus if G_1 and G_2 are the galvanometer deflections observed with beams of wavelengths λ_1 and λ_2, the relative quantum intensity is given by:

$$Q_1/Q_2 = G_1\lambda_1/G_2\lambda_2 \tag{231}$$

3. Vacuum Photocells

The vacuum photocell consists of an evacuated bulb of glass or silica containing two electrodes. The cathode consists of a coating of a mixture of metals—sometimes containing oxides also—supported on a metal plate, or on the surface of the bulb itself. When light impinges on the cathode, photons carrying energy greater than the work function of the surface eject photo-electrons. If the anode is charged positively with respect to the cathode, the photo-electrons are collected by the anode, a current proportional to the incident light intensity flows through the external circuit and may be measured with a galvanometer or microammeter (see Fig. 69A). At low applied voltage the electric field near the cathode is not sufficient to overcome the negative space charge, and not all photo-electrons are collected by the anode. Above about 25 volts all electrons are collected and the current is substantially independent of this applied voltage. Operated under these conditions the vacuum photocell provides a very reproducible method for measuring light intensities.

Even with photon energies above the threshold corresponding to the value of the work function, the quantum efficiency of the photo-emission (electrons emitted per quantum received), is less than unity because some light is absorbed at depths great enough to prevent the emergence of the photo-electron. Above the threshold the quantum efficiency varies in the manner shown by the curves in Fig. 70 (see reference 138). It will be observed that the quantum efficiency of some types of photo cathode (e.g., tri-alkali, type S20) rises to a value of about 0.2 in the wavelength region of maximum sensitivity. The sensitivity of such a photocell to this spectral region can be simply calculated, since the absorption of 1 einstein of radiation at a quantum efficiency of unity will eject 1 Faraday (96,500 coulombs) of electrons. Thus for the photocell in question:

$$1\ \mu\text{amp of photo-current corresponds to } 10^{-6}/0.2 \times 96{,}500 \text{ or } 5 \times 10^{-11} \text{ einstein sec}^{-1}.$$

Using a galvanometer having a sensitivity of 2,000 mm per microamp, light intensities of the order 10^{-12} einstein sec^{-1} can be measured simply and with precision.

Unlike the thermopile, the vacuum photocell is a high impedance source of *current* and the latter can be simply amplified by passing it through a high resistance load connected to the grid of an electrometer valve. In the simple battery-operated circuit shown in Fig. 69 B, the valve anode circuit with galvanometer, serves as a sensitive null point detector for balancing the voltage developed by the photocurrent across the load L, against the voltage applied by the calibrated potentiometer P. With the light off and P set to zero the galvanomer is brought to zero by adjusting the anode potentiometers. The light shutter is then opened and the galvanometer again brought

A. Working principle

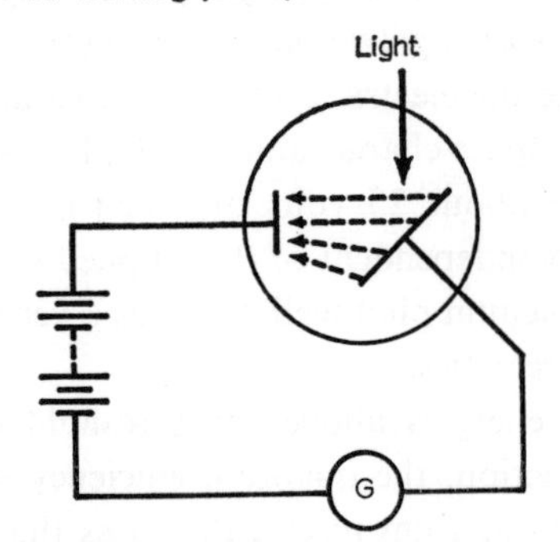

B. Circuit for measuring low light intensities

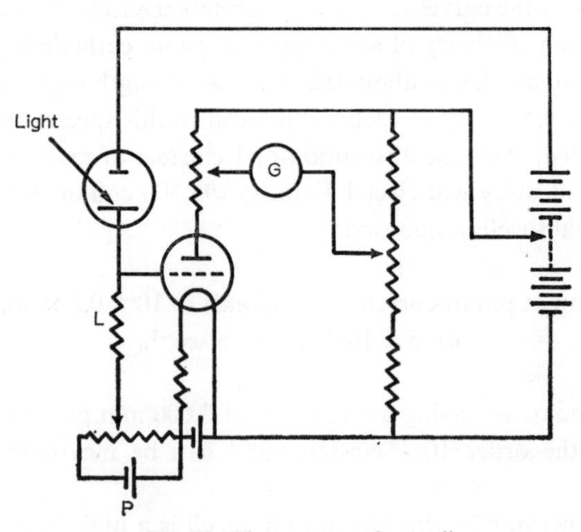

Fig. 69. The vacuum photocell.
A, working principle. B, simple circuit for measuring low light intensities.

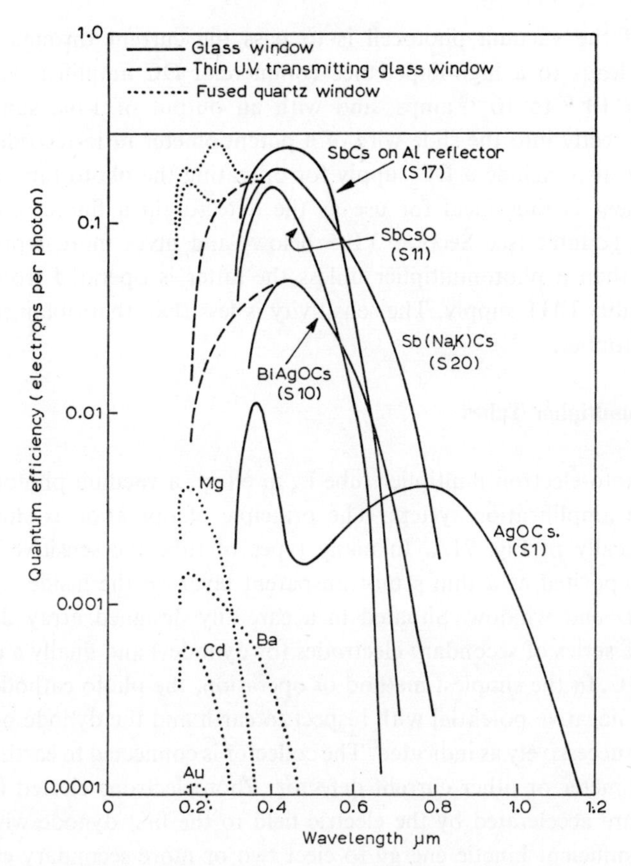

Fig. 70. Spectral response of various photocathodes (from Sharpe[138]).

to zero by adjustment of the calibrated potentiometer P. The voltage reading of P is proportional to the light intensity. For the measurement of very low light levels, load resistors as high as 10^{12} ohms are used. To avoid variation in sensitivity and excessive dark current due to surface leakage, it is necessary to seal the photocell, load resistor and electrometer valve in an evacuated glass bulb. Once calibrated, such photometers retain their calibration for long periods. The limiting sensitivity with a load of 10^{12} ohms is of the order 10^{-15} amperes. Photometers are however more frequently used with loads of 10^{10} ohms or less and measurement of photocurrents of 10^{-13} amperes or more is simple and precise.

A more convenient (though more expensive) method of amplifying the output of the vacuum photocell is to pass the current through suitably screened leads to a high-impedance commercial DC amplifier with input ranges of 10^{-6} to 10^{-12} amps, and with an output of 1 ma suitable for feeding directly into the slide wire of a potentiometer ratio-recorder. Some amplifiers also include a DC supply for operating the photo tube. Such an arrangement is thus ideal for use as the detector in a fluorescent screen quantum counter (see Section 3 E 5 below) and gives more reproducible readings than a photomultiplier unless the latter is operated from a very high-stability EHT supply. The sensitivity is less than that obtained with a photomultiplier.

4. Photomultiplier Tubes

The photo-electron multiplier tube is, in effect, a vacuum photocell with a built-in amplification system. The principle of operation is shown diagrammatically in Fig. 71A. In many types of tube the sensitive material (PC) is deposited as a thin semi-transparent layer on the inside surface of the quartz end window. Situated in a carefully designed array down the tube are a series of secondary electrodes (or dynodes) and finally a collector electrode C. In the simplest method of operation, the photo cathode is held at a high negative potential with respect to earth and the dynode potentials decrease successively as indicated. The collector is connected to earth through a galvanometer or other current detector. Photoelectrons ejected from the cathode are accelerated by the electric field to the first dynode which they hit with sufficient kinetic energy to eject two or more secondary electrons. These in turn are accelerated to the next dynode where they give rise to still more secondary electrons. The process continues down the dynode chain which may have up to 15 amplification stages so that one photoelectron may give rise to as many as 10^8 electrons emitted from the last dynode. These electrons are finally attracted to the collector (at earth potential when the tube is in the dark) and a current flows through the galvanometer to earth. Discussion of the factors governing the design of photomultiplier tubes is beyond the scope of this book; four types of electrode arrangements are shown in Fig. 72 (see reference 138). The connections to the electrodes are brought out to a series of pins at the base of the tube and the voltage supplies are obtained from a chain of resistors connected as shown in Fig. 71 B. Optimum values of resistors and details of methods of wiring for any particular type are best obtained from the manufacturer of the tube.

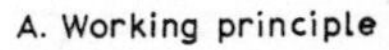

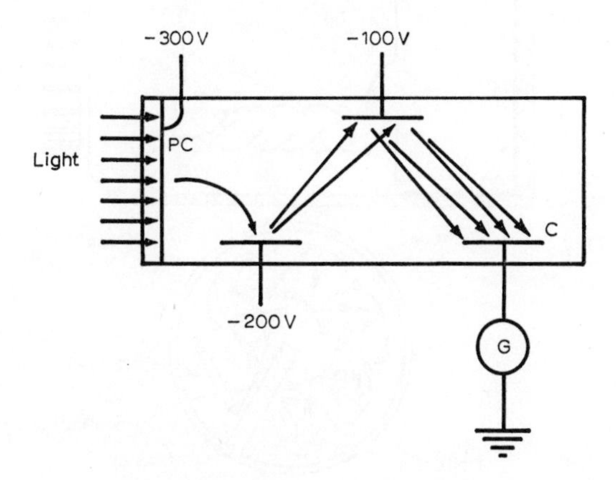

B. Method of wiring

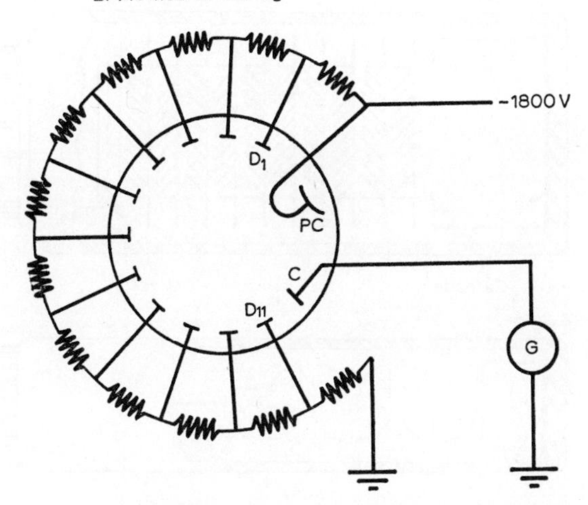

Fig. 71. The photomultiplier.
A, working principle. B, method of wiring. (All dynode resistors are 1 megohm.)

The dark current is the factor that limits the ultimate sensitivity attainable. It is caused by the ejection of electrons from the cathode by thermal activation, or by traces of radioactivity in the surroundings causing luminescence of the envelope. It can be reduced considerably by cooling the tube in solid CO_2 or even liquid air. This is particularly advantageous with tubes sensitive to the near infra-red region (type S1 cathodes, see Fig. 70). Such tubes of necessity have cathodes coated with material of low work function so that

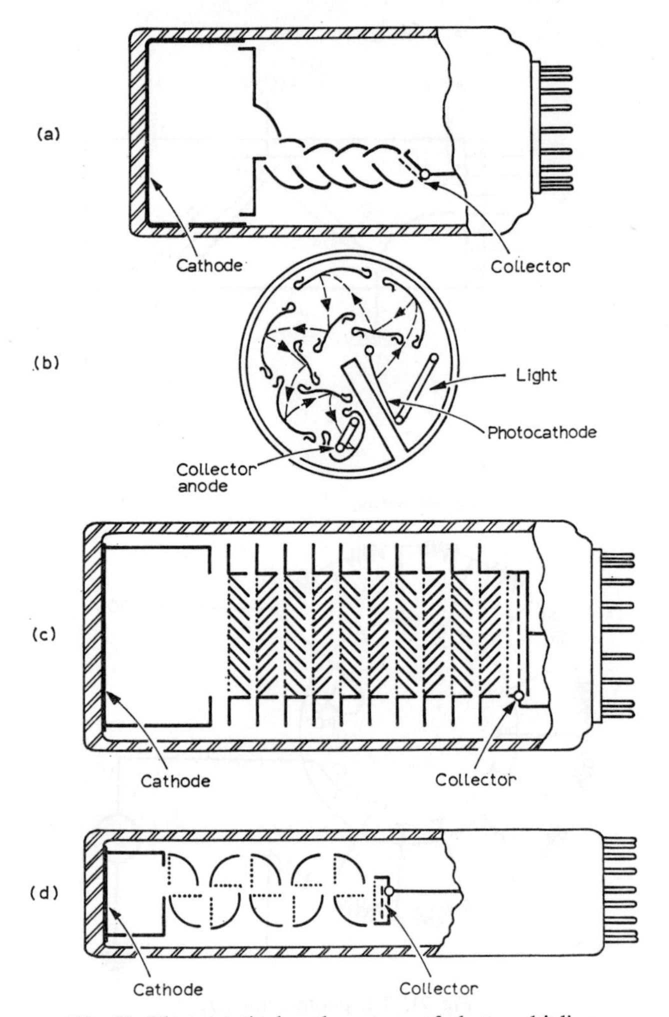

Fig. 72. Electrostatic dynode systems of photomultipliers.
(a) focused structure, (b) compact focused structure, (c) venetian-blind structure, (d) box-and-grid structure (from Sharpe[138]).

quanta of long wavelengths are capable of ejecting electrons, but the low work function automatically results in greater thermal activation. It would clearly be desirable to have one photomultiplier for use at wavelengths throughout the ultra-violet, visible and the near infra-red regions. Unfortunately type S1 photo cathodes have a much lower quantum efficiency compared with the other types, and this, added to the inconvenience of cooling, means that in high sensitivity work they are only worth using when measurements beyond 800 nm are required. For general work at wavelengths up to 800 nm the E.M.I. 9558Q tube with type S20 photo cathode is recommended. This has eleven stages and high quantum efficiency. With an overall applied voltage of 1800 volts the sensitivity is about 200 amperes per lumen, and the dark current only 0.002–0.01 microamp. The latter can of course be greatly reduced by cooling, if desired, but in the author's experience many problems in high sensitivity photoluminescence can be tackled using this tube without cooling. If a cut-off at shorter wavelengths can be accepted the 13-stage 6256 tube with type S11 photo cathode (Fig. 70) is also valuable. This has a maximum sensitivity of 2,000 amperes per lumen and a dark current of 0.1 microamp under these conditions.

The sensitivity of a photomultiplier can be conveniently varied over a wide range of values by simply changing the overall voltage applied to the resistor chain (Fig. 71 B). Typical relative sensitivity values for a 9558Q tube are shown in Fig. 73. It will be observed that increasing the applied voltage from 800 to 1800 volts increases the sensitivity by a factor of about 1,000. A further sensitivity factor of 10 or more can conveniently be built into the final amplifier that feeds the photomultiplier signal to the recorder,

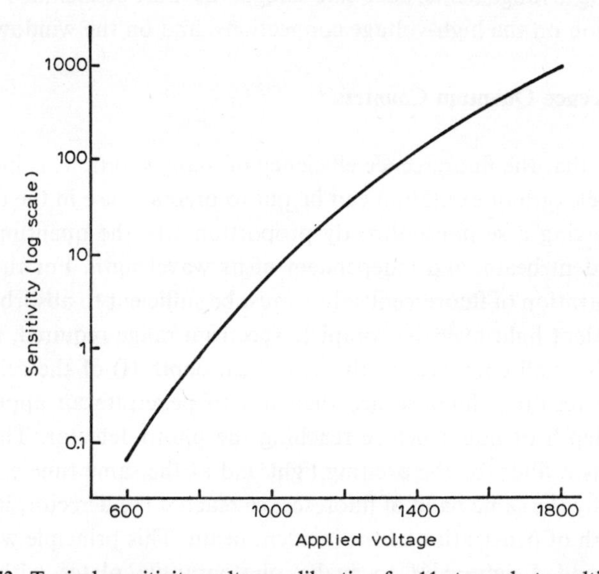

Fig. 73. Typical sensitivity–voltage calibration for 11-stage photomultiplier.

so that light intensities varying by factors of more than 10^4 can be directly compared. This facility is extremely valuable in investigations of phosphorescence and delayed fluorescence in fluid solutions. It should be remembered however that there is a limit to the output current that can be drawn from the tube whilst still maintaining the linearity of response, and that this limit is lower when the applied voltage is lower. However the outputs required in spectro-luminescence measurements will rarely exceed 10 microamps and this is well within the limit for most tubes. It is a wise precaution to check linearity at various voltages by inserting filters of known transmission in the beam of luminescence.

Some precautions are necessary in the design of the photomultiplier housing. If the tube is operated with the cathode at high negative potential (the simplest arrangement for DC operation, as described in Fig. 71 B), contact between the wall of the tube and a part of the housing at earth potential may result in high dark current due to the extraction of electrons from the dynode system and the production of phosphorescence of the glass. The multitude of pins at the base of the tubes are generally sufficient to hold the tube in place in its socket, and contact with other parts of the tube is thus avoided. Alternatively it may be arranged that any material contacting the tube is at the same potential as the cathode. In the design of photomultiplier cooling arrangements, adequate safeguards must be included to prevent condensation on the high-voltage connections, and on the window.

5. Fluorescence Quantum Counters

The fact that the fluorescence efficiency of many solutions is independent of the wavelength of excitation can be put to practical use in the design of a detector having a response directly proportional to the quantum intensity of the incident beam, and independent of its wavelength. For this purpose the concentration of fluorescent solute must be sufficient to absorb the whole of the incident light over the complete spectrum range required, in a thickness that is small compared with the optical depth (l) of the cell (see Fig. 74A). The resulting fluorescence then has to penetrate an approximately constant depth of liquid before reaching the photo-detector. The solution thus acts as a filter for the exciting light and at the same time ensures that substantially the same band of fluorescence reaches the detector, irrespective of the depth of penetration of the incident beam. This principle was used by Harrison and Leighton[139] to make photographic plates with constant

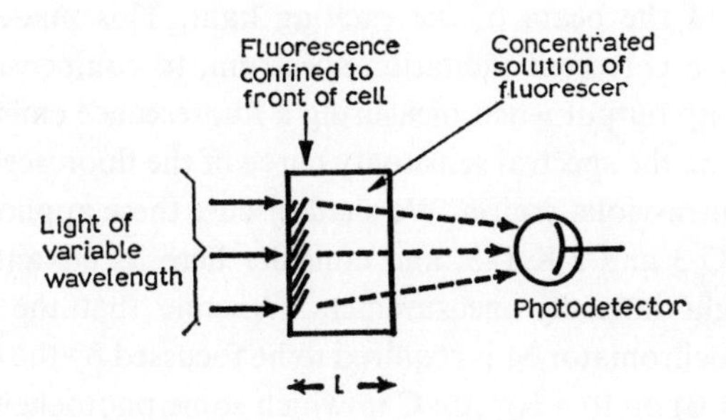

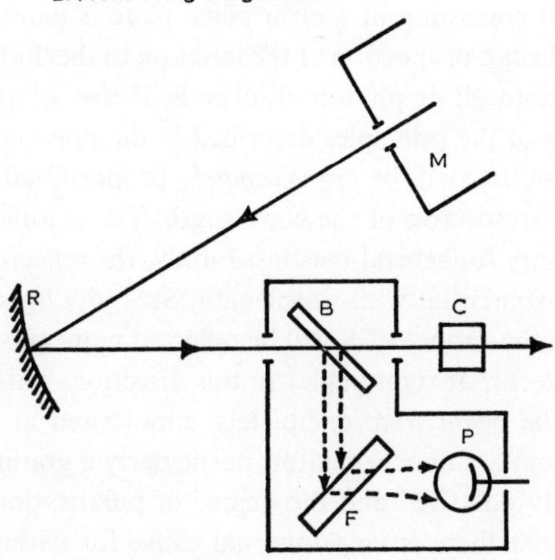

Fig. 74. The fluorescence quantum counter.
A, working principle. B, monitoring a light beam.

spectral sensitivity by coating them with fluorescent machine oil or with aesculin in gelatine. It was first applied to photocells by Bowen[140] who recommended a solution of aesculin in water (1 g per litre) in a 1 cm cell for the region 252–366 nm. Melhuish[141] has recommended a solution of rhodamine B in ethylene glycol (3 g per litre) as a wide range quantum counter, and found its quantum efficiency to be constant to within $\pm 5\%$ over the range 220–600 nm. Weber and Teale[142] have used a 2×10^{-2} M solution of 1-dimethylamino-naphthalene-5-sodium sulphonate in water for the range 200–400 nm.

Parker[143] incorporated a fluorescent screen quantum counter as a permanent feature in a spectrofluorimeter, to measure continuously the quantum intensity of the beam of the exciting light. This made it possible to record directly a corrected excitation spectrum, to compensate for fluctuations in the lamp output when measuring a fluorescence emission spectrum, and to determine the spectral sensitivity curve of the fluorescence monochromator in the ultra-violet region. We shall discuss these applications in detail in Sections 3 G 3 and 3 K 1–3, and consider here its advantages and limitations for light intensity measurement. Assume that the beam of light from the monochromator M is required to be focussed by the concave mirror R (see Fig. 74 B) on to a cuvette C in which some photochemical or photophysical process is to be observed. To monitor the light flux reaching C, a beam splitter B consisting of a clear silica plate is placed at an angle in the beam and reflects a proportion of the latter on to the fluorescent screen F, viewed by the photocell or photomultiplier P. If the solution F is suitably chosen according to the principles described in the previous paragraph, the photomultiplier output will be *approximately* proportional to the quantum flux reaching C, irrespective of the wavelength. The qualification "approximately" is necessary for several reasons. Firstly, the reflectivity of the beam splitter will vary somewhat with wavelength. Secondly light with its electric vector parallel to the surface of B will be reflected more effectively than light with its electric vector at right angles to this direction. This would be of no consequence if the beam were completely unpolarised at all wavelengths. However light from a monochromator, particularly a grating monochromator, is appreciably polarised and the degree of polarisation may vary with wavelength, so that there is an additional cause for variation in the total reflectivity of the beam splitter.

It will be noted that in Parker's quantum counter the cell is placed at an angle of 45° to the beam, and the fluorescence is viewed through the back face of the cell in a direction at right angles to that of the incident beam. This arrangement has the advantage that if the incident beam contains a small proportion of "impure" wavelengths not absorbed by the quantum counter, these are not registered by the photocell. In the more usual arrangement (Fig. 74 A) such impurities would give rise to a large photocell response and hence a considerable error. Parker's arrangement also has the advantage that if the monochromator is inadvertently scanned beyond the absorption limit of the quantum counter, the photocell or photomultiplier is not subjected to the full intensity of the unabsorbed beam. Placing the screen F at an angle has the disadvantage that the response will vary with the polarisation of the incident light. However since the incorporation of the beam splitter has already introduced such an error, the additional error is of little consequence, and could indeed be arranged partially to compensate for the former. Because of these errors it is desirable to calibrate the quantum counter at a series of wavelengths. This is most simply done by means of the ferrioxalate actinometer (see Section F). As fluorescent screen, Parker used a cuvette of 5 mm optical depth containing 4.4×10^{-3}M fluorescein in a mixture of sodium carbonate and bicarbonate (both 0.1 N). Over the range 2.2 to 4.3 μm^{-1} ($\sim$ 450–230 nm) he found the apparent relative fluorescence efficiency to vary by $\pm$ 10%, when using a prism monochromator to isolate the beam of exciting light. Once calibrated in this way the quantum counter will measure directly in absolute units the total flux received by the contents of C. As an alternative, the calibration may be done by means of a thermopile substituted for the cuvette C. However for very low light intensities the actinometer is better in view of the tedious nature of thermopile measurements under these conditions. The fluorescein solution transmits beyond about 500 nm. A rhodamine B solution of appropriate concentration may of course be substituted, although it is still desirable to calibrate it if precise results are required, for the reasons just explained. For calibration, the ferrioxalate actinometer can be used for wavelengths up to about 500 nm. To calibrate the rhodamine B quantum counter at longer wavelengths than this the thermopile must be used. It need not however be calibrated independently in absolute units because the actinometer can be used to calibrate the thermopile at a shorter wavelength via the intermediate calibration of the quantum counter. Provided that a mirror or an *achromatic* lens is used for focussing the beam, the spatial distribution of rays to the thermopile will remain unchanged when the wavelength setting of the monochromator is changed, and hence the thermopile will still retain its calibration at the longer wavelength. It is important to note that an appreciable error may be introduced in this procedure if a non-achromatic lens is used.

Finally two further points should be noted. First, if it is desired to use a quantum counter (e.g. rhodamine B) for the direct measurement of *relative* spectral distribution of the beam, without the use of a beam splitter and without calibration (i.e. as in Fig. 74A), it is possible to avoid the effect of transmitted impurity wavelengths by moving the photocell or photomultiplier slightly off the optic axis, so that it cannot receive the direct transmitted beam. Provided that the face of the fluorescent screen remains perpendicular to the incident beam no error will be introduced by variation in the polarisation of the latter. Secondly to obtain reproducible measurements, the rhodamine B or other fluorescent compound used, must be carefully purified. If small amounts of impurities are present that absorb in the wavelength region where the absorption of the chosen compound is low, the impurity will absorb an appreciable proportion of the incident beam and the observed quantum efficiency will no longer be independent of wavelength.

F. MEASUREMENT OF LIGHT BY CHEMICAL METHODS

1. Advantages of the Chemical Actinometer in Radiant Energy Measurements

The observation of rate of photochemical change has frequently been suggested as a means of measuring actinic light in absolute units, and in principle such a chemical method has important advantages over physical methods employing calibrated thermopiles or photoelectric devices. Using an ideal actinometric system, the main advantages are as follows. (a) The results are readily reproducible in different laboratories. The actinometer can be repeatedly re-standardised by different workers so that accurate values of its quantum efficiency can be firmly established. It then provides a simple laboratory reference standard for absolute radiation measurement. (b) The actinometer is an integrating system and automatically records the total dose of light received during an exposure, whether the intensity remains constant or shows wide fluctuations. (c) By choice of suitable shape and size of containing vessel the actinometer can be made to integrate almost any beam, no matter how widely dispersed or non-uniform. (d) With a quantum efficiency independent of wavelength, the ideal actinometer will act as a quantum counter for a polychromatic beam.

2. The Ferrioxalate Actinometer

The principal requirements of an ideal general purpose chemical actinometer are: (a) constant quantum efficiency and high absorption factor over a wide range of wavelengths, of intensities and of total radiation dose; (b) high sensitivity and precision coupled with simplicity of operation and ready availability of the photochemical material. The ferrioxalate actinometer developed by Parker and Hatchard[144–147] largely fulfils these conditions for wavelengths between 500 and 250 nm, and the indications are that it would be suitable down to at least 200 nm. The photolyte consists of a solution of potassium ferrioxalate, $K_3Fe(C_2O_4)_3.3H_2O$, in 0.1 N sulphuric acid. In acid solution the trioxalato-ferric ions are largely dissociated into monoxalato and dioxalato complexes. On exposure to light the following reactions occur:

$$[Fe(C_2O_4)]^+ \xrightarrow{h\nu} Fe^{2+} + (C_2O_4)^- \qquad (232)$$

$$(C_2O_4)^- + [Fe(C_2O_4)]^+ \rightarrow 2CO_2 + Fe^{2+} + (C_2O_4)^{2-} \qquad (233)$$

After photolysis the ferrous ion formed is converted to its 1,10-phenanthroline complex and the latter determined absorptiometrically. The quantum efficiency of ferrous ion formation has been accurately determined and varies smoothly and slowly over the wavelength range 500–250 nm (see Table 22). The minimum detectable amount of radiation is about 2×10^{-10} einstein per ml using standard measuring methods, and smaller quantities can be detected using micro-methods. Quantities of 2×10^{-8} einstein per ml and greater can be measured with an accuracy of $\pm 2\%$. The maximum amount of radiation that can be accepted without alteration of quantum efficiency is 5×10^{-6} einstein per ml of solution and very wide ranges of total radiation dose can thus be measured. The quantum efficiency remains constant over a range of intensities from 5×10^{-11} to 2×10^{-4} einstein cm^{-2} sec^{-1}. The actinometer can thus be used equally well to measure very low intensities isolated by a monochromator or very high intensities such as those obtained from a flash tube.

Both the photolyte and photolysis products are stable in the dark and the procedure is simple and versatile. The versatility of the system depends on the fact that the photolyte itself absorbs strongly, but gives rise to weakly absorbing photolysis products. Accumulation of the latter does not therefore disturb the linearity of response, even when the proportion of photolyte decomposed is considerable. Large or small amounts of decomposition can be measured with equal accuracy by simple dilution before absorptiometric reaction.

The ferrioxalate actinometer is being increasingly adopted as the primary standard for actinic light measurement in photochemical laboratories throughout the world. It has many applications in connection with the measurement of photoluminescence and we therefore give detailed instructions for its use.

TABLE 22

QUANTUM EFFICIENCY OF THE FERRIOXALATE ACTINOMETER AT 22°C

Wavelength (nm)	By comparison with uranyl oxalate	By reference to calibrated thermopile	By comparison with ferrioxalate at 436 nm using thermopile	Recommended value
577/9 (0.15 M)	–	–	0.013	0.013
546	–	–	0.15	0.15
509	–	–	0.86	0.86
480	–	–	0.94	0.94
468	–	–	0.93	0.93
436 (0.15 M)	–	–	1.01	1.01
436 (0.006 M)	1.12	1.11	[1.11]	1.11
	1.07[a]			
405	1.14	1.16	1.13	1.14
365/6 (0.15 M)	–	1.20[b]	1.16[c]	1.18
365/6 (0.006 M)	1.16	1.26	1.21	} 1.22
	1.20[a]	1.26[b]		
361/6	–	–	1.22	
334	1.25	–	1.21	1.23
313	1.23	–	1.24	1.24
297/302	1.24	–	–	1.24
254	1.22	–	1.29	1.25
	1.25[a]			

Notes

Most of the results are taken from Hatchard and Parker[146] for which the concentrations of potassium ferrioxalate were 0.15 M for wavelengths 579 to 436 nm and 0.006 M for wavelengths 436 to 254 nm. Results at both concentrations are reported for 436 and 365/6 nm.

The results marked [b] were obtained by Lee and Seliger[148] using 0.15 M and 0.006 M potassium ferrioxalate. Those marked [a] were obtained by Baxendale and Bridge[149] using a somewhat different procedure.

The recommended values at 361/6 have been modified slightly from those originally recommended[146] to take into account the additional data of Lee and Seliger. The value marked [c] was determined by reference to 0.006 M ferrioxalate at 366 nm (Φ assumed to be 1.22).

3. Preparation of Actinometer Liquid and Choice of Concentration and Optical Depth

Pure potassium ferrioxalate can be readily prepared by mixing three volumes of 1.5 M AR potassium oxalate with 1 volume of 1.5 M AR ferric chloride with vigorous stirring. The precipitated potassium ferrioxalate should be recrystallised three times from warm water and the crystals sucked dry and dried in a current of air at 45°C. The whole procedure should be carried out in a room lit only by a yellow safelight. The composition corresponds to $K_3Fe(C_2O_4)_3.3H_2O$. The 0.006 M actinometer solution is prepared by dissolving 2.947 g of the crystals in 800 ml of water. 100 ml of 1.0 N sulphuric acid are added, the solution diluted to 1 litre and mixed. The 0.15 M solution is made in a similar manner using 73.68 g of ferrioxalate crystals.

The actinometer solutions could be made up by mixing standard solutions of potassium oxalate and ferric sulphate. The quantum efficiency at 366 nm is little affected by the presence of excess neutral electrolyte and this probably applies at the other wavelengths also. However, it is preferred to use the pure potassium ferrioxalate, because this forms a very convenient standard substance. It is readily prepared in a pure state, it has an accurately known composition, and it can be stored in the dark indefinitely without change.

A 1 cm depth of 0.006 M solution absorbs 99% or more of the light of wavelengths up to 390 nm. It can be used conveniently for wavelengths up to about 430 nm (absorption about 50% per cm). For longer wavelengths the 0.15 M solution is usually more convenient. The fractions of light absorbed by various thickness of the two solutions are shown in Fig. 75 and can be used as a guide to indicate the required depth and concentration for any particular purpose. When very accurate results are required, and light absorption is not complete, it is desirable to measure directly the fraction of light absorbed by the actual layer of liquid used, preferably

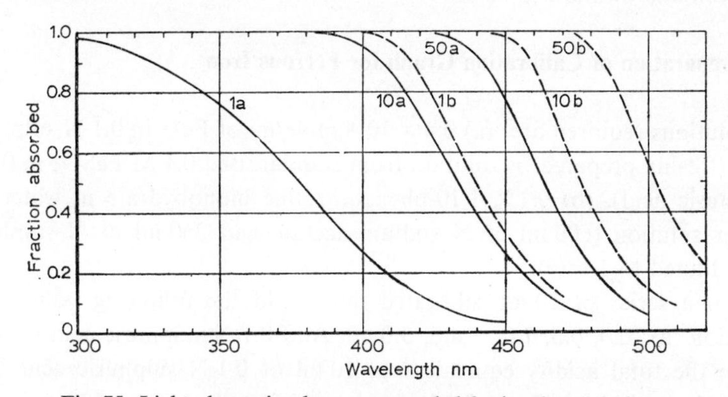

Fig. 75. Light absorption by recommended ferrioxalate solutions.
Curves 1a, 10a and 50a are for 0.006 M potassium ferrioxalate in 0.1 N sulphuric acid with optical depths of 1, 10 and 50 mm respectively. Curves 1b, 10b and 50b are for 0.15 M potassium ferrioxalate in 0.1 N sulphuric acid at the same optical depths respectively (from Hatchard and Parker[146]).

using the same source of light, because the fraction absorbed will in general vary with temperature and of course with the spectral purity of the light.

If the fraction of light transmitted by the solution is large, a small correction should be applied for the light reflected back into the liquid from the rear face of the cell (approximately 4–5%) as well as for the light reflected by the front surface of the cell. Thus if t is the observed transmission of the ferrioxalate as measured with an absorption spectrophotometer by comparison with a similar cell filled with water, the fraction of *incident* light absorbed by the ferrioxalate is given by

$$f_a = 0.95[(1 - t)(1 + 0.05t)] \quad (234)$$

In many applications the quantity required is the dose rate delivered to the *contents* of the cell. The appropriate value of the fraction absorbed is then equal to $(1 - t)(1 + 0.05t)$. If t is large and the proportion of ferrioxalate decomposed is large, a mean transmission factor must be used. It would however then be best to use the more concentrated solution of ferrioxalate or a greater optical depth so that the proportion of decomposition and the fraction transmitted are reduced to low values.

When stored in an amber bottle in the dark room the actinometer solutions can be kept unchanged for long periods. The "blank" values obtained by the method described below correspond respectively to optical densities of approximately 0.01 and 0.025 per cm for 10 ml of 0.006 M or 0.15 M after reaction and dilution to 20 ml.

4. Preparation of Calibration Graph for Ferrous Iron

Solutions required are: (a) 0.4×10^{-6} mole/ml of Fe^{2+} in 0.1 N sulphuric acid (freshly prepared by dilution from standardised 0.1 M $FeSO_4$ in 0.1 N sulphuric acid), (b) 0.1% 1,10-phenanthroline monohydrate in water, (c) buffer solution (600 ml of N sodium acetate and 360 ml of N sulphuric acid diluted to 1 litre).

Into a series of 20 ml calibrated flasks add the following volumes of solution (a): 0.0, 0.5, 1.0 ... 4.5, 5.0 ml. Add 0.1 N sulphuric acid so as to make the total acidity equivalent to 10 ml of 0.1 N sulphuric acid. Add 2 ml of solution (b), 5 ml of solution (c) (mixing after each addition) make up to mark, mix and allow to stand for at least one half hour. Measure the optical densities at 510 nm in a 1 cm cell with an absorption spectrophotometer. Correct each optical density for the value obtained with the solution to which no ferrous ion was added and plot the results against the amount of ferrous ion added.

Using a spectrophotometer of type similar to the Unicam 500 or 600 it has been found that 1.81×10^{-6} mole of ferrous ion in the 20 ml test solution produces an optical density of 1.0 per cm at 510 nm. The optical density of the "blank" solution was less than 0.01 per cm, and a linear calibration graph was obtained. Using either of these spectrophotometers the calibration graph was found to remain constant during several years. In view of this and the fact that the graph is linear, it is possible to use a spectrophotometric factor to convert optical density to amounts of ferrous ion, and only occasional checks of the calibration graph are required. With a filter photometer somewhat different values of "blank" and graph slope may be obtained. Whatever the instrument, the user must of course prepare his own calibration graph to allow for instrumental variation. The "difference optical density" at 510 nm is independent of the presence of potassium ferrioxalate and of the volume of buffer added, provided that the latter is equal to at least one half of the volume of the 0.1 N sulphuric acid present in the solution (final pH equals 3.5). Half an hour is more than sufficient to allow for complete reaction after addition of the phenanthroline. If desired the solutions may be kept for several hours (in the dark) before measurement.

5. General Procedure for Actinometry

Irradiate the appropriate actinometer solution for a period sufficient to produce a concentration of ferrous ion between 0.005×10^{-6} and 3×10^{-6} mole per ml. (During irradiation it is desirable to stir the liquid by means of a current of nitrogen.) After irradiation and mixing, pipette an aliquot of the solution into a 20 ml calibrated flask. Add in succession 2 ml of phenanthroline solution (b), and a volume of buffer (c) equal to one half the volume of photolyte taken (mixing after each addition). Make up to the mark with water, mix and allow to stand for at least half an hour. Measure the optical density at 510 nm in either a 1 or 4 cm cell. Repeat with the same volume of unexposed actinometer solution. Convert the "difference optical density" to quantity of ferrous iron using the calibration graph or spectrophotometric factor as appropriate. Convert the quantity of ferrous iron formed in the total volume of the irradiated solution to a radiation dose using the recommended quantum efficiency recorded in Table 22 (see equation 5). If necessary allow for the fraction of light absorbed to obtain the dose incident on the surface of the liquid. For concentrations up to 0.02×10^{-6} mole per ml a 10 ml aliquot should be taken and measurement made in a 4 cm cell. For higher concentrations an aliquot should be taken such that the optical density in a 1 cm cell is between 0.1 and 0.8. The amount of ferrioxalate decomposed should correspond to a reduction in concentration of not more than 0.005 M.

For the measurement of very low light intensities it is desirable to keep the volume of solution as small as possible. The following modified procedure has been found suitable for this purpose. 3 ml of actinometer liquid are irradiated in a 1 cm spectrophotometer cell. To the irradiated liquid is added 0.5 ml of a solution that is 1.8 N in sodium acetate and 1.08 N in sulphuric acid, and contains 0.1% w/v 1,10-phenanthroline monohydrate. Measurement is carried out in the same cell and the limiting sensitivity (corresponding to a difference optical density of 0.002) corresponds to 6×10^{-10} mole of ferrous ion. For low intensity beams of very small cross section, the volumes can be reduced still further and measurement made in a micro-absorption cell.

6. Comments on Procedure

The procedures described above are those that have been found most satisfactory and convenient by the author. They can of course be modified in some degree to suit individual requirements. In this respect the comparative insensitivity of the quantum efficiency to wide changes in solution composition, light intensity and total radiation dose is a particularly attractive feature of the ferrioxalate actinometer. If however major modifications to the procedure are contemplated, reference should be made to the original papers for details of changes in quantum efficiency that are likely to be produced.

For the most accurate work provision should be made for stirring the liquid during irradiation by means of a current of nitrogen. However the error introduced by using an aerated and unstirred solution is only a small percentage provided that absorption is complete throughout the irradiation.

It is most important to carry out all manipulation of ferrioxalate solutions in a dark room. A Kodak OB yellow safe light, which does not transmit below 550 nm, provides adequate and safe illumination for short periods, but even dim electric light decomposes the solution at an appreciable rate.

G. AMPLIFICATION AND RECORDING OF PHOTOMULTIPLIER OUTPUT

1. Manual Operation

The simplest method of reading the intensity of fluorescence emission is to pass the output from the collector electrode of an 11- or 13-stage photomultiplier to a sensitive galvanometer (2,000 mm per microamp) as shown in Fig. 71B. The dark current from an E.M.I. 9558Q photomultiplier operated at 1800 volt will then give a deflection of about 10 mm (the exact value varies from tube to tube). If the galvanometer is critically damped by means of a shunt in the usual way, its time constant will damp out much of the dark current fluctuation so that light intensities corresponding to about one fifth of the dark current will produce a significant deflection. Such an arrangement is simple, reliable and remarkably sensitive. Provided that the source used for excitation gives a reasonably constant intensity it is quite adequate for many applications involving intensity measurements at a few wavelengths only (e.g. the determination of inorganic constituents after chemical separation—see Chapter 5). As in all fluorescence measurements it is desirable to compare the intensity from the "unknown" specimen with that from a fluorescence standard, e.g. quinine bisulphate (see Section 3 L 4). Long term fluctuations in the intensity of the exciting light or in the EHT supply to the photomultiplier are then automatically compensated. It is desirable to use a galvanometer shunt with accurate tappings corresponding to sensitivities of, say, 1.0, 0.3, 0.1, 0.03, 0.01. A maximum scale deflection of 500 mm on the lowest sensitivity will then correspond to a photomultiplier output of 25 micro-amp. To measure higher light intensities the voltage supply to the photomultiplier should be reduced: nearly all commercial EHT units are fitted with a control knob for this purpose. A complete fluorescence emission spectrum is measured by taking galvanometer readings at a series of wavelength settings of the fluorescence monochromator (Fig. 44). An excitation spectrum is measured similarly by varying the wavelength setting of the excitation monochromator. These measurements will give "uncorrected" spectra (see Section 3 K 3).

2. Single Beam Recording

Manual measurement of complex spectra is time-consuming, and most spectrofluorimeters are fitted with some form of recorder. If the light source has been adequately stabilised, single beam recording gives quite satisfactory results. If a pen recorder is to be used, an amplifier capable of converting the photomultiplier output (generally 0.002 to 1.0 micro-amp) to a current of 1 to 2 milliamp, will be required. A variety of DC amplifiers are available commercially and these are generally equipped with various current sensitivity ranges and a variable time constant. The latter is particularly valuable in high sensitivity work; if a large value of time constant is used, e.g. 2 to 3 seconds, much of the dark current fluctuation is damped out and a smoother record obtained. Of course the spectrum must then be scanned more slowly if full resolution is to be obtained.

Either a direct reading 2 milliamp recorder may be used, or preferably (but more expensively) a potentiometer recorder. The latter is preferred because it generally gives a more positive movement and has the additional advantage that it can be converted to a ratio recorder as described in the next section. The principle of the potentiometer recorder is illustrated in Fig. 76. It consists of a potentiometer P, the sliding contact of which is attached to the recording pen and is driven by a servo motor. The voltage to be measured is fed in at point I to the sliding contact via the servo control. The latter detects the out-of-balance between the applied voltage at I and the voltage from the potentiometer P, and the servo motor drives the contact carrying the pen to the point on P at which a balance is attained. If the

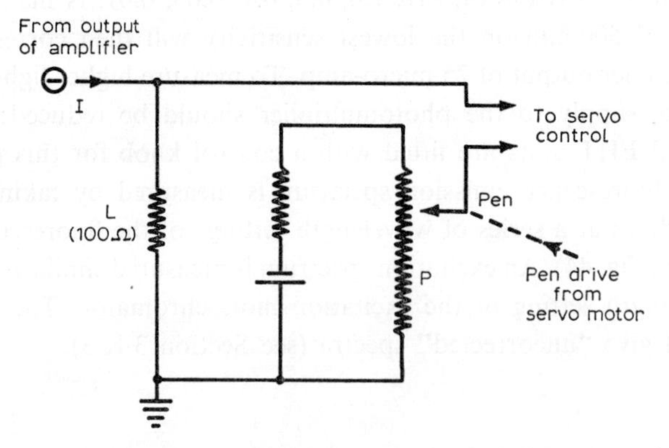

Fig. 76. Principle of potentiometer recorder in single beam operation.

photomultiplier output is amplified with an amplifier giving a maximum output of, say, 1 milliamp, this is passed through 100 ohm load resistor L to provide a maximum of 100 mv to give full scale deflection of the recorder. It is of course necessary to arrange that the polarity of the recorder is the same as that of the amplifier output. Many amplifiers give an output with *positive* at earth potential when the input is negative with respect to earth, as it is with the photomultiplier shown in Fig. 71 B.

For some experiments it is necessary to record spectra more rapidly than is possible with a pen recorder of 1 second pen speed. Suppose that it is required to scan the spectrum from 200 to 700 nm in 5 seconds with a resolution of 1 nm. The recorder must then be capable of giving full scale deflection in about 0.01 second and an oscilloscope must be used. DC oscilloscopes generally have a built-in amplifier of variable gain and it is simply necessary to feed the output from the collector of the photomultiplier direct to the input terminal of the Y amplifier on the oscilloscope. Because of the requirement for fast response, a much greater noise level from the photomultiplier must be accepted and much lower sensitivities are achieved by this method than with a pen recorder and amplifier having a time constant of 1 to 2 seconds. It is of course possible to damp out the noise by shunting the input with a capacitor (see Section 3 N 6), but the advantage of fast recording is then lost without gaining the advantage of a permanent chart record provided by a pen recorder.

3. Double Beam Recording

Many types of discharge lamp give somewhat unstable light output and to make precise measurements of fluorescence excited by such lamps it is necessary to incorporate some form of compensating device in the equipment. The use of a fluorescent screen quantum counter (see Section 3 E 5) for this purpose was first described by Parker[143]. It has the great advantage that it also permits the direct recording of *corrected* fluorescence excitation spectra. The latter function is discussed in Section 3 K 2. We describe here (Fig. 77) a suitable input circuit for a potentiometer recorder to permit ratio recording. The battery supplying the slide wire potentiometer P in Fig. 76 is disconnected and the slide wire activated by passing through it the output (approximately 1 milliamp) from the amplifier (A_2 in Fig. 77) of the fluorescent screen quantum counter. (The circuit can simply be re-converted to single beam recording by reconnecting the battery via switch S.) The

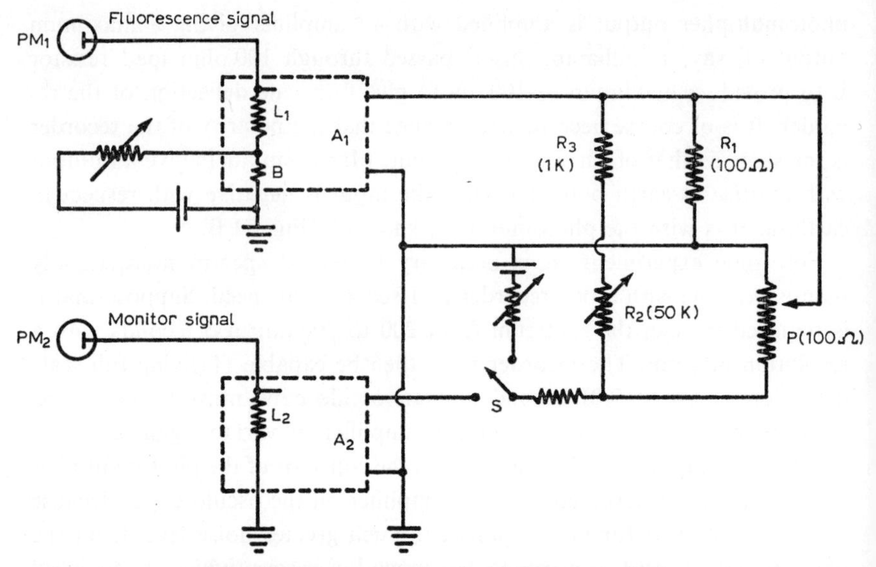

Fig. 77. Arrangement for double beam recording.

output from the fluorescence signal amplifier (A_1) is passed to earth through the 100 ohm resistor (R_1) and the voltage across the latter (maximum 100 mv) fed to the slide wire contact of the recorder. With this arrangement the position of the servo-operated slide wire contact on P gives a direct measure of the ratio of the voltages across R_1 and P, and hence a measure of the ratio of the currents flowing from the collectors of the two photomultipliers (PM_1 and PM_2). Fluctuation in the intensity of the exciting light produces the same fractional change in the output of PM_1 and PM_2 and the position of the recorder pen on P is unaffected.

Two simple additional devices are incorporated in the circuit. The first consists of a small resistor (B) placed in series with the load resistor L_1 of the fluorescence amplifier. A small current passed through this resistor from its supply battery provides a voltage acting in the reverse sense to the voltage developed across L_1 by the output from PM_1, and serves to "back off" the photomultiplier dark current when working at high sensitivity. No backing off is required for the monitor amplifier because PM_2 always works at comparatively low sensitivity and the zero control incorporated in A_2 provides sufficient adjustment. The second device consists of the resistors R_2 and R_3 that feed a small proportion of the output from A_2 to the slide wire contact. Adjustment of R_2 permits the true chart zero to be shifted along P without interfering appreciably with the ratio recording characteristics. This is useful when there is considerable noise shown by the pen at high sensitivity.

An AC amplifier can be used for A_1 and it is then necessary to chop the exciting light at the frequency to which the amplifier is tuned. This arrangement has several advantages. Firstly, the backing off circuit can be dispensed with because the DC component of the dark current will not be registered by the amplifier. Secondly, the instrument will be insensitive to stray "DC" light that may leak into the sample compartment of the fluorescence monochromator. Thirdly, the instrument will register short-lived fluorescence but will not record long-lived phosphorescence, unless the chopper is placed in the beam of phosphorescence instead of in the beam of the exciting light. It has some disadvantages also. One is the inconvenience of having to interpose a chopper, and the second is that half of the exciting light is wasted in the chopping process so that, in principle at least, the overall sensitivity is somewhat reduced. It is, however, a great convenience to incorporate choppers in both beams so that the instrument can be used both as a spectrofluorimeter and as a spectrophosphorimeter (see Section 3 N 1). There is then no disadvantage in using AC amplification.

4. Calibration of Sensitivity

If is often required to compare the intensities of two emissions differing by several orders of magnitude, for example when measuring the delayed and prompt fluorescence from the same fluid solution. The overall sensitivity of the equipment can be varied by changing the amplifier gain or by varying the voltage applied to the dynode resistor chain feeding the photomultiplier (see Section 3 E 4). Large variations in sensitivity are best achieved by a combination of both methods. Thus, suppose we have an amplifier with gain control positions corresponding to factors of 1, 3 and 10. The photomultiplier operates reasonably satisfactorily at light levels corresponding to output currents between 0.002 and 20 micro-amp or more, i.e. it will cover a sensitivity range of about 10^4. (Reference to Fig. 73 indicates that this corresponds to a variation in EHT supply from about 600 to 1800 volt.) The relative values of sensitivity settings may be conveniently determined as follows. With a strongly fluorescent solution in the sample compartment, and the analysing monochromator set to a wavelength far removed from the maximum in the fluorescence spectrum, and with narrow slits on the analysing monochromator, the amplifier gain is set to 10 and the photomultiplier voltage to 1800 volts. These settings are allocated the overall sensitivity value of 10,000. The slit width is adjusted so as to produce a full scale deflection on the recorder. The amplifier gain is then reduced to 3 and the recorder deflection noted. The ratio of the two recorder readings gives the precise value of the new sensitivity setting (approximately 3,000). With this new setting the slit width is increased, or the wavelength setting is adjusted to a value nearer to the absorption maximum, so as to give full scale deflection on the recorder. The amplifier gain is now reduced to 1 and from the recorder reading the new sensitivity value is calculated (approx. 1,000). This procedure is repeated at lower photomultiplier voltages so that a series of precise sensitivity settings is obtained corresponding approximately to 300, 100, 30, 10, 3 and 1. These values should be reproducible to within a few per cent over quite long periods if a stable amplifier and EHT supply is used. It is desirable to check linearity on selected ranges (particularly the lowest sensitivity) by inserting filters of known transmission in the beam of exciting light, or by measuring the fluorescence intensity from a series of solutions containing known concentrations of a fluorescent substance under conditions for which the fluorescence is known to be proportional to concentration.

H. FACTORS AFFECTING THE CHOICE OF GEOMETRICAL ARRANGEMENT OF THE SPECIMEN

1. Inner Filter Effects and Quenching

The geometrical arrangement of the beam of exciting light and the direction of viewing the fluorescence light in relation to the specimen has been one of the most controversial points in the design of a spectrofluorimeter. This is partly because of the different uses to which instruments have been put, partly because of the different sensitivities of the instruments used, and partly to confusion by some workers between *solute quenching* and *inner filter* effects. Lack of appreciation of the elementary principles governing the latter has in the past led to the use of inappropriate geometrical arrangements, or incorrect interpretation of the results obtained, and was one of the main reasons why the technique of fluorescence measurement had been regarded with suspicion by some workers. We shall therefore define what we mean by the terms *quenching* and *inner filter* effects. By quenching we mean all those processes that result in the *true* fluorescence or phosphorescence efficiency being reduced to below unity. These processes have already been discussed in Sections 1B1, 1C2, 2A, 2B and 2C. They are those processes that divert the light energy absorbed by a potentially luminescent molecule into channels other than the emission processes responsible for luminescence. They include internal conversion, intersystem crossing, energy transfer etc., as well as deactivation caused by collision with other molecules of solute. Luminescence quenching is thus a *fundamental* effect characteristic of the system under the particular conditions concerned, and independent of the experimental arrangement used to observe it. On the other hand inner filter effects are instrumental artifacts. They have no influence on the primary process of emission from the molecules originally excited, but simply reduce the observed intensity of luminescence, either by absorption of the exciting light, or of the luminescence, within the material being tested. The two kinds of inner filter effects have been discussed

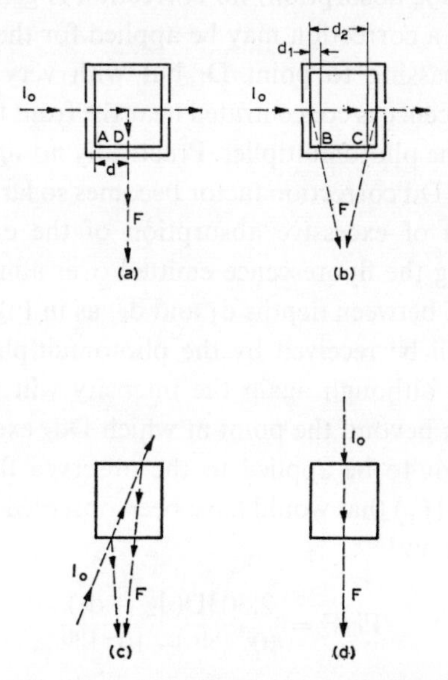

Fig. 78. Methods of illumination and viewing.
(a) and (b), right angle method; (c) frontal method; (d) in-line method (from Parker and Rees[150]).

by Parker and Rees[150]. They are: (a) excessive absorption of the exciting light, and (b) absorption of the luminescence emitted. We shall now discuss them in relation to the three basic specimen arrangements shown in Fig. 78.

2. Inner Filter Effects with Right Angle Illumination

In the method of right angle illumination (*a* and *b* in Fig. 78) the exciting light (I_0) passes through the specimen in a direction at right angles to that along which the fluorescence is viewed, and the cell compartment is usually masked in such a way that the photomultiplier "sees" only the illuminated liquid and not the illuminated cell faces (*a* in Fig. 78). The advantage of this arrangement is that interference by stray light arising from reflection at the faces of the container, or fluorescence of the container itself, is minimised. Since the exciting light has to pass through a depth of liquid AD (= d) before reaching the region viewed by the photomultiplier, its effective intensity will be reduced by a factor 10^{-Dd} where D is the total optical density per cm of the solution at the wavelength used for excitation. The arrangement (*a*) is clearly suitable only for weakly absorbing solutions. If Dd is less than 0.02 (4.6% absorption) no correction is generally necessary. At higher values of Dd a correction may be applied for the amount of exciting light adsorbed in passing to point D, but with very strongly absorbing solutions the fluorescence is concentrated near the front face of the container, and out of view of the photomultiplier. Practically no signal is then recorded by the latter and the Dd correction factor becomes so large as to be valueless. Some of the effects of excessive absorption of the exciting light can be mitigated by viewing the fluorescence emitted over almost the whole width of the container, i.e. between depths d_1 and d_2, as in Fig. 78b. If d_1 is small, fluorescence will still be received by the photomultiplier even with highly absorbing solutions although again the intensity will not be proportional to the concentration beyond the point at which Dd_2 exceeds a few per cent. The correction factor to be applied to the observed fluorescence intensity (F) to give the value (F_0) that would have been observed in the absence of the inner filter effect is now[131]:

$$F_0/F = \frac{2.303D(d_2 - d_1)}{10^{-Dd_1} - 10^{-Dd_2}} \tag{235}$$

In practice, correction becomes inaccurate for values of F_0/F greater than about 3.

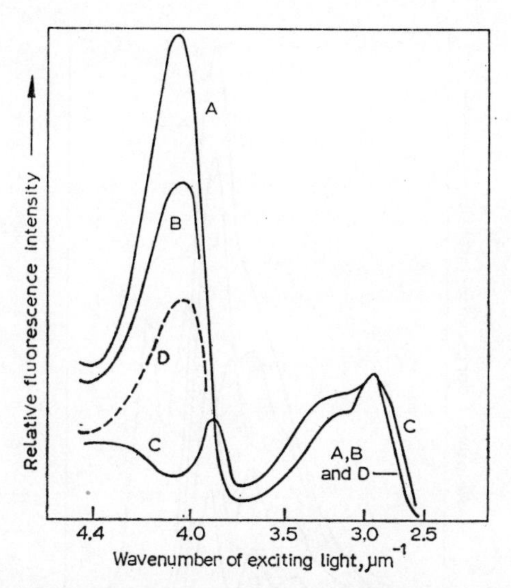

Fig. 79. Effect of excessive absorption of the exciting light on the excitation spectrum. Quinine bisulphate in 0.1 N sulphuric acid at 20°C in a cell 1 cm square. A, 1 µg per ml; B, 10 µg per ml (sensitivity decreased); C, 100 µg per ml (sensitivity further decreased). Curves A, B and C with right angle illumination. Curve D, 10 µg per ml with frontal illumination and optical depth 15 mm (from Parker and Rees[150]).

The use of too concentrated a solution will result in a distortion of the observed fluorescence excitation spectrum. Thus in Fig. 79, as the concentration of quinine bisulphate was increased from 1 to 10 microgram per ml (curves A and B) the height of the main excitation maximum (corresponding to maximum absorption) decreased relative to the weaker maximum at the lower wavenumbers. At a still higher concentration (curve C) the absorption of exciting light at the main excitation maximum became so great that much of the fluorescence moved out of view of the photomultiplier when this wavenumber was used for excitation, and the observed excitation spectrum actually passed through a minimum at this point. The remainder of the spectrum was also distorted.

Distortion of the excitation spectrum can equally well be produced by a second solute having a strong absorption at some wavelengths. This is a particularly insidious effect and can easily lead to false interpretation of excitation spectra by the unwary operator. The effect of various concentrations of benzene on the main excitation band of anthracene (Fig. 80) is good example. Benzene emits fluorescence at high wavenumbers far removed

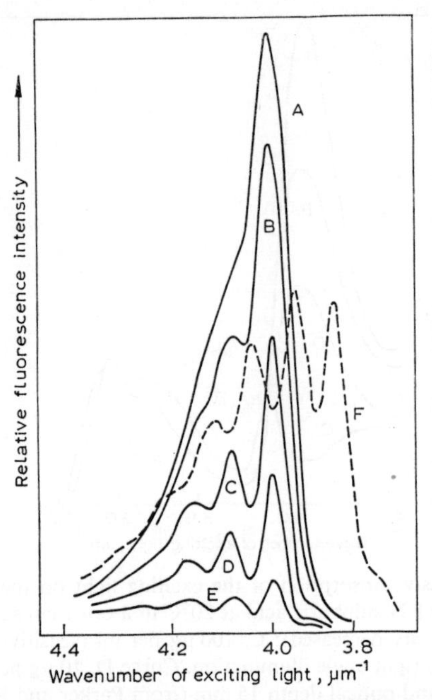

Fig. 80. Effect of a second absorbing solute on the excitation spectrum. Excitation spectra (with right angle illumination in a cell 1 cm square) of anthracene in ethanol (0.1 μg per ml) with benzene at concentrations: A, nil; B, 0.05%; C and F, 0.125%; D, 0.25%; E, 0.5%. For curves A–E the fluorescence monochromator was set to receive anthracene emission (2.5 μm^{-1}); for curve F it was set to receive benzene emission (3.6 μm^{-1}) (from Parker and Rees[150]).

from the emission band of anthracene, but its first absorption band overlaps the second absorption band of anthracene as indicated by curve F in Fig. 80, which is the fluorescence excitation spectrum of *benzene*. With no benzene in the solution, the dilute anthracene produced its correct excitation spectrum (curve A). As the concentration of benzene was increased, it absorbed an increasing proportion of the exciting light and this absorption was of course greatest at the absorption maxima in the benzene spectrum. At these wavenumbers therefore the emission from the anthracene was least and in fact the benzene superimposed a kind of inverted absorption spectrum on the excitation band of the anthracene. Curve C, for example, could well have been mistaken for the excitation of another "unidentified" component of the solution. To avoid such errors it is essential to check the *absorption* spectrum of a solution before taking an excitation spectrum, and if necessary to dilute the solution until its optical density is sufficiently small at all wavelengths of interest.

The second type of inner filter effect is that produced by absorption of *fluorescence* light; it may be absorption by an excessive concentration of the fluorescent solute itself (self-absorption) or it may be absorption by other solutes. For right angle illumination its magnitude can be approximately calculated if the path length of the fluorescence beam through the liquid is known, and re-emission of fluorescence by absorption of the original fluorescence is neglected. *Self-absorption* mainly affects the high wavenumber side of the fluorescence emission band because it is in this region that overlap with the first absorption band occurs (see Section 1 B 1 and Fig. 2). The presence of a second solute absorbing strongly in the region where the

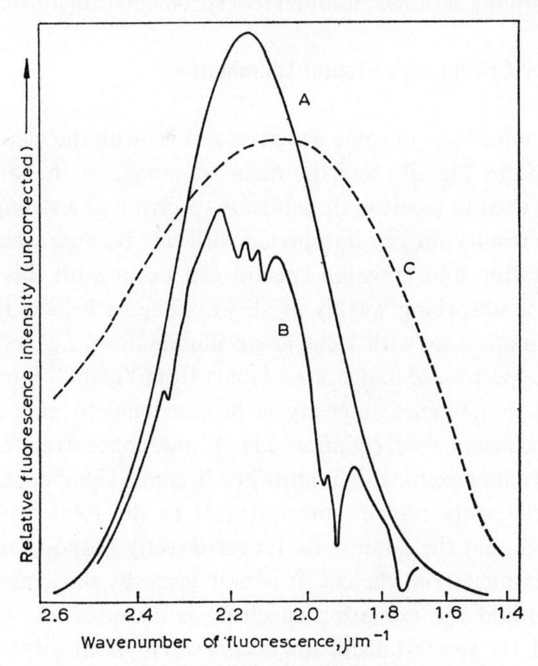

Fig. 81. Effect of a second absorbing solute on the emission spectrum (right angle illumination).

Excitation at 2.73 μm^{-1} (366 nm). Optical depth for fluorescence light was 2.5 cm. Band width of the quartz prism fluorescence monochromator was 0.01 μm^{-1} at 2.0 μm^{-1}. Curve A, 80 μg per ml of quinine bisulphate in 0.1 N sulphuric acid at 20°C; curve B, as for A but with 0.12 M neodymium chloride and sensitivity increased to compensate for quenching; curve C, spectral sensitivity of analysing system (9558 photomultiplier) (from Parker and Rees[150]).

first fluoresces, will naturally produce distortion of the emission spectrum of the latter. This is illustrated in Fig. 81 which shows the distortion produced in the emission spectrum of quinine bisulphate as a result of excessive absorption of the fluorescence by neodymium chloride. The main absorption peaks of the latter are clearly evident, in inverted form, in the emission spectrum B. Admittedly this particular example is a hypothetical one because the measurement of quinine bisulphate in such a concentrated solution of neodymium chloride would not be attempted, if for no other reason than that the neodymium chloride strongly *quenches* the fluorescence, quite apart from its inner filter effect. Neodymium was chosen to demonstrate the general principle because it has a series of sharp absorption bands. A high concentration of a substance showing a broad absorption band would produce an equally serious, although less spectacular, distortion.

3. Inner Filter Effects with Frontal Illumination

Inner filter effects are in some ways less serious with the method of frontal illumination (*c* in Fig. 78) and the main advantage of this arrangement is that it can be used to measure the *emission* spectrum of a strongly absorbing solution. The results must be interpreted with care because serious distortion of both excitation and emission spectra can occur with this arrangement also. Solutions absorbing weakly at all wavelengths behave the same with frontal illumination as with right angle illumination, i.e., both excitation and emission spectra are undistorted (apart from "blank" interference) and the observed fluorescence intensity is proportional to concentration and extinction coefficient (see equation 21). If the concentration is increased until the fraction of exciting light absorbed becomes significant, the observed fluorescence intensity remains proportional to the total rate of emission of fluorescence, but the latter is no longer directly proportional to concentration and extinction coefficient. It is now given by the general expression (equation 18) and the excitation spectrum is distorted. As with the right angle method, the greatest distortion occurs where the absorption is greatest, i.e. at the maxima in the excitation spectra, which are observed to be lower than their true values relative to the minima. At the limit, with complete absorption of exciting light at all wavelengths, and with only one absorbing species present, the observed fluorescence intensity is independent of wavenumber (if the fluorescence efficiency is constant), and the excitation spectrum then consists of a horizontal straight line. It is clearly of no value as a criterion for identification. The effect of excessive absorption with frontal illumination can be seen in curve D of Fig. 79, in which, for a concentration of 10 microgram per ml of quinine bisulphate, the main excitation maximum is lower than that observed (with right angle illumination) from a dilute solution (curve A). The degree of distortion depends also on the optical depths used. Thus with right angle illumination in a 1 cm cell, the effective optical depth for the exciting light was 0.5 cm, while for frontal illumination in a 15 mm cell the effective optical depth was 15 mm. With the same concentration of quinine bisulphate the latter produced a more distorted spectrum (curve D) than the former (curve B of Fig. 79).

If a second absorbing solute is present, it will not interfere so long as the total fraction of exciting light absorbed is small, i.e., the situation is the same as in right angle illumination. For complete absorption however each solute will absorb a fraction of the exciting light proportional to its own value of D, and if the value of D for the second solute is much higher (and with no energy transfer), the fluorescence of the first solute will be almost extinguished (*not* quenched). A simple example of this is observed in the measurement of an excitation spectrum in a solvent such as carbon tetrachloride that shows a cut off in the middle of the ultra-violet region. Beyond the cut-off the excitation spectrum falls to zero.

Frontal illumination is the best method for observing the fluorescence *emission* spectra of concentrated solutions, but some remarkable distortions of the spectra can be produced. For example Fig. 82 shows the effect of adding naphthacene to a solution of N-phenyl-2-naphthylamine. Exciting light of wavenumber 2.73 μm^{-1} (366 nm) was comparatively weakly absorbed and penetrated a great distance into the solution. As a result, much of the observed fluorescence had to pass back again through a thick layer of solution and was thus partly absorbed by the naphthacene which therefore superimposed its absorption spectrum on the emission spectrum of the phenyl naphthylamine (compare curves A and B in Fig. 82). Exciting light of wavenumber 3.19 μm^{-1} (313 nm) was more strongly absorbed and did not penetrate so far. Hence the fluorescence had to travel back through a small depth of solution and was less strongly absorbed by the naphthacene (curve C in Fig. 82).

It is possible to observe similar effects with concentrated solutions containing only one solute, if the absorption and fluorescence emission spectra of the latter overlap, since the amount of self-absorption will depend on the depth of penetration of the exciting light. Thus, when excited by frontal illumination with light of wavenumber 2.04 μm^{-1} (490 nm) a concentrated solution of fluorescein shows a normal emission spectrum because the exciting light is all absorbed in a very thin layer at the front of the cell (Fig. 83 curve A) and little self-absorption of fluorescence occurs. If however a wavenumber of 2.73 μm^{-1} (366 nm) is used for excitation, considerable penetration of the exciting light into the solution occurs. Most of the fluorescence then has to pass back through a large depth of solution which absorbs strongly in the high wavenumber region of the fluorescence emission band. As a result the high wavenumber end of the band may now be completely absent from the observed spectrum, and a considerable degree of distortion is detectable, well beyond the wavenumber corresponding to the maximum in the true spectrum. The maximum value of the observed spectrum is therefore shifted towards lower wavenumbers by an appreciable amount (compare curve B with curve A in Fig. 83). Change in emission spectrum with change of excitation wavelength is normally taken as evidence

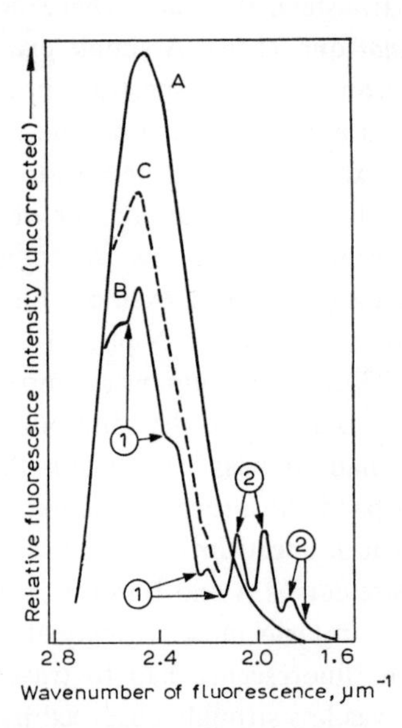

Fig. 82. Effect of second absorbing solute on the emission spectrum (frontal illumination). Curve A, 100 µg per ml of N-phenyl-2-naphthylamine in benzene and excitation with either 2.73 μm^{-1} or 3.19 μm^{-1} light; curve B, as for curve A but 40 µg per ml of naphthacene present (excitation with 2.73 μm^{-1} light); curve C, as for curve B, but with more strongly absorbed exciting light (3.19 μm^{-1}). Symbols 1 and 2 indicate, respectively, absorption and fluorescence of naphthacene. Band width of quartz prism analysing monochromator was 0.1 μm^{-1} at 2.0 μm^{-1} (from Parker and Rees[150]).

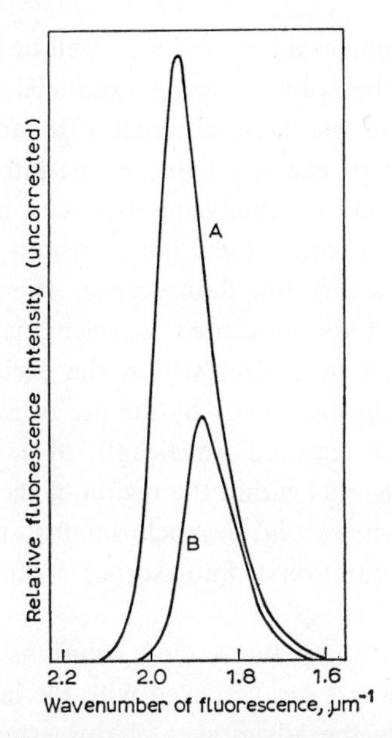

Fig. 83. Effect of self-absorption on the fluorescence emission spectrum (frontal illumination).
4×10^{-3} M fluorescein solution at pH 11; curve A, excitation at 2.04 μm^{-1} (490 nm); curve B, excitation at 2.73 μm^{-1} (366 nm). Sensitivity and intensity of exciting light were the same for both curves (from Parker and Rees[150]).

that there is more than one emitting species present, but this rule is clearly not applicable to concentrated solutions, and when applying the rule care must be taken to check by absorption measurements that an observed change in emission spectrum cannot be accounted for by inner filter effects.

4. The Method of In-Line Illumination

With dilute solutions the in-line method (*d* in Fig. 78) gives undistorted excitation and emission spectra, and a response proportional to the concentration of fluorescent solute. In this respect it is similar to the right angle and frontal methods. For measuring very weak fluorescence under these conditions it suffers from two great disadvantages. The first is that the unabsorbed exciting beam enters the analysing monochromator directly and if the latter is a single monochromator, some exciting light scattered within the monochromator will pass the exit slit at all wavelength settings, and the resulting instrumental "blank" may well be sufficient to swamp the weak emission from the solution being examined. This difficulty can be avoided by using a double monochromator to analyse the fluorescence emission, or the beam of exciting light can be intercepted by interposing between the specimen and the analysing monochromator a suitable cut-off filter that absorbs the exciting light but transmits the fluorescence. This filter must itself have a very low fluorescence. The second disadvantage is the interference caused by unwanted wavelengths in the exciting light. These can be produced by scatter within the excitation monochromator (if it is a single monochromator) or by the use of inefficient filters, if these are used to isolate the required wavelength of exciting light. Unwanted exciting light of wavelengths within the region of the fluorescence band will pass directly through the second monochromator and produce a very high "blank" reading. The question of fluorescence "blank" is discussed in more detail in Chapter 5.

When measuring strongly fluorescing solutions the interference from scattered light is much less serious, even with the in-line arrangement, and we shall discuss briefly the advantages of this arrangement and the inner filter effects that occur with it. So far as distortion of the excitation spectrum is concerned, the effects are substantially the same as those observed with frontal illumination, i.e. with solutions absorbing strongly, the excitation maximum is reduced in intensity relative to the minimum, and with complete absorption at all wavelengths the observed excitation spectrum (for wavelengths of fluorescence not absorbed by the solution) becomes a horizontal straight line. This is of course the principle on which the fluorescence quantum counter (Section 3 E 5) is based.

Distortion of the fluorescence *emission* spectrum by self-absorption or by absorption of the fluorescence by a second solute, is greater with the in-line arrangement than with the right-angle or frontal methods because with a strongly absorbing solution, all of the fluorescence light has to pass through almost the whole depth of liquid before it is observed. Nevertheless the in-line arrangement has this advantage over the frontal arrangement, that it is much easier to set up, and the observed intensity is much less critically dependent upon the exact position of the cuvette holding the specimen. For this reason it is the preferred method for the investigation of the effect of high concentrations on the fluorescence emission of the solute. When making such measurements it is necessary to ignore that region of the spectrum subject to self-absorption. For example, Fig. 84 shows the results obtained with a series of concentrations of anthracene in aerated ethanol excited by light of wavelength 248 nm isolated from a mercury lamp by means of a quartz prism double monochromator. The spectra were analysed by means of a second double monochromator so that interference from exciting light scattered within the analysing monochromator was negligible.

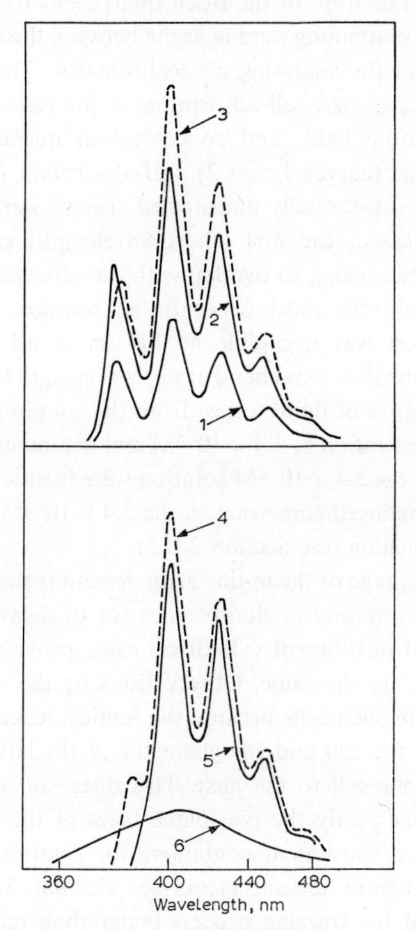

Fig. 84. Observation of self-absorption and concentration quenching with the in-line arrangement.
Anthracene in ethanol at concentrations of (1) 1×10^{-6} M, (2) 3.4×10^{-6} M, (3) 3.4×10^{-5} M, (4) 3.4×10^{-4} M, (5) 3.4×10^{-3} M; excitation at 248 nm; optical depth 15 mm. A cuvette of synthetic silica containing benzene was placed between the specimen and the entrance slit optics. Curve (6) shows the result of removing the cuvette containing benzene when pure ethanol was in the specimen cuvette.

However, one source of interference is illustrated by the curve 6 which shows the response obtained with pure ethanol in the cuvette. This emission was due to fluorescence of the *fused quartz* lens used to focus the fluorescence from the cuvette on the entrance slit of the analysing monochromator. It was eliminated by replacing the fused quartz lens by one of synthetic silica, or by preventing irradiation of the fused quartz lens by interposing a synthetic silica cuvette containing pure benzene between the sample cuvette and the focussing lens of the analysing monochromator. The remaining curves illustrate three effects, viz.: self-absorption of fluorescence, effect of high absorption of exciting light, and concentration quenching. With dilute anthracene solutions (curves 1 and 2), self-absorption of the fluorescence was negligible and substantially undistorted spectra were obtained. As the concentration increased, the first (short-wavelength) emission band was increasingly absorbed owing to overlap with the absorption band, and was completely absent in the most concentrated solution (curve 5). On the contrary, absorption was negligible at 400 nm in all solutions and the intensity of the main vibrational band at this wavelength thus gave a measure of the rate of emission of fluorescence from the solutions. Absorption was complete at a concentration of 3.4×10^{-5}M and the intensities of the 400 nm band in this and in the 3.4×10^{-4}M solution were identical (curves 3 and 4). The intensity was reduced somewhat in the 3.4×10^{-3}M solution owing to concentration quenching (see Section 2 B 2).

One further advantage of the in-line arrangement is that it permits precise comparison of the intensity of fluorescence (or of delayed emission) from solutions contained in different cylindrical cells, provided that the optical depths of the cells are the same. Observations by the right angle method can lead to errors in such cells because the luminescence is viewed through the curved face of the cell and the geometry of the latter generally varies appreciably from one cell to the next. This does not apply to the in-line arrangement in which only the two plane faces of the cell are employed. Cylindrical cells are convenient containers for solutions that have to be vacuum-degassed before measurement (see Section 3 J 2) because they generally withstand the freezing process better than rectangular cells. An application of the in-line method in the determination of triplet formation efficiencies is described in Chapter 4.

When using the in-line arrangement, it is most important to avoid scanning the fluorescence monochromator over the wavelength region of the exciting light. If this is done with a weakly absorbing solution the photomultiplier will be subjected to the full intensity of the beam of exciting light and damage may be caused. As a precaution, a protective filter such as the benzene used in Fig. 84 may be included.

5. Choice of Specimen Arrangement

It will now be clear from the previous discussion that the type of arrangement chosen will depend on the kind of specimen to be examined and the purpose for which the examination is to be made. There are three general types of specimen to be considered: (a) dilute solutions or gases, for which absorption of exciting light is small at all relevant wavelengths, (b) concentrated solutions, and (c) opaque solids and crystalline or highly-cracked rigid solutions frozen at low temperature. The choice of arrangement for opaque solids is clear; some form of frontal illumination such as that shown in Fig. 78 C must be used. Usually highly cracked or polycrystalline specimens are also best observed by frontal illumination. The choice for weakly absorbing solutions is also clear; here the sensitivity with a good spectrofluorimeter is nearly always limited not by the instrumental sensitivity itself, but by the overall "blank" due to scattered exciting light, fluorescence from the container walls, impurities in the solvent, etc. Since scattered light and fluorescence from the container are less with right angle illumination (where the photomultiplier does not "see" the illuminated walls of the container) than with all forms of frontal or in-line arrangement (where it does), the former is obviously the best choice. The choice of arrangement for moderately absorbing solutions will be decided by the nature of the investigation and the magnitude of the inner filter effects to be expected. If highly absorbing solutions *must* be measured, the choice between frontal and in-line arrangements will be decided by the degree of absorption of the fluorescence light that can be tolerated. In *all* experiments with highly absorbing solutions the distortions and errors likely to be introduced by inner filter effects must be carefully considered and the *absorption* spectrum of the solution should be measured and taken into account in deciding what exciting wavelengths to use and what the observed fluorescence means. In analytical applications it is sometimes necessary to measure concentrated solutions, e.g., when an excess reagent has to be added to convert the material to be determined into a fluorescent form. Due consideration must then be given to the interference by inner filter effects (an example is the determination of traces of boron by reaction with benzoin—see Chapter 5). In general however it is a good rule in analytical work to dilute the solution until the inner filter effects are negligible, or sufficiently small to be corrected for when right angle illumination is used. Under these conditions each fluorescent species present contributes its own spectra independently of all others.

The ideal spectrofluorimeter for general use is one in which cell compartments can readily be interchanged, so that right angle, frontal or in-line modes can be used. There is therefore a good argument for having the excitation monochromator, lamp and monitor etc. as one unit, and the analysing monochromator and photomultiplier as a second unit, so that the relative dispositions of the two units can be readily altered to suit various requirements.

J. SPECIMEN CONTAINERS AND COMPARTMENTS

1. Containers for Specimens at Room Temperature

For the measurement of dilute aerated solutions by the right angle method a rectangular cuvette with dimensions 1 × 1 cm and about 5–6 cm high is generally the most convenient. It should be optically polished on all four vertical faces to minimise light scattering. The illuminated cell faces should be screened from the analysing monochromator and the cell compartment should be designed so that the beam of exciting light after passing through the cell is dissipated as rapidly as possible (e.g. by passing into a cone lined with black velvet) and is not reflected back to illuminate parts of the cell walls visible to the analysing monochromator. To obtain the maximum sensitivity when very small amounts of material are available, a smaller cuvette may be designed of such a size as to just "fit" the focussed beam from the exit slit of the excitation monochromator. For example if the maximum width of the image of the latter is 2 mm and its height 15 mm, a cuvette 3 × 3 mm filled to a depth of 20 mm would be adequate. The volume would thus be about ten times less than that required in the 10 × 10 mm "standard" cuvette and the sensitivity correspondingly greater in terms of total weight of specimen required. Construction of very small fused silica cuvettes with plane faces is however more difficult. It is possible to use cylindrical micro-cuvettes but these introduce relatively enormous amounts of scattered exciting light and cuvette fluorescence, and for high sensitivity work they are only worthwhile if double monochromators (or single monochromators coupled with very efficient filters) are used for both the exciting light and the fluorescence light.

Glass cuvettes are reasonably transparent down to about 320 nm, depending on the wall thickness. However at wavelengths less than about 360 nm the absorption of the cell wall will introduce distortion of the excitation spectrum and below about 300 nm practically all light is cut off. Special glasses transparent down to about 250 nm can be obtained but for general purpose spectrofluorimetry some form of quartz or silica cell is best. As it is frequently necessary to work with very dilute solutions, fluorescence of the cell walls must be avoided, even with right angle illumination where it is arranged that the analysing monochromator does not view the illuminated cell faces directly. Fused natural quartz, even of "optical" quality, nearly always contains traces of impurity (e.g. aluminium) that render it fluorescent when excited with wavelengths in the region 250 nm. Synthetic silica is much purer and after suitable heat treatment it has a very low fluorescence. It is emphasised that even this pure silica is fluorescent if it has been subjected to incorrect heat treatment, and when purchasing silica cells it is desirable to specify material of low fluorescence. The wide variation in the fluorescence of various types of silica can be seen from the spectra in Fig. 85. The fluo-

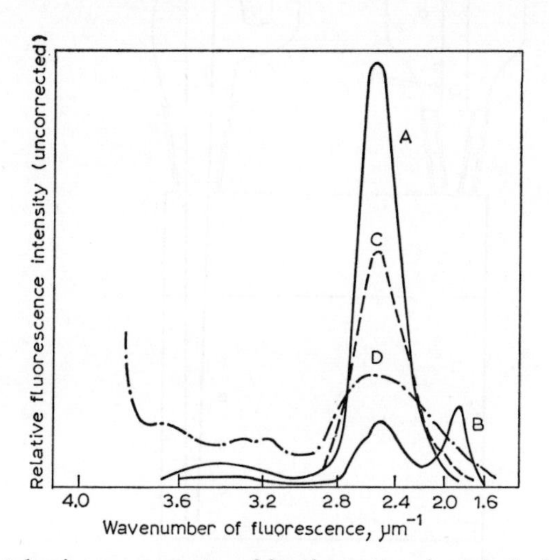

Fig. 85. Photoluminescence spectra of fused quartz and synthetic silica at room temperature.
A, optical grade fused quartz (sensitivity × 1); B, second specimen, as for A, but sensitivity × 20; C, synthetic silica (sensitivity × 540); D, second specimen of synthetic silica (sensitivity × 60,000). Excitation with 4.03 μm^{-1} (248 nm) (from Parker and Rees[150]).

rescence from the best specimen of synthetic material (curve D) was some 100,000 times lower than that from a specimen of optical quality fused quartz (curve A). An example of the effect of the fluorescence from a fused quartz cuvette when working at the limit of sensitivity is given in Section 5 C 2.

Quenching of fluorescence by dissolved oxygen has been discussed briefly in Chapter 2 and some specific examples are given in Chapter 5. Most common solvents saturated with air contain about 10^{-3}M oxygen. The degree of fluorescence quenching by this oxygen ranges from less than 1% with a compound having a strong first absorption band and low fluorescence

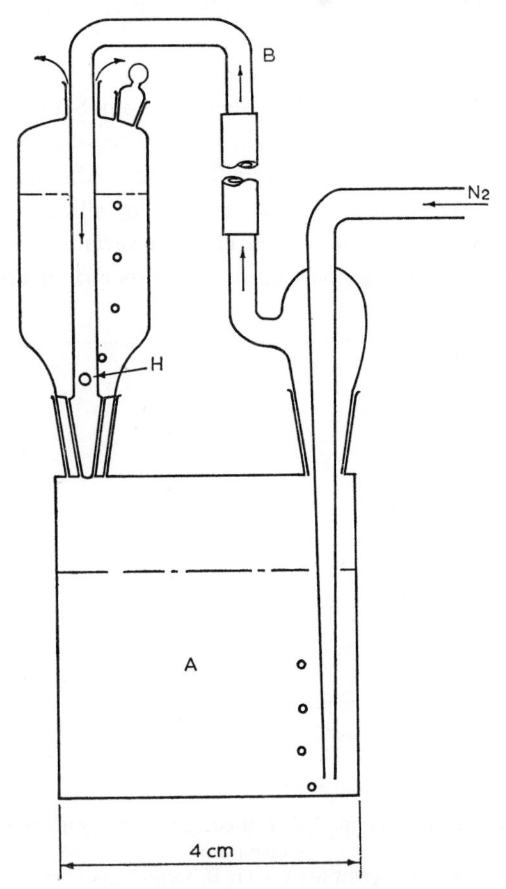

Fig. 86. Cell for reaction and measurement of photoluminescence in the absence of oxygen (from Parker and Barnes[131]).

efficiency, to 95% for a long-lived fluorescer. Oxygen quenching of phosphorescence and delayed fluorescence is very much greater than this and the special precautions required are described in Section 3 J 2. Fluorescence quenching by oxygen can be readily prevented by bubbling cylinder nitrogen (less than 10 p.p.m. oxygen) through the solution before measurement. For many purposes a loosely fitting cuvette lid carrying a glass inlet tube is sufficient. For more exhaustive de-oxygenation the cuvette should be fitted with a ground glass stopper carrying the inlet tube. Polythene or flexible metal tubing is better than rubber for connecting the cell to the gas cylinder. If precise measurements are required, or if de-aeration is prolonged, the gas must be pre-saturated with solvent vapour to prevent loss by evaporation.

In some fluorimetric reactions, e.g. the determination of traces of borate by reaction with benzoin in alkaline solution, it is necessary to de-oxygenate the reactants before mixing, to avoid the formation of fluorescent oxidation products. A cell for this purpose is shown in Fig. 86. After bubbling through the borate solution in the cuvette A, the nitrogen passes through the delivery tube B, the lower end of which is closed and acts as a liquid-tight stopper to prevent the reagent in the trap from flowing into the cuvette. The gas bubbles through the liquid in the trap from a small hole, H, in the side of the delivery tube and finally vents through the annular space at the top of the trap. When the de-oxygenation is complete the delivery tube is lifted slightly to allow the solution to flow into the cuvette.

A useful form of general purpose cell compartment for liquids at room temperature is shown in Fig. 87. It consists of a metal box with four apertures and a vertical slide by which either of two silica cells containing "standard" and "test" solutions may be moved into the beam. The contents of either cell may thus be illuminated and viewed in a variety of ways, e.g. they may be illuminated by filtered exciting light from A or by light from an excitation monochromator at F. The resulting fluorescence may be analysed by a monochromator at E or measured directly through the filters by a photomultiplier at G. By placing a photocell at A and illuminating from F, the absorption of exciting light by the solution can be measured at the same time as its fluorescence.

Precise measurement of fluorescence intensities by the frontal method requires more careful adjustment of the position of the specimen and the direction of the beam of exciting light. One suitable arrangement is shown diagrammatically in Fig. 88. The beam of light from the excitation monochromator is reflected by the front surface aluminised mirror M and focussed

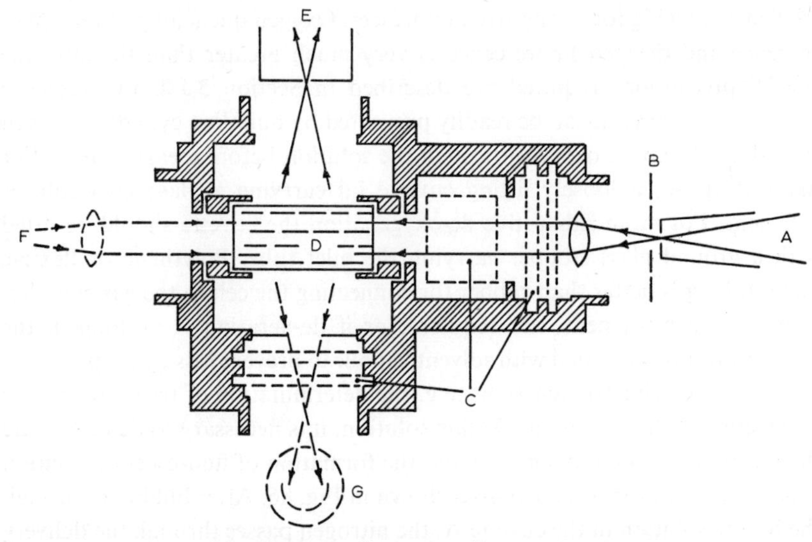

Fig. 87. General purpose cell compartment.
A, light from mercury lamp; B, chopper disc; C, glass and liquid filters; D, silica cell in vertical slide; E, monochromator for fluorescence; F, chopped exciting light from monochromator; G, photomultiplier for filtered fluorescence. For absorption measurements, G is replaced by a light source (from Parker and Barnes[131]).

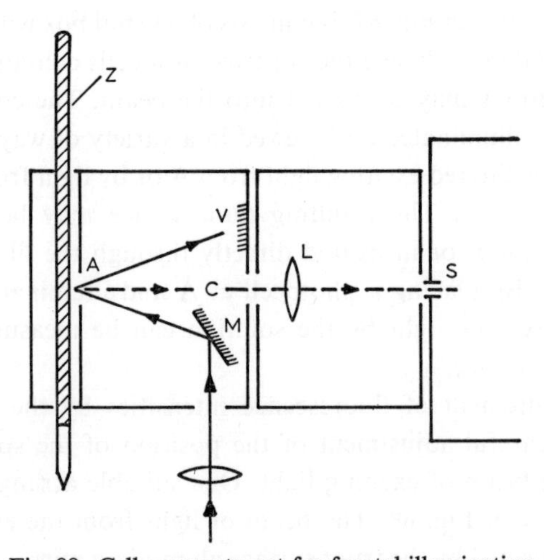

Fig. 88. Cell compartment for frontal illumination.

on to the aperture A behind which is placed the specimen to be examined. The specimen may be a cuvette of liquid, a paper chromatogram, a chromatographic column, or a column of zone refined material (e.g. Z in Fig. 88). Arrangements can be made to rotate and traverse the latter so that the distribution of segregated fluorescent impurities can be determined. The reflected beam of exciting light is absorbed by the black velvet V, or other non-fluorescent matt black surface. The fluorescence from the specimen is focussed by the lens L into the entrance slit S of the analysing monochromator. The mirror M can be rotated about a vertical axis so that the image A of the exit slit of the excitation monochromator can be adjusted slightly so as to fall on the optic axis SL. By removing the mirror M, and placing a cuvette at C, the compartment can also be used for right angle illumination, although to achieve optimum illumination of the entrance slit S the lens L must be replaced by one of shorter focal length.

As with all frontal illumination, a considerable amount of scattered exciting light passes into the analysing monochromator. For high sensitivity work both the analysing and excitation monochromators must be fitted with efficient complementary filters, or alternatively double monochromators must be used. Also the cuvette or specimen tube must have very low fluorescence.

2. Cell for Vacuum De-aeration

Both phosphorescence and delayed fluorescence in fluid solution at room temperature are quenched by exceedingly low concentrations of molecular oxygen and vacuum de-oxygenation is by far the most reliable method of achieving the necessary low oxygen concentration. The vacuum system should be kept as simple as possible, with short, wide connecting tubes to maintain an adequate pumping rate. A block diagram of a suitable system is shown in Fig. 89. The optical cell can have a variety of shapes and sizes to suit the application but for general purposes a cylindrical pyrex or silica cell, 2.5 cm in diameter and 2 cm optical depth is convenient and relatively simple to construct. It should be fitted with a long side arm of internal diameter at least 1 cm for connection to the vacuum line (see Fig. 90).

Solutions in volatile solvents are degassed as follows. The cell is first cooled in an acetone/CO_2 bath and pumped out. It is then isolated from the system by tap T_1 (Fig. 89), the contents refluxed to remove dissolved gas, cooled again in the acetone/CO_2 bath and pumped out. Tap T_1 is again

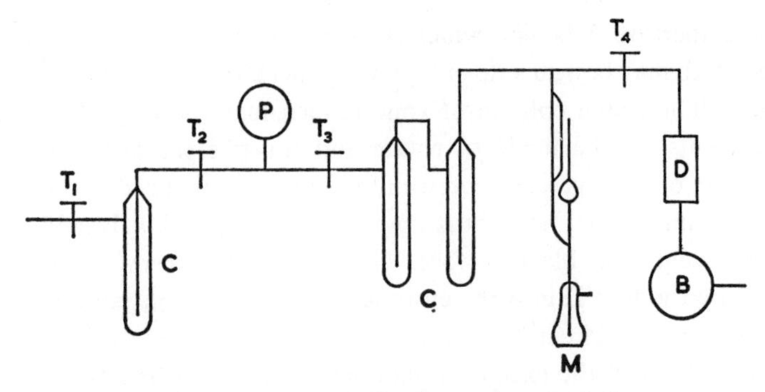

Fig. 89. Vacuum system for de-aeration of solutions.
B, D, backing and diffusion pumps; M, Macleod gauge; C, cold traps; P, Pyrani gauge; T, taps.

closed and the cell contents refluxed as before. This procedure is repeated a third time and the solution is now substantially free from CO_2 and oxygen. The last traces of oxygen are then removed by repeated cooling in an ethanol bath at $-110°C$, pumping down to less than 10^{-5} mm, warming to reflux (with tap T_1 closed) and re-cooling, etc. This is continued until the gas pressure remaining above the liquid when isolated from the pump by the tap T_3, after refluxing and recooling, is less than 2×10^{-5} mm. The cell is finally cooled, the constriction heated to softening point for several minutes while pumping, and finally sealed off. The phosphorescence or delayed

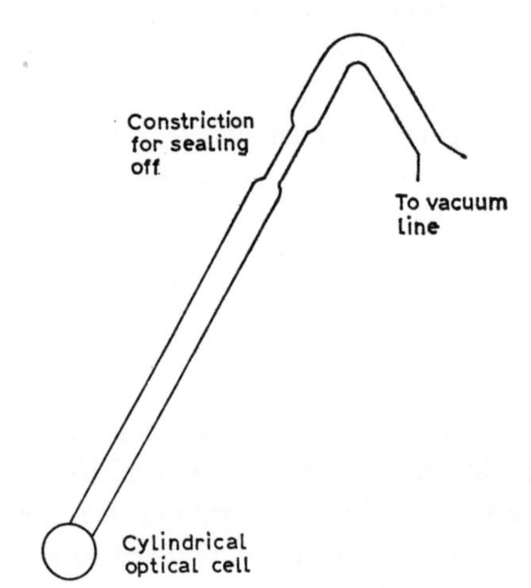

Fig. 90. Cell for vacuum de-aeration of solutions (from Parker[111]).

fluorescence should be measured immediately (see later) and again after standing overnight. Agreement between the two measurements indicates that the cell surfaces have been adequately degassed.

Relatively non-volatile liquids such as glycerol or propylene glycol cannot be refluxed satisfactorily without excessive heating and are more difficult to de-oxygenate. As much oxygen as possible is removed from the solutions before filling by passing a current of nitrogen through them at a temperature of 50°C. The same series of cooling, etc., is carried out as already described for volatile solvents, except that after each cooling and pumping, the degassing of the solution is facilitated by warming the solution to 50°C and shaking vigorously (by rotation about the greased joint).

Apiezon L grease is satisfactory for the joint on the cell, and if refluxing is carried out with care, and the cell design shown in Fig. 90 is used, there will be no detectable contamination of the solution with grease.

The cylindrical cell just described can be used for right angle, frontal or in-line illumination. For right angle measurements the cell faces can be conveniently masked by cutting a piece of carbon-black-filled polythene tube to fit round the curved surface of the cell, and having a rectangular aperture approximately 1 × 1.5 cm cut in it to let through the luminescence. The cell is irradiated through one of the plane faces and the luminescence viewed through the curved face via this aperture. The curvature will give rise to errors if it is attempted to compare the luminescence intensity of two solutions contained in different cells, but this is rarely necessary when measuring phosphorescence or delayed fluorescence, because the intensity of the latter is normally compared directly with that of the prompt fluorescence from the same solution under identical conditions. The design of cell in Fig. 90, with its long stem, is ideal for suspending in a dewar vessel when making measurements at high or low temperature.

3. Measurements at Controlled Temperature

For temperatures not too far removed from ambient, a rectangular cuvette may be seated in a hollow brass block that fits closely round the sides and bottom of the cell and has windows for the passage of exciting light and luminescence. Water or other appropriate liquid at the controlled temperature is passed through the block.

For general work at high or low temperatures a quartz dewar vessel similar to that shown in Fig. 91 is very convenient. It is furnished with

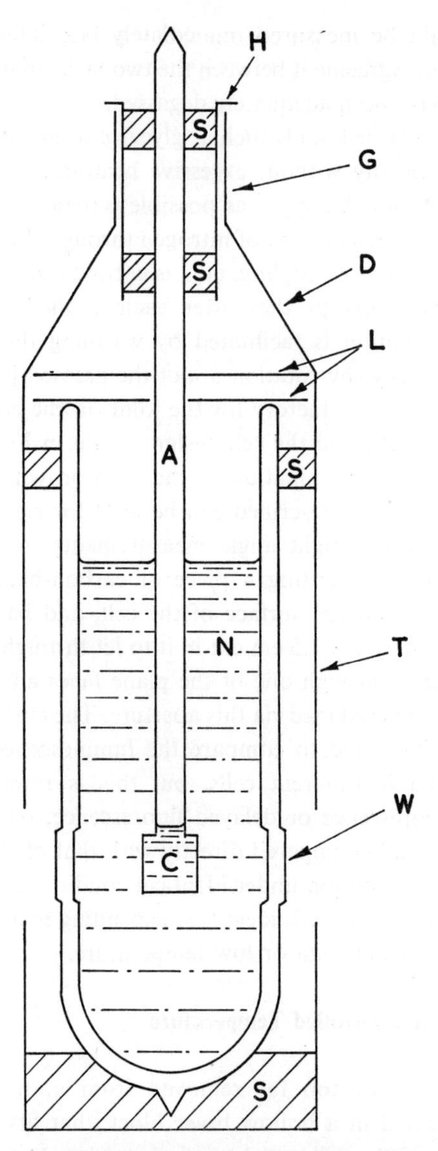

Fig. 91. Dewar vessel and cell for measurements at high or low temperatures. W, windows of quartz Dewar vessel; T, tin box; S, sponge rubber supports; D, brackets with grooves (G) to take cell holder (H); C, cuvette with long stem (A); L, loosely fitting lids; N, nitrogen or other liquid bath.

four pairs of windows situated at the same height, two pairs in line and the other two pairs also in line and in a direction at right angles to the first two pairs. It is desirable to make the dewar of comparatively large capacity so that it will take the 2 cm cylindrical cell, with room for adjustment of position. The large capacity also has the advantage that when filled with liquid nitrogen, the cell can be maintained at 77°K for several hours without replenishment of the liquid nitrogen. Such a dewar is normally purchased unsilvered and with tube attached for evacuation. It should first be carefully cleaned with chromic acid, water and ethanol. Silvering solution is then applied carefully, first to the top half and then to the bottom half, leaving an unsilvered strip at the level of the windows. After thorough washing the dewar is pumped down to a hard vacuum and heated to 280–300°C while pumping continuously for 1–2 weeks, before finally sealing off. Careful degassing at this stage to give a hard vacuum will be repaid by subsequent performance in terms of low rate of evaporation of liquid nitrogen and avoidance of misting of the windows.

The dewar is supported in a rectangular metal box (see Fig. 91) fitted at the top with brackets having grooves in which slides the Tufnol cell holder. The cell is fixed by its stem in the Tufnol holder by means of rubber gaskets. Two pairs of loose polythene covers are sufficient to prevent condensation of ice in the dewar when it is filled with liquid nitrogen.

The same arrangement can be used for controlled temperatures down to – 115°C by filling the dewar with ethanol and cooling to the required temperature by adding liquid nitrogen. The temperature is read by a thermocouple attached to the side of the cell and if the dewar has been well constructed it will be found that the temperature rises by only 2–3° during the time required to measure the fluorescence and phosphorescence spectra. For many purposes this degree of control is adequate and the arrangement is simple. A small resistance wire heater may be incorporated and the spectrum taken at a series of temperatures as the bath warms up. Above room temperature it is desirable to replace the ethanol with water. Between – 115 and – 150°C the ethanol must be replaced by isopentane.

A variety of arrangements have been described in the literature for obtaining more precise temperature control. We shall briefly describe that of Hirshberg and Fischer[151]. This was designed for absorption measurements (a more stringent requirement) and could be easily adapted for photoluminescence measurements. It consists of a quartz dewar vessel A (Fig. 92) with plane windows B and fitted at the upper end with a ground joint N. The 10 mm square quartz cuvette C is joined by means of a graded seal D to a pyrex cone. The cuvette is supported by means of the socket E of the pyrex head F. The gaseous coolant or heatant is introduced into the dewar through G by means of a hemispherical joint M and leaves the dewar through H. The cuvette is enclosed on two sides by a copper sheet envelope, and a thermocouple (inserted through another tube not shown in Fig. 92) is situated between this envelope and the cell. The thermocouple serves to measure and control the temperature by means of an indicating controller. For temperatures above 20°C, air from a blower is passed over a bare wire heating coil, the current through which is controlled by the temperature of the thermocouple. For the range – 160°C to – 10°C, liquid air is boiled by means of a small electric heater, the electrical input to which is controlled by the temperature of the thermocouple. For temperatures between – 20 and + 20°C the boiling rate of the liquid air is kept constant and the stream of cold air is passed over a heating coil, the input of which is again controlled from the thermocouple.

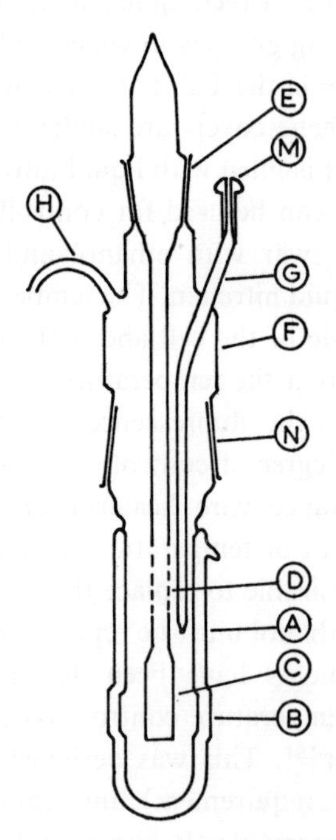

Fig. 92. Optical absorption cell for use with gaseous heating or cooling agents (from Hirshberg and Fischer[151]).

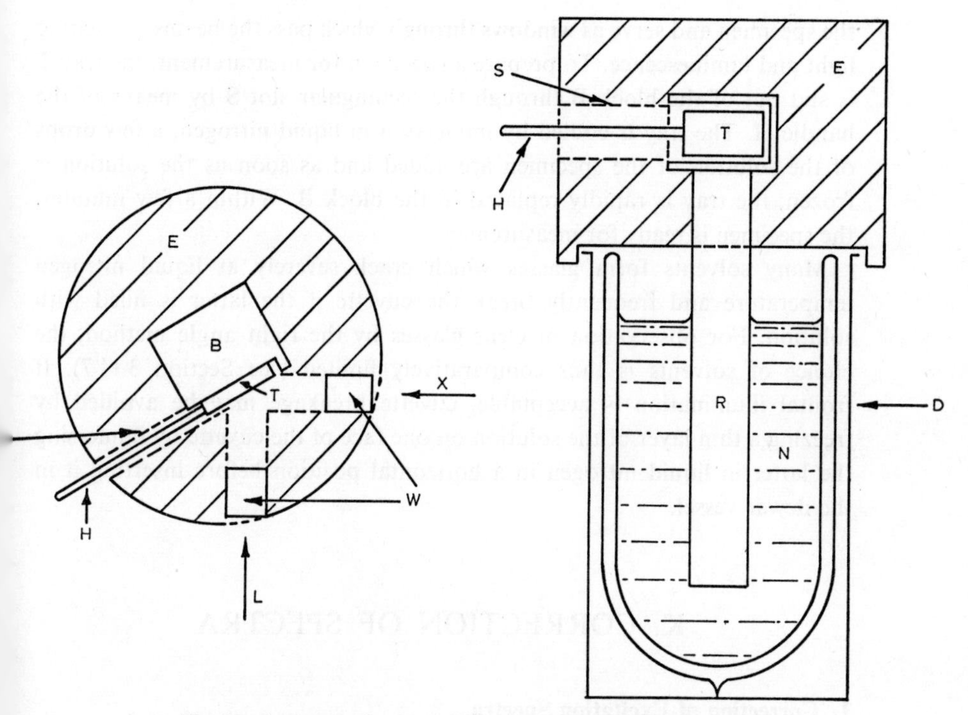

Fig. 93. Apparatus for frontal illumination at low temperature.

A device for measuring the luminescence of thin layers of rigid solutions at 77°K by frontal illumination is shown in Fig. 93. It consists of a block B, attached to a thick rod R both of high conductivity copper. A small rectangular tray T slides in grooves in the front face of the block so as to make close contact with the surface of the latter. The outer surface of the tray has a rectangular depression 2 × 3 cm and 3 mm deep, in which is placed the specimen. The tray can be removed from the block by means of the thin stainless steel handle H. The copper block is set in the middle of a large block of expanded polyurethane plastic E which serves as heat insulation. The plastic block fits over the top of a dewar vessel D containing liquid nitrogen. The copper rod dips in the latter, which abstracts heat from the copper block and the specimen in the tray T is thus cooled to near liquid nitrogen temperature. The temperature of the underside of the tray is measured by means of a thermocouple. The evacuated cylindrical silica cells W are set in the holes in the plastic block. These prevent condensation on the specimen and serve as windows through which pass the beams of exciting light and luminescence. To prepare a specimen for measurement, the tray T is slid out of the block B through the rectangular slot S by means of the handle H. The tray is cooled by immersion in liquid nitrogen, a few drops of the solution of the specimen are added and as soon as the solution is frozen, the tray is rapidly replaced in the block B. Within a few minutes, the specimen is ready for measurement.

Many solvents form glasses which crack severely at liquid nitrogen temperature and frequently break the cuvette if the latter is filled with solution. For observation of clear glasses by the right angle method, the choice of solvents is thus comparatively limited (see Section 3 N 7). If frontal illumination is acceptable, cuvette breakage may be avoided by freezing a thin layer of the solution on one face of the cuvette by immersing the latter in liquid nitrogen in a horizontal position before inserting it in the dewar vessel.

K. CORRECTION OF SPECTRA

1. Correction of Excitation Spectra

With a simple single beam spectrofluorimeter the excitation spectrum of a solution is measured by setting the fluorescence monochromator (with wide slits if necessary) to the wavelength of the peak of the fluorescence emission band of the component of the solution to be measured. The excitation monochromator is then scanned across the wavelength region of interest with the slits set to a constant value, and the variation of the intensity of the fluorescence is plotted against the wavelength or wave-number of the exciting light. The uncorrected excitation spectrum so obtained depends on the characteristics of the instrument used to measure it, and is frequently a grossly distorted version of the true spectrum. The relationship between the uncorrected excitation spectrum and the true excitation spectrum can be derived as follows. The photomultiplier output (P) is directly proportional to the total flux (**Q**) of fluorescence emitted by the specimen, and for weakly absorbing solutions the latter is given by equation 21 hence:

$$P = kQ = kI_0(2.3\varepsilon cl)\phi_f \tag{236}$$

where k is an instrumental constant. Thus for a solution of given concentration in a particular cell:

$$P \propto I_0 \varepsilon \phi_f \tag{237}$$

Now I_0 is the quantum intensity of the exciting light passed to the specimen and this will vary with the wavelength setting of the excitation monochromator according to the characteristics of the source and the monochromator. The quantity $\varepsilon\phi_f$ is a fundamental characteristic of the fluorescent solute under the particular conditions of measurement and the "true" excitation spectrum consists of a plot of $\varepsilon\phi_f$ against the wavelength or wavenumber of the exciting light. Since with many solutes the fluorescence efficiency ϕ_f is independent of the wavelength of the exciting light, the true excitation spectrum is generally a replica of the *absorption spectrum* of the solute (see Section 1 B 4 and Fig. 6).

To derive the true excitation spectrum from the recorded curve, the variation of I_0 with wavelength must first be determined by taking measurements with either a thermopile, a calibrated phototube, the ferrioxalate actinometer, or most simply with a fluorescent screen quantum counter (see Sections 3E and 3F). The method of correction is illustrated by the results in Fig. 94. The upper curve, A, represents the relative quantum intensity of the exciting light. The full curve, B, shows the absorption spectrum of quinine bisulphate, and the lower curve, C, the recorded (uncorrected) excitation spectrum. Owing to the rapidly decreasing intensity of exciting light at short wavelengths, the recorded excitation spectrum shows the second absorption band of quinine bisulphate with an intensity much lower than that of the first band, whereas in fact the second absorption band is considerably more intense. The true excitation spectrum, obtained by dividing the ordinates of curve C by those of curve A, is shown by the circles on curve B. It agrees quite closely with the absorption spectrum.

The unfavourable distribution of ultra-violet sources can often obscure important features of the excitation spectrum if corrections are not applied. For example, the absorption spectrum of alkaline fluorescein solution has an intense band in the visible region and three bands in the ultra-violet region (see full curve A in Fig. 95). The uncorrected excitation spectrum scanned at a sensitivity suitable to accommodate the visible band, did not show appreciable excitation in the ultra-violet region, and it was necessary to increase the sensitivity by a factor of 100 to record the latter. Even then the ultra-violet bands were greatly distorted. After correction to constant

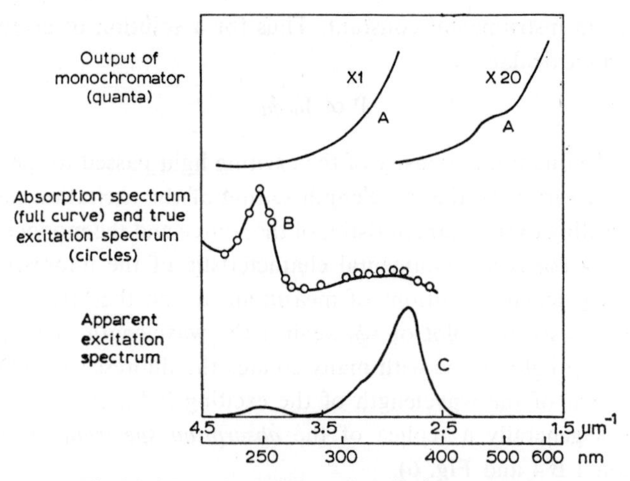

Fig. 94. Manual correction of excitation spectrum.
Quinine bisulphate in 0.1 N sulphuric acid; A, relative quantum intensity of light delivered by quartz monochromator, with wide slits, using xenon lamp; B (full line), absorption spectrum; B (circles), corrected excitation spectrum obtained by dividing the ordinates of curve C by those of curve A; C, uncorrected excitation spectrum (fluorescence intensity as a function of wavenumber setting of excitation monochromator at constant slit width) (from Parker and Rees[158]).

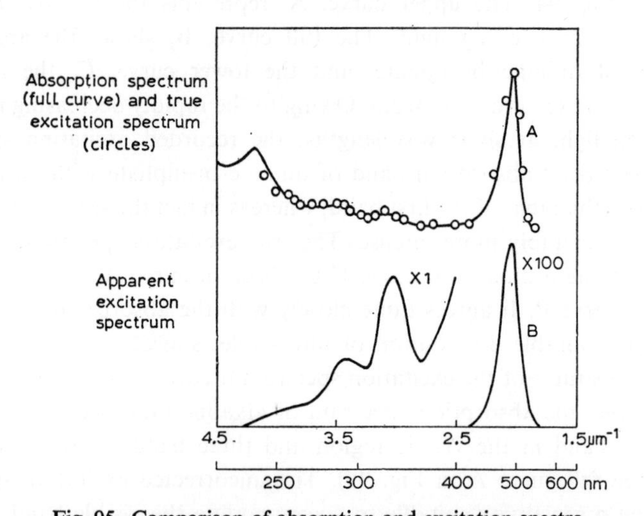

Fig. 95. Comparison of absorption and excitation spectra.
Fluorescein in carbonate–bicarbonate buffer; A (full line) absorption spectrum; A (circles) corrected excitation spectrum; B, uncorrected excitation spectrum (fluorescence intensity as a function of wavenumber setting of excitation monochromator at constant slit width) (from Parker and Rees[158]).

intensity of exciting light, the true excitation spectrum (circles on curve A in Fig. 95) was in satisfactory agreement with the absorption spectrum.

An indirect method of obtaining data for correcting the excitation spectrum has been described by Argauer and White[152]. They report that the fluorescence efficiency of an ethanolic solution of the aluminium chelate of 2,2′-dihydroxy-1,1′-azonaphthalene-4-sulphonic acid is constant over the range 250–550 nm, so that direct comparison of the uncorrected excitation spectrum of a dilute solution of this compound with its absorption spectrum provides directly the data necessary to calculate the correction curve of the excitation monochromator/lamp combination.

2. Automatic Recording of Corrected Excitation Spectra

Correction of complex excitation spectra is tedious and the accuracy depends on the spectral distribution of the source remaining constant between the time when I_0 is measured as a function of wavelength and the time when the uncorrected spectrum is recorded. An apparatus that records directly the corrected excitation spectrum (or a close approximation to it) is clearly a great advantage. The first such apparatus was described by Parker[143] and its principle is illustrated in Fig. 96. It makes use of the fluorescence quantum counter (Section 3 E 5) and the method of ratio

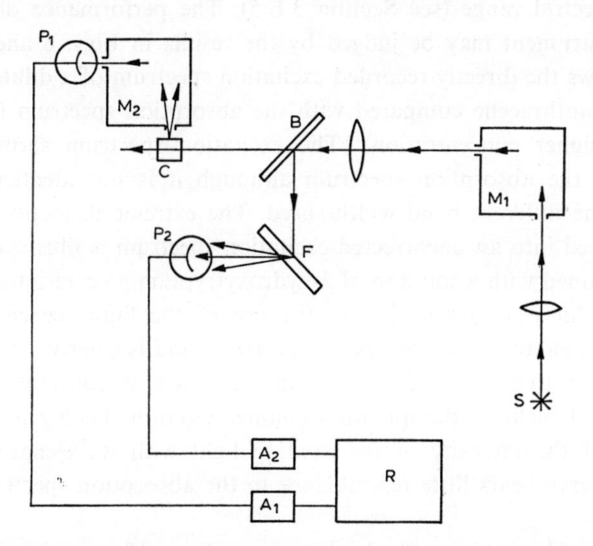

Fig. 96. Arrangement for recording corrected excitation spectrum (from Parker[143]).

recording (Section 3 G 3). The exciting light from the monochromator M_1 is focussed on the specimen in C, but a proportion of the beam is reflected by the clear silica plate B on to the fluorescence quantum counter F viewed by the photomultiplier P_2. A band of fluorescence from the specimen in C is selected by the monochromator M_2 and is received by the photomultiplier P_1. The amplified outputs of P_1 and P_2 are passed to the ratio recorder R (see Fig. 77 for details). As the wavelength setting of M_1 is varied, the slit widths are adjusted so as to maintain the output of P_2 approximately constant. Since at all wavelength settings of M_1,

$$P_2 \propto I_0 \tag{238}$$

(see Section 3 E 5) and

$$P_1 \propto I_0 \varepsilon \phi_f \tag{239}$$

the recorder measures:

$$P_1/P_2 \propto \varepsilon \phi_f \tag{240}$$

i.e. the corrected excitation spectrum. In his original experiments Parker used a solution of 4.4×10^{-3}M fluorescein in a mixture of sodium carbonate and bicarbonate (both 0.1 N) and found that this monitored the quantum intensity of the beam of exciting light to within $\pm 10\%$ over the range 230–450 nm. This accuracy is adequate for many purposes but the rhodamine B quantum counter, as recommended by Melhuish[141] is better and covers a wider spectral range (see Section 3 E 5). The performance of Parker's original instrument may be judged by the results in Figs. 6 and 97. The former shows the directly recorded excitation spectrum of a dilute solution of 1,2-benzanthracene compared with the absorption spectrum (measured at much higher concentration). The excitation spectrum shows all the features of the absorption spectrum although it is not identical with it owing to the different band widths used. The extreme distortion that can be introduced into an uncorrected excitation spectrum is illustrated by the results obtained with a solution of 5-hydroxytryptamine creatinine sulphate (Fig. 97). Curve 1, obtained with the use of the fluorescence quantum counter, is similar to the absorption spectrum, and is clearly a much more useful record than curve 2, which was recorded at constant slit width without the benefit of the quantum counter monitor. Owing to the rapid variation of the intensity of the exciting light with wavelength, the uncorrected curve bears little resemblance to the absorption spectrum of the compound.

A variety of instruments for the direct recording of fluorescence excitation

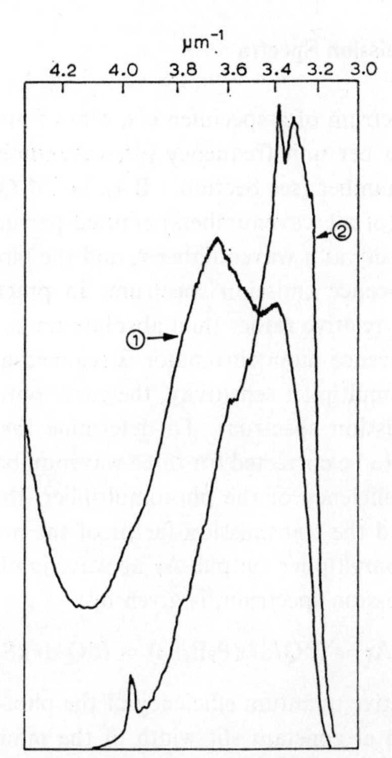

Fig. 97. Distortion of uncorrected excitation spectrum in the ultra-violet region. 5-Hydroxytryptamine creatinine sulphate in water; excitation spectrum recorded with, (curve 1), and without, (curve 2), the quantum counter monitor; fluorescence observed at 2.85–3.0 μm^{-1} (from Parker[143]).

spectra have since been described[153–157]. Some utilise quantum counters and thus record $\varepsilon\phi_f$ directly as described above. Some use a thermopile to monitor the beam of exciting light and thus record the quantity $\varepsilon\phi_f\lambda$, a less significant quantity than $\varepsilon\phi_f$. Others use a photomultiplier, thermopile or other device to monitor the beam of exciting light and apply corrections electronically for the variation of quantum efficiency of the monitoring device with wavelength, so that a precise curve of $\varepsilon\phi_f$ is recorded. For chemists who wish to construct their own apparatus, the method described by Parker is simple to set up and records corrected excitation spectra with an accuracy ($\pm$ 10%) that is sufficient for most purposes. The recorded spectra can when necessary be precisely corrected by subsequent calculation if the response of the quantum counter is determined by the methods described in Section 3 E 5.

3. Correction of Emission Spectra

The emission spectrum of a specimen is a plot of luminescence intensity, measured in quanta per unit frequency (or wavenumber) interval, against frequency or wavenumber (see Section 1 B 4), i.e., if **Q** represents the total number of quanta (of all wavenumbers) emitted per unit time, then $d\mathbf{Q}/d\bar{\nu}$ represents the intensity at a wavenumber $\bar{\nu}$, and the plot of $d\mathbf{Q}/d\bar{\nu}$ against $\bar{\nu}$ is the true luminescence emission spectrum. In practice the spectrum is normally plotted in relative rather than absolute units.

When the fluorescence monochromator is scanned at constant slit width and constant photomultiplier sensitivity, the curve obtained is the apparent or uncorrected emission spectrum. To determine the true spectrum, the apparent curve has to be corrected for three wavenumber-dependent factors, viz. the quantum efficiency of the photomultiplier, the band width of the monochromator and the transmission factor of the monochromator. Thus the observed photomultiplier output $A_{\bar{\nu}}$ at wavenumber $\bar{\nu}$, corresponding to the apparent emission spectrum, is given by:

$$A_{\bar{\nu}} = (d\mathbf{Q}/d\bar{\nu})(P_{\bar{\nu}}B_{\bar{\nu}}L_{\bar{\nu}}) = (d\mathbf{Q}/d\bar{\nu})(S_{\bar{\nu}}) \qquad (241)$$

where $P_{\bar{\nu}}$ is the relative quantum efficiency of the photomultiplier, $B_{\bar{\nu}}$ is the relative band width at constant slit width of the monochromator, and $L_{\bar{\nu}}$ is the fraction of light transmitted by the monochromator. The true emission spectrum can be calculated from the observed curve by dividing each ordinate by the corresponding value of $S_{\bar{\nu}}$. The quantity $S_{\bar{\nu}}$ is the spectral sensitivity factor of the monochromator–photomultiplier combination (including entrance slit optics): it is proportional to the photomultiplier output that would be observed if the entrance slit were illuminated by a source having constant spectral distribution. If therefore a diffuse source of known spectral distribution is used to illuminate the entrance slit of the monochromator in such a way that the collimator is filled with light, and the photomultiplier response (R_{SL}) recorded as a function of wavenumber, then $S_{\bar{\nu}}$ can simply be determined from:

$$S_{\bar{\nu}} = R_{SL}/(d\mathbf{Q}/d\bar{\nu})_{SL} \qquad (242)$$

This is the method generally used for calibration in the visible region, using a tungsten lamp run at a known colour temperature, from which the spectral distribution $(d\mathbf{Q}/d\bar{\nu})_{SL}$ can be calculated from Planck's equation (eqn. 226). In practice it is desirable to use the lamp to illuminate a magnesium oxide

TABLE 23

REFLECTIVITY OF MAGNESIUM OXIDE
(Adapted from Benford, Lloyd and Schwarz[159])

nm	0	20	40	60	80
700	0.997	–	–	–	–
600	0.994	0.995	0.995	0.996	0.996
500	0.986	0.988	0.990	0.991	0.992
400	0.974	0.976	0.979	0.981	0.984
300	0.92	0.925	0.93	0.94	0.95
200	–	–	0.955	0.94	0.91

screen placed at an angle in front of the entrance slit (as in Fig. 49A). The collimator is then filled with light and the correct (i.e. mean) value of $S_{\bar{\nu}}$ is obtained. The intensity values of the lamp emission must then be multiplied by the reflectivity of the magnesium oxide (see Table 23) before inserting into eqn. 242 as $(d\mathbf{Q}/d\bar{\nu})_{SL}$. Alternatively if a magnesium oxide screen is not used, several sets of measurements must be made with the lamp situated in various positions on either side of the optic axis of the instrument—so that different regions of the collimator are illuminated—and the mean values taken. This is necessary because the transmission factor of the monochromator may vary according to which area of the collimator is illuminated.

Calibration data for standard tungsten lamps are often provided in the form of energy units (watts) per unit wavelength interval (i.e. $dE/d\lambda$ according to equation 225). These must first be multiplied by the corresponding wavelengths to give quanta per unit wavelength interval, i.e., $(dE/d\lambda)\lambda$. They must then be converted to quanta per unit frequency interval by multiplying again by λ^2 (see equation 214). Thus:

$$(d\mathbf{Q}/d\bar{\nu})_{SL} = \lambda^3(dE/d\lambda)_{SL} \tag{243}$$

and the sensitivity factor $S_{\bar{\nu}}$ is given by:

$$S_{\bar{\nu}} = \frac{R_{SL}}{\lambda^3(dE/d\lambda)_{SL}} \tag{244}$$

The various stages in the calculation are shown in Fig. 98. Because the results at the various stages of the calculation are in relative units only, all curves are normalised to give the same value at some convenient wavelength.

The method just described is not usually applied in the ultra-violet region because ultra-violet lamps of accurately known spectral distribution are not readily available. Parker[160] has described a useful method for this spectral

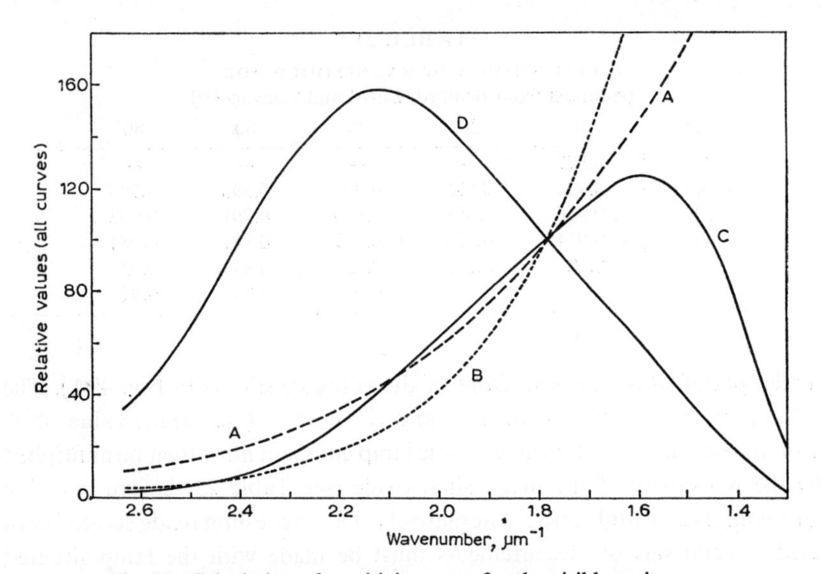

Fig. 98. Calculation of sensitivity curve for the visible region.
Glass prism spectrometer and E.M.I. 9558 photomultiplier. A, relative spectral distribution of light from standard lamp run at a colour temperature of 2856°K (energy per unit wavelength interval—$dE/d\lambda$); B, as for A, but quanta per unit wavenumber interval—$dQ/d\bar{\nu}$; C, photomultiplier output at constant slit width; D, spectral sensitivity curve—$S_{\bar{\nu}}$ (i.e., ordinates of curve C divided by ordinates of curve B). All curves are normalised to 0.56 μm (from Parker and Rees[158]).

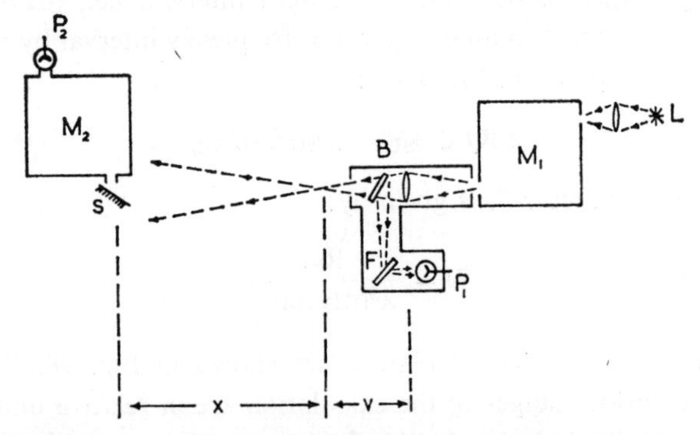

Fig. 99. Arrangement for calibrating fluorescence spectrometer in the ultra-violet region. L, xenon arc lamp; M_1, excitation monochromator; B, silica plate beam splitter; F, 0,5 mm silica optical cell containing fluorescent screen solution; P_1, monitoring photomultiplier; S, screen coated with MgO; M_2, fluorescence monochromator; P_2, fluorescence photomultiplier (from Parker[160]).

region. It makes use of the fluorescence screen monitor to measure directly the product $P_{\bar{\nu}}L_{\bar{\nu}}$. The principle is illustrated in Fig. 99. A magnesium oxide screen S is prepared by holding a piece of aluminium sheet in the smoke from a burning magnesium ribbon until it is coated uniformly with a thick layer of oxide. The coefficient of reflection of magnesium oxide in the visible and ultra-violet regions ($\sigma_{\bar{\nu}}$) has been determined by Benford and co-workers[159] and varies by only 8 % over the range 250–500 nm (see Table 23). The screen is placed at an angle in front of the entrance slit of the monochromator–photomultiplier combination to be calibrated (M_2–P_2 in Fig. 99) and situated in the centre of the diverging light from the focussed beam from the excitation monochromator M_1 which is fed by the xenon arc L. If a lens is used to focus the beam, as shown in Fig. 99, it must be an achromatic lens, otherwise a correction has to be applied for the change in focal length of the lens with wavelength[161]. Alternatively a concave mirror may be used, as shown in Fig. 74B. The distance X (Fig. 99) must be large compared with the distance V, so that the screen is uniformly illuminated with an intensity proportional to the total quantum intensity of the beam. The latter is measured by means of the fluorescence quantum counter, i.e. the monitor (B–F–P_1), which is previously calibrated by means of a thermopile or by the ferrioxalate actinomer etc. (see Sections 3 E and 3 F) to give a series of relative values of quantum efficiency of the monitor, $\phi_{\bar{\nu}}$. The procedure to be followed for the calibration will be described in the next paragraph with reference to Table 24 which shows the results obtained for a quartz prism monochromator fitted with an EMI 6256 photomultiplier.

The entrance slit of M_2 is set to a small value and the exit slit is set fully open (corresponding to a band width $\Delta\lambda$). The widths of the slits of the excitation monochromator are kept equal to one another and are adjusted as the wavelength setting of the excitation monochromator is changed, so

TABLE 24

METHOD OF CALCULATING SPECTRAL SENSITIVITY IN THE ULTRA-VIOLET REGION

μm^{-1}	$\sigma_{\bar{\nu}}$	$\phi_{\bar{\nu}}$	$R_{\bar{\nu}}$	$P_{\bar{\nu}}L_{\bar{\nu}}$	$B_{\bar{\nu}}$	$S_{\bar{\nu}}$
3.0	0.93	1.04	5.67	6.34	0.0462	0.293
3.2	0.92	1.05	6.05	6.90	0.0425	0.293
3.4	0.92	1.05	6.83	7.80	0.0390	0.304
3.6	0.91	1.06	6.78	7.90	0.0352	0.278
3.8	0.94	1.07	4.78	5.44	0.0318	0.173
4.0	0.95	1.08	3.05	3.47	0.0288	0.100

as to keep the output of the monitor constant. It must be arranged that the *full* band width passed by the excitation monochromator is always less than $\Delta\lambda$ so that the complete band of wavelengths passed by M_1 is received by the photomultiplier, P_2, when the wavelength settings of the two monochromators coincide. At each wavelength setting of the excitation monochromator, M_1, the fluorescence monochromator, M_2, is scanned across this wavelength and the maximum output ($R_{\bar{\nu}}$) of the photomultiplier P_2, is noted. Since the output of the monitor is kept constant, the entrance slit of M_2 is illuminated by a quantum intensity proportional to $\sigma_{\bar{\nu}}/\phi_{\bar{\nu}}$ of which a fraction $L_{\bar{\nu}}$ reaches the photomultiplier and gives rise to a signal $R_{\bar{\nu}}$ proportional to the photomultiplier quantum efficiency ($P_{\bar{\nu}}$). Hence:

$$R_{\bar{\nu}} \propto P_{\bar{\nu}} L_{\bar{\nu}} \sigma_{\bar{\nu}} / \phi_{\bar{\nu}} \tag{245}$$

Thus the values of $P_{\bar{\nu}}L_{\bar{\nu}}$ in column 5 are calculated from those of $\sigma_{\bar{\nu}}$, $\phi_{\bar{\nu}}$ and $R_{\bar{\nu}}$ (columns 2, 3 and 4 in Table 24). The values of $B_{\bar{\nu}}$ in column 6 are obtained by reference to the dispersion data supplied with the instrument (or they may be determined by the method described in Section 3 B 5) and hence finally the values of $S_{\bar{\nu}}$ are calculated. The same method may be used with a mercury lamp in place of the xenon lamp, by setting M_1 to the wavelengths of the principal mercury lines.

If a series of compounds are available for which the corrected fluorescence emission spectrum has been precisely determined, measurement of the uncorrected spectrum of these compounds with the instrument to be calibrated permits the direct calculation of the spectral sensitivity function by the application of equation 242, in which $(dQ/d\bar{\nu})_{SL}$ now represents the known spectral distribution of one of the standard compounds and R_{SL} the observed readings. It is of course essential that the compound be free from other fluorescent material and the fluorescence must be measured under the same conditions of temperature, solvent, concentration, pH, etc., for which the standard fluorescence curve is provided. The two compounds for which standard fluorescence spectra have been most frequently reported are quinine bisulphate in dilute sulphuric acid (either 1.0 N or 0.1 N) and anthracene (see for example Fig. 2). The former is recommended by Melhuish[162] for calibrating spectrofluorimeters in the visible region. It is most useful in the range 410–520 nm. Outside these limits the fluorescence intensity falls to rather low values compared with the maximum, and considerable care is required to obtain precise spectral sensitivity curves. In particular, "blank" fluorescence due to scattered light, Raman emission, etc. must be corrected for. Lippert and co-workers[163] have measured the corrected emission spectra of a series of compounds that between them cover almost the whole of the visible region of the spectrum. They used concentrated solutions with the method of frontal illumination. Argauer and White[152] found that three of these compounds (quinine bisulphate, 3-aminophthalimide and m-nitrodimethylaniline) gave nearly the same emission spectrum in dilute solution when measured by right angle illumination, but reported that β-naphthol shows a change in spectral shape on dilution and it was not therefore considered a good standard substance under these conditions. They recommended an ethanolic solution of the aluminium chelate of 2,2′-dihydroxy-1,1′-azonapththalene-4-sulphonic acid as an additional standard substance.

For the reasons explained in Section 1 B 4, the corrected fluorescence emission spectra are theoretically most significant when plotted on a wavenumber, or frequency basis, e.g., $d\mathbf{Q}/d\bar{\nu}$ against $\bar{\nu}$. The area under the corrected curve is proportional to **Q**, the total rate of emission of fluorescence of all wavelengths. If a grating monochromator is used to analyse the emission spectrum, the uncorrected spectrum is automatically recorded on a scale that is linear in wavelength. With such a record it is much simpler to calculate the corrected spectrum on a wavelength basis, i.e. to plot $d\mathbf{Q}/d\lambda$ against λ. Integration of the corrected spectrum will still give **Q**, the total flux of fluorescence of all wavelengths emitted. To derive $d\mathbf{Q}/d\lambda$ from the recorded curve, a different series of spectral sensitivity values, i.e., S_λ, is required. These are related to $S_{\bar{\nu}}$ by:

$$S_\lambda = S_{\bar{\nu}}\lambda^2 \tag{246}$$

Examples of spectral sensitivity curves plotted in the two ways (i.e. $S_{\bar{\nu}}$ and S_λ) are shown in Figs. 54 A and 100. It will be observed that the curve for the prism monochromator is smooth and continuous while that for the grating shows discontinuities. The latter are due to grating "anomalies". For much work, particularly if the spectra are not required to be corrected, these anomalies are of little consequence. They give rise to some additional labour in calculating a corrected spectrum and can be somewhat troublesome in instruments designed to record emission spectra directly. When comparing uncorrected spectra the presence of the anomalies must be borne in mind, because when their effect is superimposed on a broad structureless emission band, they can make it appear as though the band has some structure.

Although there are still no generally agreed methods of reporting fluo-

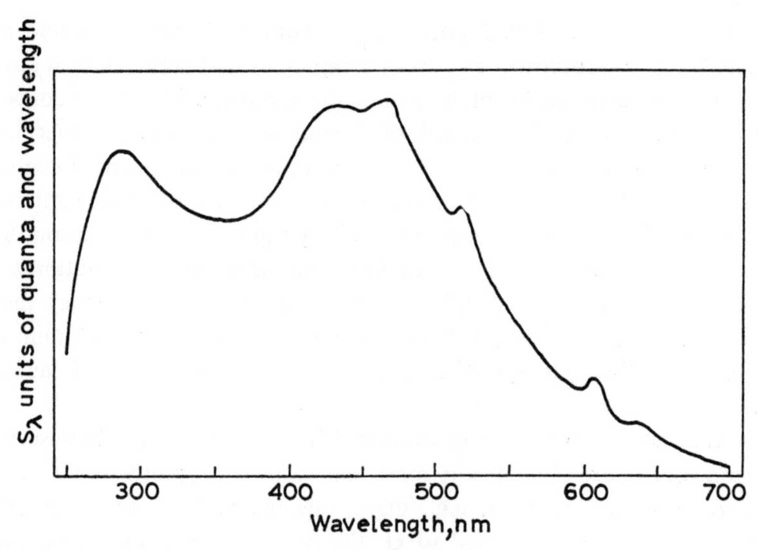

Fig. 100. Spectral sensitivity curve of grating monochromator with 9558Q photomultiplier (from Parker[111]).

rescence spectra, the proposals set forth by a group of internationally recognised experts[164] are worth consulting. Corrected spectra plotted in two of the three ways outlined in Chapter 1 are shown in Fig. 2. It should be noted that for photochemical purposes the plot of $d\mathbf{Q}/d\bar{\nu}$ against $\bar{\nu}$ is the most useful. The plot of $d\mathbf{Q}/d\lambda$ against λ is acceptable when using a grating monochromator, for reasons of convenience in calculation. Integration of the latter curve still gives the total quantum flux as already described. The method of plotting $dE/d\lambda$ against λ (see Fig. 2) has nothing to commend it for photochemical purposes because integration of the curve gives the total *energy* flux and the latter bears no simple relationship to the fluorescence efficiency.

Whatever method is used to correct the fluorescence emission spectrum, the corrected spectrum will represent the true emission spectrum only if precautions have been taken to avoid appreciable absorption of the fluorescence by the solution itself (see Section 3 H on inner filter effects).

4. Automatic Recording of Corrected Emission Spectra

Calculation of corrected emission spectra is time-consuming, particularly if the spectra show much fine structure. For this reason many spectra reported in the literature (including most in this book) are reproduced "uncorrected". For many purposes, where direct comparison, qualitative or quantitative, between a series of spectra recorded with the same instrument is all that is required, these uncorrected spectra are perfectly adequate. Nevertheless the value of the spectra to other workers—particularly when they are the spectra of pure compounds—is greatly increased if they are presented in corrected form. The labour of calculation can be avoided by the use of a computer, as described by Drushel and co-workers[165] but it is clearly a great advantage to use an instrument that records the corrected emission spectrum directly. A variety of methods for automatic correction have been reported in the literature[153–155, 166–171]. Some employ a mechanical cam coupled to the wavelength scan to provide the correction programme in the instrument; some use an electrical cam to apply the correction, either by variation of the monochromator slits, by movement of an optical wedge in the light beam, or by variation of the detector sensitivity; some workers have used a computer. We shall describe the method used by Hamilton[167] because it is a very versatile system. The correcting mechanism can be readily adjusted to fit the correction curve of the instrument, and it can be easily readjusted if the correction curve changes due to ageing of the photomultiplier etc., or if a different photomultiplier is used.

A block diagram of Hamilton's instrument is shown in Fig. 101. Light from the source (filtered as required) was chopped at 400 c/s. The fluorescence from the specimen was analysed by the monochromator–photomultiplier combination and as the slits of the monochromator were not involved in the correcting system they could be varied as required. The

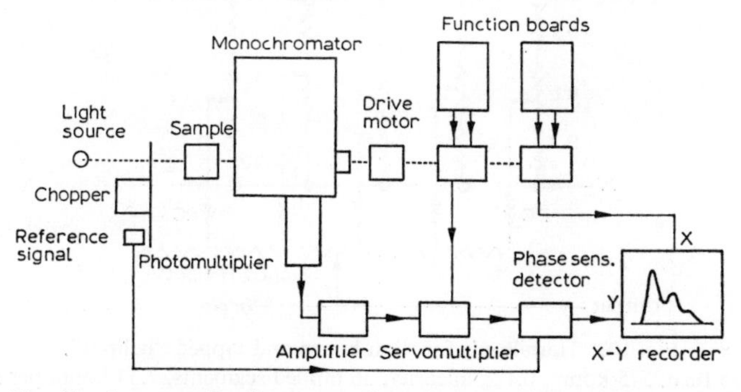

Fig. 101. Hamilton's system for direct recording of corrected fluorescence emission spectra (from Hamilton[167]).

photomultiplier output was amplified, multiplied by the sensitivity correction function and detected by means of a phase-sensitive detector, the reference signal for which was obtained from a phototransistor–lamp combination mounted on the chopper unit. The detected signal was fed to the Y axis of the recorder.

The amplified photomultiplier output had to be multiplied by the reciprocal of the spectral sensitivity curve of the monochromator–photomultiplier combination. The correction function was approximated by a series of linear segments using a tapped helical potentiometer geared to the wavelength drive (see Fig. 102). The potential of each tap was adjusted to be proportional to the correction function at the wavelength corresponding to the tap. The helical potentiometer then provided a linear interpolation between each tap. The output potential was then approximately proportional to the correction function at all wavelength settings, the accuracy depending on the spacing of the taps and the smoothness of the function. Hamilton used 36 segments and obtained a correction accuracy greater than the accuracy with which the spectral sensitivity curve itself could be determined. The voltages of the taps were derived from preset potentiometers on the function board (Fig. 102) supplied from a stabilised 5 v power supply. The function boards were printed circuit plug-in arrangements so that functions could be simply interchanged. The output from the helipot controlled a simple potentiometric servomultiplier that multiplied the photomultiplier signal by the correction function.

Hamilton used a grating spectrometer with linear wavelength drive, and by the use of a second tapped helipot coupled mechanically to the first, he

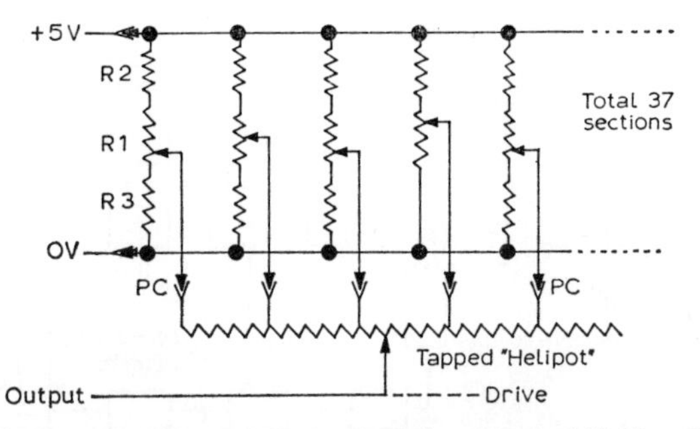

Fig. 102. Hamilton's function board and tapped "helipot".
Helipot, 15 turn, 375 kohm, 0.1% linearity, 36 tapped segments, 7.34 kohm per segment; R1, 22 ohm; R2, 3, selected to suit function; total resistance per section $\simeq$ 220 ohm; PC, printed circuit connector (from Hamilton[167]).

arranged to plot the spectra automatically with a linear wavenumber scale. The function board of the second helipot was adjusted to fit the wavelength–wavenumber conversion function and the output of the helipot was fed to the X axis of the recorder. By the use of several function boards the apparatus could be made to give scans linear in wavelength or wavenumber, with both prism or grating monochromators.

L. MEASUREMENT OF FLUORESCENCE EFFICIENCY

1. Absolute Quantum Efficiency of Fluorescence

To determine fluorescence efficiency directly it is necessary to compare the rate of absorption of exciting light with the total rate of emission of fluorescence of all wavelengths and in all directions. In principle this is simple but in practice it is a difficult experiment to perform with precision. The most direct method, and the first to be applied, was that of Vavilov[172]. This involves the comparison of the intensity of fluorescence emitted from the front surface of the specimen with the intensity of exciting light scattered from a "completely reflecting" matt surface. Many of the difficulties and correction factors in this method are avoided by enclosing the specimen or the matt surface, in an integrating sphere as described by Forster and Livingston[173]. The difficulty of comparing the quantum intensity of the exciting light with that of the polychromatic fluorescence may be overcome by using a quantum counter as detector (see Section 3 E 5) as applied by Bowen and Sawtell[174]. In view of the difficulties associated with all absolute methods, and the ease with which fluorescence efficiency may be determined by reference to a "standard" substance (see below) we shall not describe the absolute methods in detail. Some more recent publications, in addition to those already quoted are those of Förster[6] who has reviewed the methods of measurement, Melhuish[141,175], Gilmore, Gibson and McClure[176] as well as that of Pringsheim[126]. The method devised by Weber and Teale[142] is in one respect an absolute method in that comparison with a standard fluorescent solution is not required. It also has the advantage that it is applied to dilute solutions in which absorption effects are small. In this method the intensity of fluorescence emission is compared with the intensity of exciting light scattered by a dilute solution of glycogen. The proportion of the exciting beam scattered by the glycogen solution is calculated from the optical density of the latter. The method assumes dipolar scattering by the glycogen solution and in practice variations of $\pm 5\%$ in the scattering power of solutions of equal optical density were sometimes observed. This is the main source of error. Taking this and other sources of error into account the authors estimated that their values of quantum efficiency have an overall accuracy of 7–10%. This is probably as good as that attained by most other absolute methods. Table 25 shows some data taken from their paper together with that obtained by other workers using both absolute and relative methods.

2. Determination of Relative Fluorescence Efficiencies

The determination of the relative fluorescence efficiencies of two substances, both in solution at room temperature, is a relatively simple matter[158]. With a modern spectrofluorimeter of high sensitivity it is possible to work with solutions so dilute that errors due to excessive absorption of exciting light, or to self-absorption of the fluorescence are negligible. Under these conditions the total rate of emission of fluorescence is proportional to the product $I_0 \varepsilon c l \phi_f$ (see equation 21). Now the integrated area under the corrected fluorescence spectrum is also proportional to the total rate of emission of fluorescence (see Section 3 K 3) and thus, if the fluorescence emission spectra of two solutions are measured with the same instrumental geometry (this implies also the same solvent—see below) and at the same intensity of exciting light, the ratio of the two *observed* fluorescence intensities is given by:

$$\frac{F_2}{F_1} = \frac{Q_2}{Q_1} = \frac{\text{area 2}}{\text{area 1}} = \frac{I_0 \varepsilon_2 c_2 l \phi_2}{I_0 \varepsilon_1 c_1 l \phi_1} = \left(\frac{\phi_2}{\phi_1}\right) \frac{\text{optical density of 2}}{\text{optical density of 1}} \tag{247}$$

If the absolute fluorescence efficiency (ϕ_1) of one of the substances is known, that of the other is then simply calculated. In practice, it is most convenient to record the two uncorrected spectra in separate experiments, and to correct them as described in Section 3 K 3. The areas under the corrected curves are then measured and for each substance the factor (area of corrected spectrum)/(uncorrected peak height) is calculated, using the same units (e.g. recorder chart divisions) for both. The two solutions are then accurately compared by recording the fluorescence intensities at their respective peak heights. The ratio of the observed peak intensities multiplied by the ratio of the factors gives the value of F_2/F_1 for insertion in equation 247.

The ratio of fluorescence intensities measured by the instrument (F_2/F_1) is equal to the ratio of the absolute rates of emission of fluorescence by the two solutions (i.e. $\mathbf{Q}_2/\mathbf{Q}_1$) only if the geometrical arrangement of the specimen and optics is identical for the two measurements. A change in the refractive index (n) of the solution results in a variation in the angles of the rays emerging from a plane cuvette–air interface[184]. Thus if the two substances to be compared are dissolved in different solvents the observed intensities must be corrected by multiplying by n^2. When quinine bisulphate in aqueous 0.1 N sulphuric acid is used as the standard substance, the overall factor n^2(solvent)/n^2(water), amounts to a correction of 27% for benzene and 5.5% for ethanol.

3. Precautions

In measuring relative fluorescence efficiencies, care must be taken to avoid errors due to inner filter effects, non-monochromatic exciting light, fluorescence "blank", oxygen quenching and photodecomposition. Errors due to the first two factors are in principle simple to avoid, but in practice they are easily overlooked. Equation 21 applies strictly only if the proportion of exciting light absorbed by the solution is negligible. An optical density of 0.01 at the depth from which the fluorescence is viewed (see Section 3 H 2) will introduce an error of 2.3% into the intensity of fluorescence observed. Whenever possible the optical density should not exceed this value. If the fluorescence efficiency and instrumental sensitivity are low, so that higher optical densities have to be used, correction may be made for absorption of the exciting light (Section 3 H 2), or better, the optical densities of the two solutions may be adjusted to the same value. This should not exceed 0.4 per cm, if a 1 cm cell is viewed at its mid-point (i.e. effective optical density 0.2). The use of such high optical densities in the measurement of relative efficiencies of prompt fluorescence is however rarely necessary with modern sensitive equipment.

When using dilute solutions special care is required in determining the values of optical density for insertion into equation 247. To obtain sufficiently accurate results with solutions having an optical density of 0.01 per cm, a 10 cm cuvette must be used. Care must then be taken to ensure that the solutions are free from traces of turbidity or suspended dust particles, otherwise an appreciable error in the optical density measurement may be introduced in a 10 cm depth of solution. Alternatively the optical density per cm may be calculated from measurements on more concentrated solutions, but it is then necessary to confirm that Beer's law is obeyed by measuring a series of solutions of varying concentration.

If there is an appreciable overlap between the absorption and fluorescence emission spectra, care must be taken to avoid errors due to self-absorption of fluorescence (see Section 3 H 2). The simplest way is to use solutions sufficiently dilute that self-absorption is negligible. If however it is desirable to make the measurements on more concentrated solutions, it is generally possible to establish the true shape of the fluorescence emission spectrum by separate measurements with a more dilute solution. The relationship between the area of the corrected spectrum and the intensity at a wavelength (λ) outside the region of self-absorption can thus be established. This relationship can then be used to derive the corrected area of the emission from the more concentrated solution by measuring the intensity of the latter at the same wavelength (λ). Such a measurement does not however correct for the increased emission resulting from excitation by the fluorescence light absorbed in the region of overlap.

Large errors can be introduced in the determination of fluorescence efficiency if the exciting light is not monochromatic. For example, the mercury line at 313 nm is weakly absorbed by anthracene, the absorption of the mercury line at 254 nm being over 100 times as great. Thus, if the 313 nm line is isolated by filters that transmit 1% of the 254 nm line, the observed intensity of fluorescence will be about double that which would be observed if pure 313 nm radiation were used for excitation. Since the comparison substance will not in general, show the same difference in absorption of the two lines, a considerable error will be introduced into the measured ratio of fluorescence efficiencies. It should be noted that even when a monochromator is used to isolate a mercury line, the scattered light of unwanted wavelengths may account to an appreciable proportion of the total light transmitted. It is therefore always desirable to check the purity of the exciting light. This can be done by measuring the spectral distribution of the light after reflection from a magnesium oxide surface, or more simply but less precisely, by filling the cuvette with a slightly turbid, but otherwise transparent, solution and measuring the spectrum of the scattered light.

Quite large errors can be introduced in the measurement of optical density if the absorption spectrum is very steep in the region concerned. Many of the mercury "lines" consist of groups of closely spaced lines—or a broadened line if a high pressure lamp is used. Light isolated by a monochromator from a xenon lamp generally contains an even broader band of wavelengths. If therefore the absorption spectrum is very steep it is desirable to measure the optical density of the solution with the same beam of light as that used to excite the fluorescence. It should be noted however that this procedure does not fully correct for the fact that the exciting light is not completely monochromatic. In particular it does not correct for the presence of a small proportion of light of greatly different wavelength for which the extinction coefficient is many times greater, as in the example of anthracene described in the previous paragraph. Because of the difficulties of deriving satisfactory absorption factors for polychromatic light, mercury lamps are preferred to xenon lamps when measuring fluorescence efficiencies.

If the fluorescence measurement is made with a very dilute solution and the fluorescence efficiency is low, the fluorescence "blank" may contribute appreciably to the total measured emission. The factors that may contribute to the fluorescence "blank" are discussed in detail in Chapter 5. They include impurities in the solvent, Raman emission, scattered exciting light, cuvette fluorescence, etc. It is always desirable to check the fluorescence blank by making measurements on the pure solvent under the same conditions as those used to measure the solution. If appreciable, the recorded spectrum of the "blank" must be deducted from that of the solution before the latter is corrected.

With substances emitting relatively long-lived fluorescence, quenching by dissolved air may be considerable. It may generally be eliminated by passing a current of nitrogen through the solution before measurement.

4. Standard Fluorescent Substances

One of the most widely used standard solutions for fluorescence measurement is quinine bisulphate in dilute sulphuric acid. It is used as an arbitrary standard of comparison to compensate for variations in spectrofluorimeter sensitivity (Section 3 G 1), as a standard for expressing instrumental sensitivity (Section 5 B 2), as a standard for determining the spectral sensitivity curve in the short wavelength half of the visible spectrum (Section 3 K 3) and as a standard for the determination of fluorescence efficiency (see below). Both the corrected fluorescence emission spectrum (see Fig. 2) and the absolute fluorescence efficiency in 1.0 N sulphuric acid have been carefully

TABLE 25

FLUORESCENCE QUANTUM EFFICIENCIES IN SOLUTION (20–25°C)

Compound	Solvent	Method	Exciting wavelength (nm)	ϕ_f	Ref.
Acenaphthene	Ethanol	A_a(0.30)	313	0.39	104
	Cyclohexane	DPA(1.0)	303	0.60	177
Acridine orange HCl	Ethanol	A_a(0.30)	366	0.46	36
Anthracene	Benzene	Absolute	366	0.26	175
	Benzene	Absolute	366	0.29	142
	Benzene	Q(0.55)	366	0.24	103
	Ethanol	Absolute	366	0.27	175
	Ethanol	Absolute	254	0.30	142
	Ethanol	Q(0.55)	366	0.30	103
	Ethanol	Q(0.55)	366	0.30	158
	Hexane	Absolute	254	0.31	142
	Hexane	Q(0.55)	366	0.29	103
	Cyclohexane	DPA(1.0)	254	0.36	177
Anthranilic acid	Benzene	Absolute	366	0.58	175
	Ethanol	Absolute	366	0.59	175
1,2-Benzanthracene	Ethanol	A_a(0.30)	366	0.20	36
Benzene	Ethanol	A_a(0.30)	248	0.04	160, 161
	Cyclohexane	DPA(1.0)	254	0.07	177
Benz(a)pyrene	Ethanol	A_a(0.30)	366	0.42	36
Chlorophyll a	Ethanol	Absolute	644	0.23	142
	Methanol	Absolute	644	0.23	142
	Methanol	Absolute		0.24	173
	Ether	Absolute	644	0.32	142
	Ether	Absolute		0.24	173
Chlorophyll b	Ethanol	Absolute	644	0.10	142
	Methanol	Absolute	644	0.10	142
	Methanol	Absolute		0.06	173
	Ether	Absolute	644	0.12	142
	Ether	Absolute		0.11	173
Chrysene	Ethanol	A_a(0.30)	313	0.17	104
	Cyclohexane	DPA(1.0)	313	0.14	177
9,10-Dimethylanthracene	Ethanol	A_a(0.30)	366	0.89	179
9,10-Diphenylanthracene	Benzene	Absolute	366	0.84	175
	Benzene	Q(0.55)	366	0.81	103
	Benzene	A_a(0.24)	366	0.80	178
	Ethanol	Absolute	366	0.81	175
	Ethanol	Q(0.55)	366	0.76	103
	Ethanol	A_a(0.29)	366	1.0	23
	Ethanol	A_a(0.30)	366	0.89	179
Eosin	Water	Absolute	366	0.16	173
	Aq. NaOH	Absolute	366	0.19	142
	Aq. NaOH	RB(0.73)	467	0.23	158
	Aq. NaOH	RB(0.73)	492	0.21	158
	Aq. NaOH	RB(0.73)	366	0.23	158
Fluoranthene	Ethanol	A_a(0.30)	366	0.21	36
Fluorene	Ethanol	Absolute	254	0.54	142
	Hexane	Absolute	254	0.54	142
	Cyclohexane	DPA(1.0)	265	0.80	177

TABLE 25 (continued)

Fluorescein	Aq. NaOH	Absolute	366	0.92	142
	Aq. NaOH	Absolute	436	0.79	173
	Aq. NaOH	Absolute		0.84	172
	Aq. NaOH	Absolute		0.85	180
	Aq. NaOH	Absolute		0.85	181
	Aq. NaOH	RB(0.73)	467 or 492	0.85	158
1-Methoxy-naphthalene	Ethanol	A_a(0.30)	313	0.53	104
N-Methyl-acridinium chloride	Water	Absolute	366	1.01	142
2-Methylanthracene	Ethanol	A_b(0.29)	366	0.24	23
9-Methylanthracene	Ethanol	A_a(0.30)	366	0.33	179
	Ethanol	A_b(0.29)	366	0.38	23
	Cyclohexane	DPA(1.0)	254	0.35	177
Naphthalene	Ethanol	Absolute	254	0.12	142
	Ethanol	A_a(0.30)	313	0.21	104
	Hexane	Absolute	254	0.10	142
	Cyclohexane	DPA(1.0)	265	0.23	177
2-Naphthylamine	Benzene	Absolute	366	0.50	175
Perylene	Benzene	Absolute	366	0.89	175
	Benzene			0.98	182
	Ethanol	Absolute	366	0.87	175
	Cyclohexane	DPA(1.0)	254	0.94	177
Phenanthrene	Ethanol	Absolute	254	0.10	142
	Ethanol	A_a(0.30)	313	0.13	104
Phenol	Ethanol	A_a(0.30)	248	0.16	160, 161
	Water	Absolute	254	0.22	142
	Cyclohexane	DPA(1.0)	265	0.08	177
9-Phenylanthracene	Ethanol	A_a(0.30)	366	0.49	179
	Ethanol	A_b(0.29)	366	0.52	23
	Isopropanol	Q(0.55)	366	0.46	103
	Cyclohexane	DPA(1.0)	365	0.49	177
Proflavine hydrochloride	Ethanol	A_a(0.30)	366	0.40	36
Pyrene	Ethanol	A_a(0.30)	313	0.72	104
	Cyclohexane	DPA(1.0)	313	0.32	177
Quinine bisulphate	1.0 N H_2SO_4	Absolute	366	0.55	175
Rhodamine B	Ethanol	Absolute	535	0.97	142
	Ethanol	Q(0.55)	366	0.73	158
Rubrene	n-Heptane	Absolute	436	1.02	173
Thionine	0.1 N H_2SO_4	RB(0.73)	546	0.02	158
Toluene	Hexane		254	0.23	183
	Cyclohexane	DPA(1.0)	265	0.17	177
Triphenylene	Ethanol	A_a(0.30)	313	0.09	104
	Cyclohexane	DPA(1.0)	265	0.08	177

Notes

Values in column 3 refer to quantum efficiencies assumed for the following standard solutions:

- A_a anthracene in ethanol.
- A_b anthracene in benzene.
- DPA 9,10-diphenylanthracene in cyclohexane.
- Q quinine bisulphate in dilute H_2SO_4.
- RB rhodamine B in ethanol.

determined by Melhuish[162,175] who reports a quantum efficiency of 0.508 at a concentration of 5×10^{-3}M with a self quenching constant of 14.5, corresponding to an efficiency at infinite dilution of 0.546 at 25°C or 0.553 at 20°C. The latter value applies up to a concentration of about 10^{-3}M. Standard quinine bisulphate (mol. wt. 548.6) is best prepared by recrystallising the laboratory grade chemical several times from water. It crystallises with 7 molecules of water of crystallisation which may be removed by drying *in vacuo* over silica gel. The anhydrous material is hygroscopic. The heptahydrate is efflorescent.

The fluorescence efficiencies of a selection of substances in various solvents is given in Table 25. Several of these could be used in place of quinine bisulphate as the standard substance. Anthracene is attractive as a standard because it is readily available and can be purified exhaustively by zone refining. Parker and Rees[158] have investigated several other substances as possible standards for fluorescence measurements. Alkaline solutions of eosin were found to be unsuitable because of rapid decomposition. Strongly alkaline fluorescein solutions decomposed slowly but weakly alkaline solutions (pH 9.6 in carbonate–bicarbonate buffer) were much more stable. However the absorption at 366 nm is low so that fairly large errors are introduced by the presence of traces of impurities absorbing at this wavelength. Rhodamine B in ethanol is also stable and absorbs more strongly at 366 nm but less at 436 nm.

Values of relative fluorescence efficiency can be determined more accurately than absolute values, and thus, as relative values are accumulated, it should be possible to correlate them with the absolute values available, and ultimately to correct the latter where appropriate. Corrections to the absolute values would clearly be necessary if a substance were found that gave an apparent fluorescence efficiency (measured against an existing standard) considerably greater than unity. Although the values recorded for some substances under the same conditions vary appreciably (see Table 25), it is unlikely that the average values for the most widely investigated substances (e.g. quinine bisulphate or anthracene) will ultimately be found to be in error by more than 10%. Among the highest generally accepted values are those for 9,10-diphenylanthracene (0.76–1.0), alkaline fluorescein (0.79–0.92) and perylene (0.87–0.98). Some indirect confirmation of high values will be obtained as more triplet formation efficiencies are determined (see Section 4 A 4–6). For example the low triplet formation efficiency of perylene is consistent with its high fluorescence efficiency.

M. MEASUREMENT OF FLUORESCENCE LIFETIME

Apparatus for the direct measurement of fluorescence decay rates are of two general kinds—phase fluorometers and flash fluorometers. A variety of equipment of both types has been described in the literature. We shall outline briefly the principles of the methods and give sufficient references to provide a starting point for a more exhaustive study.

1. Phase Fluorometers

In phase fluorometers the fluorescence is excited by means of a light beam modulated at high frequency. The phase, or the degree of modulation, of the fluorescence is then compared with that of the exciting light. For fluorescence decaying exponentially with lifetime τ the phase ξ of the fluorescence relative to that of the exciting beam is given by:

$$\omega\tau = \tan \xi \tag{248}$$

where ω is the angular rate of the modulation[185, 186]. The signals from the source and the fluorescing specimen are fed to a circuit which indicates some fixed phase relationship between them. A known delay is introduced between the two signals so as to "zero" the indicating circuit. For example Bailey and Rollefson used a circuit that effectively subtracted the two signals and thus gave a minimum output when the two signals were in phase. Other types of circuit have also been used[185, 187–189].

An alternative method of measuring ξ is to determine the degree of modulation of the two signals[191]. Clearly the longer the lifetime of the fluorescence the lower will be its degree of modulation. The degrees of modulation in the exciting beam (m_s) and in the fluorescence beam (m_f) are related to ξ by:

$$m_f/m_s = \cos \xi \tag{249}$$

The degree of modulation, m_f or m_s, is determined by measuring the ratios of the AC (i_a) to the DC (i_d) components in the beam concerned:

$$m = i_a/i_d \tag{250}$$

For exponential decay, values of ξ obtained by the two methods give good agreement. Disagreement is taken to indicate non-exponential decay. One of the main disadvantages of the phase or modulation fluorometer is that it is difficult to interpret the results with non-exponential decay, although Birks, Dyson and Munro[192] have applied the method to such measurements.

Early phase fluorometers made use of polarised exciting light and the Kerr effect to modulate a light beam. This involves the application of a strong high frequency field across a liquid or crystal exhibiting the electro-optical effect. Later methods of modulation have used an ultrasonic standing-wave diffraction grating (formed in a liquid by means of a quartz crystal transducer), through which the steady beam of exciting light passed[186,187,189,190]. A simpler method[185] uses an air discharge tube driven by a high frequency oscillator to produce modulated light directly. An even more satisfactory method is to use a hydrogen lamp fed from a suitably modulated current source[191,193].

2. Flash Fluorometers

Pulse source fluorometers provide the most direct and perhaps simplest methods of determining fluorescence lifetimes. The specimen is illuminated with a source giving a light flash of duration less than that of the lifetime to be determined. The decay of the fluorescence is then recorded by means of a fast oscilloscope. The principle is thus similar to that used for the determination of the lifetimes of phosphorescence and delayed fluorescence in the millisecond region although, of course, mechanical interruption of the light beam is not possible at the frequencies required for the measurements of the fast rates of decay of prompt fluorescence. Two alternative procedures have been used. In the first the photomultiplier is pulsed to give high gain for a period of several lifetimes and the voltage output corresponding to the emission is presented directly on the Y axis of an oscilloscope and photographed[194]. Usually several thousand traces are superimposed to obtain sufficient intensity for photographic integration. Birks, King and Munro[195] used a sampling oscilloscope and plotted the trace on an X–Y recorder. In the alternative procedure the photomultiplier is pulsed to yield a moderate gain for a period less than one lifetime. The source and photomultiplier are triggered synchronously at a repetition frequency of a few thousand cycles per second and the decay curve of the emission is determined by changing the delay between the times of excitation and observation. The output from the photomultiplier, after amplification, is presented on a recorder as a function of time interval[196,197].

For use with the flash technique, light sources must give light pulses with a decay rate less than that of the fluorescence decay to be measured. Both hydrogen filled flash lamps[193,195,197] and spark discharges in air[198] have been used. Some applications of directly measured fluorescence decay are described later.

3. Indirect Method

Fluorescence lifetimes can also be determined indirectly from quenching experiments. In the process:

$$^1D^* + {}^1Q \rightarrow {}^1D + {}^1Q^* \qquad (251)$$

transfer of singlet energy can in general take place both by the long range dipole–dipole interaction (see Section 2 B 6) and by collisions between transferring species. With an acceptor having a low extinction coefficient such as biacetyl the long range contribution is negligible[199] and it is then possible by measuring the Stern–Volmer quenching constant for the above process (equation 75) to derive a value for $k_Q\tau_0$. In general the frequency of encounters in solution between spherical molecules of equal radii, r, is given by[181]:

$$k_c = \frac{8RT}{3000\eta}\left[1 + \frac{2r}{(\tau_0 \mathbf{D})^{\frac{1}{2}}}\right] \qquad (252)$$

D is the diffusion coefficient given by:

$$\mathbf{D} = \mathbf{k}\mathbf{T}/(3\pi\eta r) \qquad (253)$$

where η is the viscosity of the solvent.

Now the first term of equation 252 corresponds to the stationary solution of the diffusion equation (i.e., it is equivalent to equation 80). The second or transient term is small for fluid solutions at room temperature. Dubois and Van Hemert[200] made measurements of $k_Q\tau_0$ for naphthalene, benzene and 12 alkyl benzenes quenched by biacetyl, from which they calculated τ_0 from the simple diffusion equation, and also from equation 252, i.e., including the transient term. The latter gave slightly lower values which were in reasonable agreement with the values of τ_0 determined directly with a phase fluorometer by Ivanova and co-workers[201]. It seems therefore that by using biacetyl as quencher and assuming diffusion-controlled quenching, measurements of Stern–Volmer quenching constants can be used to derive values of fluorescence lifetimes. This procedure requires much simpler apparatus than that involved in the application of phase or flash fluorometer measurements.

N. MEASUREMENT OF LONG-LIVED LUMINESCENCE

Nearly all the instrumental components used for the measurement of prompt fluorescence, and many of the factors affecting their application, apply equally to the measurement of long-lived luminescence. We shall discuss here only the modifications to equipment required for the measurement of long-lived luminescence and the additional factors that must be taken into consideration.

1. The Spectrophosphorimeter

To distinguish between prompt fluorescence, and phosphorescence or delayed fluorescence, it is necessary to interrupt the beam of exciting light periodically and to view the specimen only during the periods of darkness, i.e. when the short-lived fluorescence has completely decayed. Becquerel's original phosphoroscope[202] consisted of two circular discs mounted on the same axle. The discs had holes cut round the circumference, the holes in the first disc being offset from those in the second. The specimen was placed between the discs and when the latter were rotated the specimen was illuminated intermittently by the beam of exciting light passing through the holes in the first disc, and was viewed during the periods of darkness through the holes in the second disc. Instead of using discs, Lewis and Kasha[30] mounted the specimen inside a hollow cylindrical can having a slot cut in the circumference so that as the can rotated about its axis the specimen was alternately illuminated by the exciting light passing through the slot, and viewed through the same slot. Similar arrangements are employed in some commercial spectrophosphorimeters. With both of these arrangements only the long-lived emission can be measured—to measure the total luminescence the rotating discs, or the can, must be removed. Bauer and Baczynski[18] used a different arrangement consisting of a single disc with slots, fixed to a cylindrical can having two slots situated at 180° from one another. The whole was rotated so that the specimen was illuminated intermittently through the holes in the disc and the luminescence observed through the slots in the cylinder. The relative position of disc and cylinder could be adjusted so as to place the holes in the disc either in or out of phase with the slots in the cylinder. The photomultiplier thus observed either the sum of the short- and long-lived emission, or the long-lived emission alone.

All of these devices are somewhat inconvenient because they limit the choice of geometrical arrangement of apparatus and the size of specimen compartment that can be accommodated. The limitations arise mainly because of the need to have direct mechanical coupling between the chopper in the beam of exciting light and that in the beam of luminescence. Parker overcame this difficulty by using two chopper discs driven by separate synchronous motors[19, 203]. The phase relationship between the two discs is maintained simply by running the two motors from the same 50 cycles per second supply. By turning the body of one of the motors about the axis of rotation the choppers can be put in or out of phase, without stopping them, and immediate successive readings of long-lived emission, and of prompt plus long-lived emission can be taken. With this device it is thus possible to convert a spectrofluorimeter into a spectrophosphorimeter with no other modification than placing one chopper at the exit slit of the excitation monochromator and the second chopper at the entrance slit of the analysing monochromator (see Fig. 103). This arrangement has the advantage that

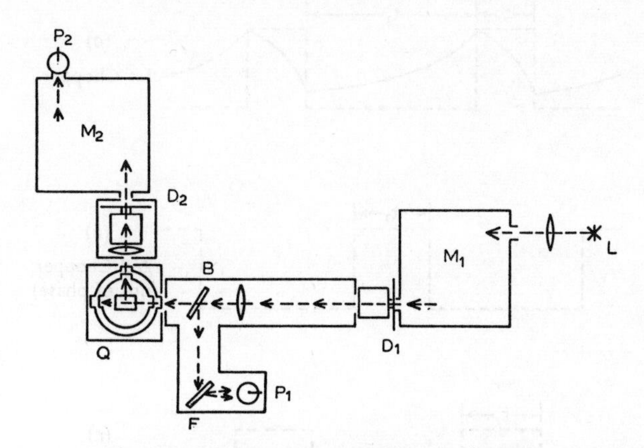

Fig. 103. Diagram of spectrophosphorimeter.

L, light source; M_1, M_2, monochromators; D_1, D_2, chopper discs driven by synchronous motors; B, silica plate beam splitter; F, 0.5 mm silica cell containing fluorescent screen solution; P_1, monitoring photomultiplier; P_2, fluorescence–phosphorescence photomultiplier; Q, fused quartz Dewar flask containing sample cell (from Parker and Hatchard[19]).

the instrument can still be used as a spectrofluorimeter by simply running the choppers in phase.

A chopping frequency of 800 cycles per second is convenient for measuring the intensity and spectrum of long-lived emission. This may be obtained by using discs having 16 slots and driven by 3000 r.p.m. motors. To avoid light leakage past the edges of the chopper blades when working out of phase, the open periods of both choppers are arranged to be somewhat less than the closed periods—an efficient arrangement is that in which the chopper in the beam of exciting light has an open period of $\frac{1}{4}$ cycle and that in the beam of luminescence an open period of $\frac{1}{3}$ cycle. The principle of operation of these choppers is shown in Fig. 104. The upper diagram (*a*) represents in ideal form the phase of the first chopper, i.e. the dotted curve depicts the periodical variation in the intensity of exciting light reaching the specimen. The dotted curve also represents the variation in intensity of prompt fluorescence, since this grows and decays in a negligible time compared with the period of the chopper. The full line in (*a*) shows the growth and decay of the long-lived luminescence during the periods of illumination and darkness. The second diagram (*b*) represents the viewing position of the second chopper when out of phase—it allows the photo-multiplier to "see" only the long-lived luminescence, for a period at the

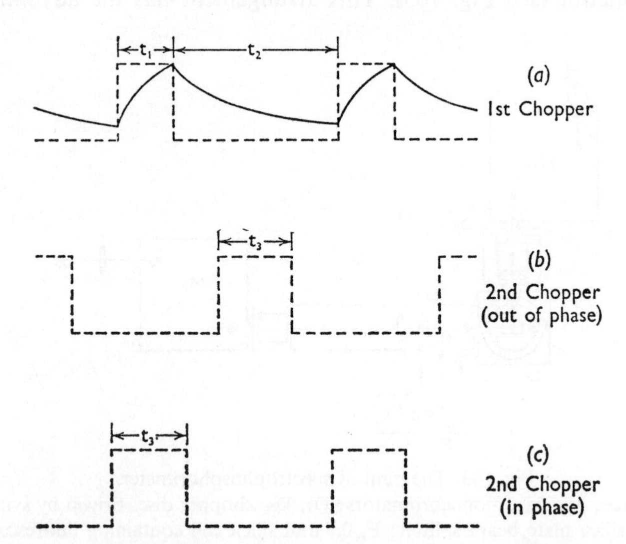

Fig. 104. Relationship between chopper phase and phosphorescence intensity. Full line represents phosphorescence intensity (from Parker and Hatchard[203]).

mid-point of the "dark" period. In the lower diagram (*c*) of Fig. 104, the second chopper is shown in phase. The photomultiplier now samples both the prompt fluorescence for a period t_1, and the long-lived luminescence for a period t_3.

2. Phosphorimeter Factor

If the lifetime of the long-lived emission is large compared with t_1, t_2 and t_3 (Fig. 104), the decay of intensity during the time t_2, will be negligible and the intensity during the complete cycle will be substantially constant. In the out-of-phase position, therefore, the photomultiplier will measure a fraction of the total long-lived emission equal to $t_3/(t_1 + t_2)$, i.e. t_3/t_c, where t_c is the time occupied by one complete cycle. With the recommended value of $t_3/t_c = \frac{1}{3}$, the observed intensity of long-lived emission must therefore be divided by a factor of $\frac{1}{3}$ to determine the total rate of its emission per cycle. Clearly, the intensity of long-lived emission observed with the choppers in phase will be equal to that observed when they are out of phase, and the intensity of prompt fluorescence can thus be simply calculated from the difference between the readings in phase and out of phase. This is illustrated by the two spectra shown in Fig. 105 which refer to a solution of impure phenanthrene at 77°K, for which the lifetime of the $T_1 \rightarrow S_0$ phosphorescence of phenanthrene is 4.3 sec. Curve A is the spectrum taken in phase, and shows both the prompt fluorescence and phosphorescence emission. Curve B is the spectrum recorded with the choppers out of phase and shows only the phosphorescence. At those wavelengths where the prompt fluorescence is negligible, the phosphorescence intensities measured in phase and out of phase are identical. The difference between curve A and curve B gives the prompt fluorescence spectrum. To determine the relative intensities of phosphorescence and fluorescence emitted, the values in curve B must be divided by the phosphorimeter factor (0.333) before comparing with the fluorescence spectrum.

If the lifetime of the long-lived emission is of the same order of magnitude as the periods of illumination and darkness, the long-lived emission will decay appreciably during the dark period before observation by the photomultiplier, and the observed intensity must be divided by a phosphorimeter factor less than t_3/t_c to obtain the total rate of emission of long-lived luminescence per cycle. The required phosphorimeter factor, i.e. the ratio of the observed intensity of long-lived emission out of phase (P_0) to the

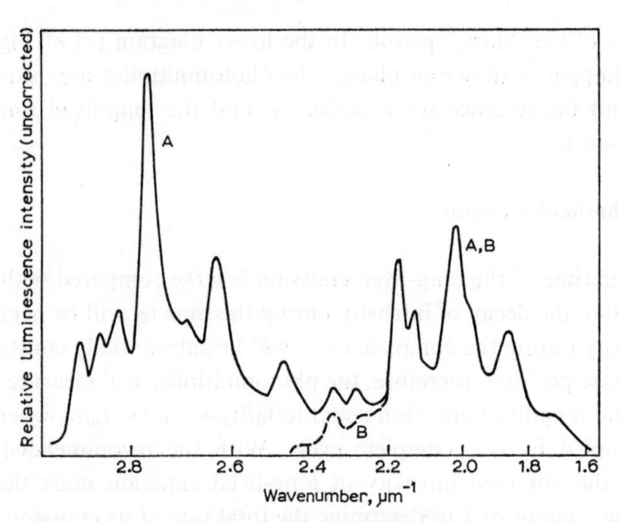

Fig. 105. Fluorescence and phosphorescence emission spectra of impure phenanthrene (10^{-3} M) in ether–pentane–ethanol glass at 77°K.
A, fluorescence plus phosphorescence (choppers in-phase); B, phosphorescence (choppers out-of-phase). The intensity of the first fluorescence band is low owing to self-absorption. Excitation at 3.19 μm^{-1} (313 nm); quartz analysing monochromator with 9558Q photomultiplier; half-band width at 2.5 μm^{-1} was 0.019 μm^{-1} (from Parker and Hatchard[203]).

total rate of its emission (P), can be calculated if two assumptions are made. The first assumption is that the long-lived emission decays exponentially, and the second is that the cut-off time of the exciting light by the first chopper is small compared with the period of the cycle. Unless the entrance slit of the excitation monochromator is very narrow, the second assumption is only approximately correct. With these two assumptions, the phosphorimeter factor is given by:

$$\frac{P_0}{P} = \frac{\tau}{t_1}\left[\frac{1-\exp(-t_1/\tau)}{1-\exp(-t_c/\tau)}\right][\exp\{-(t_2-t_3)/2\tau\} - \exp\{-(t_2+t_3)/2\tau\}] \tag{254}$$

where τ is the lifetime of the long-lived emission. For the recommended values of $t_c = 1/800$ sec, $t_1 = t_c/4$, $t_2 = 3t_c/4$ and $t_3 = t_c/3$, the phosphorimeter factor P_0/P has the following values:

τ (msec)	≥ 5	1	0.5	0.25	0.15	0.10	0.07
P_0/P	0.333	0.316	0.271	0.165	0.070	0.022	0.005

Thus with long-lived emission having a lifetime greater than 5 msec, the observed intensity out of phase is one third of the total rate of emission. For a lifetime of 0.25 msec the observed intensity in the out of phase position is reduced to about one half of this value, and for shorter lifetimes the observed intensity in the out-of-phase position decreases rapidly.

The majority of applications involving the measurement of long-lived emission having a lifetime less than 1 msec are those in which the intensity of prompt fluorescence is many times greater, e.g. measurements of delayed fluorescence in fluid solution. The intensity of the total emission observed in phase is then substantially identical to that of the prompt fluorescence. There are however compounds for which the prompt fluorescence in fluid solution is weak and the phosphorescence comparatively intense. Generally the phosphorescence band does not overlap that of the fluorescence and no correction is therefore necessary to derive the spectral curve of the latter from the in-phase measurements. An example is shown in Fig. 106. This

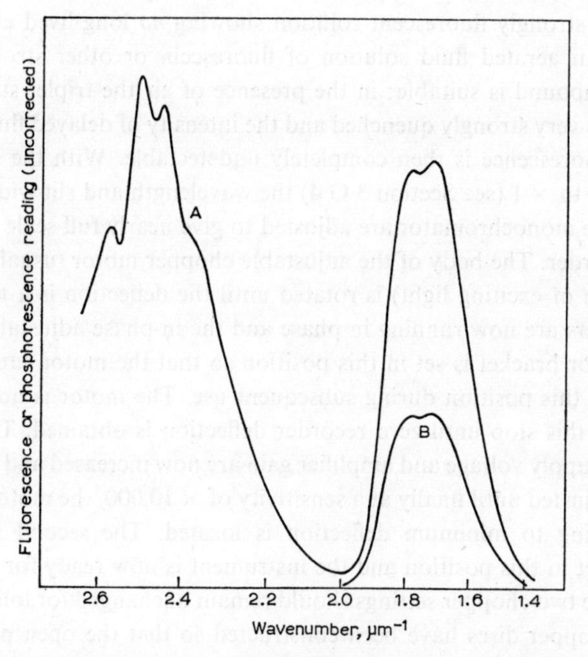

Fig. 106. Luminescence of 3,4-benzopiazselenol in de-aerated cyclohexane at 20°C. A. fluorescence plus phosphorescence (choppers in-phase); B, phosphorescence only (choppers out-of-phase); excitation at 2.73 μm^{-1} (366 nm); quartz analysing monochromator with 9558Q photomultiplier; half-band width at 2.5 μm^{-1} was 0.024 μm^{-1} (from Parker and Hatchard[203]).

refers to a solution of naphtho-1′,2′,3,4-[1,2,5-selenodiazol]—alternative name "3,4-benzopiazselenol" (see Section 5 D 2). This compound shows a weak fluorescence band with maximum at about 2.5 μm^{-1}. In de-aerated solution a second band appears in the yellow region (1.7 μm^{-1}) due to phosphorescence having a lifetime of about 0.3 msec. The observed intensity of this band is thus greater when the choppers are in phase than when they are out of phase. The ratio of the observed in-phase (P_i) to out-of-phase intensity (P_0) may be calculated if the two assumptions referring to equation 254 apply. It is as follows:

$$\frac{P_i}{P_0} = \frac{P}{P_0} - \frac{[\exp\{-(t_3 - t_1)/2\tau\} - \exp\{-(2t_2 - t_3 + t_1)/2\tau\}]}{[\exp\{-(t_2 - t_3)/2\tau\} - \exp\{-(t_2 + t_3)/2\tau\}]} \tag{255}$$

3. Method of Phasing Choppers and Determination of Chopper Leakage

To set the adjustable chopper motor at the in-phase and out-of-phase positions a strongly fluorescent solution showing *no* long-lived emission is required. An aerated fluid solution of fluorescein or other strongly fluorescent compound is suitable; in the presence of air the triplet state of the fluorescer is very strongly quenched and the intensity of delayed fluorescence and phosphorescence is then completely undetectable. With the sensitivity controls set to × 1 (see Section 3 G 4) the wavelength and slit widths of the fluorescence monochromator are adjusted to give nearly full scale deflection on the recorder. The body of the adjustable chopper motor (usually the one in the beam of exciting light) is rotated until the deflection is a maximum. The choppers are now running in phase and the in-phase adjustable "stop" on the motor bracket is set in this position so that the motor can be easily returned to this position during subsequent use. The motor is now rotated away from this stop until zero recorder deflection is obtained. The photomultiplier supply voltage and amplifier gain are now increased and the motor position adjusted until finally at a sensitivity of × 10,000, the motor position corresponding to minimum deflection is located. The second adjustable "stop" is set in this position and the instrument is now ready for use. Once adjusted the two chopper settings should remain unchanged for long periods.

If the chopper discs have been constructed so that the open periods are somewhat less than the closed periods (as described in Section 3 N 1 above) the direct beam of exciting light will be completely cut off from the specimen during the whole of the viewing period permitted by the second chopper at the entrance slit of the analysing monochromator when the angular position of the adjustable chopper motor is set to the out-of-phase position. Theoretically therefore no prompt fluorescence should be passed to the photomultiplier in this condition. In practice it will be found that a small amount of prompt fluorescence *does* pass owing to multiple reflections inside the chopper housings, but if the inside of the boxes housing the chopper motors, and the chopper blades themselves, are coated with matt black paint, the amount of light leakage should not exceed 1/200,000 of the intensity observed in the in-phase position. With care the light leakage can be reduced to 1 part in 10^6. This corresponds to a recorder deflection of 1 division at a sensitivity of $\times$ 10,000 when the prompt fluorescence observed in phase gives a deflection of 100 divisions at a sensitivity of $\times$ 1. If much larger deflections than this are observed, the inside of the chopper housing should be carefully inspected for reflecting surfaces, e.g. the faces of the monochromator slits etc., and these coated with matt black. The effect of varying the slit width will indicate whether the high reading is due to incomplete obscuration of the direct beam of exciting light during the whole viewing period. The dimensions of the chopper blades should be such that it is possible to open the slits of both monochromators to their full extent without increase in the light leakage.

In testing for light leakage care must be taken to choose an excitation wavelength at which the material of the cuvette itself does not phosphoresce. If it is desired to check the light leakage with short wavelength excitation, a synthetic silica cuvette must be used. For excitation of fluorescein with light of wavelength 436 nm, a pyrex cuvette is satisfactory provided that the right angle geometry is employed.

4. Determination of Quantum Efficiency of Phosphorescence and Delayed Fluorescence

The efficiencies of long-lived emission may be determined in precisely the same way as those of prompt fluorescence (see Section 3 L 2) by comparing the integrated area under the corrected emission spectrum with that of the prompt fluorescence of a standard compound. The intensity of the long-lived emission must, of course, be divided by the phosphorimeter factor to give the appropriate value of area for insertion in equation 247.

Frequently the compound exhibiting long-lived emission will also show prompt fluorescence. If the efficiency of the latter has already been determined by the usual method, the efficiencies of phosphorescence and delayed fluorescence under the same conditions can be derived without reference to any other solution. The ratio of the efficiency of delayed fluorescence (θ) to that of prompt fluorescence (ϕ_f) is simply calculated by comparing the intensities at one of the principal maxima in the spectra, which are of course identical in shape. The recorded emission spectrum need not be corrected, but due correction must be made for the phosphorimeter factor and also for the different instrumental sensitivities at which the two spectra are measured. The method of calculation is illustrated by the results in Fig. 107 which is taken from the paper by Parker and Hatchard[46], and was the first observation of P-type delayed fluorescence from anthracene in fluid solution. The spectra of prompt and delayed fluorescence were uncorrected and were distorted by self-absorption. Nevertheless, since they were both measured on the same solution under identical conditions the relative intensities of prompt and delayed fluorescence can be directly compared. The peak intensities at 2.5 μm^{-1} were 90.5 divisions at unit sensitivity for prompt

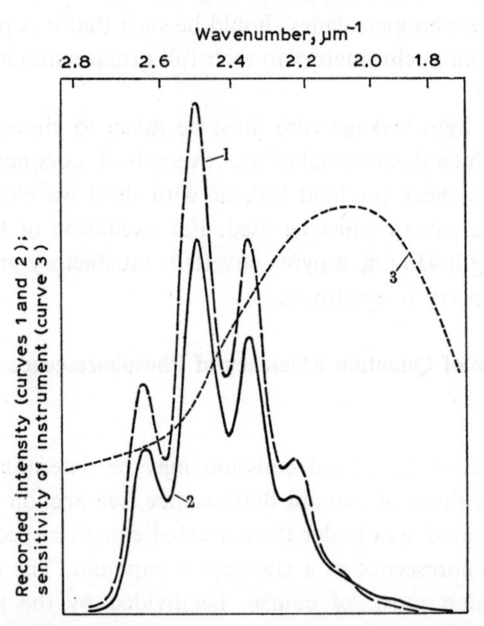

Fig. 107. Delayed fluorescence of 5×10^{-5} M anthracene in ethanol. 1, prompt fluorescence (distorted by self-absorption); 2, delayed fluorescence at 260 times greater sensitivity; 3, spectral sensitivity of instrument ($S_{\bar{\nu}}$); half-band width of analysing monochromator was 0.05 μm^{-1} at 2.5 μm^{-1}; intensity of exciting light was approximately 1.4×10^{-8} einstein cm^{-2} sec^{-1} at 2.73 μm^{-1} (366 nm) (from Parker and Hatchard[46]).

TABLE 26

APPROXIMATE FLUORESCENCE AND PHOSPHORESCENCE EFFICIENCIES AT 77°K
(From Parker and Hatchard[203])

Compound	τ (sec)	ϕ_p/ϕ_f	ϕ_p	ϕ_f
Benzene	8.0	0.89	0.19	0.21
Naphthalene	2.8	0.02	0.008	0.39
Anthracene	—	—	—	0.27
Phenanthrene	4.3	0.80	0.11	0.14
Fluorene	7.1	0.14	0.07	[0.54]
Diphenyl	5.1	1.4	0.25	0.17
Triphenylene	17.1	5.1	0.28	0.06
Phenol	2.6	0.93	0.37	0.40
Benzoic acid	2.0	> 10	0.27	Small
Acetanilide	3.6	> 10	0.05	Small
4-Nitro-N-ethylaniline	0.4	> 10	0.12	Small
N-Phenyl-2-naphthylamine	1.3	1.7	0.44	0.26
Acetophenone	0.008	> 10	1.0	Small
Benzophenone	0.005	> 10	0.71	Small
Benzoin	0.018	> 10	0.54	Small
Benzil	0.005	> 10	0.67	Small
Anthraquinone	0.004	> 10	0.41	Small

fluorescence and 66 divisions at sensitivity 260 for delayed fluorescence. The lifetime of the delayed emission was 3.8 msec corresponding to a phosphorimeter factor of 0.33. Hence:

$$\theta/\phi_f = \frac{66}{90.5 \times 260 \times 0.33} = 0.85 \times 10^{-2} \quad (256)$$

Phosphorescence efficiency is determined in a similar manner except that the spectrum is first corrected and the area under the corrected spectrum compared with that under the corrected spectrum of prompt fluorescence. The phosphorimeter and instrumental sensitivity factors are then applied as before. Determinations of absolute values of prompt fluorescence efficiency in rigid solutions at liquid nitrogen temperature are less precise than the corresponding determinations at room temperature because of the difficulties of determining values of optical density, and the errors introduced by distortion or cracking of the rigid solvent. The determination of the ratio ϕ_p/ϕ_f is not affected by these factors and this ratio can be determined much more precisely. Some values of ϕ_p/ϕ_f and approximate values of ϕ_p and ϕ_f at 77°K are shown in Table 26. The latter were calculated by assuming that the relative values of the optical densities of all the solutions at 77°K were the same as at 20°C. The fluorescence of fluorene was also assumed to be the same as its room temperature value and this value was then used as the standard to determine the values for the other compounds.

5. Determination of Rate of Light Absorption

In the measurements of prompt fluorescence and phosphorescence in fluid solutions a knowledge of the precise rate of absorption of exciting light is not normally required unless fluorescence efficiencies are being determined by an absolute method, or permanent photochemical change is also being investigated. However, certain types of delayed fluorescence show an intensity that is proportional to the *square* of the rate of light absorption (that is, the *quantum efficiency* is proportional to the first power of the rate of light absorption). The *apparent* efficiencies of prompt fluorescence and phosphorescence may also be intensity-dependent in rigid media at low temperature where triplet lifetimes are long and an appreciable proportion of the solute may be present in the triplet state during irradiation. To make quantitative investigation of such systems the rate of light absorption by the solute must be measured precisely. For this purpose the exciting light is focussed on an aperture of suitable size and shape, in such a way that the aperture is uniformly illuminated. A sharp image of the aperture is then focussed on the region of the specimen from which the luminescence is viewed. The total flux of exciting light through the aperture is then measured with the ferrioxalate actinometer. The area of the image of the aperture is measured and hence the intensity of illumination on the specimen (einstein cm^{-2}) calculated. The rate of light absorption is then derived from this value and the measured optical density per cm at the wavelength concerned. Due allowance must be made for inner filter effect by applying the factor 10^{-Dd} (see Section 3 H 2) to derive the true intensity at the point from which the emission is viewed.

6. Measurement of Lifetime of Long-Lived Luminescence

Long lifetimes may be simply determined by recording the decay of photomultiplier signal, with the choppers out-of-phase, when the exciting light is shut off, by means of a rapid mechanical shutter. For lifetimes of 5 sec or more the signal may be recorded with a fast pen recorder. For lifetimes between 0.1 and 5 sec the photomultiplier output must be passed direct to the amplifier feeding the Y plates of an oscilloscope, and the oscilloscope trace photographed, or more conveniently, if a storage oscilloscope is used, the luminescence intensity as a function of time may be read at leisure directly from the oscilloscope screen. For lifetimes less than 0.1 sec a somewhat different procedure has to be used. For lifetimes in the range 0.1–10 msec the 16-blade choppers of the 3000 r.p.m. synchronous motors are replaced by 2-blade choppers having an open cycle slightly less than the closed cycle. The beams of exciting light and luminescence are then chopped at 100 c/s. The output from the photomultiplier is fed to the Y amplifier of an oscilloscope, the time base of which is triggered at 100 c/s on the positive-going pulse from the output from the fluorescent screen monitor. With the choppers out-of-phase, the oscilloscope thus records repetitive traces of the decaying luminescence, each trace covering the 3–4 msec during which the second chopper is open to allow the photomultiplier to view the decaying luminescence. For lifetimes greater than 10 msec, the luminescence intensity does not decay sufficiently during 4 msec to allow accurate lifetimes to be derived, and slower chopping frequencies are then required. A convenient method is to use 1500 r.p.m. synchronous motors, with 2-blade choppers, and a reduction gear box giving chopping frequencies of 50, 25 and 12½ c/s. The latter is adequate for lifetimes up to 100 msec. Alternatively the whole range of lifetimes can be covered by using a single variable speed motor to drive a rotating can or rotating disc phosphoroscope, but the convenience of having separately-driven choppers is now lost and a separate phosphoroscope attachment must be employed.

For the sake of completeness it should be mentioned that lifetimes covering a whole range of values may be measured by using flash-excitation in much the same manner as has already been described for measuring prompt fluorescence lifetimes (Section 3 M). This method has the advantage that lifetimes much shorter than 0.1 msec can be measured (with appropriately chosen flash tube), and intensity measurements may be made over a long time interval after the flash. This method was used in the pioneering work of Bäckström and Sandros[33], on the phosphorescence of biacetyl in solution, and the reader is referred to their work for further details.

Whatever method is used for recording the intensity of long-lived emission (I_D) as a function of time after the exciting light is shut off, the results should be plotted in the form $\log_e (I_D)$ against time (see Fig. 108). A linear plot indicates that the luminescence decay is exponential and the slope of the plot gives the lifetime, τ, directly (see equation 25). A non-linear plot (see, for example Figs. 32, 34 and 108 *e*) indicates the operation of an unusual factor in the mechanism of the emission. For example, in fluid solution it may indicate that the intensity of the exciting light is so great that the rate of bimolecular triplet–triplet quenching contributes appreciably to the overall rate of triplet decay (see Section 2 D 4). In rigid medium it may indicate that the delayed fluorescence is produced by recombination of ionic species (see Section 1 C 6), or it may indicate the establishment of a non-random distribution of triplet molecules (see Section 2 D 3).

A few words of explanation about the time constant of the measuring circuit may be of value to the reader unfamiliar with electronics. The circuit

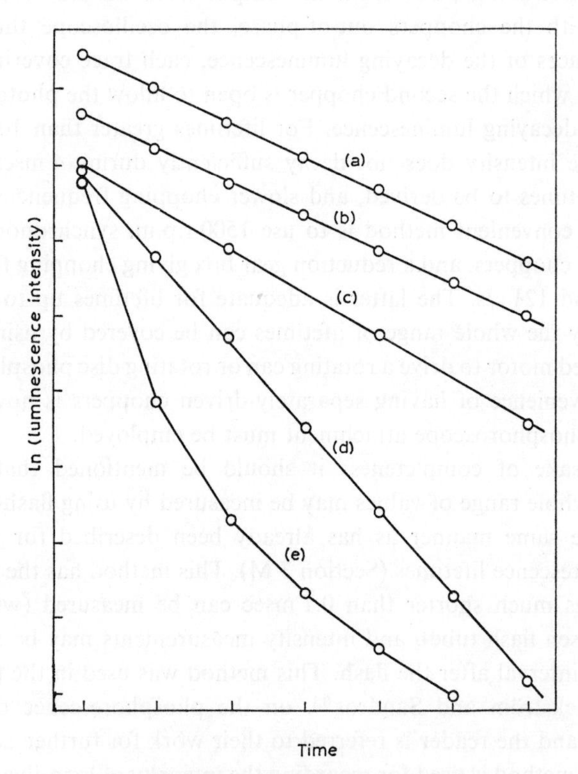

Fig. 108. Decay of luminescence with time.

Ordinate, ln (intensity), one division = 0.25; abscissa, time, one division = 0.0003 sec for curves a and b, 0.0005 sec for curve c, 1.0 sec for curve d, 0.1 sec for curve e. a, b, delayed fluorescence of pyrene monomer and dimer in ethanol at + 23°C; c, delayed fluorescence of naphthalene in ethanol at − 23°C; d, e, phosphorescence and delayed fluorescence of 10^{-1} M phenanthrene in ether–pentane–ethanol glass at 77°K (from Parker[111]).

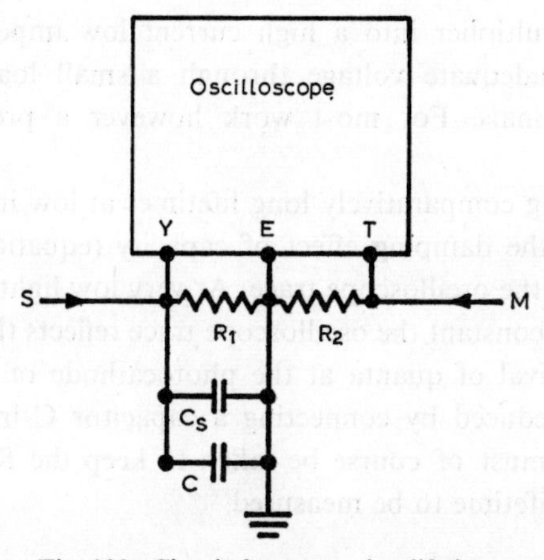

Fig. 109. Circuit for measuring lifetimes.
Y, amplifier input; T, trigger input; S, luminescence signal; M, signal from monitor; R_1, signal amplifier load (0.1–10 megohm); R_2, trigger load (to suit oscilloscope); C_s, stray capacities; C, capacitor used when measuring long lifetimes.

used to record the decay of luminescence is shown diagrammatically in Fig. 109. The current from the photomultiplier measuring the luminescence is passed to earth through the load resistor R_1 and the voltage developed across the latter is fed to the Y amplifier of the oscilloscope. The sensitivity attained will be proportional to the resistance of R_1, the maximum possible value of which will correspond to the input impedance of the Y amplifier when set on its maximum sensitivity range. A typical maximum value might be 10 megohm.

Associated with the leads from the photomultiplier there will be a stray capacity (C_s) and this will limit the rate at which the input potential can fall when the light is shut off. The decay of potential is exponential with a time constant given by:

$$\text{time constant (seconds)} = \text{megohms} \times \text{microfarads} \qquad (257)$$

With long photomultiplier leads the stray capacity C_s may be sufficient to give a time constant of several tenths of a msec with a 10 megohm load and such an arrangement could not be used for measuring short lifetimes. The remedy is to shorten the leads and/or to reduce the value of R_1, although the latter will reduce the overall sensitivity. For *very* short lifetimes a cathode follower pre-amplifier must be built into the photomultiplier housing. This effectively converts the low current high impedance signal from the photomultiplier into a high current low impedance signal that will develop an adequate voltage through a small load resistor at the oscilloscope terminals. For most work however a pre-amplifier is not required.

When measuring comparatively long lifetimes at low intensity advantage can be taken of the damping effect of capacity (equation 257) to reduce the noise level on the oscilloscope trace. At very low light levels with a very small circuit time constant the oscilloscope trace reflects the non-continuous nature of the arrival of quanta at the photocathode of the detector. The noise is greatly reduced by connecting a capacitor C in parallel with R_1 (Fig. 109). Care must of course be taken to keep the RC value less than one-tenth of the lifetime to be measured.

7. Purity of Materials

One of the virtues of fluorescence and phosphorescence measurements as analytical methods is that, properly applied they can be used to determine extremely low concentrations of impurity in the presence of much larger concentrations of a second luminescent substance (see Chapter 5). Conversely, the degree of purity of specimens used for the investigation of their luminescence must be very high, at least with respect to certain classes of impurity. Thus in the determination of the prompt fluorescence emission spectrum of a compound having a low fluorescence efficiency and low extinction coefficient at the wavelength of the exciting light, the presence of 0.1% of an impurity having a high fluorescence efficiency and high extinction coefficient can completely vitiate the results. Similar considerations apply to the determination of phosphorescence emission in rigid solution at low temperature. It is desirable therefore to check the purity by determining the fluorescence emission spectrum with a series of excitation wavelengths. If a variation in the shape of the spectrum is found, the presence of a second fluorescent substance must be suspected. The discovery of weak emission bands in unexpected spectral regions should always be treated with suspicion, and careful checks made to confirm that they are not due to impurity.

Owing to the long lifetimes of triplet states and their consequent susceptibility to quenching in fluid solution (see Chapter 2) the standard of purity of both solute and solvent required for the investigation of long-lived emission in fluid solution is far higher than that required for the investigation of prompt fluorescence. The oxygen content must be reduced to a low level, but by the application of the high vacuum method described in Section 3 J 2, this can be fairly easily achieved. The main difficulty is to remove traces of impurities capable of accepting triplet energy from the compound under investigation—such impurities can introduce errors at concentrations below 10^{-9}M. The problem is greatest when the compound under investigation has a relatively high triplet energy. Thus many impurities capable of quenching the phenanthrene triplet (situated at 2.21 μm^{-1}) have no effect on the anthracene triplet (situated at 1.48 μm^{-1}) because their triplet states are situated at higher levels than that of anthracene (see for example Section 4 B 4).

The difficulty of reducing impurity levels in solvents to less than 0.0001 p.p.m. is accentuated by the fact that the nature of the objectionable impurities cannot be predicted with certainty and there is no general method for their detection except the fact that they quench the long-lived luminescence from the substance under investigation. Measurement of optical absorption in a 10 cm cell is useful as a means of following the purification process during its early stages, and sensitive measurements of fluorescence with excitation at 250 nm or phosphorescence at 77°K, provide a useful check in the later stages of the purification process (see Section 5 C 4). Once a pure solvent has been prepared, the purity of subsequent batches may be compared with it by measuring the lifetime of the delayed fluorescence of a substance having a relatively high triplet energy, e.g. phenanthrene. There is however no way of demonstrating the presence of impurity in the original batch of "good" solvent except by producing a better one that shows a longer lifetime of delayed fluorescence, i.e. a lower degree of triplet quenching.

Some impurities can be detected not only by their effect on the lifetime and intensity of the delayed fluorescence of the substance under investigation, but also by the appearance of a new system of emission bands due to sensitised delayed fluorescence of the impurity. Such impurities sometimes undergo photodecomposition when irradiated in the presence of the sensitizer and can thus be removed. The consumption of pyrene and other impurities in a phenanthrene solution during irradiation is shown in Fig. 110. The traces of pyrene in this solution were found to have come from the ethanol used to prepare it and provide a good example of the care required in the storage and handling of purified solvents. This batch of ethanol had been previously used for preparing pyrene solutions and the

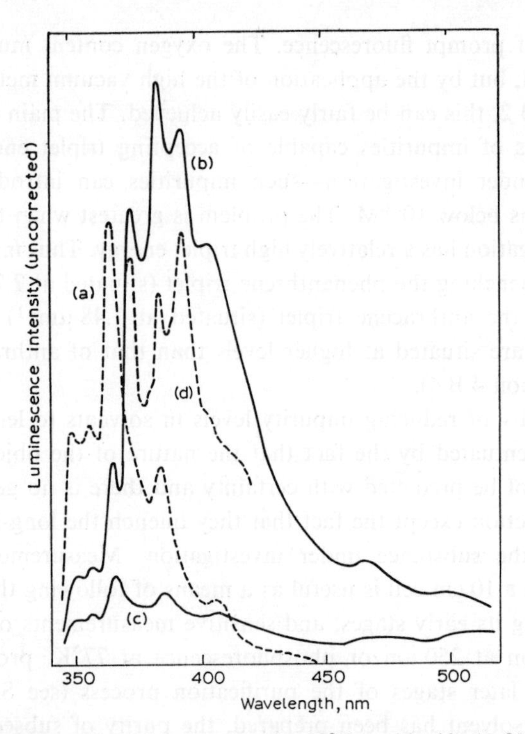

Fig. 110. Luminescence of impurities in ethanol at 20°C sensitised by 10^{-3} M phenanthrene. Rate of light absorption was 0.7×10^{-6} einstein litre^{-1} sec^{-1} at 341–362 nm. a, prompt fluorescence; b and c, delayed fluorescence at 1000 times greater sensitivity; d, prompt fluorescence of a dilute pyrene solution. Photo-decomposition of b to give c was produced by irradiation for 30 min with a rate of light absorption of 10^{-5} einstein litre^{-1} sec^{-1} (from Parker[111]).

obvious precautions against contamination had been carefully observed. The presence of 1 part in 10^9 of pyrene in the solvent was attributed to contamination by airborne dust on the many occasions when the storage bottle had been opened while pyrene was being handled in the laboratory.

In the following paragraphs we give some brief notes on the methods of preparation of some commonly used solvents.

Ethanol. 3 g of potassium hydroxide pellets are rinsed with ethanol to remove surface impurities and are then added to 3 litres of boiling laboratory grade ethanol. The solvent is immediately fractionated and the middle 50% collected. The effect of this treatment on the fluorescence emission spectrum is shown in Fig. 151 (Section 5 C 4).

Iso-pentane and Low Boiling n-Paraffins. These are subjected to extraction with oleum containing 10% SO_3 for 100 hours in a continuous extraction apparatus. The oleum is renewed periodically as it becomes exhausted. The solvent is washed with water, then with dilute alkali, then with water again. It is then dried over anhydrous sodium sulphate, fractionated, and the middle 50% collected.

Cyclohexane. Some batches of "spectroscopically pure" material are found to be sufficiently pure for use without purification. All batches received are tested and those showing the lowest fluorescence (see Fig. 150 in Section 5 C 4) are reserved for high sensitivity work. The quality of impure cyclohexane can be improved by careful treatment with sulphuric acid followed by washing, drying and fractionation.

Liquid Paraffin. The quality of this material can be greatly improved by passing over activated alumina, but the last traces of impurities absorbing in the quartz ultra-violet region are difficult to remove.

Benzene. Bäckström and Sandros[33] found that the lifetime of dilute solutions of biacetyl in high quality benzene from different sources varied considerably. By fractional distillation they were able to raise the lifetime to 1.0×10^{-3} sec at 20°C.

Glycerol and Propylene Glycol. The fluorescence and phosphorescence at 77°K from these solvents is greatly reduced by fractional vacuum distillation. Propylene glycol is preferred to glycerol if the higher viscosity of the latter is not important.

Ionising Solvents. A variety of purification methods have been reported, mainly in connection with electrochemical studies, and the quantity of luminescent impurities is not known. Coetzee and co-workers[204] recommend the following series of treatments for acetonitrile: stirring with calcium hydride for 2 days, fractionation from phosphorus pentoxide, refluxing over calcium hydride and slow fractionation. The product may contain some acrylonitrile. Maricle[205] allows dimethylformamide to stand for 12 hours over potassium carbonate before fractionating. Maricle and Hodgson[206] heat dimethylsulphoxide for 2 hours over sodium hydroxide at 90°C and then flash-distil under vacuum.

Glassy Solvents. The choice of solvent mixtures to form clear rigid glasses at 77°K without cracking is limited. Winefordner and St. John[207] tested a variety of single solvents and solvent mixtures. Of the pure solvents, pentane, petroleum ether, ethyl ether, 2-bromobutane, ethanol and n-propanol gave uncracked glasses in most tests. (It should be noted however that these tests were made in small sample tubes. If large optical cells are used the chance of cracking is greater and the cell itself may be broken. Breakage can be avoided by freezing a thin layer of solution on one face of the cell.) Of the mixtures, they found that water-free ethanol with many of the other solvents frequently gave clear glasses. The most reliable glasses were those containing pentane, e.g. ether–pentane–ethanol or methyl cyclohexane–pentane in various proportions. Greenspan and Fischer[208] have investigated the viscosity of a variety of glass-forming solvent mixtures as a function of temperature. Some of their results are shown in Table 27. These may be used to choose solvent mixtures giving clear glasses at 77°K and having various degrees of rigidity.

Solutes. Commercial specimens of many solutes—even those labelled "high-purity"—often contain sufficient impurity to render them useless for the investigation of luminescence. Each solute to be investigated presents its own purification problems and may require a combination of techniques to reduce impurities to the required level, which may be as low as 1 part per million or less. Recrystallisation of solids or fractionation of liquids, followed by liquid/solid or gas/liquid chromatography, high vacuum distillation, fractional freezing or zone refining should be tried. For solids melting

TABLE 27

VISCOSITY OF LOW TEMPERATURE GLASSES
(Adapted from Greenspan and Fischer[208])

Solvent	Approximate viscosity in poise at − 180°C
1-Propanol/2-propanol (2:3)	6×10^{12}
Ethanol/methanol	2×10^{12}
Ethanol/methanol + 4.5% water	–
Ethanol/methanol + 9% water	–
Iso-octane/isononane	3×10^{10}
Methylcyclohexane/cis/trans-decalin	1×10^{14}
Methylcyclohexane/toluene	7×10^{9}
Methylcyclohexane–isohexanes (3:2)	3×10^{6}
Methylcyclohexane/methylcyclopentane	2×10^{5}
Methylcyclohexane/iso-pentane	–
Methylcyclohexane–iso-pentane (1:3)	1×10^{3}
2-Methylpentane	7×10^{4}
2-Methyl tetrahydrofuran	4×10^{7}
Ether/iso-pentane/ethanol (5:5:2)	9×10^{3}

Note
Equal volumes of components except where indicated.

below 300°C without decomposition (under oxygen-free nitrogen) zone refining[209] (see Section 5 E 8) is a powerful method which will generally reduce most impurities to a very low level. Since, however the number of impurities present at the part per million level in an organic compound is generally large, there are often one or two impurities with segregation coefficients sufficiently close to unity to make their complete removal impracticable even with exhaustive zone refining. For example, to prepare pure phenanthrene, the anthracene impurity is best removed by treatment with maleic anhydride in benzene solution, and the fluorene impurity by melting with potassium hydroxide[210]. The phenanthrene is then recrystallised and subjected to zone refining to remove the remaining impurities. Impurities having unfavourable distribution coefficients cannot be reduced to the required level by a single exhaustive zone refining treatment, but an equilibrium distribution of impurity along the zone refined ingot is ultimately achieved[209]. Measurement of the fluorescence emission spectrum at various positions along the ingot provides a convenient method of following the course of the purification process. The observations of fluorescence or phosphorescence in rigid solution at 77°K, or sensitised delayed fluorescence in fluid solution also provide sensitive methods of detecting impurities in appropriate systems. Examples of these applications are described in Chapter 5.

8. Artifacts and Trivial Effects

Distortion of the prompt fluorescence spectrum as a result of absorption of luminescence by the specimen itself has been discussed in Section 3 H. Under certain conditions distortion of the spectra of phosphorescence or delayed fluorescence can be produced in a similar manner. There are, however, two further effects that can give rise to spurious signals in the measurement of delayed fluorescence. These were first recognised by Parker and Joyce[59] in the measurement of the delayed fluorescence of perylene solutions. When excited by light of wavelength 436 nm, solutions of perylene in ethanol emit only very weak delayed fluorescence (see Section 4 A 5), but when the same solutions were excited by light of wavelength 250 nm, Parker and Joyce observed a much stronger delayed emission. They were able to show that the additional emission was not true delayed fluorescence, but was an artifact, i.e. it was in reality *prompt* fluorescence of perylene excited by absorption of the *phosphorescence* emitted by the fused quartz container.

It therefore decayed with the same lifetime as the phosphorescence of the latter and was thus registered by the spectrophosphorimeter as delayed fluorescence. The absorption band of perylene is favourably situated for the production of this artifact because the main phosphorescence of the fused quartz is emitted in the region 370–430 nm. To avoid the possibility of spurious results due to this effect it is best when making measurements with short-wavelength excitation, to use containers made of non-phosphorescent synthetic silica.

The causes of such spurious delayed fluorescence are not restricted to phosphorescence of containers. If the "true" long-lived emission from one component of the solution (either true phosphorescence or true delayed fluorescence) is situated in a spectral region absorbed by a second component, it will excite prompt fluorescence of the latter (a "trivial effect"), and this prompt fluorescence will decay at a rate identical with that of the long-lived emission of the first component. As an example, Parker and Joyce[59] quote an experiment in which a solution of perylene in liquid paraffin at 77°K was excited by the light of wavelength 250 or 297–302 nm. The liquid paraffin contained traces of impurities emitting phosphorescence in the region 380–440 nm. This phosphorescence was partly absorbed by the perylene and excited prompt fluorescence of the latter. The prompt fluorescence decayed at the same rate as the phosphorescence of the impurities in the liquid paraffin and was thus registered by the spectrophosphorimeter as delayed fluorescence of the perylene. It is clearly important to examine all experiments in which delayed fluorescence is observed, to ensure that the trivial effect is not responsible.

P. MEASUREMENT OF POLARISATION OF LUMINESCENCE

The polarisation of photoluminescence from liquids is almost always measured in a direction at right angles to the direction of propagation of the exciting light. The equipment used consists essentially of a filter fluorimeter, or spectrofluorimeter, fitted with a device for determining the degree of polarisation of the emission. The exciting light may be completely unpolarised or completely plane polarised. The degree of complexity of the equipment depends on the type of information required, and the precision with which the polarisation must be determined. The principles and precautions already discussed for conventional spectrofluorimetry and spectrophosphorimetry will in general apply, but there are additional precautions required. These are mainly concerned with avoiding the introduction of stray polarisation produced by monochromators etc., or with compensating for them if they are present. The additional precautions will be discussed below in describing the main types of apparatus in use. It is recommended that the following sections are read in conjunction with Sections 1 D.

1. Polarising Units

(*a*) *Nicol's Prism.* A crystal of calcite about 25 × 8 mm, with its end surface ground to a particular orientation, is cut in a special way[68] and the two pieces cemented together with canada balsam. Natural light passing into the rhomb at A (Fig. 111 *a*) is split into the ordinary and extraordinary rays, which are polarised in planes at right angles. The refractive index of the balsam is intermediate between those of the calcite for the ordinary ray and the extraordinary ray (for the particular angle of incidence concerned) so that the latter is refracted at the interface and passes through to E as plane polarised light, but the ordinary ray is totally internally reflected at the interface and is absorbed at the blackened side of the prism. The direction of polarisation of the transmitted ray is parallel to the shorter diagonal of the end face. The transmission of a Nicol prism falls rapidly in the ultraviolet region owing to absorption by the canada balsam.
(*b*) *The Glan–Thompson Prism.* This consists of a square ended prism of calcite with the optic axis perpendicular to the length of the prism. It is sliced along a plane including opposite edges and the polished faces cemented together with a special resin of appropriate refractive index and transparent to the near ultra-violet. The ordinary and extraordinary rays pass undeviated through the prism until they reach the resin film. The refractive index of the latter, for the extraordinary ray, is almost the same as that of the calcite and this ray passes on undeviated while the ordinary ray suffers total internal reflection (Fig. 111 *b*).
(*c*) *The Rochon Prism.* In its usual form this consists of two right angled prisms of doubly refracting crystal cemented together. The optic axis of the left-hand prism is parallel to the normal, i.e., it is parallel to the direction of incident light beam A in Fig. 111 *c*. The optic axis of the right-hand prism is perpendicular to the plane of the paper. Light incident normally at A passes undeviated along the direction of the optic axis of the first prism,

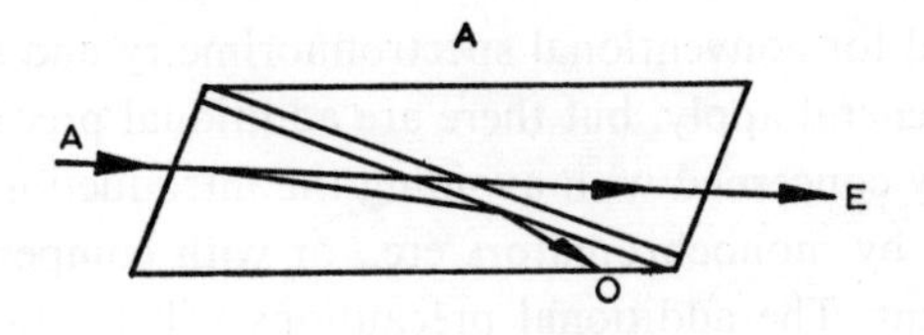

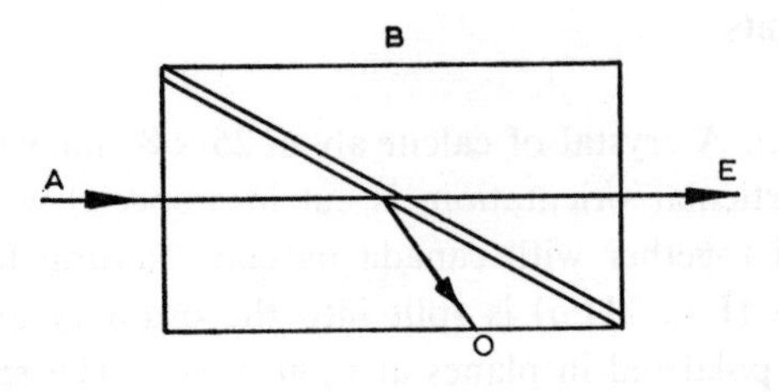

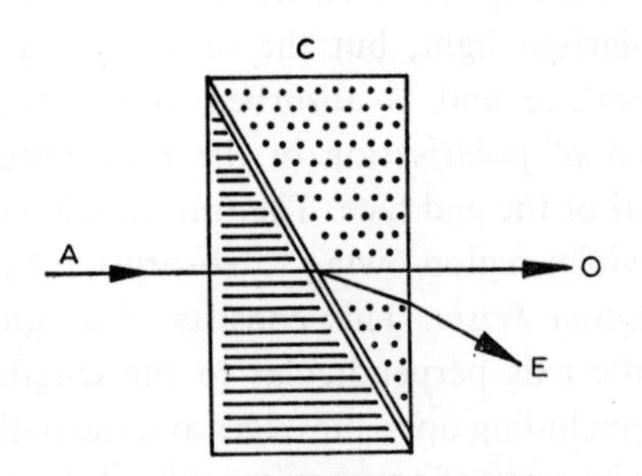

Fig. 111. Polarising elements.
a, Nicol's rhomb; b, Glan–Thompson prism (optic axis perpendicular to AE); c, Rochon prism.

but on entering the second it forms ordinary and extraordinary rays. The ordinary ray passes on undeviated but the extraordinary ray is deviated and leaves the prism at an angle to the ordinary ray. In this arrangement the doubly refracting properties of the first prism are not involved, since the light passes along the optic axis. The first prism may therefore be replaced by one of fused silica. This permits the resolution of moderately convergent beams without depolarisation, and if the two prisms are joined by a thin layer of paraffin oil the device will transmit down to a wavelength of 220 nm.

With this arrangement the extraordinary ray, which is vertically polarised, is deviated only slightly, while the ordinary ray is deviated by about 10 degrees from the extraordinary[211].

(*d*) *Polarising Films.* These were originally prepared by incorporating dichroic crystals, e.g., quinine sulphatoperiodide, in transparent polymer sheets. The crystals were oriented in the film by a magnetic field or by flow and pressure treatment. The film then has the property (like tourmaline crystal—see Section 1 D 1), of absorbing light polarised in one direction but transmitting visible light with its plane of polarisation at right angles. The "Polaroid" sheets now available give a very high degree of polarisation in the middle of the visible spectrum.

2. Polarisation Filter Fluorimeters

For the measurement of rotational depolarisation a comparatively simple arrangement can be used, with filters to isolate the required wavelengths of exciting and fluorescence light. Such an apparatus has been described by Johnson and Richards[212]. The general arrangement is shown in Fig. 112. Ultra-violet light from the mercury lamp A illuminates the slit S_1, is rendered parallel by the lens L_1, passes through the filter F_1 and the rectangular stop S_2 into the rectangular glass cell C, situated in the corner of the thermostat. The fluorescence passes out through the filter F_2 and the iris diaphragm I

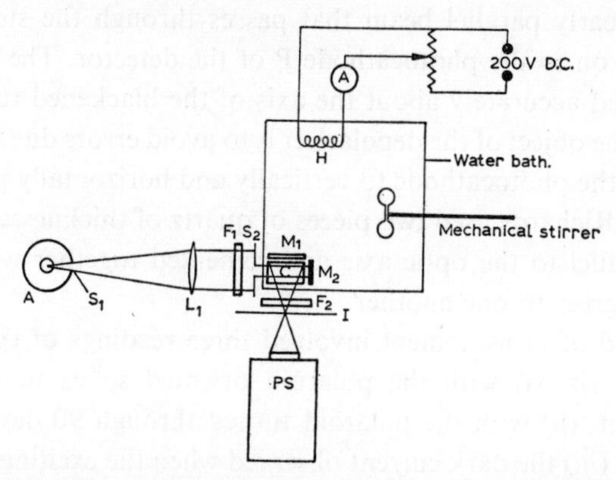

Fig. 112. Measurement of polarisation of fluorescence.
A, light source; S, illuminated slit; L_1, quartz lens; F_1, excitation filter; S_2, rectangular stop; C, cell; M_1 and M_2, silver foil mirrors; F_2, fluorescence filter; I, iris diaphragm; PS, polariscope; H, heater (from Johnson and Richards[212]).

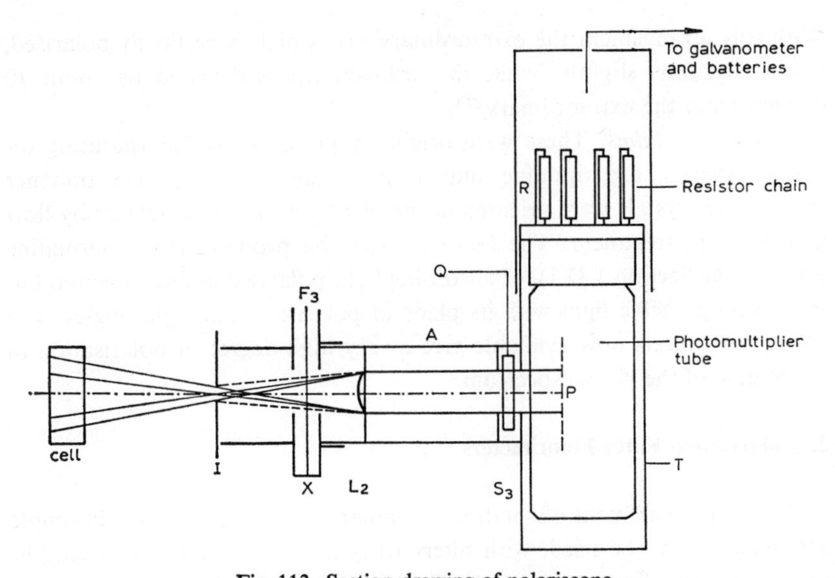

Fig. 113. Section drawing of polariscope.
X, holder for polaroid filter, F_3; L_2, lens; S_3, limiting stop; Q, quartz depolariser; A, blackened brass tube; P, photocathode; T, photomultiplier housing (from Johnson and Richards[212]).

to the *polariscope* PS, its intensity being increased by the mirrors M_1 and M_2. Details of the polariscope are shown in Fig. 113. From the diaphragm I the fluorescence passes through the polaroid film F_3 to the lens L_2. The lens produces a nearly parallel beam that passes through the stop S_3 and the *depolariser* Q on to the photocathode P of the detector. The polaroid filter may be rotated accurately about the axis of the blackened tube A through 90 degrees. The object of the depolariser is to avoid errors due to the different sensitivity of the photocathode to vertically and horizontally polarised light. Johnson and Richards used two pieces of quartz of thicknesses 2 and 4 mm, each cut parallel to the optic axis and cemented together with their optic axes at 45 degrees to one another.

The method of measurement involved three readings of the photomultiplier output, viz. (i) with the polaroid oriented so as to pass vertically polarised light, (ii) with the polaroid turned through 90 degrees from this position, and (iii) the dark current observed when the exciting light was shut off. The first two readings were corrected for dark current and the corrected values i_x and i_y substituted in the following equation to give the apparent degree of polarisation:

$$p_e = \frac{i_x - i_y}{i_x + i_y} \tag{258}$$

The true degree of polarisation (p) is in general given by:

$$p_e = pT \tag{259}$$

where T is the polarising power of the polaroid. The latter was found to be 0.997 or greater, and thus p_e and p could be equated.

Johnson and Richards discussed, amongst other factors, the magnitude of systematic errors due to the following causes: polarisation of the exciting light; orientation and rotation of the polaroid; divergence of the fluorescence beam. A correction for degree of polarisation of the exciting beam, p_L, may be calculated from the equation:

$$\frac{1}{p'} = \frac{1}{p}\left(\frac{1}{1 + p_L}\right) + \frac{p_L}{1 + p_L} \tag{260}$$

where p′ is the observed polarisation of fluorescence and p the value that would be observed if p_L were zero. Errors due to incorrect orientation or incorrect angular rotation of the polaroid were shown to be less than 0.3 %, and less than 0.15 %, respectively for all relevant values of p, if the orientation and angular adjustment were correct to within 2 degrees. Errors due to divergence of the fluorescence beam by ± 5 degrees were shown to introduce an error in p of considerably less than 1 %. Weber[213] has shown that quite large convergence or divergence of the beam of exciting light can be tolerated without the introduction of an appreciable error.

Weber[214] has described a sophisticated instrument capable of measuring small and large polarisations with an absolute precision of ± 0.001. It depends on the fact that the linearly polarised component of the fluorescence vibrating in the direction of propagation of the exciting light is independent of the plane of polarisation of the latter and may be used to give a reference signal against which the signal due to the component of fluorescence in the direction of vibration of the exciting light may be compared. The specimen is illuminated with suitably filtered light, the polarisation of which can be adjusted to the vertical or horizontal plane by means of a Glan–Thompson prism. The fluorescence is viewed by two photomultipliers, situated opposite one another and in directions perpendicular to the direction of propagation of the exciting light. A Glan–Thompson prism in front of one photomultiplier passes only horizontally polarised light and the output from this photomultiplier thus serves as the reference signal. A Glan–Thompson prism in front of the second photomultiplier (the measuring photomultiplier) is set permanently to pass only vertically polarised light. When the exciting light is horizontally polarised both beams of luminescence passing from the specimen towards the photomultipliers are unpolarised and hence the outputs of both photomultipliers are proportional to $I_\perp$. The circuit is therefore adjusted to balance their outputs under this condition. When the exciting light is vertically polarised, the reference photomultiplier still receives a signal proportional to $I_\perp$, but the measuring photomultiplier now receives a signal proportional to $I_\parallel$. To measure the ratio $I_\perp/I_\parallel$, the photomultiplier signals are again brought to balance by rotating a fourth Glan–Thompson prism through which the light to the measuring photomultiplier and its own Glan–Thompson prism has to pass. From the angle of rotation needed to balance the photomultiplier outputs the value of $I_\perp/I_\parallel$ can be calculated. The reader is referred to Weber's original paper for full details of the apparatus, and a discussion of the theory of the measurements and the magnitude of the systematic errors involved.

3. Polarisation of Fluorescence Excitation Spectrum

Weber[215] has described an apparatus for the measurement of the polarisation of the fluorescence excitation spectrum. The arrangement is shown in Fig. 114. The required wavelength of exciting light, isolated from the xenon lamp S by means of the monochromator M, is rendered parallel by the lens L and illuminates the specimen in the cuvette C. The fluorescence emerging at right angles passes through the Glan–Thompson prism P and the filter F to the photomultiplier PM. The outputs of the latter with the polariser in each of two positions are measured by means of a galvanometer viz., with the polarised component perpendicular to the directions of excitation and observation ($I_\parallel$) and with the polarised component parallel to the direction of propagation of the exciting light ($I_\perp$). The values observed at each wavelength setting of the monochromator are then substituted in equation 49 to give the degrees of the polarisation p. The values of p so obtained are then plotted against wavelength to give the polarisation of the fluorescence excitation spectrum.

To reduce rotational depolarisation to negligible proportions, glycerol, or better, propylene glycol, was used as solvent, and the solution was cooled to $-70°C$ by surrounding the cuvette with a solid copper block

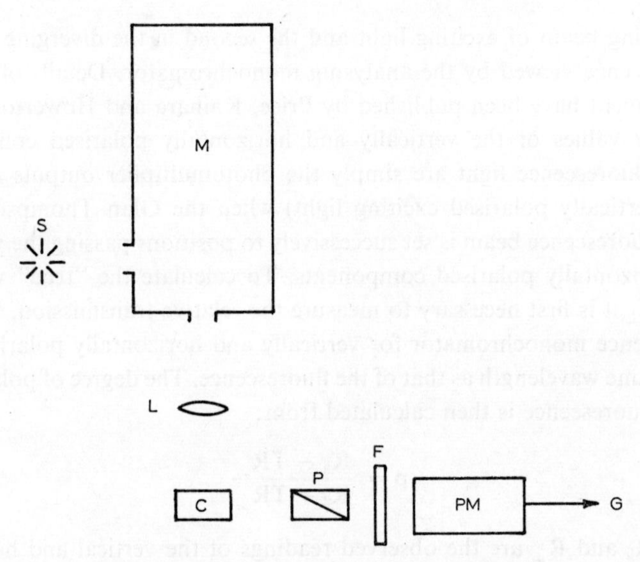

Fig. 114. Arrangement for determining the polarisation of the fluorescence excitation spectrum.
S, xenon arc; M, grating monochromator; L, quartz lens; C, spectrosil cuvette or block; P, Glan–Thompson prism; F, fluorescence filter; PM, photomultiplier; G, to galvanometer.

that in turn was placed inside a brass jacket through which cooled ethanol was rapidly circulated.

Weber found that the degree of polarisation of the exciting light varied with wavelength, but was never greater than 10%. The corresponding error introduced in the degree of polarisation of fluorescence was of the same order of magnitude as the precision of the measurements and no correction was applied. Weber also discussed the problem of depolarisation arising from absorption of fluorescence in concentrated solutions and secondary emission. To measure concentrated solutions he used thin films held between the hypotenuse faces of right-triangular prisms made of synthetic silica.

4. Conversion of Conventional Spectrofluorimeter or Spectrophosphorimeter for Polarisation Measurements

A conventional spectrofluorimeter employing right angle illumination in which the exciting light is brought to a focus at the centre of a 1 cm cuvette may be converted for polarisation measurements by placing Glan–Thompson prisms in rotatable mounts close to the specimen, one in the converging beam of exciting light and the second in the diverging beam of luminescence viewed by the analysing monochromator. Details of such an arrangement have been published by Price, Kaihara and Howerton[74]. The *apparent* values of the vertically and horizontally polarised components of the fluorescence light are simply the photomultiplier outputs observed (with vertically polarised exciting light) when the Glan–Thompson prism in the fluorescence beam is set successively to positions passing the vertically and horizontally polarised components. To calculate the "true" values of $I_{\parallel}$ and $I_{\perp}$ it is first necessary to measure the relative transmission, T, of the fluorescence monochromator for vertically and horizontally polarised light of the same wavelength as that of the fluorescence. The degree of polarisation of the fluorescence is then calculated from:

$$p = \frac{R_{\parallel} - TR_{\perp}}{R_{\parallel} + TR_{\perp}} \tag{261}$$

where $R_{\parallel}$ and $R_{\perp}$ are the observed readings of the vertical and horizontal components of fluorescence.

The value of T may be measured by placing a source of unpolarised light (e.g., light from a small tungsten filament lamp diffused through several ground glass screens) in the position normally occupied by the specimen. The ratio of the signals observed with the polariser set to the vertical and horizontal positions (with the instrument set to the appropriate wavelength) is then equal to T.

Price and co-workers used their instrument to measure the polarisation of four solutions for which precise data had previously been obtained by Weber[214]. The stability of the light source was not better than 1% and the standard deviation from the mean of several measurements was found to be about 1%. Their values of polarisation were within ± 0.03 of those reported by Weber.

Alternative methods of correction were described by Azumi and McGlynn[64]. One of these depends on the fact that at sufficiently low solvent viscosity and long emission lifetime rotational depolarisation is complete. They measured the apparent polarisation of the fluorescence from a 10^{-4}M solution of phenanthrene in methyl cyclohexane and found $R_{\parallel}$ and $R_{\perp}$ to vary with variation of the wavelength settings of both the excitation and emission monochromators. However, with a constant setting of the emission monochromator, the ratio $R_{\parallel}/R_{\perp}$ was found to be constant throughout the excitation spectrum. They therefore concluded that the fluorescence was indeed unpolarised and that the variation of $R_{\parallel}$ and $R_{\perp}$ was caused by instrumental factors. The ratio $R_{\parallel}/R_{\perp}$ then provided the required value T.

The second method of correction proposed by Azumi and McGlynn is the simplest to apply and probably the most precise. It makes use of the fact that with horizontally polarised exciting light, the emission viewed in a direction at right angles to the incident beam (i.e., in a direction parallel to the vibrations of the exciting light) must be unpolarised. Let the observed photomultiplier readings of the vertical and horizontal components of fluorescence with *horizontally* polarised exciting light be $R_{\parallel}'$ and $R_{\perp}'$, and the corresponding readings with *vertically* polarised exciting light be $R_{\parallel}$ and $R_{\perp}$. The relative transmission value of the fluorescence monochromator is then equal to $R_{\parallel}'/R_{\perp}'$ and the corrected polarisation of fluorescence is calculated from:

$$p = \frac{R_{\parallel} - R_{\perp}(R_{\parallel}'/R_{\perp}')}{R_{\parallel} + R_{\perp}(R_{\parallel}'/R_{\perp}')} \tag{262}$$

5. Automatic Recording of Fluorescence Polarisation Spectra

Ainsworth and Winter[211] have described a general purpose spectrofluorimeter that can be used to record polarisation of fluorescence excitation or emission spectra, absorption spectra and corrected fluorescence excitation and emission spectra. We shall describe briefly the first function (see Fig. 115). For full details the reader is referred to the original paper. Exciting light from the first monochromator was rendered parallel, passed through a Rochon prism and the vertically polarised light was focused on the specimen. Fluorescence emitted at right angles was dispersed by the second monochromator and was then separated into its two polarised components by the second Rochon prism. The two beams were modulated at different

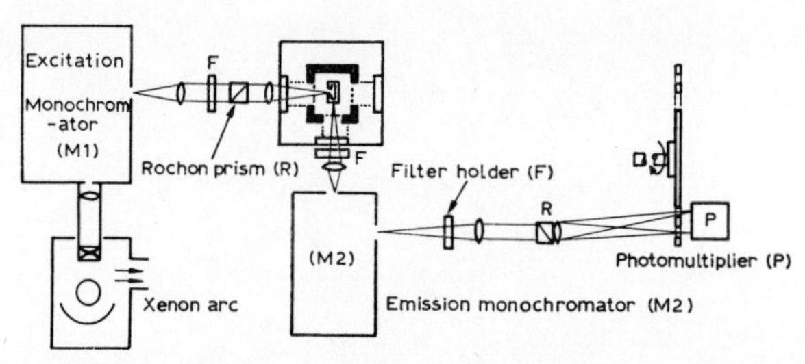

Fig. 115. Arrangement of Ainsworth and Winter's recording polarisation spectrofluori-(meter from Ainsworth and Winter[211]).

frequencies by passing through holes cut in a rotating disc, and both beams finally fell on the photocathode of the same photomultiplier. After amplification, the two AC signals were separated, rectified and the resulting voltages (proportional to $I_{\parallel}$ and $I_{\perp}$) were fed to a small analogue computer from which appropriate functions of $I_{\parallel}$ and $I_{\perp}$ could be derived and recorded. Facilities were available for recording either

$$p = \frac{I_{\parallel} - I_{\perp}}{I_{\parallel} + I_{\perp}} \tag{263}$$

or

$$\cos^2 \beta = \frac{2I_{\parallel} - I_{\perp}}{I_{\parallel} + 2I_{\perp}} \tag{264}$$

(see equations 49 and 53). Arrangements were also made to correct automatically for variations in the relative transmission factor T of the analysing monochromator for vertically and horizontally polarised light.

Chapter 4

Special Topics and Applications

A. DETERMINATION OF PARAMETERS OF THE LOWEST EXCITED SINGLET STATE

1. General Comments

The various processes by which the excited singlet state, S*, can be consumed were considered in Chapter 2 and may be summarised as follows:

radiative transition	rate = $k_f[S^*]$
internal conversion	$k_n[S^*]$
intersystem crossing	$k_g[S^*]$
solute quenching	$k_Q[Q][S^*]$
energy transfer	

Solute quenching includes self-quenching in which the fluorescent solute itself acts as the quencher, Q. Energy transfer and self-quenching can both be reduced to negligible proportions by working with moderately dilute solutions ($<10^{-4}$M). The rate constant k_Q for quenching by a second solute may be determined from the Stern–Volmer quenching constant (equation 75) and the lifetime of the fluorescence, τ_0, in the absence of added quencher. There then remain the rate constants for the first three of the processes listed above, and we shall consider these in the following sections. It should be noted that the rate constant for internal conversion includes the rate of fluorescence quenching by the solvent itself. Such quenching may result in photochemical reaction or it may correspond to an increase in the rate of intersystem crossing, as we shall see.

2. Fluorescence Lifetimes

There are two methods in general use for the determination of fluorescence lifetimes. The first and most accurate requires the use of a fluorometer (see

Section 3 M 1–2) to measure directly the actual lifetime, τ_0, under the conditions of test. In dilute solution without added quencher this gives directly the sum of the three rate constants of interest, since:

$$1/\tau_0 = k_f + k_n + k_g \tag{265}$$

Measurement of the fluorescence efficiency, ϕ_f^0, (see Sections 3 L) then provides sufficient data to calculate the radiative lifetime, τ_r, and hence k_f, since:

$$\phi_f^0 = k_f/(k_f + k_n + k_g) = k_f\tau_0 = \tau_0/\tau_r \tag{266}$$

The second method does not require the use of a fluorometer and for this reason it is more frequently used. The radiative lifetime is calculated directly from the integrated absorption spectrum using equations 27, 28 or 29. This value, together with the fluorescence efficiency, ϕ_f^0, measured under the conditions of interest, gives the actual lifetime, τ_0, via equation 266.

A third method, less frequently used, was described in Section 3 M 3. This assumes that quenching by an added solute, such as biacetyl, is diffusion-controlled and hence the lifetime, τ_0, is derived directly from the Stern–Volmer quenching constant and the calculated diffusion-controlled rate constant under the conditions of interest. The value of τ_0, together with ϕ_f^0, then yields τ_r as before.

3. Intersystem Crossing Rates

The separate values of k_n and k_g cannot be determined from fluorescence measurements alone and the value of τ_0, as determined by the methods described above, must be combined with the value of the triplet formation efficiency, ϕ_t, determined by an independent method, to obtain the value of k_g from:

$$k_g = \phi_t/\tau_0 \tag{267}$$

(see equation 93).

Until recently, the only method of determining ϕ_t was based on the assumption that at low temperature (77°K) the rate of internal conversion, k_n, is negligible so that the value of ϕ_t at low temperature is obtained directly from the fluorescence efficiency at low temperature (equation 108). The photochemist is generally more interested in the value of ϕ_t in fluid solution at room temperature. If the compound emits a measurable phosphorescence under these conditions, the value at room temperature may be derived from that at low temperature by measuring the phosphorescence efficiencies, ϕ_p, and lifetimes τ_p, at the two temperatures, and applying equation 103 so that:

$$\phi_t = \frac{\phi_p \tau_p'}{\phi_p' \tau_p} \phi_t' \sim \frac{\phi_p \tau_p'}{\phi_p' \tau_p} (1 - \phi_f') \tag{268}$$

where the primed symbols refer to low temperature.

Several direct methods for the determination of ϕ_t in fluid solution at room temperature have recently been described. Bowers and Porter[216] have devised a method to determine the absolute concentrations of triplet molecules formed in flash absorption spectroscopy and have applied this to the determination of triplet formation efficiencies. Lamola and Hammond[105] have employed a method in which the compound to be studied is used as a photosensitiser for reactions known to involve the triplet states of the substrates. Since neither of these methods depends on the measurement of photoluminescence they will not be described further. Two methods which do involve the measurement of photoluminescence are described in the following sub-sections. One of these methods combines the results of fluorescence quenching measurements with measurements of the relative concentrations of triplet molecules formed in flash photolysis. The second requires only the measurement of prompt and delayed fluorescence but uses a compound of known triplet formation efficiency as a "standard".

4. Determination of Triplet Formation Efficiencies by the Method of Medinger and Wilkinson

Fluorescence quenching by an added solute, Q, may be the result of chemical reaction, e.g. the quenching of the fluorescence of anthracene by carbon tetrachloride, the formation of dianthracene or the formation of excited dimers. It may be the result of energy transfer (Section 2 B 6):

$$^1A^* + {}^1Q \rightarrow {}^1A + {}^1Q^* \tag{269}$$

or charge transfer[217]:

$$^1A^* + {}^1Q \rightarrow A^{\pm}Q^{\mp} \rightarrow {}^1A + {}^1Q \tag{270}$$

Of a particular interest is a process suggested by Kasha[218] in which the fluorescence quenching is the result of intersystem crossing caused by collision of the excited singlet molecule with the quencher molecule:

$$^1A^* + {}^1Q \rightarrow {}^3A + {}^1Q \tag{271}$$

This process is to be expected with quenchers containing heavy atoms because of the greater spin–orbit coupling. Medinger and Wilkinson[103] investigated this process using bromobenzene as the heavy atom quencher and anthracene, and some of its derivatives, as the fluorescent substance. They measured the relative triplet formation efficiency with and without quencher, by the method of flash absorption spectroscopy, and found that the fluorescence quenching by heavy atom quenchers was indeed accompanied by an increase in triplet formation. Their results for the relative amounts of triplet formed with and without bromobenzene present (D_T/D_T^0) and the corresponding values of fluorescence quenching (F^0/F) are shown in Table 28. By making one reasonable assumption they were able to use these values to calculate the triplet formation efficiencies of the fluorescent solutes.

Using our terminology, the kinetic scheme or Medinger and Wilkinson is similar to that given in Section 2 B 1, but with two quenching reactions:

$$^1A \xrightarrow{h\nu} {}^1A^* \qquad \text{rate} = I_a \tag{272}$$

$$^1A^* \rightarrow {}^1A + h\nu' \qquad k_f[^1A^*] \tag{273}$$

$$^1A^* \rightarrow {}^1A \qquad k_n[^1A^*] \tag{274}$$

$$^1A^* \rightarrow {}^3A \qquad k_g[^1A^*] \tag{275}$$

$$^1A^* + Q \rightarrow {}^3A + Q \qquad k_Q[^1A^*][Q] \tag{276}$$

$$^1A^* + Q \rightarrow {}^1A + Q \qquad k_Q'[^1A^*][Q] \tag{277}$$

$$^3A \rightarrow {}^1A \qquad k_h[^3A] \tag{278}$$

The steady state equations for $[^1A^*]$ give:

$$\phi_f^0/\phi_f = 1 + (k_Q + k_Q')[Q]\tau \tag{279}$$

where τ is the singlet lifetime in the absence of quencher. The relative amounts of triplet formed with and without quencher are given by:

$$\frac{D_T}{D_T^0} = \frac{1 + k_Q[Q]/k_g}{1 + (k_Q + k_Q')[Q]\tau} \tag{280}$$

Elimination of [Q] between 279 and 280 gives:

$$\frac{\phi_f^0}{\phi_f} = \left(\frac{D_T\phi_f^0}{D_T^0\phi_f} - 1\right)\left(1 + \frac{k_Q'}{k_Q}\right)\phi_t^0 + 1 \tag{281}$$

TABLE 28

FLUORESCENCE QUENCHING AND INCREASE OF TRIPLET FORMATION CAUSED BY BROMOBENZENE

(Adapted from Medinger and Wilkinson[103])

Compound	Solvent	Concentration of bromobenzene (M)	F^0/F	D_{T0}/D_T	ϕ_t^0
Anthracene	Liq. paraffin	0.96	2.08	1.17	0.75
Anthracene	Liq. paraffin	1.92	3.35	1.205	
9-Methylanthracene	Liq. paraffin	0.96	1.31	1.09	0.72
9-Methylanthracene	Liq. paraffin	1.92	1.57	1.13	
9-Phenylanthracene	Isopropanol	0.96	1.50	1.30	0.51
9-Phenylanthracene	Isopropanol	1.92	1.62	1.38	
9-Phenylanthracene	Liq. paraffin	0.48	1.12	1.18	0.37
9-Phenylanthracene	Liq. paraffin	0.96	1.26	1.36	
9-Phenylanthracene	Liq. paraffin	1.92	1.44	1.64	
9-Phenylanthracene	Liq. paraffin	2.88	1.57	1.64	
9,10-Diphenylanthracene	Liq. paraffin	1.92	1.07	1.5	0.12

where ϕ_t^0 is the triplet formation efficiency in the absence of quencher ($= k_g\tau$). (It should be noted that with very high quencher concentrations 279 and 280 require the introduction of kinetic activity factors which have the effect of altering the effective value of [Q]. Since however equation 281 is independent of [Q], it is not affected by these factors.)

Medinger and Wilkinson found that plots of F^0/F against $D_TF^0/(D_T^0F) - 1$ were linear, and according to equation 281 the slopes of the lines are equal to $(1 + k_Q'/k_Q)\phi_t^0$. It was found that the sums of these slopes and the respective fluorescence efficiencies were unity within experimental error, i.e.:

$$\phi_f^0 + (1 + k_Q'/k_Q)\phi_t^0 = 1 \tag{282}$$

This means that either k_Q' is negligible, or that:

$$k_Q'/k_Q = k_n/k_g \tag{283}$$

Equation 283 follows from the fact that $1/\tau = (k_f + k_n + k_g)$. Medinger and Wilkinson argued that since there is no obvious reason why k_Q'/k_Q should equal k_n/k_g for all the compounds, it is probable that k_Q' is small in comparison with k_Q and hence ϕ_t^0 can be calculated directly from the measured values of F^0/F and D_T/D_T^0 via equation 281. The values so calculated are shown in column 6 of Table 28. The results indicated that for these four compounds, in the solvents used, the sum of ϕ_t^0 and ϕ_f^0 was unity.

In a later paper[219] Medinger and Wilkinson applied the same method to pyrene in ethanol using ethyl iodide as quencher. In dilute solutions in which the formation of excited dimer is negligible they found $\phi_t^0 = 0.38$. The sum of this and the fluorescence efficiency originally reported by Parker and Hatchard[34] was close to unity. (Compare with the values obtained by the method described in the following section—Table 30.) As the concentration of pyrene is increased, the triplet formation efficiency is expected to decrease owing to competition of the self-quenching (and accompanying excited dimer formation—see Section 4 D 3) with processes 273–275. Medinger and Wilkinson found that the triplet formation efficiency did decrease at high concentrations (see Table 29) but not so rapidly as would be predicted from the self-quenching reaction. They showed that their results could be satisfactorily interpreted on the assumption that the excited dimer dissociated to give a triplet molecule:

$$^1P_2^* \rightarrow {}^3P + {}^1P \tag{284}$$

as well as singlet excited and ground state molecules (see Section 4 D 3):

$$^1P_2^* \rightarrow {}^1P^* + {}^1P \tag{285}$$

$$^1P_2^* \rightarrow {}^1P + {}^1P \tag{286}$$

At high pyrene concentration the sum of ϕ_t, ϕ_M and ϕ_D was less than unity (see Table 29) indicating that the excited dimer decays to an appreciable extent by the non-radiative process 286.

Finally the results are interesting in showing that organic molecules containing heavy atoms can quench singlet states much more efficiently than they quench triplet states—with pyrene the ratio of the two quenching constants is about 2000.

TABLE 29

EFFICIENCIES OF TRIPLET FORMATION (ϕ_t) AND OF MONOMER (ϕ_M) AND DIMER (ϕ_D) FLUORESCENCE

(From Medinger and Wilkinson[219])

Pyrene concentration	ϕ_t	ϕ_M (ref. 34)	ϕ_D (ref. 34)
Low	0.38	0.65	0
2×10^{-5}	0.38	0.63	0.02
10^{-4}	0.33	0.52	0.08
3×10^{-4}	0.26	0.41	0.21
10^{-3}	0.185	0.21	0.38

5. Triplet Formation Efficiencies from Delayed Fluorescence Measurements

Alternative methods of determining triplet formation efficiencies have been described by Parker and Joyce[113,114,220]. These rely on comparison with a standard compound of known triplet formation efficiency. However they have the advantage of simplicity, and one of them can be applied to compounds that are themselves not fluorescent. We shall therefore describe these methods in detail.

General equations for the efficiency of sensitised P-type delayed fluorescence were discussed in Section 2 D 5, where it was shown that in general two mechanisms have to be considered, namely that involving mutual interaction of two triplet molecules of acceptor (process 170) and that involving mixed triplet quenching (process 172). If however the concentration of acceptor is made sufficiently large, the donor triplet is strongly quenched and mixed triplet interaction is negligible. Under these conditions absorption of light by the donor causes population of the triplet level of the acceptor by the processes:

$$^1D \xrightarrow{h\nu} {}^1D^* \rightarrow {}^3D \qquad \text{rate} = I_a\phi_t^D \tag{287}$$

$$^3D + {}^1A \rightarrow {}^1D + {}^3A \qquad p_e'k_c[^3D][^1A] \tag{288}$$

The probability factor p_e' is the proportion of quenching encounters that result in triplet-to-singlet energy transfer and is introduced to take account of the possible occurrence of the process:

$$^3D + {}^1A \rightarrow {}^1D + {}^1A \quad \text{rate} = (1 - p_e')k_c[^3D][^1A] \tag{289}$$

The fate of the acceptor triplets is determined by:

$$^3A \rightarrow {}^1A \qquad \text{rate} = k_j[^3A] \tag{290}$$

$$^3A + {}^3A \rightarrow {}^1A^* + {}^1A \qquad p_ck_c[^3A]^2 \tag{291}$$

$$^1A^* \rightarrow {}^1A + h\nu' \tag{292}$$

With high concentrations of acceptor, ^3D is consumed almost entirely by process 288 and 289 and the rate of formation of acceptor triplets is equal to $p_e'I_a\phi_t^D$. At low rates of light absorption, the rate of the bimolecular process 291 is negligible compared with that of process 290 and hence:

$$p_e'I_a\phi_t^D = k_j[^3A] = [^3A]/\tau \tag{293}$$

in which τ is the lifetime of the *acceptor* triplet and is equal to twice the lifetime of the sensitised delayed fluorescence. The rate of emission of delayed fluorescence via processes 291 and 292 is thus equal to:

$$\tfrac{1}{2}\phi_f p_c k_c[^3A]^2 = \tfrac{1}{2}\phi_f p_c k_c(p_e' I_a \phi_t^D \tau)^2 \tag{294}$$

where ϕ_f is the efficiency of prompt fluorescence of the *acceptor*. The relationship between the intensities, I_{DF}, of sensitised delayed fluorescence emitted by two solutions containing the same acceptor but different donors in the same solvent is thus:

$$\frac{(I_{DF})_1}{(I_{DF})_2} = \left[\frac{(I_a)_1(p_e'\phi_t^D)_1\tau_1}{(I_a)_2(p_e'\phi_t^D)_2\tau_2}\right]^2 \tag{295}$$

Thus if p_e' is the same for both donors (it is generally assumed to be unity), measurement of the relative intensities and lifetimes of the delayed fluorescence in the same apparatus at a known relative rate of light absorption allows the ratio $(\phi_t^D)_1/(\phi_t^D)_2$ to be calculated. Parker and Joyce[104, 220] chose perylene as acceptor and used a wavelength of 313 nm for excitation. At this wavelength perylene has a low extinction coefficient and a low triplet formation efficiency (see below) so that the amount of delayed fluorescence produced by direct excitation of the perylene was negligible. They used anthracene as the standard donor for which they assumed $(\phi_t^D)_2 = 0.70$. Their results are shown in Table 30 from which it will be observed that the sum $(\phi_f + \phi_t)$ was close to unity with the seven compounds investigated.

The experimental procedure is simple. Solutions containing the two donors to be compared, at concentrations giving equal optical densities

TABLE 30

TRIPLET FORMATION EFFICIENCIES FROM DELAYED FLUORESCENCE MEASUREMENTS

(From Parker and Joyce[104])

Donor	ϕ_t	ϕ_f	$\phi_t + \phi_f$
Anthracene	[0.70]	[0.30]	—
Acenaphthene	0.45	0.39	0.84
Chrysene	0.82	0.17	0.99
1-Methoxynaphthalene	0.46	0.53	0.99
Naphthalene	0.71	0.21	0.92
Phenanthrene	0.80	0.13	0.93
Pyrene	0.27	0.72	0.99
Triphenylene	0.89	0.09	0.98

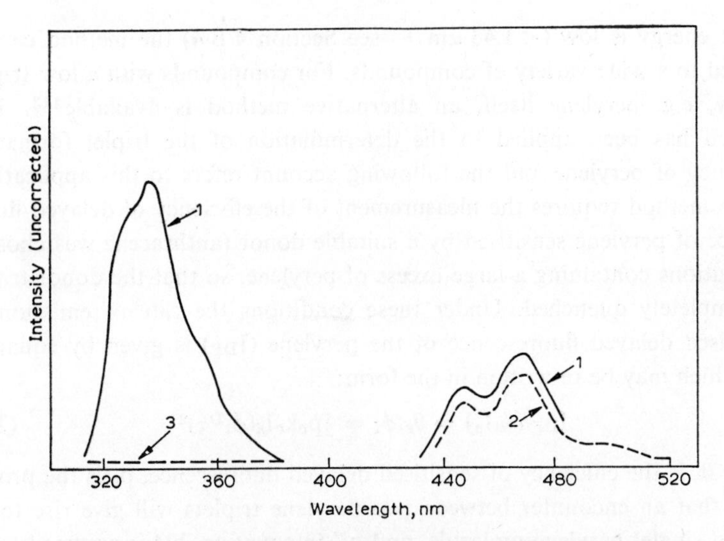

Fig. 116. Perylene delayed fluorescence sensitised by acenaphthene. 1, prompt fluorescence at sensitivity × 3, showing bands due to acenaphthene (left) and perylene (right); 2, delayed fluorescence at sensitivity × 150; 3, delayed fluorescence at sensitivity × 1500, indicating complete quenching of acenaphthene triplet. Rate of light absorption was about 10^{-6} einstein litre^{-1} sec^{-1}; band width was 10 nm. Spectra are distorted by self-absorption and low resolution (from Parker and Joyce[104]).

(e.g. 0.200) at the excitation wavelength, and containing in addition the same concentration of perylene (e.g. 4×10^{-5}M), are vacuum-degassed in optical cells (see Section 3 J 2) and the intensities of sensitised delayed fluorescence of perylene compared directly to give $(I_{DF})_1/(I_{DF})_2$. In a separate experiment the lifetimes of the sensitised delayed fluorescence from the two solutions are measured. Substitution in equation 295 ($(I_a)_1 = (I_a)_2$) gives the required ratio $(\phi_t^D)_1/(\phi_t^D)_2$. Typical prompt and delayed fluorescence spectra are shown in Fig. 116. The complete absence of delayed fluorescence from the donor (curve 3) indicates that the concentration of perylene was sufficient to quench the donor triplet almost completely so that equation 294 could be applied. One of the main sources of error with donors having a steep absorption curve in the region of the excitation wavelength is the measurement of the precise optical density of the solution. It is necessary to use monochromatic exciting light and to measure the optical density with the same beam as that used to measure the intensity of delayed fluorescence.

The method just described can be applied to all compounds that have a triplet level situated above that of the chosen acceptor. Since the perylene triplet energy is low ($< 1.48\ \mu m^{-1}$—see Section 4 B 4) the method can be applied to a wide variety of compounds. For compounds with a low triplet energy, e.g. perylene itself, an alternative method is available[113]. This method has been applied to the determination of the triplet formation efficiency of perylene and the following account refers to this application.

The method requires the measurement of the efficiency of delayed fluorescence of perylene sensitised by a suitable donor (anthracene was chosen) in solutions containing a large excess of perylene, so that the donor triplet is completely quenched. Under these conditions the rate of emission of sensitised delayed fluorescence of the perylene (I_{DF}) is given by equation 294 which may be re-written in the form:

$$I_{DF}/(\phi_f I_a) = \theta_s/\phi_f = \tfrac{1}{2} p_c k_c I_a (\phi_t{}^D \tau)^2 \qquad (296)$$

where θ_s is the efficiency of sensitised delayed fluorescence, p_c is the probability that an encounter between two perylene triplets will give rise to an excited singlet perylene molecule, and p_e' in equation 294 is assumed to be unity. All parameters except p_c in equation 296 are known ($\phi_t{}^D$ for anthracene assumed to be 0.70) or can be measured, and hence p_c can be determined. Values of p_c for perylene in three solvents are given in Table 31. This value of p_c is then used to derive the triplet formation efficiency of perylene from measurements of its directly excited delayed fluorescence. The efficiency of the latter is related to the triplet formation efficiency by equation 138. Values of θ/ϕ_f at measured values of I_a are shown in Table 32 and using the value of p_c from Table 31, the values of ϕ_t for perylene were calculated. The low values of ϕ_t are consistent with the large fluorescence efficiency of perylene reported by Bowen[182] (0.98 in benzene).

Parker and Joyce[114] have applied the same method to the determination

TABLE 31

SENSITISED DELAYED FLUORESCENCE OF PERYLENE

(From Parker and Joyce[113])

Solvent	I_a (einstein litre^{-1} sec^{-1})	θ/ϕ_f	$\tau_t{}^P$ (sec)	p_c
Ethanol	1.39×10^{-6}	0.93×10^{-3}	4.34×10^{-3}	0.026
n-Hexane	1.33	1.86	4.84	0.012
Cyclohexane	1.33	0.42	3.92	0.012

Note

All Solutions contained 10^{-5} M perylene and 5×10^{-5} M anthracene. Rates of light absorption (I_a) refer to anthracene at 366 nm.

TABLE 32

TRIPLET FORMATION EFFICIENCY OF PERYLENE
(From Parker and Joyce[113])

Solvent	I_a	θ/ϕ_f	τ_t^P	ϕ_t^P
Ethanol	1.97×10^{-5}	2.1×10^{-6}	4.34×10^{-3}	0.0088
n-Hexane	2.02	12.8	4.84	0.015
Cyclohexane	2.07	2.6	3.92	0.014

Note

All solutions contained 10^{-5} M perylene. Rates of light absorption (I_a) refer to 436 nm.

of the triplet formation efficiencies of chlorophylls *a* and *b* in ethanol, using anthracene or naphthalene as donor. Corrections had to be applied for the small proportion of delayed fluorescence excited by direct light absorption by the chlorophylls in the sensitised solutions. For details of these corrections the reader is referred to the original paper. The results obtained are included in Table 33. The sum $(\phi_f + \phi_t)$ falls far short of unity, suggesting that internal conversion from the excited singlet state is considerable with the two chlorophylls. In this respect the results differ from those obtained in ethereal solution by Bowers and Porter[216] by the method of flash absorption spectroscopy.

6. Triplet Formation Efficiencies from Phosphorescence Measurements

Borkman and Kearns[221] have devised a method of determining the approximate value of ϕ_t/ε_{ST}, where ε_{ST} is the extinction coefficient of the direct $T_1 \leftarrow S_0$ transition at a suitable wavelength. Since ε_{ST} is difficult to measure directly, the method is more appropriate to the determination of this parameter with compounds for which ϕ_t has been determined by another method. For this reason the principle of the method is dealt with in Section 4 B 5.

B. TRIPLET STATE PARAMETERS

1. Rate of $T_1 \rightarrow S_0$ Intersystem Crossing

The total rate of disappearance of triplet molecules in solution is made up of the following processes:

$$d[T]/dt = (k_p + k_m + k_e + k_v + k_q[q] + k_G[S_0])[T] + k_a[T]^2 \quad (297)$$

The rate constant k_G for quenching by the ground state of the solute is often small, and the rate of this process can be reduced to a negligible value by working in sufficiently dilute solutions. Some triplet molecules react rapidly and irreversibly with the solvent: the rate constant, k_v, for this reaction then dominates and the triplet lifetime of, for example, benzophenone in isopropanol at room temperature is less than 10^{-7} sec[107, 222]. With some triplets however, k_v can be neglected. Early investigations of such "unreactive" triplet molecules in fluid solution, by the method of flash absorption spectroscopy, gave triplet lifetimes of the order 10^{-4} sec and these lifetimes were found to be strongly viscosity dependent. These results were in many cases due to neglect of bimolecular triplet–triplet quenching, the rate constant for which, k_a, is often diffusion-controlled, and hence at the triplet concentrations used for flash absorption measurements, the bimolecular reaction proceeded at a rate comparable with or greater than the sum of the first order reaction rates. Because the measurement of phosphorescence or delayed fluorescence in fluid solution can be made at rates of light absorption for which the bimolecular triplet–triplet quenching is negligible, it has considerable advantages as a method for investigating the rates of the first order triplet decay processes. The composite first order rate constant for triplet decay:

$$k_h = k_p + k_m + k_e + k_q[q] \quad (298)$$

has a value between 10^2 and 10^3 sec^{-1} (see column 3 of Table 33) for many "unreactive" triplets in fluid solution at room temperature. It generally decreases with temperature although chlorophylls *a* and *b* seem to be exceptions to this rule[114, 223]. The degree to which k_m contributes to this composite rate constant has still not been decided. The difficulty is that k_q for impurity quenching is often diffusion-controlled ($k_q \sim k_c \sim 10^{10}$ in solvents such as ethanol or hexane), and hence to give a lifetime of 10 msec, [q] must not be greater than about 10^{-8}M. As discussed in Section 3 N 7 the only method of proving directly that a particular batch of highly purified solvent contains this amount of quenching impurity is to produce another specimen of solvent that gives an even longer triplet lifetime. However, Stevens and Walker[115] have presented indirect but convincing evidence that the temperature-dependence of the triplet decay of several aromatic hydrocarbons in liquid paraffin is due largely to the presence of quenching impurities either in the solvent or in the solute. The impurity concentrations

TABLE 33

TRIPLET DATA IN ETHANOL AT ROOM TEMPERATURE

(Singlet energies are estimated from short wavelength peaks of emission spectra)

Compound	Triplet energy (μm^{-1})	Triplet lifetime at 20° (msec)	ϕ_f	ϕ_t	p_c	Singlet energy (μm^{-1})
Triphenylene	2.38[a]	—	0.09	0.89	—	2.90
Phenanthrene	2.21[a]	2.0	0.13	0.80	0.05	2.87
Naphthalene	2.12[a]	3.6	0.21	0.71	0.56	3.16
Acenaphthene	2.09[a]	3.3	0.39	0.45	0.31	3.14
1-Methoxy-naphthalene	2.10[a]	5.5	0.53	0.46	0.35	3.05
Chrysene	2.00[a]	—	0.17	0.82	—	2.77
Fluoranthene	1.86[a]	8.5	0.21	0.79[d]	0.04[d]	2.46
Proflavine	1.71	6.1	0.40	0.60[d]	0.0001[d]	2.04
Pyrene	1.69[a]	11	0.72	0.27	0.27	2.69
1,2-Benz-anthracene	1.65[a]	9.4	0.20	0.80[d]	0.21[d]	2.59
Acridine 0	1.60	8.3	0.46	0.54[d]	0.0001[d]	1.91
Anthracene	1.48[a]	8.8	0.30	0.70	0.08	2.65
9-Methyl-anthracene	~ 1.47[c]	10	0.33	0.67	0.03	2.56
9,10-Dimethyl	~ 1.47[c]	8	0.89	0.032	0.07	2.48
9-Phenyl	~ 1.47[c]	15	0.49	0.47	0.05	2.54
9,10-Diphenyl	< 1.47[c] > 1.26	22	0.89	0.024	0.13	2.48
Benz(a)pyrene	1.47[a]	8.8	0.42	0.58[d]	0.02[d]	2.48
Eosin	1.42	1.5	0.45	—	—	1.80
Perylene	1.26	5.0	0.98[b]	0.01	0.03	2.28
Chlorophyll *b*	1.14	1.5	0.10	0.50	0.06	1.50
Chlorophyll *a*	0.99	0.8	0.23	0.24	0.10	1.49

[a] These values from phosphorescence at 77°K.
[b] In benzene.
[c] From sensitisation experiments with anthracene and perylene.
[d] These values estimated by assuming $(\phi_f + \phi_t) = 1$, and the corresponding values of p_c are minimum values.

involved here were considerably higher than 10^{-8}M because of the relatively high viscosity of liquid paraffin and the correspondingly lower value of k_c. Their work provides a good example of the information that can be obtained from measurements of phosphorescence and P-type delayed fluorescence, and we shall therefore consider it in some detail.

Stevens and Walker measured the lifetimes of both phosphorescence (τ_t) and P-type delayed fluorescence (τ_{DF}) of solutions of acenaphthene, pyrene, 1,2-benzanthracene and fluoranthene in liquid paraffin over a wide range of temperature. Over a considerable part of the temperature range they were able to measure both types of emission, and in this range they found that the lifetime of delayed fluorescence was equal to one half of that of the phosphorescence, thus confirming that equation 140 was obeyed. At high temperatures (low viscosity) phosphorescence was weak and they derived triplet lifetimes from the observed lifetimes of the P-type delayed fluorescence (i.e. $\tau_t = 2\tau_{DF}$). At very low temperatures delayed fluorescence was weak and the triplet lifetimes were then derived directly from the lifetimes of phosphorescence. Their results, plotted in the form:

$$\log_{10}(k_h) = \log_{10}(1/\tau_t) = \log_{10}(1/2\tau_{DF}) \quad (299)$$

against $1/T$ are shown in Fig. 117 and we shall now give their interpretation of these curves in terms of the presence of quenching impurities in solvent and/or solute.

At low temperatures k_h approaches a limiting value, k_h^0, to which they assigned the value:

$$k_h^0 = k_p + k_m \quad (300)$$

i.e. they assumed k_m was independent of temperature. With increasing temperature, k_h rises in a manner consistent with increasing diffusion-controlled quenching by impurities as a result of decreasing solvent viscosity[224]. In this intermediate viscosity range the diffusion-controlled rate constant is given by:

$$k_c = \frac{8\mathbf{RT}}{3000\eta^0} \exp(-E/\mathbf{RT}) \quad (301)$$

where E represents the "activation energy" of solvent fluidity ($1/\eta$). Hence ignoring thermal activation ($k_e[T]$), equation 298 becomes:

$$k_h = k_p + k_m + \frac{8p_e\mathbf{RT}}{3000\eta^0} [q] \exp(-E/\mathbf{RT}) \quad (302)$$

in which p_e is the probability that an encounter between triplet molecule and impurity will result in quenching of the triplet. The log plot for acenaphthene (open and solid circles in Fig. 117 a) could be accounted for by an equation of form similar to 302 with $E = 12.7$ kcal, i.e. close to the value for the activation energy of solvent fluidity (11 ± 2 kcal). By making certain assumptions Stevens and Walker estimated the concentration of quenching impurity to be 10^{-5}M (i.e. the equivalent of 1 mole % in the solute or 10^{-5} mole/litre in the solvent).

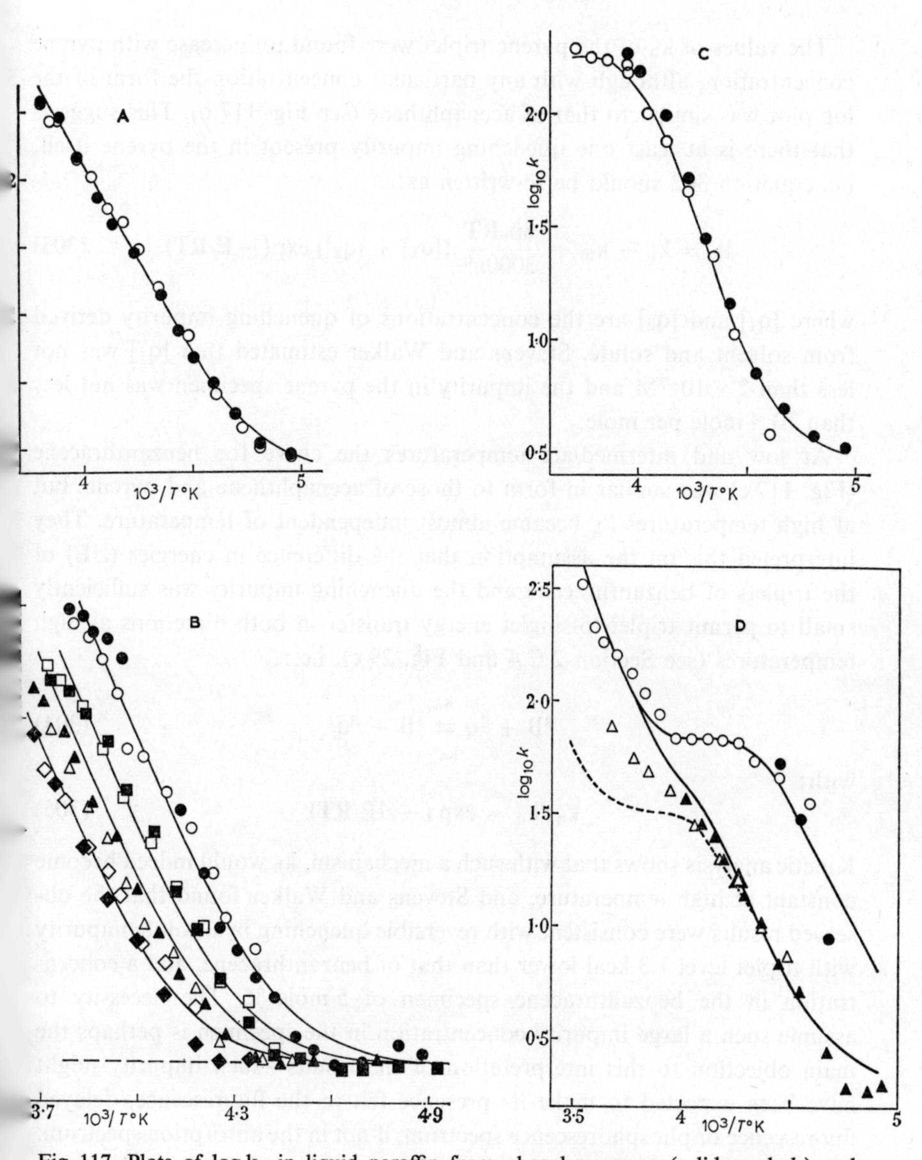

Fig. 117. Plots of log k_h in liquid paraffin from phosphorescence (solid symbols) and delayed fluorescence (open symbols).

A, acenaphthene, 10^{-3} M; B, pyrene, 3×10^{-2} M, 10^{-2} M, 5×10^{-3} M, 10^{-3} M; C, 1,2-benzanthracene, 2×10^{-3} M; D, fluoranthene, 5×10^{-2} M, 5×10^{-3} M.

The values of k_h of the pyrene triplet were found to increase with pyrene concentration, although with any particular concentration the form of the log plot was similar to that of acenaphthene (see Fig. 117 b). This suggests that there is at least one quenching impurity present in the pyrene itself, i.e. equation 302 should be re-written as:

$$k_h = k_p + k_m + \frac{8p_e\mathbf{RT}}{3000\eta^0}([q_1] + [q_2])\exp(-E/\mathbf{RT}) \tag{303}$$

where $[q_1]$ and $[q_2]$ are the concentrations of quenching impurity derived from solvent and solute. Stevens and Walker estimated that $[q_1]$ was not less than 2×10^{-7}M and the impurity in the pyrene specimen was not less than 10^{-4} mole per mole.

At low and intermediate temperatures the curve for benzanthracene (Fig. 117 c) was similar in form to those of acenaphthene and pyrene, but at high temperatures k_h became almost independent of temperature. They interpreted this on the assumption that the difference in energies (ΔE) of the triplets of benzanthracene and the quenching impurity was sufficiently small to permit triplet-to-singlet energy transfer in both directions at high temperatures (see Section 2 C 4 and Fig. 29 c), i.e.:

$$^3B + {}^1q \underset{k_q'}{\overset{k_q}{\rightleftarrows}} {}^1B + {}^3q \tag{304}$$

with:

$$k_q'/k_q = \exp(-\Delta E/\mathbf{RT}) \tag{305}$$

Kinetic analysis shows that with such a mechanism, k_h would indeed become constant at high temperature, and Stevens and Walker found that the observed results were consistent with reversible quenching by a solute impurity with triplet level 1.3 kcal lower than that of benzanthracene, and a concentration in the benzanthracene specimen of 5 mole %. The necessity to assume such a large impurity concentration in the specimen is perhaps the main objection to this interpretation of the results—such impurity might have been expected to make its presence felt in the fluorescence, delayed fluorescence or phosphorescence spectrum, if not in the absorption spectrum.

The results with 5×10^{-2}M fluoranthene (Fig. 117 d) show the same characteristics as those with benzanthracene, but in addition, at the highest temperatures k_h increases sharply with increase of temperature. Stevens and Walker interpreted this by assuming the presence of a second quenching impurity with a triplet level *higher* than that of the fluoranthene (i.e. corresponding to case D in Fig. 29). The reader is referred to the original paper for full details of the kinetic arguments used.

The photoluminescence measurements of Stevens and Walker are the most detailed that have been made of the effect of viscosity and temperature, and they have led, as we have seen, to the conclusion that the increase in k_h with temperature is due to impurity quenching. The implication from this work is that the variation of k_m with temperature is small, at least in comparison with the variation in $k_q[q]$. However, the values of [q] in the systems investigated were relatively high and it would clearly be desirable to carry out such an investigation with both solutes and solvents purified to the greatest possible degree. (For this purpose liquid paraffin is not very suitable.) To minimise the number of impurities capable of quenching the triplet of the solute, the latter should have a low triplet energy. If indeed impurity quenching is the only cause for variation of k_h with temperature, this variation would be expected to be least with solutes having very low triplet energies. Chlorophylls *a* and *b* are such compounds and k_h is in fact found to vary little with temperature[114, 223]. However, the value of k_h is relatively high and the small effect of impurity quenching with the chlorophylls may be partly due to the high value of $(k_p + k_m)$. For triplets having low values of $(k_p + k_m)$ at 77°K, e.g. many aromatic hydrocarbons, the limited amount of data available for fluid solutions at room temperature (see Table 33) show a very rough correlation between triplet lifetime and triplet energy, as would be expected if impurity quenching were the factor controlling the lifetime. Unfortunately triplet lifetimes for compounds with high triplet energies are scarce, either because the specimens so far prepared give sensitised impurity delayed fluorescence (e.g. triphenylene), or because the triplet cannot be observed at all under these conditions (e.g. benzene).

TABLE 34

TRIPLET LIFETIMES OF EOSIN DI-ANION
(From Parker and Hatchard[19])

Temperature (°C)	Lifetime (msec)	
	in glycerol	in ethanol
− 196	10.7	9.1
− 70	5.7	3.7
− 20	3.0	2.7
+ 25	2.6	1.5
+ 70	0.9	0.7

Parker and Hatchard[19] have measured the lifetimes of triplet eosin in both glycerol and ethanol at various temperatures (see Table 34). They observed an increase in k_h as the temperature increased but at any one temperature the values in the two solvents differed by less than a factor of two in spite of the considerable difference in viscosity. The implication is that diffusion-controlled quenching contributed comparatively little to the values of k_h in these solutions. It is possible however that the approximate agreement between the two solvents was fortuitous and that the higher viscosity of the glycerol was compensated by the larger concentration of quenching impurities that it contained. This system is worth more detailed investigation.

2. Rate of $S_1 \leftarrow T_1$ Intersystem Crossing

Intersystem crossing from the lowest triplet level to the first excited singlet level can only occur after thermal activation of T_1 to an upper vibrational level, T_1^*, having an energy equal to or greater than S_1. It is thus observed only with those compounds for which T_1 lies fairly close to S_1, e.g. with dyestuffs such as eosin or proflavine (see Sections 1 C 4 and 2 D 1). The rate constant, k_{ga}, for the process $T_1^* \rightarrow S_1$ can be derived from the efficiencies of phosphorescence (ϕ_p), of E-type delayed fluorescence (ϕ_e) and of triplet formation (ϕ_t), together with the triplet lifetime (τ). Comparatively little data is at present available—the most extensive is that for eosin[19] and we shall use the latter to illustrate the principles involved.

As explained in Section 2 D 1, the plot of $\log_e(\phi_e/\phi_p)$ against $1/T$ is linear with intercept equal to $\tau_R \phi_f A$ where τ_R is the radiative lifetime of the triplet and A is the frequency factor for the rate constant:

$$k_e = A \exp(-\Delta E/RT) \tag{306}$$

The value of A can be calculated from the intercept of the log plot as indicated in Table 35. The significance of A can be seen by considering the separate steps involved in the process of thermal activation from T_1 to S_1. This process involves population of the upper vibrational levels (T_1^*) of the triplet state followed by the actual process of intersystem crossing:

$$T_1 \rightleftarrows T_1^* \overset{k_{ga}}{\rightarrow} S_1 \tag{307}$$

in which

$$[T_1^*]/[T_1] = \exp(-\Delta E/RT) \tag{308}$$

If we make the reasonable assumption that the rate of thermal equilibration

TABLE 35

INTERSYSTEM CROSSING RATES FOR EOSIN IN GLYCEROL AT ROOM TEMPERATURE

ϕ_f	0.45
$\phi_t \leq (1 - \phi_f)$	≤ 0.55
ϕ_p/τ	4.8 sec^{-1}
$\tau_R = \phi_t\tau/\phi_p$	≤ 0.11 sec
Intercept of log plot ($= \tau_R\phi_f A$)	0.5×10^7
$k_{ga}(T_1{}^* \rightarrow S_1) = A$	$\geq 1.0 \times 10^8$ sec^{-1}
τ_r (singlet) $= 1/k_f$	3×10^{-9} sec
τ (singlet) $= \phi_f\tau_r$ (singlet)	1.4×10^{-9} sec
$k_g(S_1 \rightarrow T_1{}^*) = \phi_t/\tau$ (singlet)	$\leq 4 \times 10^8$ sec^{-1}

between T_1 and $T_1{}^*$ (i.e. $\sim 10^{12}$ sec^{-1}) is much greater than k_{ga}, the rate of population of S_1 from T_1 is given by:

$$k_e[T_1] = k_{ga}[T_1{}^*] = k_{ga}[T_1] \exp(-\Delta E/RT) \tag{309}$$

and hence:

$$k_{ga} = A \tag{310}$$

The value of k_{ga} calculated in this way is shown in Table 35 together with the rate constant k_g calculated from the prompt fluorescence efficiency and the singlet radiative lifetime as described in the previous section. It will be observed that the rate constants k_{ga} and k_g for the processes $T_1{}^* \rightarrow S_1$ and $S_1 \rightarrow T_1{}^*$ have the same order of magnitude.

3. Probability of Triplet-to-Triplet Energy Transfer

Encounter between two triplet molecules may in general give rise to the following products, provided that energy requirements are satisfied:

$$^3A + {}^3A \rightarrow {}^1A^* + {}^1A \tag{311}$$

$$\rightarrow {}^1A + {}^1A \tag{312}$$

$$\rightarrow {}^5A + {}^1A \tag{313}$$

$$\rightarrow {}^3A + {}^3A \tag{314}$$

$$\rightarrow {}^1A_2{}^* \tag{315}$$

With dilute solutions of many compounds the delayed dimer fluorescence (via equation 315) is small or negligible compared with the delayed monomer fluorescence, and it is convenient to express the quantum efficiency of the latter (see equation 138) in terms of the diffusion-controlled rate constant, k_c, and the probability, p_c, that an encounter between two triplet molecules will ultimately give rise to an excited singlet molecule. Equation 138 may be re-written:

$$p_c = \frac{2\theta/\phi_f}{k_c I_a (\phi_t \tau)^2} \quad (316)$$

It is a relatively simple matter to measure θ/ϕ_f, I_a and τ, and thus, with k_c calculated from equation 80, and ϕ_t determined by one of the methods described in Section 4 A, the value of p_c can be derived. The values of p_c for a variety of triplets in ethanol are given in column 6 of Table 33.

The very low values for the two cationic compounds proflavine and acridine orange may be explained qualitatively in terms of mutual electrostatic repulsion inhibiting close approach of the triplets. The values of p_c for the 9-substituted and 9,10-di-substituted anthracenes[179] are particularly interesting. If triplet-to-triplet energy transfer requires the close approach of the two triplets with their molecular planes and axes parallel, the steric hinderance caused by substitution at the 9,10 positions would be expected to reduce the value of p_c. Substitution at one position does indeed reduce p_c, but with di-substitution the value increases again. It seems likely therefore that the triplet-to-triplet energy transfer takes place to a large extent at distances greater than the encounter distance—this certainly seems to be the case with pyrene (see Section 4 E 3). It may also be relevant that unsubstituted anthracene forms a stable photo-dimer but diphenylanthracene does not; formation of stable dimer may thus compete with triplet-to-triplet energy transfer and hence reduce the value of p_c for unsubstituted anthracene, although even the latter forms an excited dimer at low temperature[179].

Among the remaining compounds there is no obvious correlation of p_c with the other triplet parameters. If it is assumed that for these compounds the *total* rate of triplet–triplet quenching is close to the diffusion-controlled value, the value of p_c may be expressed in terms of the rates of reactions 311–315 viz.:

$$p_c = k_{311}/(k_{311} + k_{312} + k_{313} + k_{314} + k_{315}) \quad (317)$$

It is conceivable that process 313 could contribute to the production of delayed fluorescence by:

$$^5A \rightarrow {}^1A^* \quad (318)$$

and dissociation of excited dimer would also contribute via process 315. The value of p_c would then be correspondingly greater.

With some compounds, notably naphthalene, p_c determined from equation 316 is high. This may be interpreted in two ways. It may mean that p_c as defined by equation 317 is indeed high, or it may mean that the effective rate constant for bimolecular triplet–triplet quenching is considerably greater than that defined by equation 80, as would be the case if a considerable amount of triplet-to-triplet energy transfer took place at distances greater than the encounter distance. This question would be resolved by precise measurements of the total triplet–triplet quenching rate, e.g., by the method of flash absorption spectroscopy.

4. Determination of Triplet Energies

The simplest method of determining the energy of the triplet state is to measure the phosphorescence spectrum in rigid solution at low temperature. The energy of the triplet state is given with reasonable precision from the wavenumber of the 0–0 band (strictly it is equal to the mean of the wavenumbers of the 0–0 bands in absorption and emission). This method is applicable to any compound that has an appreciable phosphorescence efficiency under these conditions, provided that the high wavenumber limit of the emission lies within the wavenumber region to which the spectrophosphorimeter is sensitive. There are two precautions required. First, location of the 0–0 vibrational band requires care because with some compounds it is weak compared with the 0–1 band and may then be overlooked. Second, with compounds giving weak phosphorescence, purity is important for the obvious reason that if the phosphorescence emission of a strongly phosphorescent impurity appears just to the high wavenumber side of the emission from the substance under investigation, the impurity emission may be mistaken for the weak 0–0 band. Hammond and coworkers[225] have measured the triplet energies of 41 organic compounds in this way using hydrocarbon solvents (see Table 36). Some values in ethanol are included in column 2 of Table 33. A further selection of triplet energies is given in Table 37.

There are some important compounds for which the method of phosphorescence measurement at 77°K is not applicable, either because the phosphorescence efficiency is too low, or because the phosphorescence is situated in the infra-red region where spectrophosphorimeters are insensitive. The triplet energies of such compounds can be determined by the measurement of either their E-type or P-type delayed fluorescence. For example, chlorophylls *a* and *b* in propylene glycol at moderate rates of light absorption emit only E-type delayed fluorescence[114], and measurement of its efficiency and lifetime as a function of temperature provides sufficient data to calculate the energy difference, ΔE, between the triplet and the first excited singlet states as follows. The efficiency of E-type delayed fluorescence is given by equation 123 which may be written:

$$\phi_e/\phi_f = \tau_R \phi_p A \exp(-\Delta E/\mathbf{R}\mathbf{T}) \tag{319}$$

Substituting the value of ϕ_p given by equation 103 we have:

$$\phi_e/(\tau\phi_f) = \phi_t A \exp(-\Delta E/\mathbf{R}\mathbf{T}) \tag{320}$$

Thus if ϕ_t is independent of temperature over the range considered, a plot of ln $[\phi_e/(\tau\phi_f)]$ against $1/\mathbf{T}$ will give a straight line with slope $\Delta E/\mathbf{R}$. Parker

TABLE 36

TRIPLET ENERGIES FROM PHOSPHORESCENCE MEASUREMENTS IN HYDROCARBON SOLVENT AT 77°K

(Adapted from Herkstroeter, Lamola and Hammond[225])

Compound	E_T (μm^{-1})	Compound	E_T (μm^{-1})
Propiophenone	2.61	Phenylglyoxal	2.19
Xanthone	2.60	Anthraquinone	2.18
Acetophenone	2.58	Phenanthrene	2.18
1,3,5-Triacetylbenzene	2.57	α-Naphthoflavone	2.18
Isobutyrophenone	2.56	Flavone	2.17
1,3-Diphenyl-2-propanone	2.53	Ethyl phenylglyoxalate	2.17
Benzaldehyde	2.52	4′,4-Bis(dimethylamino)-benzophenone	2.14
Triphenylmethyl-phenyl ketone	2.48	Naphthalene	2.13
Carbazole	2.45	β-Naphthyl phenyl ketone	2.09
Diphenylene oxide	2.45	β-Naphthaldehyde	2.08
Triphenylamine	2.45	β-Acetonaphthone	2.08
Dibenzothiophene	2.44	α-Naphthyl phenyl ketone	2.01
o-Dibenzoylbenzene	2.41	α-Acetonaphthone	1.97
Benzophenone	2.40	α-Naphthaldehyde	1.97
4,4′-Dichlorobenzophenone	2.38	5,12-Naphthacenequinone	1.95
p-Diacetylbenzene	2.37	Biacetyl	1.92
Fluorene	2.37	Acetylpropionyl	1.92
9-Benzoylfluorene	2.34	Benzil	1.88
Triphenylene	2.33	Fluorenone	1.87
p-Cyanobenzophenone	2.32	Pyrene	1.70
Thioxanthone	2.29		

TABLE 37

SELECTION OF TRIPLET ENERGIES

Compound	E_T (μm^{-1})	Ref.	Compound	E_T (μm^{-1})	Ref.
Benzene	2.94	226	β-Naphthylamine	2.09	227
Toluene	2.88	226	α-Naphthoate ion	2.08	227
Phenol	2.86	226	α-Chloronaphthalene	2.07	30
Anisole	2.82	226	α-Bromonaphthalene	2.07	30
Benzoate ion	2.77	226	α-Iodonaphthalene	2.05	228
Benzoic acid	2.72	226	α-Naphthol	2.05	30
Aniline	2.68	226	β-Naphtholate ion	2.04	227
Fluorene	2.38	226	α-Naphthoic acid	2.04	227
Quinoline	2.20	227	N,N-Dimethyl-β-naphthylamine	2.03	227
Quinolinium ion	2.16	227	β-Naphthoic acid	2.02	227
β-Naphthylammonium ion	2.13	227	β-Naphthoate ion	2.02	227
N,N-Dimethyl-β-naphthylammonium ion	2.13	227	Biacetyl	1.97	30
β-Naphthol	2.11	227	α-Nitronaphthalene	1.92	30
β-Bromonaphthalene	2.11	30	Coronene	1.91	226
α-Fluoronaphthalene	2.10	226	α-Naphthylamine	1.90	226
β-Chloronaphthalene	2.10	30	1,2,5,6-Dibenzanthracene	1.83	226
β-Iodonaphthalene	2.10	30	Naphthacene	1.03	122
α-Methylnaphthalene	2.09	226			

and Joyce[114] found that the plots for chlorophylls *a* and *b* were indeed linear and they thus determined ΔE, and hence, from the known value of the singlet energy (determined from the maxima of the absorption and prompt fluorescence spectra) they were able to calculate the triplet energies shown in column 2 of Table 33.

It should be noted that solutions of the chlorophylls exhibit both E-type and P-type delayed fluorescence[114]: the former predominates at low rates of light absorption in viscous solvents such as propylene glycol, and the latter predominates at high rates of light absorption in less viscous solvents such as ethanol. It is also of interest that photo-excited delayed fluorescence (probably that of chlorophyll *a*) has been observed from the living leaves of a variety of plants[229]. Parker and Joyce[229] suggest that this is E-type delayed fluorescence originating from chlorophyll molecules situated at those centres in the photosynthetic units that are not easily available for reaction with the substrate.

In general, the triplet energy of a compound may be located by reference to the known triplet energies of two other "test" compounds. If the compound under investigation can act as a donor to one of the "test" compounds (e.g. if it can sensitise its P-type delayed fluorescence), and as an acceptor to the other, its triplet energy must lie between those of the two test compounds (see Section 2 D 5). Such selective sensitisation of P-type delayed fluorescence has been demonstrated for a series of aromatic hydrocarbons using proflavine, acridine orange and eosin as sensitisers[110].

A slightly different method has been used to determine limits for the triplet energy of perylene[113]. Perylene gives rise to a weak P-type delayed fluorescence by direct excitation and hence its triplet energy must be at least one half of that of the excited singlet, i.e. $\geqq 1.14\ \mu m^{-1}$ (see Section 2 D 5). Its P-type delayed fluorescence is strongly sensitised by anthracene, and hence its triplet energy must be $< 1.48\ \mu m^{-1}$. These limits are consistent with the calculated value[230] shown in Table 33.

5. Triplet Data from $T_1 \leftarrow S_0$ Excitation

We have seen (Section 1 C 1) that the probability of the $T_1 \leftrightarrow S_0$ transition is exceedingly small for triplet molecules containing light atoms. The corresponding extinction coefficients (ε_{TS}) are therefore very small and difficult to measure. They are increased by substitution of a heavy atom in the molecule[226] or by working in a solvent containing heavy atoms[218], and the second effect can be used to facilitate the measurement of triplet–singlet absorption spectra[231, 232]. An alternative method of measuring the triplet–singlet absorption spectrum has been used by Kearns and co-workers[221, 233]. This involves the measurement of the phosphorescence excitation spectrum over wavelength regions covering both the $T_1 \leftarrow S_0$ and the $S_1 \leftarrow S_0$ absorption spectra, and provides data from which the ratio ϕ_t/ε_{TS} can be calculated. The principle of the method is as follows.

Let ε_{TS} be an extinction coefficient within the triplet–singlet absorption band at which the intensity of exciting light is I_{TS}. By a treatment similar to that used to derive equation 21 (see also Section 2 C 2) it is easily seen that, for low optical densities, the observed intensity of phosphorescence, P_{TS}, is:

$$P_{TS} = wI_{TS}\varepsilon_{TS}c_{TS}\phi_p/\phi_t \tag{321}$$

where w is an instrumental constant and c_{TS} is the concentration of solute at which the phosphorescence excitation spectrum is measured. Clearly, if the wavelength of the exciting light is varied while its quantum intensity is kept constant, P_{TS} will be proportional to ε_{TS}, i.e. the shape of the phosphorescence excitation spectrum will be identical with the $T_1 \leftarrow S_0$ absorption spectrum.

Now consider the phosphorescence spectrum excited within the $S_1 \leftarrow S_0$ absorption band for which the extinction coefficient is ε_{SS}, the concentration c_{SS} and the intensity of exciting light I_{SS}. The observed intensity of phosphorescence (P_{SS}) will again be given by an equation analogous to equation 21 viz.:

$$P_{SS} = wI_{SS}\varepsilon_{SS}c_{SS}\phi_p \tag{322}$$

Hence from equations 321 and 322:

$$\frac{\phi_t}{\varepsilon_{TS}} = \frac{P_{SS}I_{TS}c_{TS}}{P_{TS}I_{SS}c_{SS}\varepsilon_{SS}} \tag{323}$$

All quantities on the right hand side of equation 323 can be easily determined. Thus, if ϕ_t can be determined by an independent method, ε_{TS} can be calculated. In practice the values of C_{SS} and C_{TS} must be chosen to give optical densities for which inner filter effects are negligible. This implies small values of c_{SS}, but very large values of c_{TS} can be used to obtain a measurable value of P_{TS}.

In fluid solution phosphorescence efficiencies are too small to allow this method to be applied. P-type delayed fluorescence is often much more intense than phosphorescence under these conditions, and in principle it could be excited by light absorbed within the triplet–singlet band and its intensity as a function of wavelength of the exciting light could then be used to derive the triplet–singlet absorption spectrum. Since the intensity of P-type delayed fluorescence is proportional to the square of the rate of light absorption, the equivalent of equation 323 would be:

$$\left(\frac{\phi_t}{\varepsilon_{TS}}\right)^2 = \frac{(I_{DF})_{SS}}{(I_{DF})_{TS}}\left[\frac{I_{TS}c_{TS}\tau_{TS}}{I_{SS}c_{SS}\tau_{SS}\varepsilon_{SS}}\right]^2 \tag{324}$$

where I_{DF} and τ (with appropriate subscripts) are the intensities and lifetimes of the P-type delayed fluorescence observed by $T_1 \leftarrow S_0$ and $S_1 \leftarrow S_0$ excitation in the two solutions containing concentrations c_{TS} and c_{SS} respectively. To the author's knowledge such measurements have not yet been made in fluid solution but Avakian and co-workers[234] have measured the triplet–singlet excitation spectrum of the delayed fluorescence of crystalline anthracene.

C. CHEMICAL EQUILIBRIA IN THE EXCITED STATE

1. General Principles

Since a chemical compound is recognised by its observable chemical and physical properties, which are largely determined by the orbital arrangements of its electrons, it is reasonable to regard a molecule in one of its electronically excited states as a new compound, quite distinct from the molecule in the ground state, i.e., as a metastable "isomer" of the ground state molecule, although it differs from ordinary isomers in that the latter can exist indefinitely in the dark under appropriate conditions. This new molecule may exhibit a whole range of chemical reactions not shown by the ground state molecule—these are normally referred to as "photochemical reactions" of the ground state molecule. It may also take part in some reactions analogous to those undergone by the ground state molecule, and if these reactions are reversible, we should expect the equilibrium constants in the excited state to differ from those in the ground state. In this section we deal with one important group of such reactions, namely protolytic equilibria.

Consider the equilibrium between an acid RH and its conjugate base R^- for which the pK value in the ground state is 7:

$$RH + H_2O \rightleftarrows R^- + H_3O^+ \tag{325}$$

but for which the pK value (pK*) in the excited state is 3:

$$(RH)^* + H_2O \rightleftarrows (R^-)^* + H_3O^+ \tag{326}$$

Suppose that RH and R^- show characteristic absorption spectra in the visible or near ultra-violet region and emit characteristic fluorescence bands, each of which will be situated just to the long wavelength side of the corresponding absorption spectrum. In strongly alkaline solution only R^- will be present and excitation will yield only the fluorescence band due to the transition:

$$(R^-)^* \rightarrow R^- + h\nu' \tag{327}$$

In strongly acid solution (pH $\ll$ 3) only RH will be present and the solution will emit only the second fluorescence band due to the transition:

$$(RH)^* \rightarrow RH + h\nu \tag{328}$$

Now at the intermediate pH value of say 5, essentially only RH will be present in the dark and excitation will yield $(RH)^*$. At this pH value $(RH)^*$ will attempt to dissociate and we may therefore expect to observe the fluorescence band of R^- produced by excited state dissociation:

$$(RH)^* + H_2O \rightarrow (R^-)^* + H_3O^+ \qquad (329)$$

followed by process 327. To observe an appreciable emission from $(R^-)^*$ at a pH value of 5, two further conditions must be satisfied:

(a) The loss of the proton from $(RH)^*$ must leave the electronic excitation energy intact, i.e. process 329 must be fast enough to compete with dissociation involving direct formation of ground state molecules:

$$(RH)^* + H_2O \rightarrow R^- + H_3O^+ + \text{heat} \qquad (330)$$

(b) Process 329 must be fast enough to compete with the first order radiative and non-radiative conversions of $(RH)^*$ to the ground state, i.e.:

$$k\ (\text{dissociation}) \gtrsim 1/\tau \qquad (331)$$

where τ is the lifetime of $(RH)^*$.

Both these conditions are satisfied for many fluorescent protolytic reaction systems, although frequently the equilibrium in the excited state is not *completely* established before fluorescence occurs.

The first observation of "abnormal" fluorescence due to protolytic dissociation was reported by Weber[235] who found that the blue-violet fluorescence of neutral solutions of 1-naphthylamine-4-sulphonate changed to green in strongly alkaline solutions although there was no change in the absorption spectrum. The effect was interpreted by Förster[236] who first demonstrated the occurrence of excited state dissociation with solutions of hydroxy- and amino-pyrene sulphonates. He later obtained similar results with naphthalene derivatives[237, 238] and we shall consider two of the latter, one of which goes substantially to equilibrium in the excited state before fluorescence occurs, and one that does not achieve equilibrium before fluorescing.

The pK value of β-naphthylamine in the ground state is 4.07, corresponding to the equilibrium:

$$NpNH_3^+ + H_2O \rightleftarrows NpNH_2 + H_3O^+ \qquad (332)$$

and hence at pH values greater than about 6 it is present in solution almost entirely as the free base $NpNH_2$. This gives a fluorescence emission band in the violet with maximum at 420 nm (see Fig. 118 curve 1). This same violet band appears also in solutions of pH value 2 (Fig. 118 curve 2), although the solution is now 2 pH units below the pK value, and therefore contains substantially only protonated molecules. Clearly the $NpNH_3^+$ dissociates after excitation and hence the pK value in the excited state must be considerably lower than 2. The naphthylammonium ion fluoresces in the ultra-violet region, i.e. at a wavelength a little longer than its long wavelength absorption band, and to observe the ultra-violet fluorescence free from the blue fluorescence of the free base, Förster had to use very strongly acidic solutions characterised by negative pH values (Hammett's acidity scale) as indicated by curve 4 in Fig. 118. The change over from the basic to the acidic spectrum with changing pH value was quite rapid, indicating that the equilibrium in the excited state was almost completely established before the two forms fluoresced. The pH value at the change over point (Fig. 118 curve 3) i.e. the pK* for the equilibrium in the excited state, was about − 1.5. Thus pK* was some 5–6 units less than pK, indicating that $NpNH_3^+$ is a very much stronger acid in the excited state.

In very strongly alkaline solution (pH value 14) Förster found the fluorescence spectrum to undergo a further change with the appearance of a green emission band (Fig. 118 curve 5), although the solution had the same absorption spectrum as that at a pH value of 9. The same green fluorescence band is emitted by solutions of β-naphthylamine in liquid ammonia containing sodamide, where β-naphthylamide anions are known to be produced

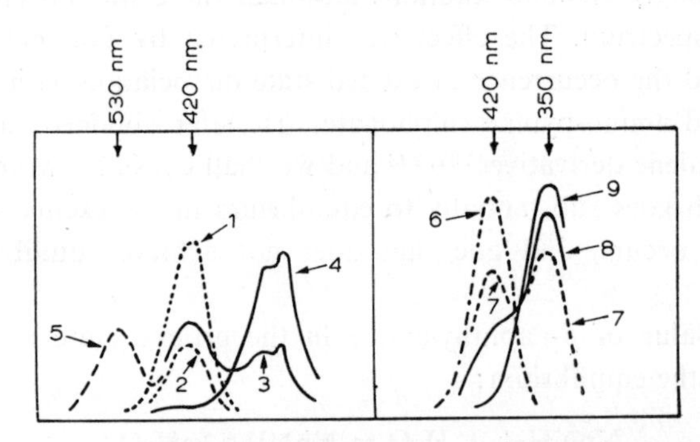

Fig. 118. Changes in the fluorescence emission spectra of solutions of β-naphthylamine (left) and β-naphthol (right) with pH value.
Curves 1–5, β-naphthylamine at pH values of 9, 2, − 1.5, − 5 and 14; curves 6–9, β-naphthol at pH values of 13, 9, 3 and 0 (adapted from Förster[238]).

by ionisation. In aqueous solutions these ions do not exist at all in the ground state, but it must be concluded from the appearance of their characteristic emission, that they are formed in the excited state by the process:

$$(NpNH_2)^* + OH^- \rightarrow (NpNH^-)^* + H_2O \qquad (333)$$

Confirmation is obtained from the observations that N-methyl-β-naphthylamine shows a similar change in fluorescence spectrum in alkaline solution, but N,N-dimethyl-β-naphthylamine does not because it has no proton available.

Förster found that phenols were also more strongly acidic in the excited state, although with these compounds the reaction rates are generally not sufficiently rapid to establish equilibrium before radiative and radiationless depopulation of the excited states occurs. Accordingly the pH-dependence of the fluorescence spectra are determined both by the pK* of the excited state equilibrium, and the rates of reaction. β-Naphthol is an example. At a pH value of 13 only naphtholate ions are present in the ground state and these give rise to a blue fluorescence band due to $(NpO^-)^*$ (Fig. 118 curve 6). At a pH value of zero, only undissociated β-naphthol is present and at this pH value only the violet fluorescence band due to $(NpOH)^*$ is observed (Fig. 118 curve 9). The ground state equilibrium:

$$NpOH + H_2O \rightleftarrows NpO^- + H_3O^+ \qquad (334)$$

has a pK value of 9.6 and if complete equilibrium in the excited state were achieved one would expect the change over from the blue $(NpO^-)^*$ fluorescence band to the $(NpOH)^*$ violet fluorescence band to occur within a narrow pH range (i.e. from 1% to 99% of each form within 4 pH units) at some point between pH values of 13 and zero. In fact the fluorescence emissions of both forms are present over a wide range of pH values from above 9 (see Fig. 118 curve 7) to below 3 (Fig. 118 curve 8). This is due to the fact that in solutions of intermediate pH value (~ 6) the excited $(NpOH)^*$ molecules start to ionise, but before this change is complete the remaining $(NpOH)^*$ molecules return to the ground state with the emission of their characteristic fluorescence. The pK* value in the excited state is about 3 and as the pH of the solution is reduced below this value the excited state equilibrium shifts in favour of $(NpOH)^*$, and at pH values < 1.0 practically no blue fluorescence due to $(NpO^-)^*$ is emitted.

2. Relationship between Spectra and Equilibrium Constants

As explained in Section 1 B 3 a molecule immediately after undergoing a 0–0 transition in absorption relaxes to a state of slightly lower energy corresponding to the stable configuration in the lowest vibrational level of the excited state. A similar relaxation occurs immediately after emission of fluorescence via the 0–0 transition. As a result of these relaxations the energy difference between the stable configurations of the ground state and the first excited state (i.e. between states a and c in Fig. 4) is slightly less than hν for absorption, but more than hν for emission, and the best estimate for the energy of the first excited singlet state is obtained from the mean of the two values, i.e.

$$E_S = \tfrac{1}{2}(h\nu_{abs} + h\nu_{em}) \qquad (335)$$

When the absorption and emission bands show no vibrational structure, it is usually not possible to locate the position of the 0–0 transition. The best estimate for E_S, is then obtained from equation 335 where $h\nu_{abs}$ and $h\nu_{em}$ now refer to the maxima in the spectra. Using these values of E_S we can set up the energy level diagram shown in Fig. 119 for systems involving equilibria in both ground and excited states, as first suggested by Förster[236]. We take as an example a compound such as β-naphthol for which the molar enthalpy, ΔH^*, is less for the reaction in the excited state than for that in the ground

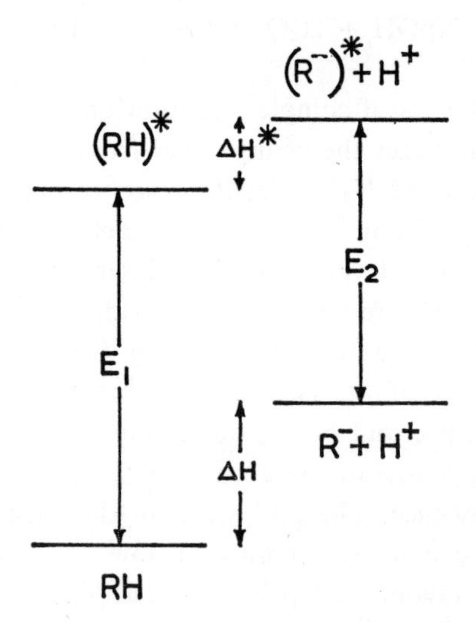

Fig. 119. Energy levels of acidic and basic forms.

TABLE 38

LIST OF ACID DISSOCIATION CONSTANTS (pK*) FOR THE LOWEST EXCITED SINGLET STATE

Acid	pK	pK*: Spectroscopic energy differences	pK*: Change of fluorescence spectrum	Reference
Phenol	10.0	5.7	–	239
α-Naphthol	9.23	2.0	2.5	240
β-Naphthol	9.46	2.5	2.81	240
2-Naphthol-5-sulphonate	9.18	0.53	0.73	241
2-Naphthol-6-sulphonate	9.10	1.6	1.66	240
1,2-Naphthalenediol	8.1[a]		2.4	242
1,5-Naphthalenediol	8.1[a]		2.9	242
1,7-Naphthalenediol	10.3[a]		3.1	242
1,8-Naphthalenediol	9.8[a]		2.9	242
2,3-Naphthalenediol	9.9[a]		3.2	242
2,6-Naphthalenediol	9.3[a]		3.4	242
2,7-Naphthalenediol	9.2[a]		2.5	242
3-Hydroxypyrene-5,8,10-trisulphonate	7.30	1.0	1.38	240
3-Acetylaminopyrene-5,8,10-trisulphonate	13.90	6.7	6.9	241
Acridinium cation	5.45	10.3	10.65	243
β-Naphthoic acid cation	∼ – 6.9	–	∼ 0	238
β-Naphthylamine	≫14	–	∼14	238
β-Naphthylamine cation	4.07	–	∼ – 1.5	238

[a] Calculated from column 4 and the spectroscopic value of pK – pK*.

state, ΔH. The reverse is true for some systems as we shall see later. It is obvious from the diagram that:

$$E_1 - E_2 = \Delta H - \Delta H^* \tag{336}$$

This applies of course whether the actual values of ΔH and ΔH^* are positive, zero or negative. If we make the assumption that the reaction entropies in ground and excited state are equal (this assumption is borne out by the results actually obtained), then since:

$$\Delta G = \Delta H - T\Delta S = -\mathbf{R}T \ln K \tag{337}$$

we have:

$$\log (K^*/K) = \frac{E_1 - E_2}{2.3\mathbf{R}T} = \frac{Nh(\Delta \nu)}{2.3\mathbf{R}T} \tag{338}$$

Inserting the values of Planck's constant, and converting to units of reciprocal microns, equation 338 becomes:

$$pK^* = pK - \frac{6.25 \times 10^3}{T}(\Delta\bar{\nu}) \tag{339}$$

or, at 20°C:

$$pK^* = pK - 21.4(\Delta\bar{\nu}) \tag{340}$$

Further, a good approximation for $\Delta\bar{\nu}$ can be obtained from the difference in the *absorption* maxima (or the fluorescence maxima) of the acidic and basic forms, and hence pK* values can be calculated approximately from absorption measurements alone. A useful rule to remember from equation 340 is that a shift in the spectrum of the basic form of 0.1 μm^{-1} to lower wavenumbers compared with the acid form corresponds to a decrease in pK value of about 2 units for the dissociation of the acid form.

Table 38 includes values of pK* determined both from equation 340 and from the pH changes of the fluorescence spectra. The close agreement shows that the assumption that the reaction entropies in the ground and excited state are equal is a reasonable one.

3. pH-Dependence of Fluorescence Intensities

The implications of Fig. 119 and equations 336 or 339 are worth considering in more detail. Firstly, whatever form is *more* stable in the excited state, this form will show an absorption band at *longer* wavelengths. It is thus possible to choose an excitation wavelength that will excite only this form (e.g. R^- in Fig. 119). Now reference to Table 38 shows that the decrease in pK value on excitation often amounts to as much as 6–8 pH units. At all pH values where such systems contain an appreciable proportion of R^- the equilibrium ratio $[RH]^*/[R^-]^*$ in the excited state will be negligible, i.e. excitation in the band of the long-wavelength form will produce fluorescence *only* from this form. In dilute solution the intensity of this fluorescence will be proportional to the concentration of this form (i.e. $[R^-]$ in the example shown in Fig. 119) and hence we can use such fluorescence measurements to determine the pK value of the system in the *ground* state.

We shall illustrate the effect of pH value on the observed intensities of fluorescence of the basic and acidic forms by means of the curves shown in Fig. 120. These were calculated for solutions sufficiently dilute that inner filter effects can be neglected, and they thus correspond to conditions fre-

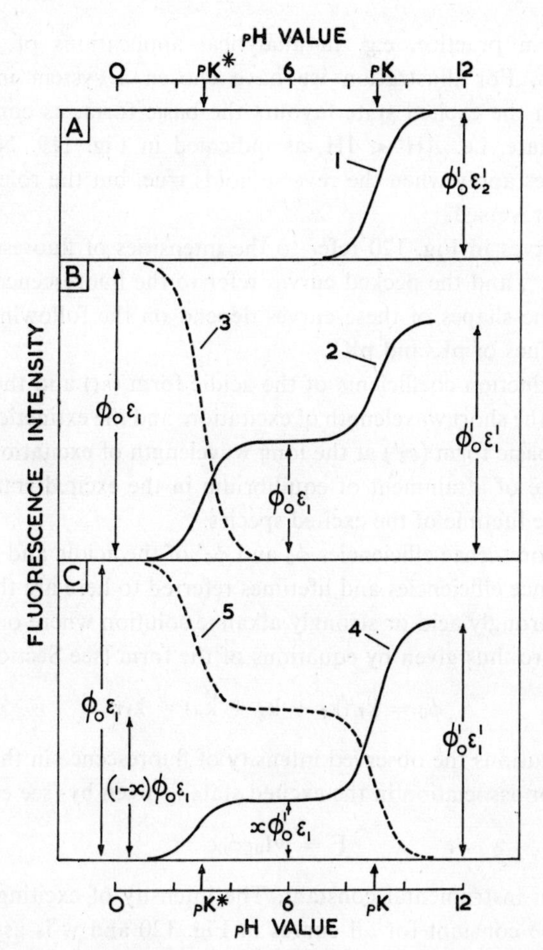

Fig. 120. Variation of fluorescence intensity with pH value for system $RH \rightleftarrows R^- + H^+$. The basic form R^- (indicated by primed symbols and full curves) is assumed to have the lowest excited singlet state, and absorbs at λ_2 where the acid form RH (indicated by unprimed symbols and pecked curves) is transparent. ε_2' is the extinction coefficient of R^- at λ_2. ε_1' and ε_1 are the extinction coefficients of R^- and RH at a shorter wavelength λ_1. ϕ_0' and ϕ_0 are assumed to be independent of wavelength. The curves are drawn for a system with: $\varepsilon_1' = 1.6\varepsilon_2' = 2\varepsilon_1$ and $\phi_0 = 2.5\phi_0'$.

Block A is with excitation at λ_2.

Blocks B and C are with excitation at λ_1.

In B, equilibrium is established at all pH values before fluorescence occurs to an appreciable extent.

In C, the proportion x of the acid that dissociates at pH 6 before fluorescence occurs is assumed to be 0.5.

quently met in practice, e.g. in analytical applications of fluorescence measurements. For illustration we have chosen a system in which the equilibrium in the excited state favours the basic form, as compared with the ground state, i.e. $\Delta H^* < \Delta H$, as indicated in Fig. 119. Naturally the same principles apply when the reverse holds true, but the roles of R^- and RH are then reversed.

The full curves in Fig. 120 refer to the intensities of fluorescence of the basic form, R^-, and the pecked curves refer to the fluorescence of the acid form, RH. The shapes of these curves depend on the following factors:

(a) The values of pK and pK*.

(b) The extinction coefficients of the acidic form (ε_1) and the basic form (ε_1') at the short wavelength of excitation, and the extinction coefficient of the basic form (ε_2') at the long wavelength of excitation.

(c) The rate of attainment of equilibrium in the excited state compared with the lifetime of the excited species.

(d) The fluorescence efficiencies ϕ_0 and ϕ_0' of the acidic and basic forms.

The fluorescence efficiencies and lifetimes referred to here are those that are observed in strongly acid or strongly alkaline solution where only one form emits. They are thus given by equations of the form (see Section 2 B 2):

$$\phi_0 = k_f/(k_f + k_g + k_n) = k_f\tau \tag{341}$$

For dilute solutions the observed intensity of fluorescence in the absence of dissociation or association in the excited state is given by (see equation 21):

$$F = wI_0\varepsilon\phi_0 c \tag{342}$$

where w is an instrumental constant. The intensity of exciting light, I_0, is assumed to be constant for all curves in Fig. 120 and w is assumed to be equal to $1/I_0c$.

The upper diagram A, refers to excitation in the long wavelength band of R^- (extinction coefficient $= \varepsilon_2'$), where RH does not absorb. As already explained, only fluorescence from $(R^-)^*$ is observed under these conditions with an intensity $f'\phi_0'\varepsilon_2'$, where f' is the fraction present as R^-. The fluorescence intensity thus follows the usual concentration–pH curve with inflection at a pH value equal to pK, and maximum equal to $\phi_0'\varepsilon_2'$. This same curve will be observed, whatever the value of pK* (remembering our assumption that pK* < pK).

Sections B and C refer to excitation with a shorter wavelength absorbed by both forms. In Section B the equilibrium in the excited state is assumed to be completely established before fluorescence occurs. The inflection in curve 2 at a value pK simply reflects the change in optical density of the solution as R^- is converted to RH. Since $(RH)^*$ dissociates completely at this pH value, only fluorescence from $(R^-)^*$ is observed, and when the conversion to RH is complete, this fluorescence has an intensity $\phi_0'\varepsilon_1$. At lower pH values approaching pK^* we observe a further fall in the fluorescence of $(R^-)^*$ and a corresponding growth in that of $(RH)^*$ (curve 3), both having an inflection at pK^*. Clearly with compounds having higher values of pK^* these inflections will move to the right in block B of Fig. 120 and with $pK^* = pK$, the horizontal section in curve 2 will disappear. Note that the pH values at which the fluorescence from both forms can be observed simultaneously are restricted to a range of about $\pm$ 2 pH units centred on pK^*.

Block C of Fig. 120 refers to the situation in which equilibrium in the excited state is not completely established before fluorescence occurs. The curves are calculated for a system in which a fraction $x = 0.5$ of $(RH)^*$ dissociates at a pH value of 6. The right hand inflections in curves 4 and 5 are situated at precisely the value pK, but the left hand inflections are situated at slightly lower pH values than pK^*. The important feature is the centre section where the fluorescence intensities of both forms remain approximately constant. It is also obvious that the range of pH values at which the fluorescence from both forms can be observed simultaneously extends from above pK to below pK^*, and this wide range of pH values is characteristic of a system in which protolytic reactions in the excited state take place without the establishment of equilibrium. The heights of the flat sections are equal to $x\phi_0'\varepsilon_1$ and $(1-x)\phi_0\varepsilon_1$, where x is the fraction of $(RH)^*$ that dissociates at intermediate pH values. Clearly as x increases the heights of the flat sections of curves 4 and 5 increase and decrease respectively, and ultimately block C becomes identical with block B.

4. Effect of Added Buffer

In general the solution will contain at least one other acid or base in addition to H_2O and H_3O^+ (or OH^-). The presence of other acids or bases will affect the *equilibrium* concentrations of $(R^-)^*$ and $(RH)^*$ *only* in so far as it affects the pH value and activity coefficients. However, the presence of other acids and bases can produce profound changes in the *rates* of the reactions involved in the equilibrium, and hence it can alter the intensities of fluorescence of $(R^-)^*$ and $(RH)^*$ if fluorescence occurs before equilibrium is established. The kinetics of protolytic reactions in the excited state have been worked out by Weller[244, 245] and we shall follow his methods here.

Consider an aqueous solution of the fluorescent acid RH and its conjugate base R^- containing also a buffer acid HB and its conjugate base B^-. The various protolytic reactions in which the excited states $(RH)^*$ and $(R^-)^*$ can take part may be represented as follows:

$$(RH)^* + H_2O \underset{k_2[H_3O^+]}{\overset{k_1}{\rightleftarrows}} (R^-)^* + H_3O^+ \tag{343}$$

$$(RH)^* + B^- \underset{k_a[HB]}{\overset{k_b[B^-]}{\rightleftarrows}} (R^-)^* + HB \tag{344}$$

$$(RH)^* \overset{k_f}{\rightarrow} RH + h\nu \tag{345}$$

$$(RH)^* \overset{k_0}{\rightarrow} RH \tag{346}$$

$$(R^-)^* \overset{k_f'}{\rightarrow} R^- + h\nu' \tag{347}$$

$$(R^-)^* \overset{k_0'}{\rightarrow} R^- \tag{348}$$

Let ϕ and ϕ' be the efficiencies of fluorescence emission from $(RH)^*$ and $(R^-)^*$ at any pH value, and ϕ_0 and ϕ_0' the corresponding values in strongly acid and strongly alkaline solutions respectively, i.e. where emission from only one species is observed. Then

$$\phi_0 = k_f/(k_f + k_0) = k_f\tau \tag{349}$$

$$\phi_0' = k_f'/(k_f' + k_0') = k_f'\tau' \tag{350}$$

Let us further assume that $pK^* < pK$, i.e. a situation similar to that depicted in Figs. 119 and 120, so that at all pH values less than pK minus 2 only RH is present in the solution. By setting up steady state equations for $(RH)^*$ and $(R^-)^*$ and substituting by means of equations 349 and 350, it can be shown that:

$$\frac{\phi}{\phi_0} = \frac{1}{1 + \tau(k_1 + k_b[B^-])/\{1 + \tau'(k_2[H_3O^+] + k_a[HB])\}} \tag{351}$$

and

$$\frac{\phi'}{\phi_0'} = \frac{1}{1 + \{1 + \tau'(k_2[H_3O^+] + k_a[HB])\}/\tau(k_1 + k_b[B^-])} \tag{352}$$

Clearly also:

$$\phi/\phi_0 + \phi'/\phi_0' = 1 \tag{353}$$

so that the pH–fluorescence curves will still have the general form shown in Fig. 120.

Weller applied these equations to the investigation of β-naphthol in formate, acetate, propionate and butyrate buffers at various temperatures. He chose a buffer ratio $[B^-]/[HB]$ of 10, and concentrations of $[HB] \leq 1.5 \times 10^{-3}$M. Under these conditions both $\tau' k_2[H_3O^+]$ and $\tau' k_a[HB]$ are negligible compared with unity and hence equations 351 and 352 become:

$$\phi_0/\phi - 1 = \tau k_1 + \tau k_b[B^-] = \frac{1}{\phi_0'/\phi' - 1} \tag{354}$$

In terms of Fig. 120 this equation means that when the total rate of dissociation of $(RH)^*$ is small, i.e. when $(k_1 + k_b[B^-]) \to 0$, $\phi \to \phi_0$ and the system behaves as though $pK^* = pK$. When $(k_1 + k_b[B^-]) \to \infty$ the system will behave like Fig. 120 block B, and with intermediate values of $(k_1 + k_b[B^-])$ it will behave as in Fig. 120 block C. If the latter behaviour is shown by solutions of low buffer concentration (i.e. with an intermediate value of k_1), increase in buffer concentration will cause the system to change in the direction from Fig. 120 block C to Fig. 120 block B. Weller found precisely this effect with solutions of β-naphthol. Plots of $(\phi_0/\phi - 1)$ against $[B^-]$ at a pH value intermediate between pK and pK* were linear and from the slopes of the plots, together with independently determined values of τ, he was able to calculate values of k_b. These values varied comparatively little among the different buffers (log k_b between 9.16 and 9.27 at 25°C), but the values were considerably lower than the maximum rate calculated from the diffusion coefficients. Weller interpreted this in terms of a steric factor with values between 0.14 and 0.21.

It should be noted that the above treatment has assumed that a steady state concentration is fully established in the excited state. This assumption leads to negligible errors in the above system, but non-stationary conditions must be taken into account when buffer is absent, as described in the following section.

5. Reaction in the Absence of Buffer

Weller[246] investigated the β-naphthol system in solutions of low pH value in which the hydrogen ion concentration was adjusted only by the addition of perchloric acid. In the absence of buffer, equation (351) reduces to:

$$\frac{1}{\phi_0/\phi - 1} = \frac{1}{\tau k_1} + \frac{\tau'[H_3O^+]}{\tau(K')^*} \tag{355}$$

where

$$(K')^* = k_1/k_2 = K^*/f^2 \tag{356}$$

and f is the mean activity coefficient and K^* is the thermodynamic equilibrium constant in the excited state. Weller found that plots of $1/(\phi_0/\phi - 1)$ against $f^2[H_3O^+]$ deviated from linearity at high acidities. This deviation was due to the fact that equation 355 is strictly valid only when a steady state is established. He calculated the factors required to correct for the transient effect and thus obtained satisfactory linear plots from which values of τk_1 and $\tau'/\tau K^*$, and hence also of k_1, k_2 and K^*, were derived. From the variation of K^* with temperature, the values of ΔH^* and ΔS^* were also calculated. The value of k_2 indicated that the reaction between $(R^-)^*$ and H_3O^+ is diffusion-controlled with a steric factor of about 0.7. Weller discussed the protolytic reaction in terms of the formation of an intermediate complex:

$$(NpOH)^* + H_2O \rightleftarrows [(NpO^-)^*.H_3O^+] \rightleftarrows (NpO^-)^* + H_3O^+ \tag{357}$$

6. Relationship between (pK – pK*) and Chemical Structure

Our discussion so far has been concerned mainly with systems which show increased acidity on excitation, i.e. for which $pK > pK^*$. This has been done partly for convenience and partly because this is the type of system with which excited state dissociation was first discovered by Förster. We should also expect there to be systems for which $pK^* > pK$, i.e. which are more strongly *basic* in the excited state, and this is indeed found to be so. Thus, while all phenols and aromatic amines so far investigated are found to have $pK > pK^*$, the aromatic carboxylic acids and aromatic ketones become much more strongly basic in the excited state, i.e. $pK < pK^*$ for their protonated forms. Thus β-naphthoic acid in the ground state will *accept* a proton only in solutions of extremely high acidity and the equilibrium:

$$Np{-}C\begin{matrix} \diagup O \\ \diagdown OH \end{matrix} + H^+ \rightleftarrows Np{-}C\begin{matrix} \diagup \overset{+}{O}H \\ \diagdown OH \end{matrix} \tag{358}$$

has a pK value of -6.9. The pK^* value as indicated by the change over point in the fluorescence spectrum is about zero. The shift in pK value with many systems is as much as $\pm$ 6 units as indicated by the values given in Table 38. This is not always the case however. Thus, pyridine has pK $\sim$ pK* although the related base acridine has (pK − pK*) equal to − 5.2. Weller[243] has suggested that the difference in this case is due to the fact that in acridine the lowest excited singlet state is L_b and this level shifts to longer wavelengths on protonation, while in pyridine the lowest excited singlet state is L_a which is little displaced by protonation.

Weller[243] has made a detailed study of the excited state protolytic reactions of acridine in aqueous solutions. The value of pK and pK* are 5.45 and 10.65, and the free base and the acridinium cation fluoresce blue and green respectively. However, like β-naphthol, the equilibria are not completely established, and indeed at a pH value of 8 (i.e. 4.65 units below pK*) solutions of acridine containing low concentrations of buffer emit mainly the blue fluorescence of the free base because the reaction:

$$(A)^* + H_2O \rightarrow (AH^+)^* + OH^- \quad (359)$$

proceeds only to a small degree before fluorescence occurs. With increasing concentration of ammonia buffer the proportion of green acridinium fluorescence increases owing to the additional reaction:

$$(A)^* + NH_4^+ \rightarrow (AH^+)^* + NH_3 \quad (360)$$

In principle this system in alkaline solution may be treated kinetically in the same way as was described for β-naphthol in acid solution, using the following equilibria in place of those given in equations 343 and 344:

$$(A)^* + H_2O \underset{k_4[OH^-]}{\overset{k_3}{\rightleftarrows}} (AH^+)^* + OH^- \quad (361)$$

$$(A)^* + NH_4^+ \underset{k_b[NH_3]}{\overset{k_a[NH_4^+]}{\rightleftarrows}} (AH^+)^* + NH_3 \quad (362)$$

General equations analogous to 351 and 352 can then be derived. From measurements of ϕ/ϕ_0 and ϕ'/ϕ_0' in alkaline solutions without buffer, Weller derived the values of k_3 and k_4 and hence also pK*. From measurements at a pH value of 8.3 in the presence of various concentrations of ammonium/ammonia buffer he derived values of k_a.

In a review article[241] Weller lists a variety of kinetic and thermodynamic data for such rapid acid–base reactions and discusses them in relation to other fast reactions of excited molecules.

7. Isomerisation in the Excited State

The fact that phenolic groups become more strongly acidic in the excited state while aromatic carboxylic acid groups become more basic gives rise to an interesting effect when both groups are present in the same molecule and situated in ortho positions. Weller[247] has found that although

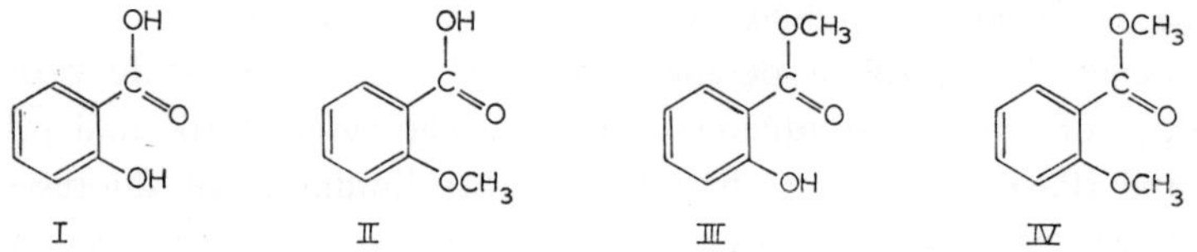

the maxima of the long-wavelength absorption bands of salicylic acid (I) and o-methoxy benzoic acid (II) differ in wavenumber by only about 0.1 μm^{-1}, the maxima of their fluorescence bands differ by 0.5 μm^{-1}—that of the acid is situated in the blue at 2.3 μm^{-1} but that of the ether is in the ultra-violet at 2.8 μm^{-1}. He interpreted the greatly increased Stokes' shift of salicylic acid in terms of an intramolecular proton transfer in the excited state:

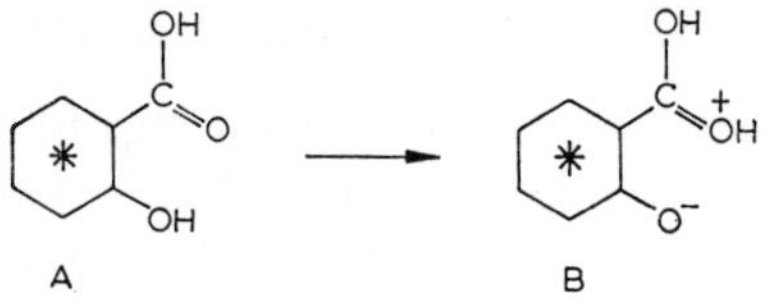

(363)

so that the absorption spectrum (maximum 3.3 μm^{-1}) corresponds to form A but the fluorescence spectrum (maximum 2.3 μm^{-1}) corresponds to form **B**. The methyl *ester* (III) behaved in a similar manner but the o-methoxy-benzoic ester (IV) gave a normal Stokes' shift similar to that observed with o-methoxy benzoic acid.

In hydrocarbon solutions Weller found that the salicylic ester (III) showed a weak short wavelength component in the emission spectrum (corresponding to form A) in addition to the intense blue emission (corresponding to form B). The relative intensity of the short wavelength component decreased as the temperature was reduced. All these effects can be explained in terms of the following protolytic equilibrium in the excited state:

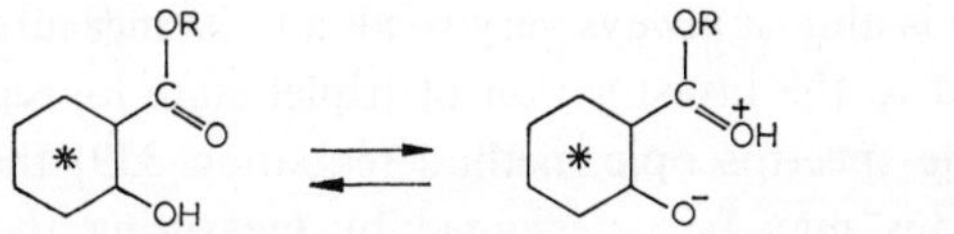

(364)

From the results of quenching experiments Weller concluded that the equilibrium at room temperature is fully established in a time much shorter than the lifetime of the excited molecules. From the temperature dependence of the ratio of the intensities of the two bands in hydrocarbon solvents he calculated $\Delta H^* = -1.0$ kcal mole^{-1} and hence, from the spectral shift he deduced that $\Delta H = 13.5$ kcal mole^{-1}. Because of the dipolar character of form B, ΔH^* is expected to be even more negative in polar solvents and this is probably the reason why no short wavelength component at all is observed with compounds I and III in alcohol or ether solvents.

Such abnormally large Stokes' shifts due to process 363 occurring in the excited state are to be expected with all compounds that contain two groups situated ortho to one another, one of which becomes strongly acidic in the excited state, and the other strongly basic.

8. Protolytic Equilibria in the Triplet State

In principle both of the methods used to determine pK* values in the excited singlet state can be used to determine pK^T values, i.e. the corresponding values in the triplet state, by observing the $T_1 \rightarrow S_0$ phosphorescence spectra of the two forms. Phosphorescence in fluid solution at room

TABLE 39

ACID DISSOCIATION CONSTANTS IN THE TRIPLET AND SINGLET STATES
(From Jackson and Porter[227])

Compound	pK	pK*	pK^T By flash absorption spectroscopy	pK^T By phosphorescence measurement
β-Naphthol	9.5	3.1	8.1	7.7
β-Naphthoic acid	4.2	10–12	4.0	4.2
α-Naphthoic acid	3.7	10–12	3.8	4.6
β-Naphthylamine	4.1	−2	3.3	3.1
N,N-Dimethyl-α-naphthylamine	4.9	–	2.7	2.9
Acridine	5.5	10.6	5.6	–
Quinoline	5.1	–	6.0	5.8

temperature is almost always very weak and its measurement has not so far been applied to the investigation of triplet state ionisation. For the application of the spectroscopic method (equation 339) the triplet energies of the two forms may be determined by measuring their phosphorescence spectra at low temperature in acid and alkaline rigid solvents, where phosphorescence is strong and easily measured. For the direct determination of pK^T the concentrations of the triplets of the two forms in fluid solutions of various pH values may be compared by flash absorption spectroscopy. Jackson and Porter[227] have applied both methods to some naphthols, naphthylamines, naphthoic acids and heterocyclic amines (see Table 39). They found that the pK^T values were intermediate between pK and pK^*, although generally much closer to the former.

D. EXCITED DIMERS AND PROMPT FLUORESCENCE

1. Types of Dimers

Association of the molecules of a solute by processes not involving primary valencies can in general give rise to polymeric complexes containing two or more monomer molecules. We shall be concerned here only with the formation of dimeric complexes. The stability of a dimer is dependent on its free energy of formation and this in turn is governed mainly by the heat of formation (enthalpy), $-\Delta H$. From the photochemical viewpoint we must distinguish between several kinds of dimer. The first is formed by the reversible combination of two *unexcited* monomer molecules:

$$A + A \rightleftarrows (AA) \qquad (365)$$

Dimers of this type (ground state dimers) are well known and are responsible for the change in the shape of the *absorption* spectrum of many dyestuff solutions at high concentrations. Such dimers are often non-fluorescent. Excitation of such a dimer will give an excited dimer $(AA)^*$ which may be envisaged as the combination of an excited monomer and a ground state molecule, i.e. as A^*A. If the heat of formation of the excited dimer $(-\Delta H^*)$ is much less than **RT**, the excited dimer will dissociate rapidly:

$$A^*A \rightarrow A^* + A \qquad (366)$$

If however $-\Delta H^*$ is large enough to allow the formation of a stable excited dimer, we can then envisage the formation of the latter by an alternative route, i.e. by the reverse of process 366, in which the *monomer* is first excited and then reacts with a ground state molecule:

$$^1A \xrightarrow{h\nu} {}^1A^* \quad (367)$$

$$^1A^* + {}^1A \rightarrow A^*A \quad (368)$$

Process 368 has particular relevance if $-\Delta H$ for the ground state dimer is zero or negative, i.e. if the ground state dimer is completely unstable. This corresponds to our *second type* of dimer, i.e. a dimer that is stable in the excited state only. As explained in Section 1 C 5 many workers have adopted the name "excimer" for excited dimers of this type, in order to differentiate them from the first type which have a stable ground state. Although the term "excimer" was originally coined for the hypothetical "long-lived excited dimers", it must be clearly understood that as now used it refers to dimeric species having short lifetimes characteristic of a singlet state.

It is with "excimers" that we shall mainly be concerned here, but we shall have occasion to refer to a *third type* of dimer, namely the photo-dimer. This is produced in its *ground state* when the interaction between A* and A results in "irreversible" chemical change, in which new covalent bonds are formed. The best known case is dianthracene, formed when concentrated solutions of anthracene are irradiated. It has the structure shown in Fig. 121 (a). Experimentally, photo-dimer formation is simply distinguished from

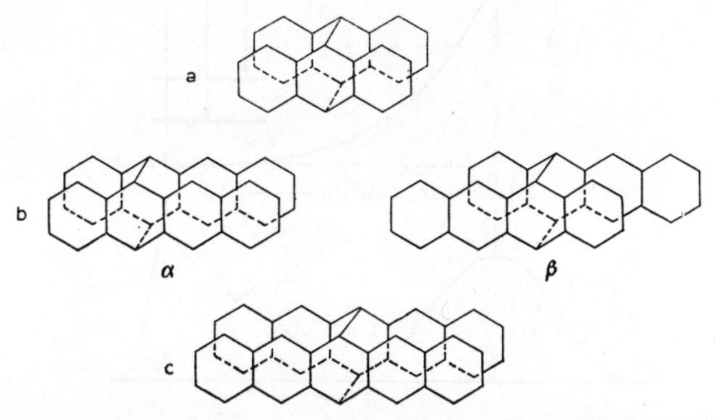

Fig. 121. Proposed molecular structures of (a) dianthracene, (b) dinaphthacene (α and β isomers), and (c) dipentacene (from Birks, Appleyard and Pope[248]).

excited dimers of types 1 and 2 by the fact that continued irradiation of the solution results in the progressive and permanent consumption of the monomer.

The three types of dimer formation may thus be summarised as follows:

$$A + A \rightleftarrows (AA) \xrightarrow{h\nu} (AA)^* \quad \text{“ordinary dimers”} \tag{369}$$

$$A + A \xrightarrow{h\nu} A^* + A \rightarrow (AA)^* \quad \text{“excimers”} \tag{370}$$

$$A + A \xrightarrow{h\nu} A^* + A \rightarrow A_2 \quad \text{“photo-dimers”} \tag{371}$$

2. Stability of Excited Dimers

Consider a solution of a compound that forms a dimer that is stable in the excited state only, and for which radiative transitions occur from the excited singlet states of both the monomer (A^*) and the dimer (A_2^*). These two radiative transitions will give rise to two fluorescence bands characteristic of the excited species concerned. The system may be represented by the diagram shown in Fig. 122 which is due to Stevens and Ban[249]. The

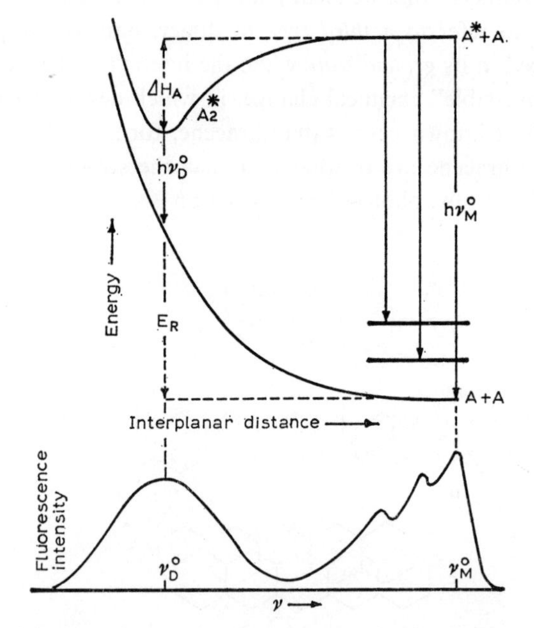

Fig. 122. Potential energy diagram for dimer in ground and excited states (from Stevens and Ban[249]).

upper part of the figure shows the potential energies of the excited and unexcited systems, $A^* + A$ and $A + A$, as a function of the distance of separation of the pairs of monomers. The upper curve has a minimum of depth $-\Delta H_A$ corresponding to the positive heat of formation of the excited dimer (excimer binding energy). The lower potential energy curve shows a steeply rising potential energy with a high negative value of $-\Delta H$ (E_R in Fig. 122), for the ground state system at the separation distance equal to the equilibrium distance of the stable excited dimer. There is clearly a strong repulsive force between the two components of the ground state "dimer", and the latter will thus dissociate immediately at all temperatures.

It is obvious from Fig. 122 that the energies of the quanta emitted by the monomer in its radiative transitions to the ground state ($h\nu_M{}^0$) must be greater than the corresponding dimer emission ($h\nu_D{}^0$). The former shows the usual vibrational structure typical of most aromatic hydrocarbons, and this is indicated diagrammatically in Fig. 122 by the transitions to higher vibrational levels of the ground state corresponding to the various maxima in the fluorescence emission spectrum of the monomer shown in the right hand (high frequency) section of the lower part of Fig. 122. The radiative transition of the dimer ends in a repulsive ground state and hence produces a continuous emission spectrum in the form of a broad structureless band having a maximum corresponding to $h\nu_D{}^0$ with:

$$h(\nu_M{}^0 - \nu_D{}^0) = -\Delta H_A + E_R \qquad (372)$$

3. The Excited Dimer of Pyrene

The most striking example of excimer fluorescence is that exhibited by concentrated solutions of pyrene, which were first investigated by Förster and Kasper[40, 250]. They found that dilute solutions of pyrene in benzene showed only the structured fluorescence spectrum characteristic of the monomer (similar to curve 4 of Fig. 131 in Section 4 E which refers to much later measurements made in ethanol), but as the concentration was increased this band was progressively quenched and a new structureless emission band appeared at longer wavelengths (see for example curves 3, 2 and 1 in Fig. 131). Förster and Kasper found that there was no change in the shape of the absorption spectrum, even in the most concentrated solutions, which still obeyed Beer's Law. They therefore concluded that the change in fluorescence spectrum was not due to the association of pyrene molecules before excitation. They confirmed the absence of association by measurements of freezing point depression and thus concluded that the change in fluorescence spectrum was due to excimer formation (equation 370) rather than to excitation of dimers already present (equation 369). In these and later papers[251, 252] Förster and co-workers elucidated all the essential features of the phenomenon and proposed the following scheme:

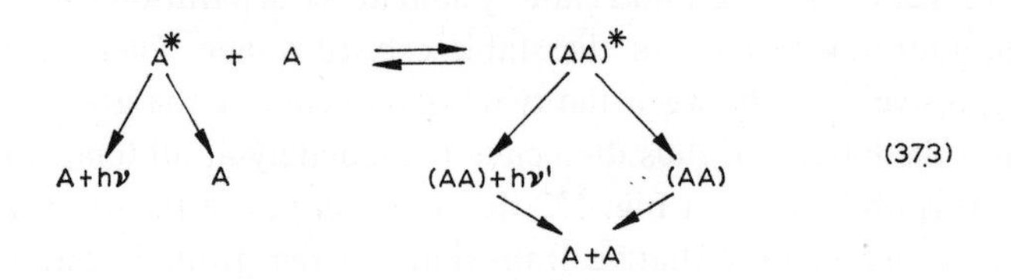

in which irradiation of the solution results in the establishment of a photo-stationary state with excited monomer and excited dimer both present, and both undergoing radiative and radiationless transitions with the ultimate formation of ground state monomer molecules.

We shall re-write the separate processes involved in this scheme, using a terminology for the rate constants consistent with that in Chapter 2, and we shall then use this scheme as a basis for discussing Försters results and those of later workers. We have therefore:

$$A \xrightarrow{h\nu} A^* \qquad \text{rate} = I_a \qquad (374)$$

$$A^* \rightarrow A + h\nu' \qquad k_f[A^*] \qquad (375)$$

$$A^* \rightarrow A \qquad k_n[A^*] \qquad (376)$$

$$A^* \rightarrow {}^3A \qquad k_g[A^*] \qquad (377)$$

$$A^* + A \rightarrow A_2^* \qquad k_A[A^*][A] \qquad (378)$$

$$A_2^* \rightarrow A^* + A \qquad k_d[A_2^*] \qquad (379)$$

$$A_2^* \rightarrow A + A + h\nu'' \qquad k_f'[A_2^*] \qquad (380)$$

$$A_2^* \rightarrow A + A \qquad k_n'[A_2^*] \qquad (381)$$

By solving the stationary state equations in a manner similar to that described in Section 2 B 1 and writing c for the total concentration of solute, it is found that:

$$\frac{\phi_D}{\phi_M} = \frac{k_f'[A_2^*]}{k_f[A^*]} = \frac{k_f' k_A c}{k_f(k_f' + k_n' + k_d)} = K_1 c \qquad (382)$$

where ϕ_D and ϕ_M are the fluorescence efficiencies of dimer and monomer at a concentration c. It is also found that:

$$\begin{aligned}\frac{\phi_M^0}{\phi_M} &= 1 + \left[\frac{k_A}{(k_f + k_n + k_g)}\right]\left[1 - \frac{k_d}{(k_f' + k_n' + k_d)}\right]c \\ &= 1 + k_A\tau_M^0(1 - k_d\tau_D)c \qquad (383)\end{aligned}$$

where ϕ_M^0 and τ_M^0 are the fluorescence efficiency and lifetime of the excited monomer and τ_D is the lifetime of the dimer at infinite dilution. Thus according to equation 383 the monomer quenching should follow a Stern–Volmer Law (Section 2 B 2). The relationship between the efficiency of dimer fluorescence and concentration can likewise be derived to give:

$$\frac{\phi_D^\infty}{\phi_D} = 1 + \frac{1}{k_A\tau_M^0(1 - k_d\tau_D)c} \qquad (384)$$

where ϕ_D^∞ is the maximum fluorescence efficiency of the dimer, i.e. that observed at infinitely high concentration. Equations 383 and 384 may be re-written in the forms first used by Förster and Kasper, viz.:

$$\phi_M = \phi_M^0\left[\frac{1}{1 + c/c_h}\right] \qquad (385)$$

and

$$\phi_D = \phi_D^\infty\left[\frac{1}{1 + c_h/c}\right] \qquad (386)$$

where c_h is the "half-value concentration", i.e. the concentration at which the fluorescence of the monomer has been 50% quenched, and at which the fluorescence efficiency of the dimer has risen to one half of its maximum value obtained in very concentrated solutions.

Förster and co-workers[40, 251, 252] showed that equations 385 and 386 were indeed obeyed by pyrene in a variety of solvents. A typical plot of ϕ_M/ϕ_M^0 and ϕ_D/ϕ_D^∞ against c, taken from their work[252] is given in Fig. 123 which shows that the intensities of monomer and dimer fluorescence passed through their "half-values" at the same concentration. Now the "half value" concentration is given by:

$$1/c_h = k_A\tau_M^0(1 - k_d\tau_D) \qquad (387)$$

If the association reaction 378 is diffusion-controlled, $1/k_A$ should be proportional to the viscosity and hence c_h should be critically dependent

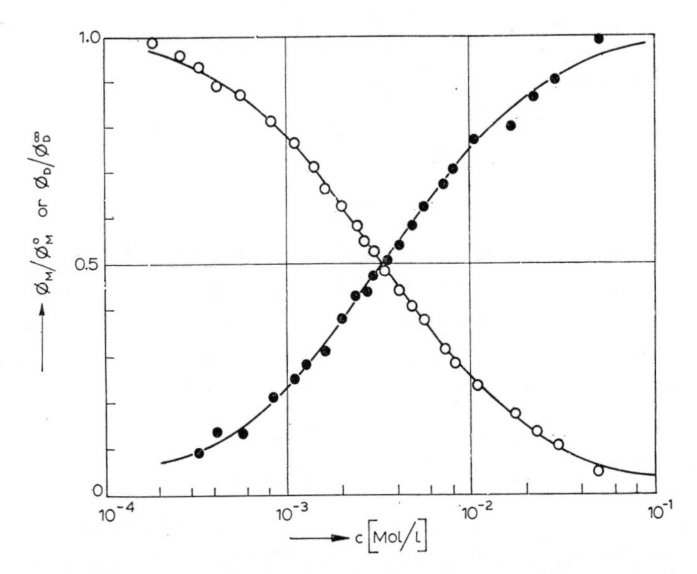

Fig. 123. Change in the fluorescence spectrum of pyrene with concentration. Solvent was a mixture of n-octylalcohol and n-dodecylalcohol; open circles, relative quantum yield of violet fluorescence; solid circles, same for blue fluorescence. The curves represent calculated values with $c_h = 3.4 \times 10^{-3}$ M (from Döller and Förster[252]).

on viscosity. Förster and co-workers found this to be the case for a variety of solvents, and they concluded therefore that k_A was indeed diffusion-controlled. Values of k_A close to the diffusion-controlled value have since been found by other workers (see below).

4. Lifetime of the Pyrene Monomer and Excited Dimer

Direct measurements of the lifetimes of the pyrene monomer and excimer were carried out by Birks, Dyson and Munro[192] using flash and phase fluorometers. The curves obtained with the former instrument provide a graphic demonstration of the reality of the excimer formation process already described. In Fig. 124 the curve p(t) represents the duration of the light pulse. The monomer fluorescence (curve $f_M(t)$) decays continuously after the end of the light pulse, as is predicted from equations 375–378. During the early stages of this decay, the excimer fluorescence (curve $f_D(t)$) rises and reaches a maximum at a time considerably later than the end of the light pulse, before finally decaying. Curve $f_D(t)$ thus represents the formation of the dimer (process 378, competing with 379–381) and its

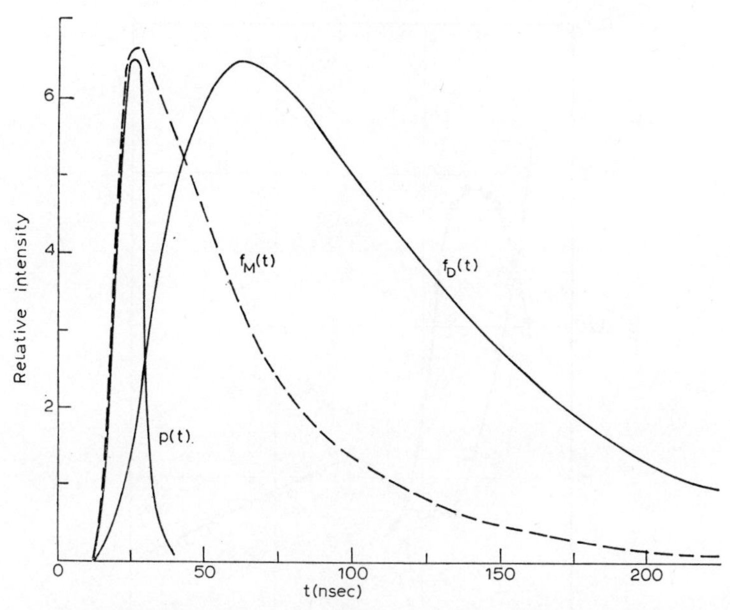

Fig. 124. Monomer fluorescence response, $f_M(t)$, and excimer fluorescence response, $f_D(t)$, to excitation light pulse, p(t) (5×10^{-3} M pyrene) (from Birks, Dyson and Munro[192]).

subsequent decay by processes 379–381. By combining the lifetime measurements with measurements of fluorescence efficiency Birks and co-workers were able to determine all the rate constants in equations 375–381 (except that only the sum of k_g and k_n was determined). The radiative lifetimes of the monomer and dimer were found to be 6.8×10^{-7} and 0.9×10^{-7} sec. Excimer formation was found to be a diffusion-controlled process with k_A given by equation 80.

5. Thermodynamic Data for Excited Dimers

By observing the fluorescence efficiencies of the pyrene monomer and dimer in liquid paraffin solution as a function of temperature, Döller and Förster[252] were able to demonstrate by direct observation that dimer formation is indeed a reversible reaction. This is illustrated by the curves in Fig. 125 taken from their paper. In a dilute solution in which dimer formation is negligible they found that the fluorescence efficiency decreases as the temperature increases owing to the increased rates of radiationless deactivation processes. With a solution containing a higher concentration sufficient to produce both

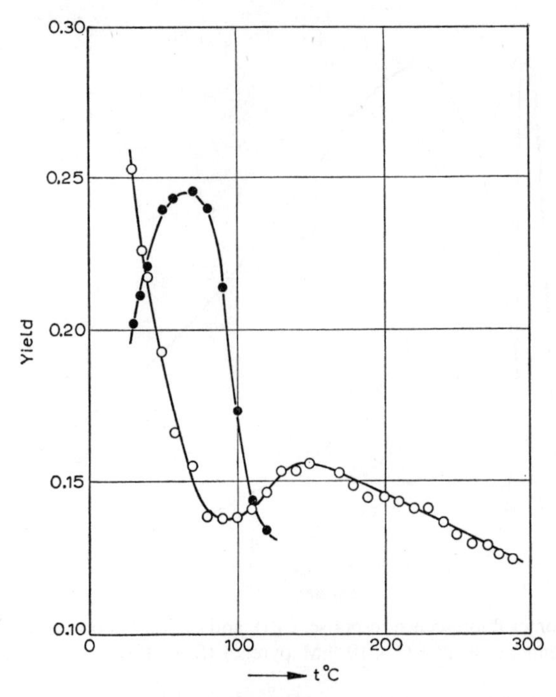

Fig. 125. Temperature dependence of the quantum yield of 5×10^{-3} M pyrene in liquid paraffin.
Open circles, monomer component; closed circles, dimer component (from Döller and Förster[252]).

monomer and dimer emission they found the results shown in Fig. 125. With increasing temperature, the quantum yield of the monomer fluorescence at first falls sharply, and that of the dimer increases, but above 80°C the quantum yield of the dimer begins to decrease and at the same time that of the monomer *increases*. The latter increase is due to thermal dissociation of the dimer at the higher temperature with re-formation of the monomer. They found the increase in monomer emission at high temperatures to be even more striking with a more concentrated solution (not shown in Fig. 125).

To interpret the effect of temperature quantitatively Döller and Förster assumed that the rate constants ($k_n + k_g$), k_A, k_d and k_n' (equations 376–379 and 381) included an activation energy term, and they therefore expressed them in the form:

$$k = N \exp(-E/RT) \tag{388}$$

(The radiative processes (k_f and k_f') were assumed to be temperature independent.) They showed that from the respective slopes of the plots of the logarithms of $(1/\phi_M^0 - 1)$, $(1/\phi_D^\infty - 1)$ and c_h against $1/T$, it was possible to calculate values for all four activation energies. In particular they found:

$$E_A = 6.5 \pm 1 \text{ kcal/mole} \tag{389}$$

$$E_d = 17.5 \pm 1 \text{ kcal/mole} \tag{390}$$

and hence the heat of formation of the dimer:

$$\Delta H^* = E_A - E_d = -11.0 \text{ kcal/mole} \tag{391}$$

Further, since the formation constant for the dimer, K^*, is equal to k_A/k_d, and,

$$-\Delta G = \mathbf{RT} \ln K^* = -\Delta H^* + T\Delta S^* \tag{392}$$

it is found that:

$$\Delta S^* = \mathbf{R} \ln (N_A/N_d) \tag{393}$$

where ΔS^* is the entropy of formation of the excited dimer and N_A and N_d are the temperature independent factors corresponding to equation 388, which were determined from the intercepts of the log plots already referred to. From 393 Döller and Förster found ΔS^* to be -20 cal deg^{-1} mole^{-1}.

Birks, Lumb and Munro[253] determined the binding energy of the pyrene dimer ($-\Delta H^*$) in a variety of solvents. Their method using only spectroscopic data is illustrated by the curves in Fig. 126 representing plots of $\log_{10}(\phi_D/\phi_M)$ against $1/T$. The curves tend to constant limiting slopes at both high and low temperatures and this may be interpreted in terms of equation 382. Thus at low temperatures, the rates of dissociation and radiationless decay of the dimer (k_d and k_n') tend to zero and hence from equation 382:

$$\phi_D/\phi_M \rightarrow k_A c/k_f \rightarrow (N_A c/k_f) \exp(-E_A/\mathbf{RT}) \tag{394}$$

Values of E_A derived from the low temperature slopes were in close agreement with the activation energy (E_c) of the diffusion-controlled rate constant, expressed as:

$$k_c = 8\mathbf{RT}/3000\eta = N_c \exp(-E_c/\mathbf{RT}) \tag{395}$$

thus confirming that excimer formation is diffusion-controlled. At high temperatures k_d becomes large compared with k_f' and k_n', and hence from equation 382:

$$\phi_D/\phi_M \rightarrow \frac{k_f'k_{A}c}{k_f k_d} \rightarrow \left(\frac{k_f'N_{A}c}{k_f N_d}\right) \exp\left(-\Delta H^*/RT\right) \tag{396}$$

The values of $-\Delta H^*$ derived from the high temperature slopes of Fig. 126 were within experimental error independent of the nature of the solvent. Birks and co-workers also combined the spectroscopic data in three solvents (ethanol, acetone and cyclohexane) with lifetime data to derive more precise values of $-\Delta H^*$. Their mean value was 0.32 μm^{-1}.

Stevens and Ban[249] have described a somewhat different method of treating the spectroscopic data to derive values of the heats of formation and entropies of formation of excited dimers. This method also depends on the application of equation 382, in the temperature range where:

$$k_d \gg (k_f' + k_n') \tag{397}$$

and makes use of the fact that if two independently emitting species in a

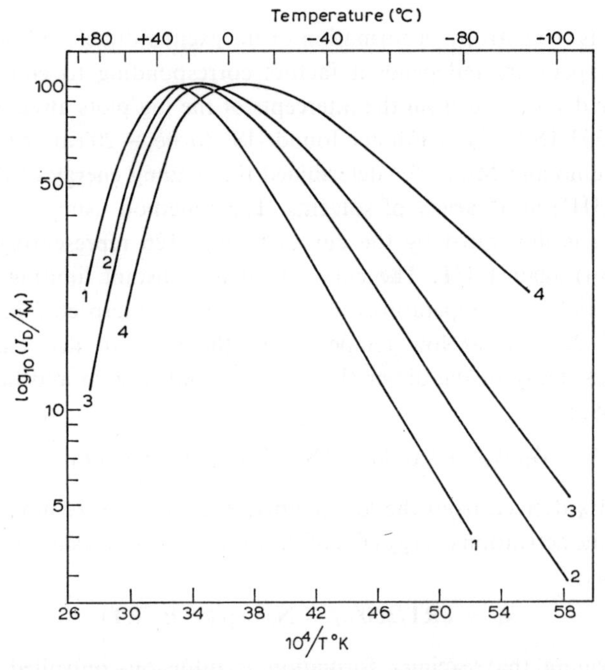

Fig. 126. Ratio of dimer to monomer emission in the fluorescence spectrum of 10^{-3} M pyrene solutions at various temperatures.
1, butanol; 2, ethanol; 3, toluene; 4, hexane (from Birks, Lumb and Munro[253]).

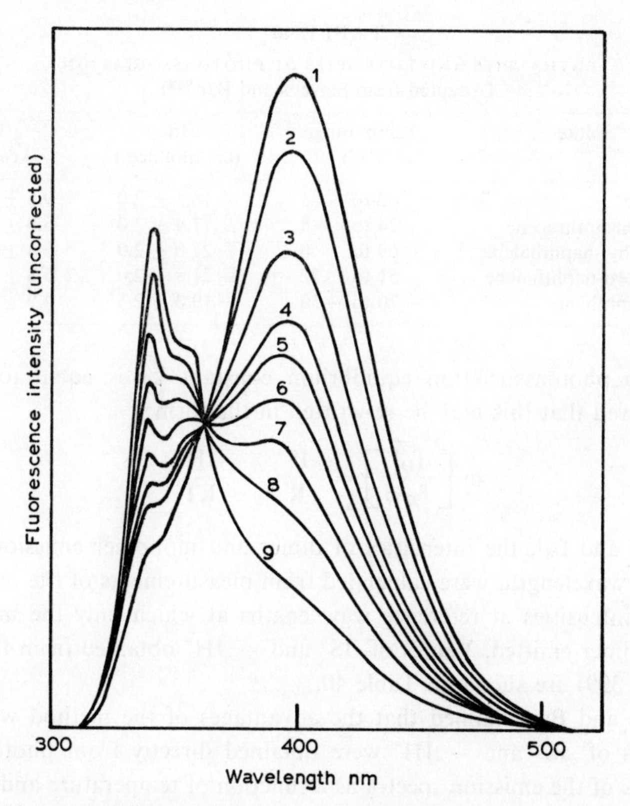

Fig. 127. Uncorrected fluorescence spectra of 0.5 M solution of 1-fluoronaphthalene in toluene at various temperatures.
1, − 51°C; 2, − 40°C; 3, − 33°C; 4, − 25°C; 5, − 19°C; 6, − 8°C; 7, − 4°C; 8, + 7°C; 9, + 22°C (from Stevens and Ban[249]).

closed system have a common wavelength region of emission, the overall spectrum recorded as a function of some variable (e.g. temperature) will exhibit an isosbestic point provided that the concentrations of emitting states are linearly related. Stevens and Ban used the presence of an isosbestic point in the recorded emission spectra (see for example the recordings shown in Fig. 127 which is taken from their paper) together with a linear dependence of log (ϕ_D/ϕ_M) on $\mathbf{T}^{-1}$ to define the temperature range over which equation 397 applies. In this region equation 382 may be written:

$$\ln\left[\frac{\phi_D}{\phi_M}\cdot\frac{k_f}{k_f'c}\right] = \ln K_A = \frac{\Delta S^*}{\mathbf{R}} - \frac{\Delta H^*}{\mathbf{RT}} \qquad (398)$$

TABLE 40

ENTHALPIES AND ENTROPIES OF PHOTOASSOCIATION
(Adapted from Stevens and Ban[249])

Solute	Temp. range (°C)	ΔS^* (cal/mol deg.)	$-\Delta H^*$ (kcal)
Pyrene	60 to 85	-18.5 ± 2.0	9.5 ± 1.0
1,2-Benzanthracene	-24 to -8	-17.4 ± 2.0	6.0 ± 0.5
2-Methyl-naphthalene	-69 to 0	-21.0 ± 2.0	5.8 ± 0.5
1-Fluoro-naphthalene	-51 to 22	-21.8 ± 2.0	5.8 ± 0.6
Acenaphthene	-70 to -39	-19.0 ± 2.5	4.9 ± 0.7

where the photoassociation equilibrium constant k_A is equal to k_A/k_d. They showed that this may be re-written in the form:

$$\ln\left[\frac{f_D{}^i}{f_M{}^i c}\right] = \frac{\Delta S^*}{R} - \frac{\Delta H^*}{RT} \tag{399}$$

where $f_D{}^i$ and $f_M{}^i$, the intensities of dimer and monomer emission at the isosbestic wavelength, were calculated from measurements of the respective emission intensities at reference wavelengths at which only the monomer and the dimer emitted. Values of ΔS^* and $-\Delta H^*$ obtained from the plots (equation 399) are shown in Table 40.

Stevens and Ban claimed that the advantages of the method were that the values of ΔS^* and $-\Delta H^*$ were obtained directly from photoelectric recordings of the emission spectra as a function of temperature and did not involve independent measurements of lifetimes, integrated spectral intensities, or corrections for detector sensitivity and reabsorption of fluorescence.

6. Excited Dimers of Other Compounds

At the time of Förster and Kasper's original experiments with pyrene, the only other compounds known to produce a concentration change in the fluorescence spectrum were some pyrene derivatives, and benz(a)pyrene[254]. Döller[255] later measured the relevant parameters relating to the formation of excimers in solutions of 4-methyl, 3-cyan-, 3-chlor- and 3-brom-pyrene, and pyrene-3-sulphonate and 3,5,8,10-tetrasulphonate. Subsequently, dimer formation in concentrated solutions of naphthalene was described by Berlman and Weinreb[256], and by Döller and Förster[257] who showed that the excimer emission band intensity increased greatly on lowering the temperature. Many papers have since been published, particularly by Birks and his school, showing that excimer formation in solution is a widespread phenomenon. A partial list of compounds for which excimer formation has been observed is given in Table 41 which also includes some references in which excimer formation has been looked for but not found. It will be observed that some compounds form stable photo-dimers, and one forms both an excimer and a stable photo-dimer.

With many compounds, the excimer emission is weak and only observed in very concentrated solutions or in the pure liquid. To detect such weak excimer bands, Birks and co-workers plot the series of emission spectra of solutions of increasing concentration in the form shown in Fig. 128, by normalising the spectra at one of the peak wavelengths of the monomer that is not affected by self-absorption and is outside the dimer emission band.

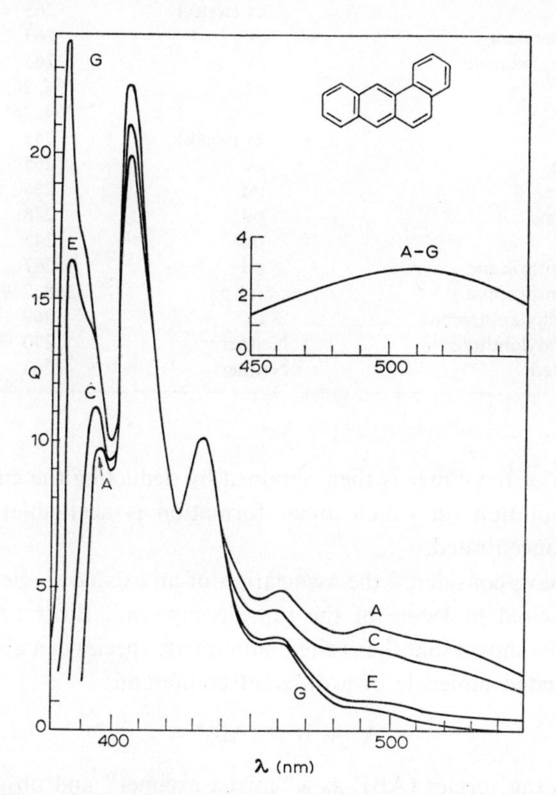

Fig. 128. Illustrating method of deriving fluorescence spectrum of excited dimer. 1,2-benzanthracene; A, 10^{-2} M; C, 5.5×10^{-3} M; E, 10^{-3} M; G, 10^{-4} M (from Birks and Christophorou[262]).

TABLE 41

DIMER FORMATION

Compound	Excimer (ex) or photodimer (pd)	Reference
Pyrene	ex	See text
Pyrene derivatives	ex	255, 258
Naphthalene and alkyl derivatives	ex	256, 257, 259, 260, 261
1,2-Benzanthracene and alkyl derivatives	ex	262, 263
1,2,3,4-Dibenzanthracene	ex	263
Benz(a)pyrene	ex	254, 258
1,2-Benzpyrene	ex	258
1,2,4,5-Dibenzpyrene	ex	258
Anthanthrene	ex	263
Perylene	ex (weak)	263
1,12-Benzperylene	ex	263
2,5-Diphenyloxazole	ex	263
Benzene	ex	264, 265
Toluene	ex	264, 265
p-Xylene	ex (weak)	265
Mesitylene	ex	265
Anthracene	pd	266
Naphthacene	pd	248
Pentacene	pd	248
2-Methylanthracene	pd	267
9-Methylanthracene	ex and pd	268, 269
9,10-Dimethylanthracene	ex	269
9,10-Diphenylanthracene	Neither	270
Others tested	Neither	263

The spectrum of the dimer is then obtained by deducting the curve for the most dilute solution (in which dimer formation is negligible) from that of the most concentrated.

So far we have considered the association of an excited singlet monomer with an unexcited molecule of the same compound. Birks and Christophorou[271] have shown that the excited monomeric species can also associate with an unexcited molecule of a different compound:

$$A^* + B \rightleftarrows (AB)^* \tag{400}$$

They refer to the species $(AB)^*$ as a "mixed excimer" and observed fluorescence emission bands due to mixed excimers formed between compounds having similar values of their lowest singlet excitation energies (e.g. pyrene and 3,4-benzphenanthrene). Birks has recently proposed the term "exciplex" for such excited complexes.

7. Factors Affecting Excited Dimer Formation

Of the two lowest excited singlet states of the aromatic hydrocarbons, classified by Platt[272] as ^1L_a and ^1L_b, the former has a stronger transition moment from the ground state (log $\varepsilon \sim 3.5$–5.0) than the latter (log $\varepsilon \sim$ 2.4–3.4). The radiative lifetime thus depends critically on whether ^1L_a or ^1L_b has the lower energy. Since excimer formation is a diffusion-controlled process which competes with monomer fluorescence, it will be favoured when the lifetime of the excited monomer is large, i.e. when ^1L_b is the lowest excited singlet state, and this is the situation with many compounds showing excimer fluorescence. For example, in Fig. 129, the ^1L_b bands of pyrene are situated at the longer wavelength and are considerably weaker than the ^1L_a bands.

Stabilisation of the excited dimer may be explained in two ways[263]. The

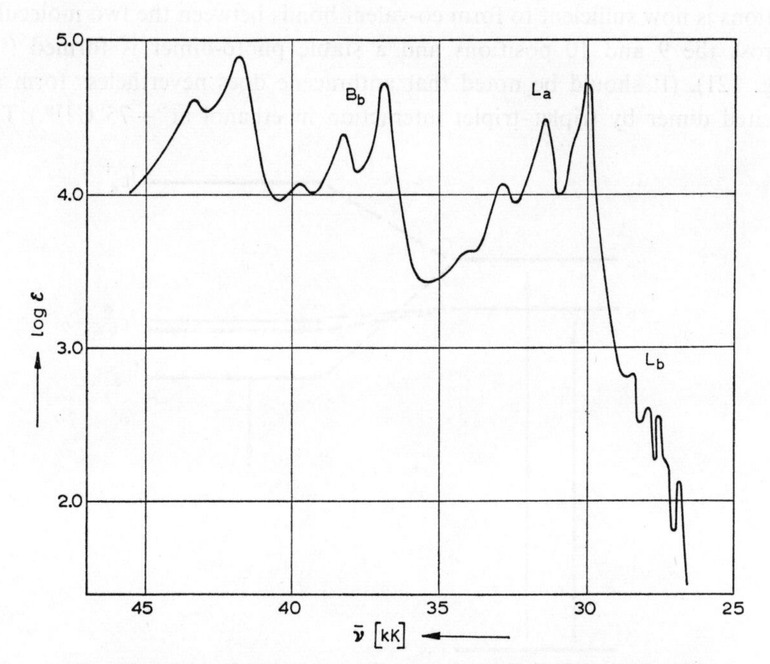

Fig. 129. Absorption spectrum of pyrene in ethanol (from Döller[255]). (Note: 1 kK = 0.1 μm^{-1}).

first depends on the coupling of the electron systems of the excited and unexcited monomer molecules (A*A) which leads to a splitting of the energy levels of the ^1L_b and ^1L_a states. The splitting of the former is much smaller than that of the latter (see Fig. 130) so that in the dimer the lowest state is of the ^1L_a type and has a lower energy than the monomer. The second method of stabilisation depends on electron transfer between the two monomers (A^+A^-). Quantum mechanical mixing between the two states AA* and A^+A^- accounts best for the experimental data, and for excimer formation to occur, the resulting state of the dimer must lie below the ^1L_b state of the monomer [263].

Barnes and Birks [269] have extended the theory to cover the case of stable photo-dimer formation, and excimer formation by some substituted anthracenes. They argue that in anthracene, tetracene and pentacene, the lowest excited singlet state is ^1L_a with dipole polarised across the short molecular axis. The ^1L_a dipole–dipole interaction potential between molecules is large and this potential orients A* and A with their molecular planes parallel and the 9,10 positions opposite one another. The magnitude of the interactions is now sufficient to form co-valent bonds between the two molecules across the 9 and 10 positions and a stable photo-dimer is formed (see Fig. 121). (It should be noted that anthracene does nevertheless form an excited dimer by triplet–triplet interaction in ethanol at −75°C [179].) The

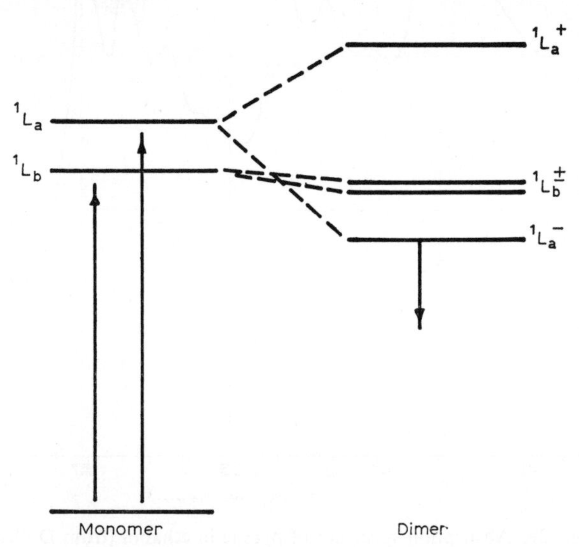

Fig. 130. Energy levels of monomer and excited dimer.

introduction of substituent groups limits the distance of approach of A^* and A, thus limiting the magnitude of the interactions. Thus 9-methyl anthracene still forms a photo-dimer but with lower quantum yield than anthracene, but in addition forms a fluorescent excimer. 9,10-Dimethyl anthracene does not form a photo-dimer but still forms an excimer, while 9,10-diphenylanthracene does not form either, owing to the non-coplanarity of the phenyl groups with the main ring system. When the lowest excited singlet state is ^1L_b as in pyrene, the effect of the ^1L_a dipole–dipole interaction potential is reduced but is still sufficient to orient A^* and A with their molecular planes and axes parallel so that polarisation of the dimer fluorescence is the same as that of the ^1L_a state, but the interaction is insufficient to convert the π-electrons into σ-electrons to form stable covalent bonds between the two monomers.

8. Intramolecular Excited Dimer Formation

In dilute solution the average separation distance of solute molecules is too great to allow an appreciable proportion of excited molecules to encounter a second solute molecule within the lifetime of the excited state. Hence excimer formation in dilute solutions is very small. If however pairs of aromatic rings are tied together by a string of saturated carbon atoms, the probability of encounter between the rings is greatly increased and it is to be expected that excimer emission will be observed at much higher dilutions than for unattached rings. Hirayama[273] found that this was indeed the case. He investigated the fluorescence spectra of dilute solutions of diphenyl and triphenyl alkanes and found that those compounds in which the phenyl groups along the main alkyl chain are separated by three carbon atoms (e.g. 1,3-diphenylpropane and 1,3,5-triphenylpentane) give rise to excimer emission. He attributed this to the fact that in these compounds the benzene rings can most easily take up the required configuration for excimer formation. Yanari, Bovey and Lumry[274] have observed a similar effect with solutions of styrene polymers. Thus the absorption spectra of both isotactic and atactic polystyrene in 1,2-dichloroethane were substantially identical with that of toluene. The emission spectra of dilute solutions of toluene and the atactic polymers were also identical and showed only a monomer band in the region of 290 nm. However the emission spectrum of the isotactic polystyrene showed a band at 324 nm close to the excimer band at 320 nm and found in concentrated toluene solutions. They suggested that each phenyl group could form an excimer with either of its nearest neighbours. If so, however, congruent plane-to-plane contacts between phenyl rings are not necessary for excimer formation, since such contacts are not possible in isotactic polystyrene. Whatever the explanation, the measurement of fluorescence emission seems to provide a useful method of distinguishing between atactic and isotactic polystyrene.

9. Other Methods of Producing Excited Dimer Emission

According to Barnes and Birks[269] the reason that anthracene does not show excimer emission is that the dipole–dipole interaction between the molecules is so great that a stable photo-dimer is formed instead. If however it were possible to hold the two species A^* and A at a separation distance too great to allow stable photo-dimer formation, but close enough to allow some interaction, it should be possible to observe excimer emission from anthracene. Just such an experiment has been carried out by Chandross[275]. He found that a solution of dianthracene in rigid methyl-cyclohexane solution at 77°K was non-fluorescent. When the specimen was irradiated at 77°K with light of wavelength 254 nm, a broad structureless fluorescence appeared. No fluorescence of monomeric anthracene was detectable although after warming and re-freezing a strong anthracene fluorescence appeared. He interpreted the results on the basis of photolytic cleavage of the dianthracene to give two anthracene molecules which are forced to remain close to each other by the rigid matrix. Hence on excitation of one of them the excimer spectrum is produced.

Although not strictly in the field of photoluminescence, it is relevant to mention the work of Chandross and co-workers on the electrochemiluminescence of aromatic hydrocarbons[276]. They generated positive and negative radical ions A^+ and A^- by alternating current electrolysis of the hydrocarbons in acetonitrile or dimethylformamide and observed luminescence which in general was of two kinds. The first was that characteristic of the singlet excited monomer of the hydrocarbon, and the second was a broad band which they attributed to the dimer. They suggested that the luminescence was produced by the two processes:

$$A^+ + A^- \rightarrow A^* + A \tag{401}$$

$$A^+ + A^- \rightarrow A_2^* \tag{402}$$

(see also Section 4 E 4).

E. EXCITED DIMERS AND DELAYED FLUORESCENCE

1. Pyrene Solutions at Room Temperature

To explain the characteristics of P-type delayed fluorescence emitted from solutions of anthracene and phenanthrene, Parker and Hatchard suggested that the encounter between two triplet molecules (see equations 127–132) resulted in the formation of a short-lived excited species, X, which then gave rise to an excited singlet molecule. They concluded[46] that the species X might be a short-lived dimer:

$$^3A + {}^3A \rightarrow {}^1A_2^* \rightarrow {}^1A^* + {}^1A \qquad (403)$$

or a singlet molecule in the quintuplet state:

$$^3A + {}^3A \rightarrow {}^5A + {}^1A \rightarrow {}^1A^* + {}^1A \qquad (404)$$

The question was partly resolved by their subsequent work with solutions of pyrene in ethanol[34]. They found that the spectra of prompt fluorescence varied with concentration in a manner similar to that observed by Förster and Kasper[40], i.e. the monomer emission showed concentration quenching, following the Stern–Volmer law (see equation 383) and this was accompanied by an increase in the intensity of the dimer emission. The relative intensity of the dimer emission (ϕ_D/ϕ_M) decreased according to equation 382 as the pyrene concentration decreased, and it tended to zero at low concentrations as shown in Fig. 131. The relative intensity of the dimer band in the delayed fluorescence spectrum (Fig. 132) was in each solution greater than that in the prompt fluorescence spectrum and was still large in the most dilute solution, although this showed no dimer in prompt fluorescence.

The intensities of delayed fluorescence of both the monomer and the dimer were proportional to the *square* of the rate of light absorption, and Parker and Hatchard therefore proposed the following mechanism in which the delayed fluorescence is produced as a result of triplet–triplet interaction. The sequence of reactions giving rise to singlet excited monomer and excited dimer as observed by *prompt* fluorescence is:

$$^1P \xrightarrow{h\nu} {}^1P^* \xrightarrow{+{}^1P} {}^1P_2^* \qquad (405)$$

When the exciting light is shut off, both excited states rapidly decay to lower concentrations that are then maintained by triplet–triplet interaction:

$$^{3}P + {^{3}P} \rightarrow {^{1}P_2^*} \rightarrow {^{1}P^*} + {^{1}P} \tag{406}$$

Thus, in *prompt* fluorescence it is the singlet excited *monomer* that is first formed, but in *delayed* fluorescence it is the excited *dimer* that is first formed. Obviously, if $^{1}P^*$ and $^{1}P_2^*$ fluoresce before equilibrium between them is established, the proportion of dimer observed in delayed fluorescence will in all solutions be greater than that observed in prompt fluorescence, and as the concentration of pyrene decreases, the relative intensity of the *delayed* fluorescence of the dimer will decrease to a constant finite value. This reaction scheme is shown in Fig. 133 and accounts qualitatively for all the features shown by the spectra in Figs. 131 and 132. Parker and Hatchard found that it was also sufficient to provide a quantitative interpretation of their measurements at room temperature, and we shall now give a brief account of their kinetic treatment.

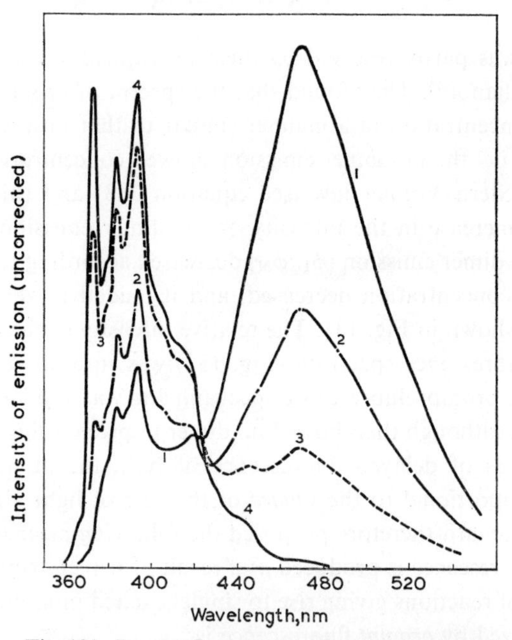

Fig. 131. Prompt fluorescence of pyrene in ethanol.
1, 3×10^{-3} M; 2, 10^{-3} M; 3, 3×10^{-4} M; 4, 2×10^{-6} M; the instrumental sensitivity settings for curves 1 and 4 were approximately 0.6 and 3.7 times that for curves 2 and 3; the short wavelength ends of the spectra in the more concentrated solutions are distorted by self-absorption (from Parker and Hatchard[34]).

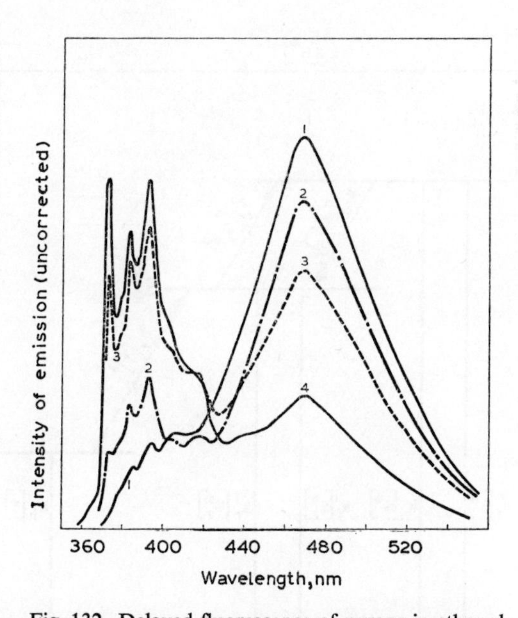

Fig. 132. Delayed fluorescence of pyrene in ethanol.
1, 3×10^{-3} M; 2, 10^{-3} M; 3, 3×10^{-4} M; 4, 2×10^{-6} M; the instrumental sensitivity settings were approximately 1000 times greater than those for the corresponding curves in Fig. 131; the short wavelength ends of the spectra in the more concentrated solutions are distorted by self-absorption (from Parker and Hatchard[34]).

During the measurement of delayed fluorescence, i.e. when the exciting light is shut off, the population of excited singlet states, S^* and S_2^* is maintained by the processes given in equation 406 and since the lifetime of the delayed fluorescence is about 3–5 msec, it decays comparatively little during the dark periods of the 800 c/s choppers. We can thus write an equation for the establishment of a steady state concentration of S^* (see Fig. 133 for rate constants):

$$k_d[S_2^*] = (k_f + k_n + k_g + k_A c)[S^*] \tag{407}$$

Hence the ratio of the efficiencies of delayed fluorescence of dimer and monomer is given by:

$$\frac{\theta_D}{\theta_M} = \frac{k_f'[S_2^*]}{k_f[S^*]} = \frac{k_f'(k_f + k_n + k_g)}{k_f k_d} + \frac{k_f' k_A}{k_f k_d} c \tag{408}$$

i.e.,

$$\theta_D/\theta_M = K_2 + K_3 c \tag{409}$$

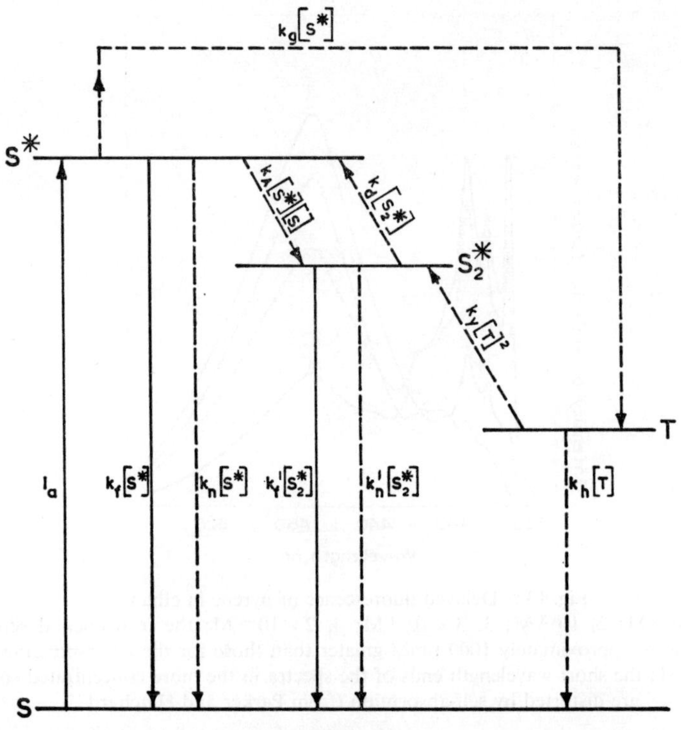

Fig. 133. Proposed mechanism for prompt and delayed fluorescence of pyrene in ethanol (from Parker and Hatchard[34]).

This expresses the fact that at low pyrene concentrations θ_D/θ_M tends to a constant finite value, K_2, which from measurements of the most dilute solution was found to be 0.706. Values of K_3, and also of K_1 (see equation 382), calculated from the results with the more concentrated solutions were found to be constant within experimental error. These calculations, together with others reported in the original paper[34] gave a complete quantitative explanation of the data then available on the pyrene system. As will be described in Section 4 E 3, they had to be modified in the light of later experiments.

2. Naphthalene in Ethanol at Low Temperature

If absence of dimer emission in delayed fluorescence (e.g. of anthracene) is an indication that the dimer dissociates very rapidly, it is to be expected that with some compounds at least, lowering the temperature will reduce the rate of dissociation sufficiently for the dimer emission to appear. Recent work[179] has indeed shown that dilute solutions of anthracene in ethanol show an appreciable dimer band in the delayed fluorescence spectrum at $-75°C$. More extensive measurements have been made with solutions of naphthalene in ethanol[112]. Thus at $-105°C$ a $3 \times 10^{-2}M$ solution gave a *prompt* fluorescence spectrum showing both monomer and dimer bands as indicated in Fig. 134 (curve a) (i.e. similar to the results already obtained by Döller and Förster[257] in toluene solutions at low temperature). The spectrum of delayed fluorescence also showed both bands, but the intensity of the dimer band was relatively much greater (Fig. 134 curve (b)). At the lower concentration of $3 \times 10^{-3}M$ the intensity of the dimer band was very

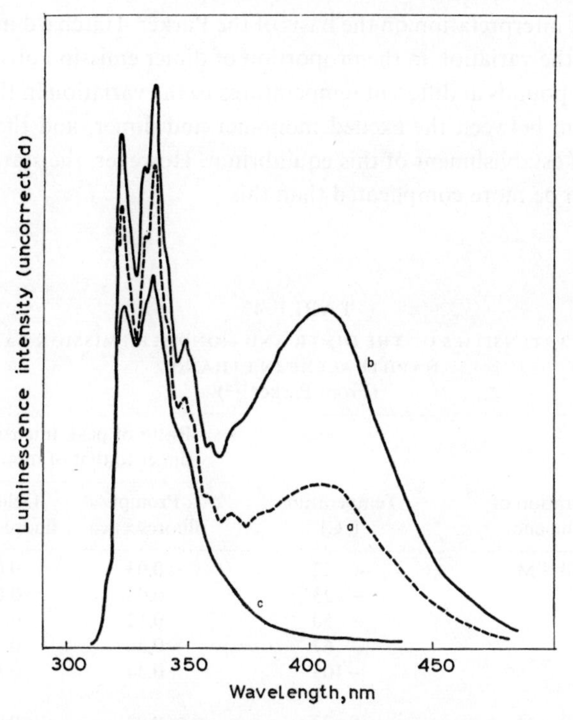

Fig. 134. Prompt and delayed fluorescence of 3×10^{-2} M naphthalene in ethanol. a, prompt fluorescence at $-105°C$ at sensitivity $\times 1$; b, delayed fluorescence at $-105°C$ at sensitivity $\times 60$; c, delayed fluorescence at $+22°C$ at sensitivity $\times 140$; excitation at 313 nm by frontal illumination (from Parker[112]).

small in prompt fluorescence but was still quite large in delayed fluorescence (see Table 42). Thus the behaviour of naphthalene solutions at −105°C was qualitatively similar to the behaviour of pyrene solutions at room temperature. At higher temperatures, the proportion of dimer observed in delayed fluorescence was almost the same as that in prompt fluorescence (Table 42) and it thus seemed that at these temperatures establishment of the equilibrium between the excited dimer and excited monomer was largely complete before fluorescence was emitted. As the temperature was raised, the equilibrium shifted in favour of the monomer and the proportion of dimer observed in either prompt or delayed fluorescence decreased (see Table 42) until at room temperature little dimer was observed in the delayed fluorescence spectrum even from the more concentrated solution (Fig. 134 curve c).

At this stage therefore it seemed that all P-type delayed fluorescence might be capable of interpretation on the basis of the Parker–Hatchard mechanism, by ascribing the variation in the proportion of dimer emission observed with different compounds at different temperatures to the variation in the position of equilibrium between the excited monomer and dimer, and the variation in the rate of establishment of this equilibrium. However, the state of affairs turned out to be more complicated than this.

TABLE 42

RELATIVE INTENSITIES OF THE DIMER AND MONOMER EMISSION BANDS OF NAPHTHALENE IN ETHANOL

(From Parker[112])

		Ratio of peak intensity of dimer to that of monomer	
Concentration of naphthalene	Temperature (°C)	Prompt fluorescence	Delayed fluorescence
3×10^{-2} M	+ 22	< 0.03	0.03
	− 23	0.04	0.05
	− 50	0.12	0.13
	− 67	0.26	0.28
	− 105	0.34	0.60
3×10^{-3} M	+ 22	< 0.03	< 0.03
	− 50	< 0.03	0.03
	− 77	< 0.03	0.09
	− 105	< 0.03	0.17

3. Pyrene Solutions at Low Temperature

Assuming that the Parker–Hatchard mechanism is the only one that operates, we can use equation 408 to predict the effect of lowering the temperature on the spectrum of delayed fluorescence. At low concentrations of pyrene such that dimer emission in prompt fluorescence is negligible, equation 408 reduces to:

$$\frac{\theta_D}{\theta_M} = \frac{k_f'(k_f + k_n + k_g)}{k_f k_d} = \frac{k_f'}{k_d \phi_M{}^0} \tag{410}$$

Lowering the temperature will cause k_d to decrease and $\phi_M{}^0$ to increase. Since however $\phi_M{}^0$ for pyrene at room temperature is already as high as 0.7, the increase in $\phi_M{}^0$ with reduction in temperature cannot cause θ_D/θ_M to decrease to less than 70% of its room temperature value. Since k_d decreases continuously as the temperature is lowered it would be expected that lowering the temperature would cause θ_D/θ_M to increase, and that at very low temperatures k_d would be negligible and the whole of the delayed fluorescence would be dimer emission.

Experiments with solutions sufficiently dilute to test the prediction from equation 410 were made by Parker[86] following previous work by Stevens and co-workers[85] with more concentrated pyrene solutions. The results obtained with pyrene and benz(a)pyrene are shown in Fig. 135. As the

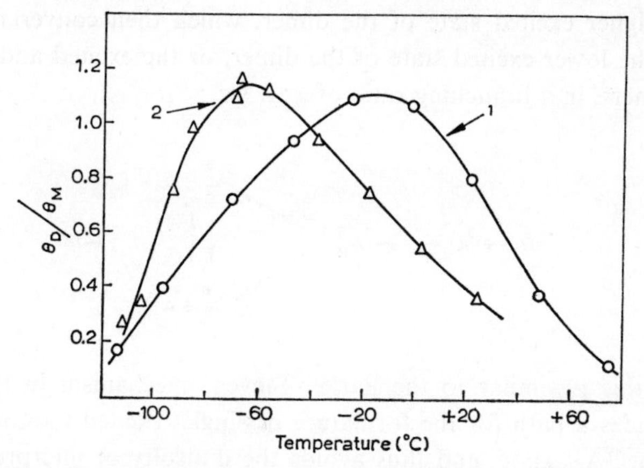

Fig. 135. Effect of temperature on the ratio of dimer emission to monomer emission in the delayed fluorescence spectra.
1, 2 × 10^{-5} M pyrene in ethanol, and 2, 10^{-5} M benz(a)pyrene in ethanol (from Parker[86]).

temperature was reduced θ_D/θ_M at first increased but then, contrary to prediction, decreased again and at $-110°C$ the proportion of dimer emission from both compounds was small. Stevens and co-workers[85] had interpreted their results by assuming that at low temperature there intervened a second mechanism involving triplet–triplet interaction at distances greater than that required for excited dimer formation. They also suggested that the formation of the excited dimer at low temperatures might be limited by the requirement of an activation energy. The results of Parker supported this view, and indeed all measurements of delayed fluorescence in fluid solution made to date[179] support the combined Parker–Stevens mechanism (see equation 411) which involves the transition from a collisional mechanism of triplet–triplet interaction at high temperature (equation 411 (a)) through an intermediate temperature range where both collisional and long range triplet interactions occur, to low temperature where only the long range interaction (equation 411 (b)) operates:

$$
{}^3A + {}^3A \begin{cases} \xrightarrow{(a)} {}^1A_2^* \longrightarrow {}^1A^* + {}^1A \\ \xrightarrow{(b)} {}^1A^* + {}^1A \end{cases} \qquad (411)
$$

Birks and co-workers[277] proposed a somewhat different combination of mechanisms. They supposed that the triplet–triplet interaction yields initially a higher excited state of the dimer, which then converts rapidly into either the lower excited state of the dimer, or the excited and ground state monomers, in a branching ratio of α:1, viz.:

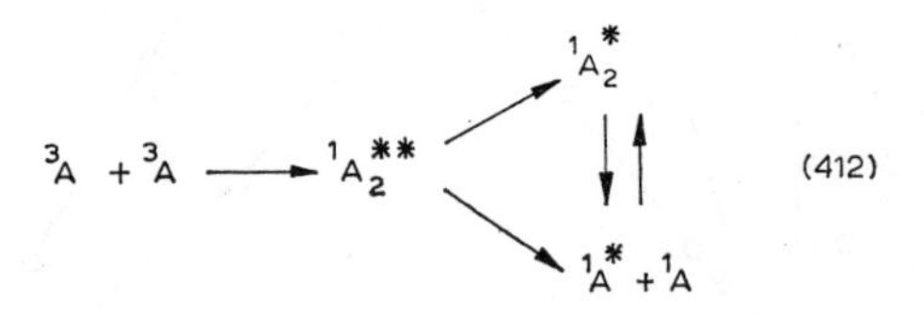

Kinetically this is similar to the Parker–Stevens mechanism in the sense that it provides a path for the formation of singlet excited monomer that by-passes the ${}^1A_2^*$ state, and thus avoids the difficulty of interpreting the results in Fig. 135 in terms of equation 410. At any given temperature the value of α determined by Birks and co-workers can be interpreted equally well by either of the mechanisms, since on the Parker–Stevens mechanism it would represent the ratio of the rates of processes 411(a) and 411(b). The Parker–Stevens mechanism gives an acceptable qualitative explanation of why the ratio of dimer to monomer should decrease at low temperatures. The rate of diffusion is then lower, the triplet molecules spend a relatively longer time within the triplet–triplet transfer distance, and a greater proportion of them would be expected to interact without the production of excited dimer. The requirement of an activation energy for the formation of the latter would further operate against its formation at low temperatures. Parker[278] criticised the Birks mechanism on the ground that it requires the rate of dissociation of the dimer in its upper excited state to be *greater* at low temperature relative to the competing reaction, and he concluded that on present evidence the Parker–Stevens mechanism, involving triplet-to-triplet energy transfer at a distance, is to be preferred. There can in any case be little doubt that triplet-to-triplet energy transfer over distances greater than the encounter distance takes place in rigid media (see Section 2 D 3). The results of experiments with substituted anthracenes also support this contention[179]. Birks[443] has since proposed a mechanism involving the formation of a quintuplet state of the dimer having a lower energy than $^1A_2^*$, and its thermal activation and intersystem crossing to $^1A_2^*$.

4. General Mechanism for Excitation of Monomer and Dimer Emission

Mechanisms such as those proposed by Stevens and co-workers[85] or Birks and co-workers[277, 279] involving direct population of both the $^1A^*$ and $^1A_2^*$ excited states (see equations 411 and 412) can be extended to cover the general case of a system involving a variety of processes, some populating $^1A^*$ and some $^1A_2^*$. The system may derive its energy by initial photo-excitation, by absorption of ionising radiation, by chemical reaction or by any other means of excitation. Whatever the mode of excitation, the processes that populate the system of excited states represented by:

$$^1A^* + {}^1A \rightleftarrows {}^1A_2^* \tag{413}$$

may be divided into two groups, viz., those giving rise initially to $^1A^*$, and those giving rise initially to $^1A_2^*$. We may represent the total rate of the former by R_M and the total rate of the latter by R_D. With the terminology used in equations 375–381, the general kinetic scheme then becomes:

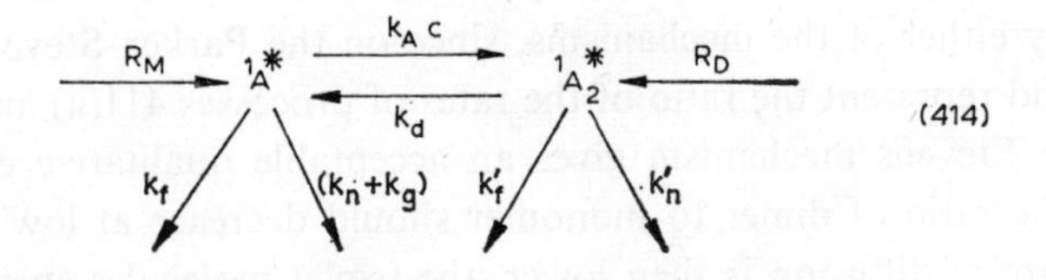

To derive an expression for the ratio of dimer emission to monomer emission (Φ_D/Φ_M) in terms of the rate parameters, we assume that triplet molecules produced *from* $^1A^*$ (rate $k_g[^1A^*]$) or *from* $^1A_2^*$ (rate included in $k_n'[^1A_2^*]$) are not present in sufficient quantity for their mutual interaction to contribute appreciably to R_M or R_D. That is, we assume that the steady state concentrations of $^1A^*$ and $^1A_2^*$ are limited to values for which P-type delayed fluorescence would obey the square law. We do not, of course, exclude processes (e.g. ion–electron recombination) that lead directly to the formation of molecules in the triplet state. With these assumptions the steady state equations for $^1A^*$ and $^1A_2^*$ are:

$$(k_f + k_n + k_g + k_A c)[^1A^*] = R_M + k_d[^1A_2^*] \tag{415}$$

and

$$(k_f' + k_n' + k_d)[^1A_2^*] = R_D + k_A c[^1A^*] \tag{416}$$

Hence:

$$\frac{\Phi_D}{\Phi_M} = \frac{k_f'[^1A_2^*]}{k_f[^1A^*]} = \frac{k_f'\tau_D(\alpha/\tau_M^0 + (1+\alpha)k_A c)}{k_f(1+\alpha\tau_D k_d)} \tag{417}$$

where α is R_D/R_M, and τ_D and τ_M^0 are the lifetimes of the dimer and the monomer at infinite dilution, and are given by:

$$\left.\begin{aligned} 1/\tau_D &= k_f' + k_n' + k_d \\ 1/\tau_M^0 &= k_f + k_n + k_g \end{aligned}\right\} \tag{418}$$

To make use of the general equation 417 the dependence of α on the concentration of solute, c, must be known or assumed. For systems in which α is independent of concentration, i.e. in which R_M and R_D vary with concentration in the same way, equation 417 becomes:

$$\Phi_D/\Phi_M = K' + K''c \tag{419}$$

Thus for such systems, the plot of Φ_D/Φ_M against c should give a straight line, from which:

$$K''/K' = (1 + 1/\alpha)k_A\tau_M^0 \tag{420}$$

Equation 417 is equivalent to that first derived by Birks[279] for photoexcited delayed fluorescence. The constants K_2 and K_3 (see equation 409) derived by Parker and Hatchard for pyrene in ethanol are equivalent to K' and K''. Birks and co-workers have applied the method to the determination of α for the P-type delayed fluorescence of pyrene in ethanol ($\alpha = 2.0$)[279] and of pyrene, 1,2-benzanthracene and 5-methyl-1,2-benzanthracene in cyclohexane ($\alpha = 2.0$, 0.9 and 0.9)[277]. Parker and Short[280] found that the monomer and dimer emission in the electrochemiluminescence of 9,10-dimethylanthracene in dimethylformamide obeyed equation 419, and that both K' and K'' increased with decrease in temperature or voltage applied. The value of K' at low temperature and low applied voltage was much greater than the value of K_2 (see equation 409) found by Parker and Joyce[179] for the photo-excited delayed fluorescence under the same conditions, and this indicated that triplet–triplet annihilation was not the main source of the electrochemiluminescence.

F. EFFECTS OF SOLVENT

1. General Comments

The solvent can affect the spectrum and efficiency of fluorescence and phosphorescence in a variety of ways, some of them complex and still imperfectly understood. Most of the effects may be grouped under the headings, polarisation and hydrogen bonding, viscosity effects, heavy atom effect, compound formation and photo-reaction. Chemical reaction of the excited singlet or triplet state with the solvent provides an additional pathway for consumption of the excited state and clearly will result in quenching of the corresponding luminescence. Examples are anthracene in carbon tetrachloride (fluorescence quenching) or benzophenone in ethanol (triplet quenching). Such reaction is often greatly reduced at 77°K. Compound formation, i.e. reaction of the ground state of the solute with the solvent, involves the formation of a new species with its own absorption spectrum, and hence the luminescence is expected to be affected—it may be reduced in intensity compared with that observed in an "unreactive" solvent, or a new fluorescence band may appear. The more subtle effects of hydrogen bonding, which may also involve marked changes in the absorption spectrum—and hence in effect, compound formation—are discussed under a separate heading below. Solvents containing heavy atoms may increase the rate of intersystem crossing as a result of increased spin–orbit coupling. They then produce fluorescence quenching and increased triplet formation (see Section 4 A 4). If such solvents do not quench the triplet state (e.g. when rigid at low temperature), they are expected to increase the phosphorescence efficiency. Finally there is one "trivial" effect which is nevertheless of practical importance: if the solvent absorbs strongly at the wavelength of the exciting light it will act as an inner filter, and little or no luminescence from the solute will be observed (see Section 3 H). This automatically limits the choice of solvents available for short-wavelength excitation.

2. Solvation by Polarisation

In considering the effects of solvation and hydrogen bonding we must distinguish carefully between the effects on the absorption spectrum and the effects on the emission spectrum. In Section 1 B 3 we discussed the reason why the 0–0 bands in the absorption and fluorescence spectra do not coincide. The difference of energy arises from the fact that immediately after light absorption the nuclear configuration of the solvated molecule is identical with the equilibrium configuration in the ground state (Franck–Condon principle) and this is generally not the same as the equilibrium configuration in the excited state. The unexcited and excited molecules may be solvated both by induced and permanent dipolar attraction, and by hydrogen bonding where appropriate. All these effects can produce differences between the equilibrium configurations of the ground and excited states and, as explained in Fig. 4, the greater the energy difference between the configurations, the greater will be the difference in the frequencies of the 0–0 bands in absorption and emission. In so far as change of solvent alters the energy difference between the two configurations, it will alter the 0–0 band separation. Increasing polarity of the solvent increases the 0–0 separation slightly, even with molecules that do not have a permanent dipole moment, owing to change in polarisability of the molecule on excitation. Molecules having a permanent dipole moment show an additional effect. Hydrogen bonding between solvent and solute can produce more profound effects and we shall discuss these in the following section.

3. Hydrogen Bonding

Compounds containing groups such as CO, –OH, –NH_2, etc. may in general form hydrogen bonds with appropriate solvents both in the ground state and in the excited state. Hydrogen bonded excited states can thus be produced via two routes as shown by the following scheme in which S represents a solvent molecule:

$$A + S \rightarrow AS \xrightarrow{h\nu} (AS)^* \tag{421}$$

$$A^* + S \rightarrow (AS)^* \tag{422}$$

The effect of hydrogen bonding in the ground state was investigated by Nagakura and Baba[281] who observed that compounds such as phenol or aniline showed abnormally high apparent dipole moments and abnormally large red shifts of their absorption spectra when dissolved in proton acceptor solvents such as dioxane, as compared with the spectra in aliphatic hydrocarbons. Comparatively small concentrations of dioxane in heptane were sufficient to produce the effect and since the corresponding O- or N-methylated derivatives did not show it, they attributed the abnormal spectral shift to the formation of hydrogen-bonded complexes of the type:

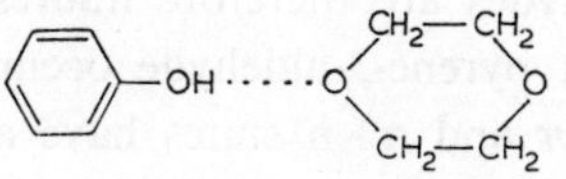

C_6H_5—OH·····O(CH_2—CH_2)$_2$O

Weller[241] found that the absorption spectrum of 3-hydroxypyrene in methyl cyclohexane was shifted some 0.03 μm^{-1} to longer wavelengths on the addition of hydrogen bond acceptors such as triethylamine or pyridine. He found that the fluorescence efficiency decreased on hydrogen bond formation. He found similar results with β-naphthol and concluded that the excited hydrogen bond complex with this compound could be formed by both mechanisms given in equations 421 and 422. From the reduction in fluorescence efficiency he calculated the rate constants for hydrogen bond formation in the excited state. Hydrogen bond formation in the excited state can be regarded as the preliminary stage of acid dissociation (in which the proton is transferred completely to the base—see Section 4 C), for which the shifts in absorption bands are considerably greater.

It will be remembered (Section 1 B 6) that increasing solvent polarity causes a shift in the π^*–n absorption bands to shorter wavelengths, and a smaller shift in the π^*–π absorption bands to longer wavelengths. With simple molecules having atoms with lone pair electrons, the π^*–n state is often the state of lowest energy, and owing to the long lifetime of this state such compounds give little fluorescence. With compounds having their lowest π^*–π and π^*–n states of similar energy it may happen that in inert solvents the π^*–n state is lowest, but in polar solvents or admixtures, this state is shifted to shorter wavelengths so that the π^*–π state is lower in energy. Such compounds are expected to be non-fluorescent in hydrocarbon solvents, but fluorescent in polar solvents. This may explain the fact that chlorophylls *a* and *b* are non-fluorescent in pure dry hydrocarbon solvents, but fluoresce on the addition of small amounts of water [282]. Similar behaviour has been reported for quinoline and acridine [283], which show increasing fluorescence as the polarity of the solvent increases.

The complexities that can arise when π^*–n shifts are superimposed on the other hydrogen bonding effects are illustrated by the interesting work of Bredereck, Förster and Oesterlin [284], which we shall discuss in some detail. The lower aldehydes have π^*–n lowest excited singlet states and are non-fluorescent. With larger aromatic rings, both types of state are situated at lower energies, but the π^*–π states move much more than do π^*–n states so that they finally become the lowest singlet excited states, and the most complex aromatic aldehydes are therefore fluorescent. Bredereck and co-workers considered that pyrene-3-aldehyde occupied an intermediate position in which the π^*–π and π^*–n states have about equal energy. This aldehyde is not fluorescent in heptane or diethyl ether, weakly fluorescent in chlorobenzene or acetonitrile, and fairly strongly so in ethanol, acetic acid or water–acetonitrile mixtures. With increasing fluorescence efficiency (with change of solvent) the fluorescence band shifts from the violet to the blue, although the absorption maximum (i.e. π^*–π, under which the much weaker π^*–n absorption is not visible) shows comparatively little red shift. The large red shift of fluorescence indicates that a considerable degree of solvation of the (lowest) π^*–π state takes place in polar solvents during the lifetime of this state. Bredereck and co-workers consider that it is this increased solvation of the π^*–π state in polar solvents that is responsible for altering the relative populations of the two types of excited state and hence the ratio of the competing radiative and non-radiative processes.

With solutions of the above moderately activating solvents in an inactive solvent, they found that (in contrast to the chlorophylls) the fluorescence efficiencies increased continuously and only reached their maximum values in the pure activating solvent. However, with a strongly activating solvent, trichloracetic acid, in heptane, they observed appreciable fluorescence at 10^{-3}M and the intensity increased rapidly with increasing concentration of trichloracetic acid. At low concentrations the character of the spectra resembled those in moderately activating solvents in that they showed structure and were situated at comparatively high wavenumbers, although they did not show a red shift. However with high concentrations of trichloracetic acid the spectra developed an additional maximum at lower wavenumbers and this shifted towards the red with increasing concentration, e.g. in 1 M trichloracetic acid the fluorescence appeared green in colour. They concluded that since the activation by trichloracetic acid in the low concentration range is not accompanied by an appreciable red shift, it is not due to a decrease in the energy of the fluorescing π^*–π state but to an increase in energy of the π^*–n state relative to the ground state. Furthermore it cannot result from reaction in the excited state because of the low concentrations that are effective. It is therefore probably due to ground state interaction giving a separate hydrogen-bonded species. Changes in the absorption spectra seemed to confirm this.

Bredereck, Förster and Oesterlin summarised their work by saying that

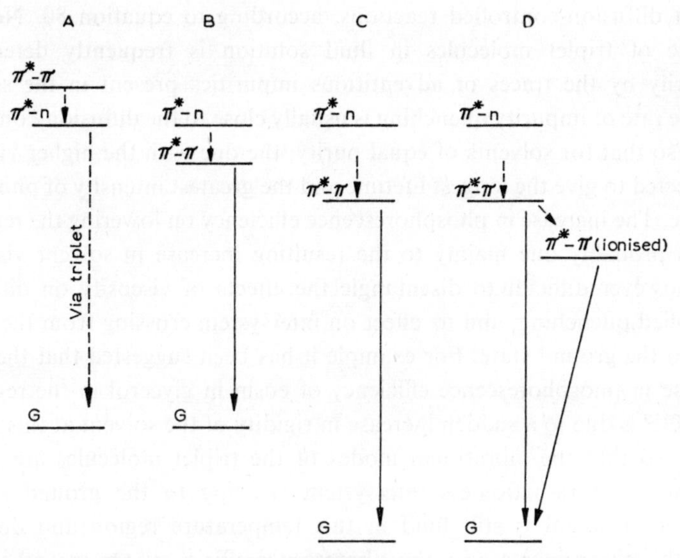

Fig. 136. Effect of solvent on the relative energies of ground and excited states of pyrene-3-aldehyde.
A, in non-polar solvent; B, in moderately activating solvent producing H-bonding of π^*–π excited state; C, with low concentrations of a strongly activating solvent producing H-bonding of ground and π^*–π excited states; D, with high concentrations of a strongly activating solvent producing the additional effect of complete proton transfer.

the separation of the π^*–n and π^*–π singlet states determines the ability of the compound to fluoresce. If this separation is small enough, its sign may be changed by change of solvent and this change may be produced in two ways. Weakly or moderately activating solvents such as chloroform or ethanol act mainly by lowering the π^*–π state energy by increasing the interaction in the excited state (i.e. with additional molecules not involved in the ground state hydrogen bonding). Strongly activating solvents such as trichloracetic acid on the other hand act at low concentration by raising the π^*–n state energy by hydrogen bonding in the ground state. The extremely large red shift at high concentrations of trichloracetic acid may involve actual proton transfer in the excited state, assisted by other activator molecules. We have attempted to interpret these conclusions by means of the diagram in Fig. 136.

4. Viscosity Effects

The viscosity of the solvent is the main parameter that determines the rate of diffusion-controlled reactions, according to equation 80. Now the lifetime of triplet molecules in fluid solution is frequently determined primarily by the traces of adventitious impurities present in the solvent, and the rate of impurity quenching is usually close to the diffusion-controlled value, so that for solvents of equal purity, the one with the higher viscosity is expected to give the longest lifetime and the greatest intensity of phosphorescence. The increase in phosphorescence efficiency on lowering the temperature is probably due mainly to the resulting increase in solvent viscosity. It is however difficult to disentangle the effects of viscosity on diffusion-controlled quenching, and its effect on intersystem crossing from the triplet state to the ground state. For example it has been suggested that the sharp increase in phosphorescence efficiency of eosin in glycerol in the region of $-60°C$[19] is due to a sudden increase in rigidity of the solvent at this temperature, so that the vibrational modes of the triplet molecules are greatly restricted and radiationless intersystem crossing to the ground state is inhibited. Ethanol is still fluid in this temperature region and does not show this sharp increase in phosphorescence efficiency. On the other hand Stevens and Walker[115] have shown that triplet lifetimes of some aromatic hydrocarbons in liquid paraffin vary with temperature in a manner that would be expected from diffusion-controlled impurity quenching (see Section 4 B 1). An obvious example of the effect of viscosity on diffusion-controlled quenching is the comparison of phosphorescence efficiency in an aerated solvent such as ether–pentane–ethanol at room temperature, and at 77°K. At room temperature the phosphorescence is completely quenched by the oxygen, but in the rigid solution at 77°K the phosphorescence efficiency is high and has the same value whether or not the solvent is de-aerated before cooling.

The concentrations of quencher required to quench prompt fluorescence appreciably are so high (see Section 2 B 2) that impurity quenching rarely presents a difficulty. The most likely source of fluorescence quenching is dissolved oxygen. Oxygen quenching may be considerable in air-saturated solutions of compounds having an exceptionally long fluorescence lifetime. For example, saturation of a dilute solution of pyrene in cyclohexane with air quenches the fluorescence by about 20 times[43]. In liquid paraffin or glycerol the degree of quenching is much less owing to the higher solvent viscosities. It is obviously important to eliminate oxygen quenching in experiments designed to elucidate more specific solvent effects.

Bowen and Seaman[24] have discussed the effects of both temperature and viscosity on the efficiency of fluorescence in solution. They found that with flexible molecules such as di-9-anthryl-ethane, the activation energy for fluorescence quenching was close to that for solvent fluidity, suggesting that such an excited molecule can degrade its energy by relative thermal diffusional movements of the two large rings joined by the flexible chain. The effect of viscosity on the activation energy of rigid molecules such as anthracene was much less.

The effect of solvent viscosity on the efficiency of P-type delayed fluorescence is more complicated. On the one hand triplet lifetimes are increased at higher viscosities owing to the lower rate of impurity quenching. On the other hand the rate of triplet–triplet encounters is also decreased, although triplet-to-triplet energy transfer can apparently occur over distances greater than the encounter distance. With naphthalene in ethanol[112] the overall effect results in a moderate increase in θ/ϕ_f as the temperature is reduced from $+20$ to -80°C, and some other aromatic hydrocarbons show a similar behaviour.

5. Resolution of Vibrational Bands and the Shpol'skii Effect

The molecules that produce the most highly resolved vibrational bands in the fluorescence emission spectrum at room temperature are those having rigid ring structures with the minimum of substituent groups. In this section we shall consider how the resolution of these bands is affected by solvent and temperature. At room temperature in fluid solution the widths of the vibrational bands are several hundredths of a μm^{-1}. An example is the spectrum of pyrene in ethanol shown in Fig. 137 curve 2. This was taken with a double silica prism monochromator at a slit width of 0.04 mm corresponding to a half-band width of 0.38 nm (0.0024 μm^{-1}) at 400 nm. Thus the resolution of this spectrum is not limited by the performance of the instrument. In saturated hydrocarbon solvents at room temperature the widths of the bands are only marginally narrower. The structure is sharpened quite considerably in a rigid glassy medium at 77°K (see Fig. 137 curve 1) but the bands are still some 0.01–0.02 μm^{-1} wide, and the widths of the bands of other compounds under these conditions are rarely less than this.

By the use of a selected n-paraffin as solvent, the finely crystalline solid solutions obtained by freezing at 77°K, or lower, give quite remarkable fluorescence emission spectra[285]. The relatively broad vibrational bands observed in glassy media split into a series of lines which are as narrow as 0.0002–0.0003 μm^{-1} in favourable cases and up to 0.001 μm^{-1} in un-

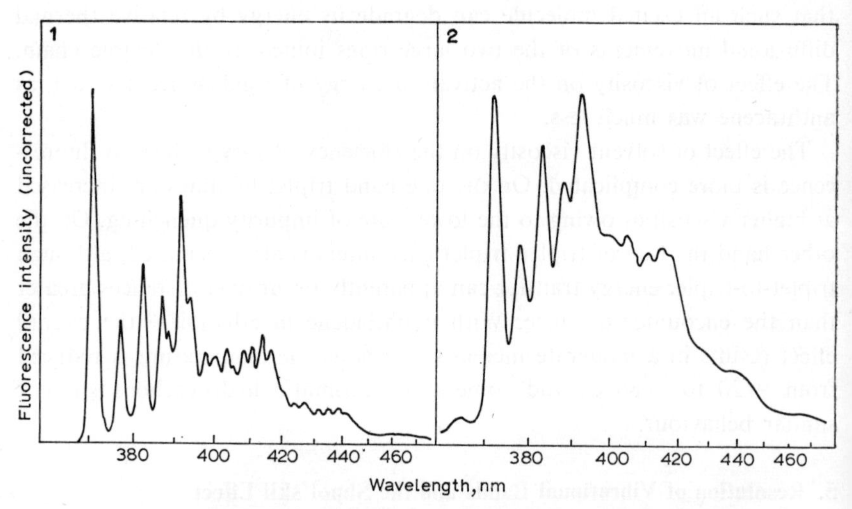

Fig. 137. Prompt fluorescence emission spectra of 2×10^{-5} M pyrene.
1, in ether–pentane–ethanol at 77°K; 2, in ethanol at room temperature; excitation at 313 nm; half-band width of silica prism analysing monochromator was 0.38 nm (0.0024 μm^{-1}) at 400 nm.

favourable ones. Regularly associated groups of lines appear and an interesting dependence on solvent is found. These sharp spectra were first observed by Shpol'skii and co-workers, using the photographic method of recording, with solutions of coronene[286], and later pyrene and benz(a)-pyrene[287]. Bowen and Brocklehurst[288] repeated the work with coronene, using a glass prism photoelectric spectrometer, and confirmed in general the results of Shpol'skii and co-workers. They showed however that the specimen of coronene used by the latter must have contained a trace of 1,12-benzoperylene, which was preferentially excited at 366 nm, and they thus provided the first demonstration of the potentialities of the technique as an analytical method. In a later paper, Bowen and Brocklehurst[289] found that the sharpness of the spectra of coronene, 1,12-benzoperylene, triphenylene and 9,10-dichloranthracene varied markedly with solvent, and they attributed these solvent effects to specific interactions of the solute molecules with the crystal lattice of the solvent, the degree of interaction depending on the relative sizes of solvent and solute. In a later review paper[285] Shpol'skii reported that the results with various substances indicated that the most favourable conditions for the appearance of the quasi-linear spectra are when the dimensions of the solute molecule are approximately equal to that of the solvent molecule. He concluded that in these circumstances the solute molecule replaces a paraffin molecule in the crystalline matrix. When the solute–solvent "fit" is not good the spectrum becomes diffuse. Thus, pyrene (molecular length 0.7 nm) gives the sharpest spectrum in n-hexane (0.88 nm) while benz(a)pyrene (1.0 nm) gives the sharpest spectrum in n-heptane (1.0 nm) or n-octane (1.1 nm).

The various applications of this technique have since been much investigated, particularly by the Russian workers. Shpol'skii[290] reviewed the analytical applications up to 1959. Dikun[291], Muel and Lacroix[292] and Personov[293] have applied it to the determination of traces of benz(a)pyrene. In addition to their analytical applications, the spectra can be used to obtain data of theoretical interest. Since in condensed media all radiative transitions take place from the lowest vibrational levels of the first excited singlet or triplet states, the band structures in the fluorescence and phosphorescence emission spectra reflect the vibrational frequencies of the *ground* state of the molecule. Thus the analysis of the highly resolved Shpol'skii spectra provides a means of measuring these molecular vibration frequencies. For example, vibrational analyses have been made of the spectra of phenanthrene (phosphorescence)[294], diphenylpolyenes and polyphenyls[295], azulene[296] and pyrene[297]. The spectrum of the latter in n-hexane at 4°K is remarkable in showing more than 220 lines compared with about 60 at 77°K.

The quasi-linear spectra are sensitive to the influence of intramolecular and intermolecular forces and are considerably affected by hydrogen bonding both of the solvent and the solute. Thus, for example, the fluorescence emission spectrum of pyrene is diffuse in ethanol at 77°K and this has been attributed to the formation of a network of hydrogen bonds between the solvent molecules[285]. In hexanol the influence of the OH groups relative to that of the hydrocarbon chain is less, and the pyrene spectrum in this solvent at 77°K is more highly resolved[285]. Some interesting effects due to substituent groups of the solute have been observed by Shigorin and co-workers[298]. Thus anthraquinone gives a sharp phosphorescence spectrum in hexane, heptane and octane at 77°K showing a series of bands, the components of which are separated from one another by an average distance of 0.166 μm^{-1}, i.e. a frequency characteristic of the fully symmetrical vibration of the C=O group in the ground state. Similar fine structure with the same frequency interval was also observed in the spectra of the β-chloro-, β-methyl- and β-amino-derivatives. With certain substituents in the α-position, in particular OH, the spectra become diffuse and this was attributed to intramolecular hydrogen bonding:

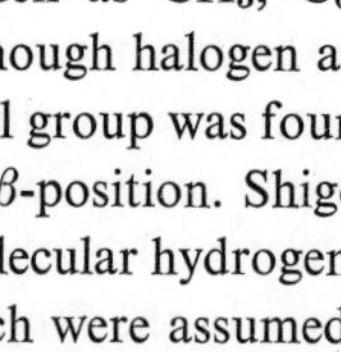

In contrast, α-substituents such as CH_3, C_6H_5 and CH_3O caused little change in the spectrum[285] although halogen atoms at the α-position caused some diffuseness. The hydroxyl group was found to cause a diffuse spectrum even when substituted in the β-position. Shigorin and co-workers[298] interpreted this in terms of intermolecular hydrogen bonding between β-hydroxyanthraquinone molecules which were assumed to be associated into dimers and polymers even in dilute solution. Later work[299] discussed the spectra of polyhydroxy anthraquinones, among which it appears that the 1,4-derivative gives a *sharp* vibrational structure but without showing the carbonyl frequency difference.

Grajcar and Leach[300] have shown that photolysis of solutes can be accomplished in the crystalline matrices used for measuring Shpol'skii spectra, and have used this technique for observing the highly resolved spectra of the trapped radical photolysis products. For example they have analysed the vibrational frequencies in the emission spectrum of the benzyl radical produced by photolysis of toluene in a monoclinic cyclohexane matrix. Toluene itself emits in the region below 300 nm but the benzyl radical spectrum is situated in the blue-green region.

Measurement of the most highly resolved Shpol'skii spectra with a photoelectric instrument is difficult, particularly if low concentrations have to be dealt with. It requires a monochromator of high resolving power

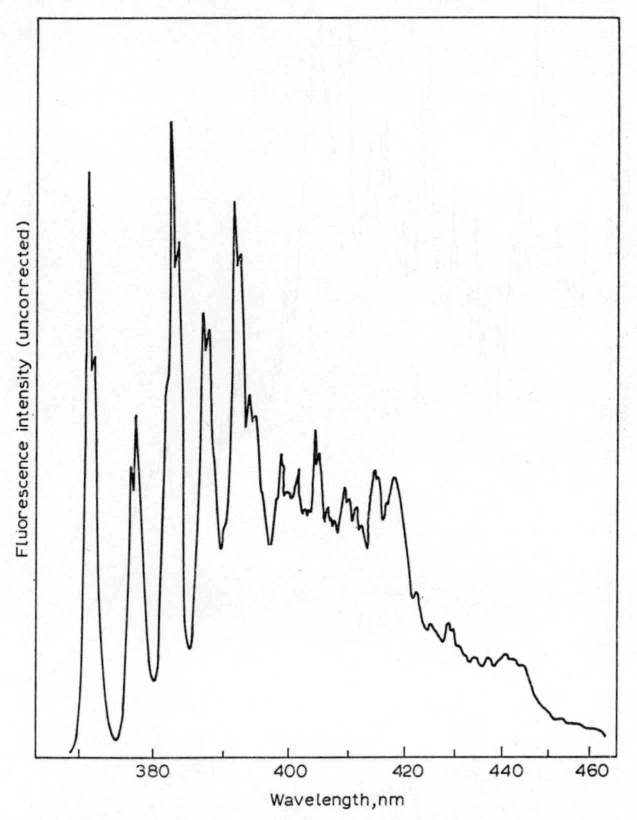

Fig. 138. Prompt fluorescence emission spectrum of 10^{-4} M pyrene in n-hexane, containing 10% cyclohexane, at 77°K.
Excitation at 313 nm; half-band width of silica prism analysing monochromator was 0.38 nm (0.0024 μm^{-1}) at 400 nm.

and large LGP value because of the narrow slit widths that have to be used. Very intense exciting light is generally needed and the spectrum has to be scanned slowly. Muel and Lacroix[292] used a specially designed double glass prism monochromator for measuring the spectrum of benz(a)pyrene

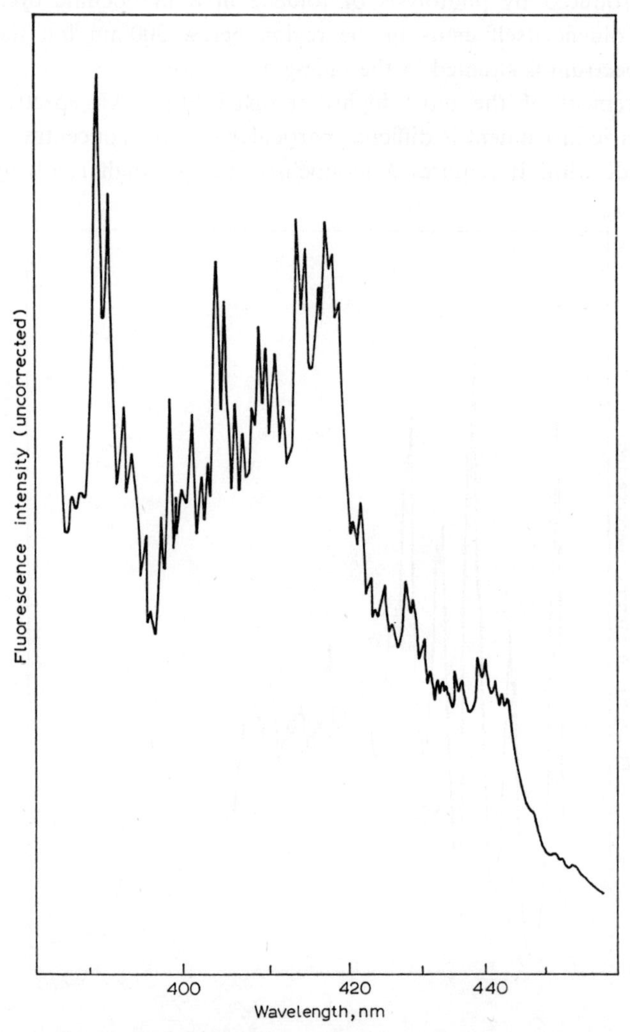

Fig. 139. Prompt fluorescence emission spectrum of 10^{-4} M pyrene in n-hexane, containing 10% cyclohexane, at 77°K.
Excitation at 313 nm; half-band width of glass prism analysing monochromator was 0.17 nm (0.0011 μm^{-1}) at 400 nm.

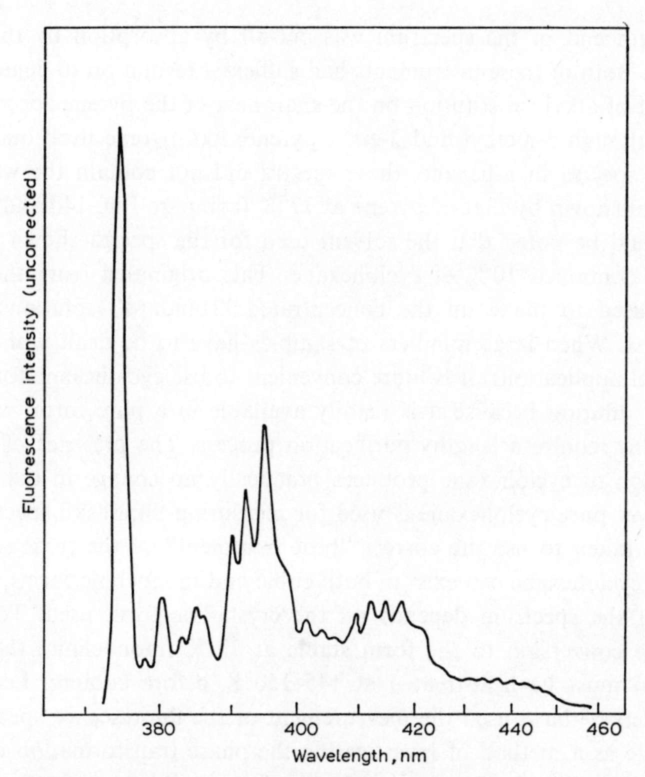

Fig. 140. Prompt fluorescence emission spectrum of 10^{-4} M 3-ethylpyrene in n-hexane, containing 10% cyclohexane, at 77°K.
Excitation at 313 nm; half-band width of silica prism analysing monochromator was 0.38 nm (0.0024 μm^{-1}) at 400 nm.

in n-octane. They located 47 of the 50 lines measured photographically by Shpol′skii as well as 10 additional lines. Most commercial monochromators of the type normally used in spectrofluorimeters do not have adequate performance to resolve the finest Shpol′skii spectra completely, but a good commercial instrument will provide sufficient resolution to show the difference between spectra measured in glassy solvents and those measured in crystalline n-paraffin matrices. For example Fig. 138 shows the spectrum of pyrene measured in a crystalline matrix with the same instrument (double silica prisms) and at the same band width, as that used to obtain curve 1 in Fig. 137. A greater resolution with the same slit widths was obtained with a single glass prism monochromator (see Fig. 139) although the short wavelength end of the spectrum was cut-off by absorption by the prism material. Both of these instruments had sufficient resolution to demonstrate the effect of alkyl substitution on the sharpness of the pyrene spectrum[301]. Thus, although 3-methyl and 3-ethyl pyrene, like pyrene itself, gave their sharpest spectra in n-hexane, these spectra did not contain the wealth of fine detail shown by that of pyrene at 77°K (compare Fig. 140 and 138).

It should be noted that the solvent used for the spectra shown in Figs. 137–140 contained 10% of cyclohexane. This originated from the cyclohexane used to make up the concentrated "standard" solutions of the specimens. When large numbers of samples have to be dealt with (e.g. in analytical applications) it is more convenient to use cyclohexane for all but the final dilution because it is readily available in a pure form, while the n-paraffins require a lengthy purification process. The presence of a small proportion of cyclohexane produces practically no change in the spectra. If however pure cyclohexane is used for measuring Shpol'skii spectra, care must be taken to use the correct "heat treatment" of the frozen mixture because cyclohexane can exist in both cubic and monoclinic forms, and the shape of the spectrum depends on the crystalline form used. To obtain complete conversion to the form stable at 77°K (monoclinic) the frozen specimen must be heat-treated at 145–150°K before cooling. Leach and co-workers[302] have used the measurement of the fluorescence spectrum of the solute as a method of investigating the phase transformation in cyclohexane, including the formation of a third metastable modification.

G. PHOTOLUMINESCENCE MEASUREMENTS IN THE STUDY OF IRREVERSIBLE PHOTOCHEMICAL REACTIONS

1. Analysis of Photochemical Products

Since the measurement of photoluminescence provides a sensitive and versatile method of chemical analysis, it will clearly be considered along with other analytical techniques, when devising methods for the identification of photochemical products, or investigating their quantum yield under various conditions. The direct measurement of fluorescence is often a convenient means of monitoring the consumption of reactant or accumulation of product, for example the formation of carbazole from diphenylamine[131]. In some systems separation of the mixture of products may be required before techniques of identification are applied, and here again, if the products can be made to fluoresce or phosphoresce, the measurement of this photoluminescence is applicable to the identification of the small quantities that can be separated by such techniques as thin-layer or gas chromatography. Since such measurements are in essence analytical applications of fluorescence and phosphorescence they are dealt with in Chapter 5. We shall consider here some applications of photoluminescence measurement that are particularly relevant to the investigation of photochemical reaction systems.

2. Measurement of Fluorescence Quenching

A frequently encountered photochemical system is that containing two solutes, one of which is excited by light absorption and then reacts with the second:

$$A \text{ (excited)} + B \rightarrow \text{products} \tag{423}$$

One of the first questions that arises is whether the reaction involves the lowest excited singlet state of A, or the triplet state. Consideration of the concentration of B required to produce an appreciable rate of reaction will sometimes provide an indication: if this concentration is less than 10^{-4}M the reaction is unlikely to involve the singlet state because, even with a diffusion-controlled reaction, the rate will be low at these concentrations. However, there is always the possibility that the singlet state is abnormally long-lived, or that a weak complex between A and B is formed in the ground state. Measurement of the fluorescence quenching of A by B will often settle the question. We shall take as an example the photoreaction of the dyestuff thionine with ferrous sulphate in 0.1 N sulphuric acid, in which the primary product is semithionine. The relative quantum yield of semithionine can be determined by flash absorption spectroscopy[303] and this yield rises to a maximum value when the ferrous sulphate is present in large excess. If the semithionine is formed by reaction of Fe^{2+} with singlet thionine, its yield should be 50% of the maximum value at the concentration of Fe^{2+} required to quench the fluorescence to 50%. In fact it was found that the yield of semithionine was 50% of its maximum value at a ferrous iron concentration of about 0.002 M, while 50% quenching of the fluorescence of thionine required about 2 M ferrous sulphate. It was thus clear that the singlet state of thionine was not involved directly in the photoreaction, at least at these low concentrations of ferrous sulphate.

3. Investigation of Photochemical Reactions in Rigid Media

Photochemical reactions in rigid glasses at low temperature are generally investigated by the method of absorption spectroscopy. Comparatively high concentrations of solute and high intensities of filtered exciting light are used so as to produce sufficient product to measure the absorption spectrum. To determine quantum efficiencies of photodecomposition, and quantum yields of products, it is necessary to produce a high and uniform intensity of monochromatic exciting light covering an area of the reaction cell sufficient to allow absorption measurements to be made. If a variety of excitation wavelengths have to be investigated it is difficult to find sources intense enough to provide the necessary light fluxes through a monochromator. The difficulty is accentuated if the quantum efficiency happens to be small, and a further difficulty is that with high concentrations, the products of the photolysis act as inner filters for the exciting light and thus introduce a considerable error in the determination of quantum efficiencies.

The techniques of fluorescence or phosphorescence measurement, capable as they are of measuring very low concentrations, thus have considerable advantages over absorption measurements in the investigation of low temperature systems, where reactants and products often show fluorescence or phosphorescence. Very dilute solutions can be used so that only a small proportion of exciting light is absorbed, and inner filter effects can be reduced to manageable proportions. Another advantage is the possibility of focussing the light from a monochromator on a small area of the face of the reaction vessel so that a variety of wavelengths of monochromatic light can be obtained at adequate intensities. After irradiation, the proportion of photolyte decomposed can be determined directly by comparing the intensity of its fluorescence (or phosphorescence) in the irradiated area with that from adjacent unirradiated areas on either side. Since right angle illumination and low concentrations are used, the spectrometer can also be employed to measure the emission and excitation spectra of the photochemical products. This technique was first used by Parker and Rees[304, 305] to determine the quantum efficiency of photodecomposition of thionine in EPA glass at 77°K. They were able to determine quantum efficiencies as low as 10^{-6} and found values that varied from $< 10^{-6}$ at 578 nm up to 3×10^{-3} at 248 nm. They also observed the fluorescence emission and excitation spectra of some of the products. Some results taken from their paper are shown in Fig. 141. Although Parker and Rees first outlined the advantages of this method in 1959[304] comparatively little use has yet been

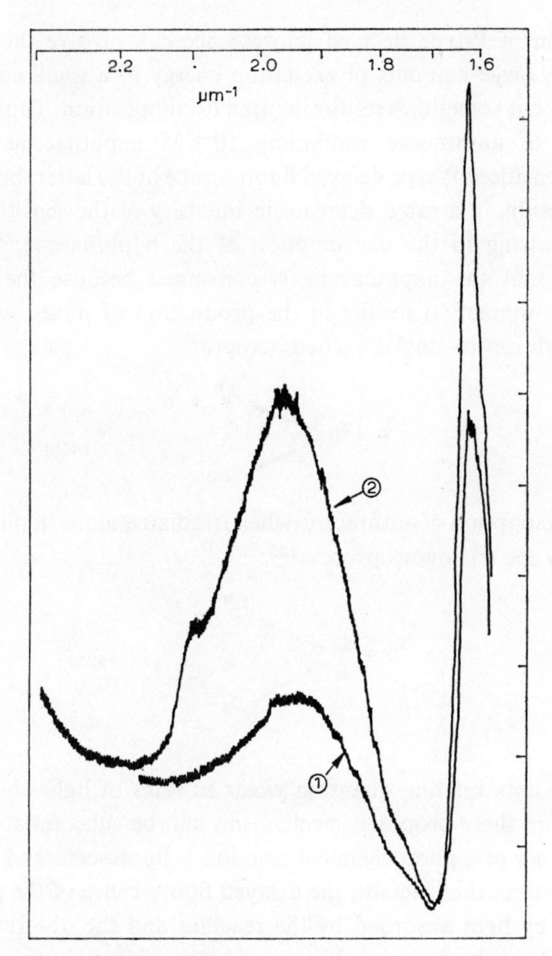

Fig. 141. Fluorescence emission spectra recorded during the photolysis of thionine in ether–pentane–ethanol at 77°K.

1, after 5 min irradiation; 2, after 40 min irradiation; 2×10^{-5} einstein litre^{-1} min^{-1} absorbed at 248 nm. The maxima at low wavenumber are due to thionine fluorescence; the broad maxima at higher wavenumber are due to photolysis products (from Parker and Rees[305]).

made of it, at least for the determination of quantum yields. Leach[306] has used the photographic technique to record and analyse the spectra of trapped radical products such as benzyl and triphenylmethyl.

4. Sensitised Delayed Fluorescence of Photochemical Reactants and Products

Sensitisation of P-type delayed fluorescence can involve the transfer of comparatively large amounts of excitation energy to a small concentration of acceptor and it sometimes results in rapid decomposition. Thus irradiation of solutions of anthracene containing 10^{-8} M naphthacene causes the emission of sensitised P-type delayed fluorescence of the latter, but continued irradiation results in a rapid decrease in intensity of the sensitised delayed fluorescence owing to the consumption of the naphthacene. It has been suggested[123] that the naphthacene is consumed because the process of mixed triplet interaction results in the production of mixed *stable photo-dimers*, in addition to singlet excited acceptor:

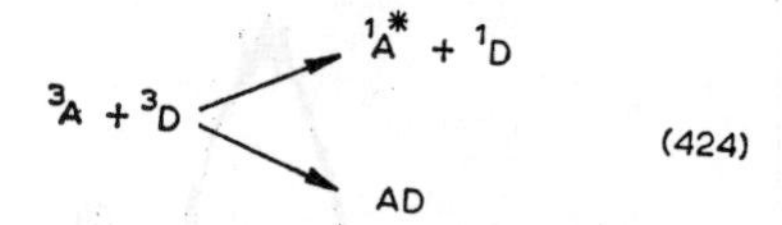

The slow consumption of anthracene when irradiated alone in dilute solution may occur by the analogous process[123]:

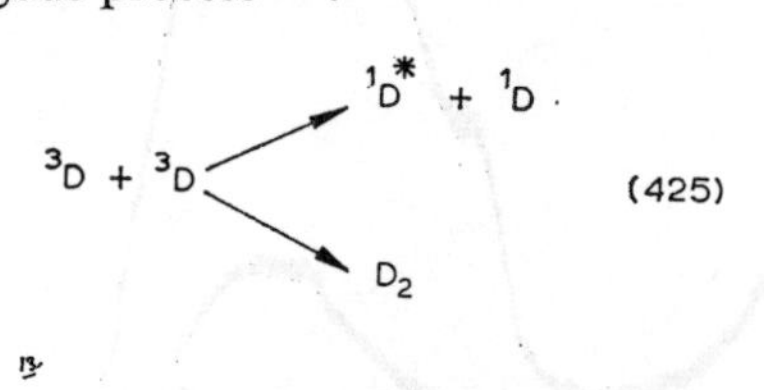

More precise data relating quantum yields to rates of light absorption are required before these proposed mechanisms can be substantiated.

If the product of a photochemical reaction is fluorescent and has a lower triplet energy than the reactant, the delayed fluorescence of the product will be sensitised by light absorbed by the reactant and the observation of the sensitised delayed fluorescence then provides an extremely delicate method for detecting the occurrence of the photochemical reaction. Parker and Hatchard[307] have observed such a reaction in solutions of benz(a)pyrene sensitised to visible light by means of proflavine hydrochloride, and containing polyvinylpyridine n-butyl bromide. Exposure of such solutions to the exciting light for short periods at room temperature produced a considerable distortion of the delayed fluorescence spectrum (measured at low temperature) although the amount of benz(a)pyrene decomposed was not sufficient to affect the prompt fluorescence spectrum (see Fig. 142 curves 1 and 2). The same photo-reaction took place in the absence of proflavine hydrochloride if the solution was irradiated with light absorbed by the benz(a)-pyrene. Measurements of the intensity and lifetime of the delayed fluorescence of solutions of benz(a)pyrene with and without the polymer present showed that the benz(a)pyrene *triplet* was quenched by the polymer, and it was

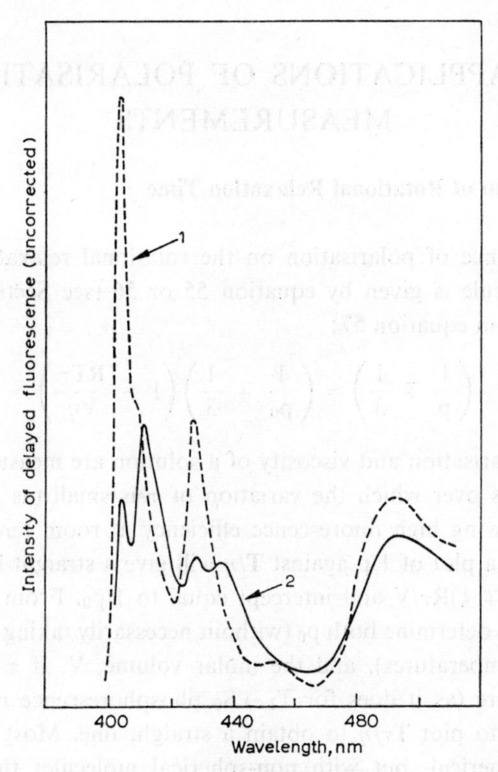

Fig. 142. Delayed fluorescence from de-oxygenated ethanolic solutions containing 10^{-6} M benz(a)pyrene, 8×10^{-6} M proflavine hydrochloride, and polyvinyl pyridine-N-butyl bromide equivalent to 10^{-4} M monomer.

Exciting light, 436 nm; spectra measured at $-75°$C; 1, before irradiation; 2, after irradiation for 80 sec at 20°C with 10^{-4} einstein litre^{-1} absorbed; prompt fluorescence spectra excited by 366 nm were substantially the same for both solutions and were similar to curve 1 (from Parker and Hatchard[307]).

therefore concluded that the photo-reaction proceeded via the triplet state of the hydrocarbon. This was confirmed by the fact that as the photo-product accumulated the benz(a)pyrene delayed fluorescence (and hence the benz(a)-pyrene triplet) was quenched by the product and the rate of reaction decreased. It is probable that there are many such photoreactions, the occurrence of which is at present unsuspected because the reactions are "shut off" in their early stages by product-quenching of the triplet state of the reactant. Measurement of the sensitised delayed fluorescence of the product is a means of observing these reactions in their very early stages before complete "shut off" occurs.

H. APPLICATIONS OF POLARISATION MEASUREMENTS

1. Determination of Rotational Relaxation Time

The dependence of polarisation on the rotational relaxation time of a spherical molecule is given by equation 55 or 56 (see Section 1 D 6). By substitution from equation 57:

$$\left(\frac{1}{p} \mp \frac{1}{3}\right) = \left(\frac{1}{p_0} \mp \frac{1}{3}\right)\left(1 + \frac{\mathbf{RT}\tau}{V\eta}\right) \tag{426}$$

Thus if the polarisation and viscosity of a solution are measured at a series of temperatures over which the variation of τ is small (as it is for many substances showing high fluorescence efficiency at room temperature—see Section 2 B 5), a plot of $1/p$ against $\mathbf{T}/\eta$ will give a straight line with slope equal to $(1/p_0 \mp \frac{1}{3})\mathbf{R}\tau/V$ and intercept equal to $1/p_0$. From this data it is thus possible to determine both p_0 (without necessarily taking measurements at very low temperatures), and the molar volume, V. If τ varies rapidly with temperature (as it does for $T_1 \rightarrow S_0$ phosphorescence in fluid media) it is necessary to plot $\mathbf{T}\tau/\eta$ to obtain a straight line. Most molecules are not strictly spherical, but with non-spherical molecules the plot of $1/p$ against $\mathbf{T}/\eta$ generally gives an approximate straight line from which an effective molar volume, V, may be derived[308].

The range of temperatures and viscosities over which the polarisation will show a measurable variation will clearly depend on the ratio $\mathbf{RT}\tau/V\eta$. With long emission times (τ) at high temperatures in solvents of low viscosity the emission will be almost completely depolarised and p too small to measure. With low values of τ and **T** in solvents of high viscosity p will be substantially constant and equal to p_0. It is instructive therefore to tabulate values of p for a variety of conditions and to consider what values of the four parameters **T**, τ, V, η give values of p that are measurably different from p_0 and zero. Some values are given in Table 43 for a typical "small" fluorescent molecule, e.g. fluorescein, for which V ~ 500 cc, and for a large polymer molecule (e.g. a protein) for which V ~ 5×10^4 cc. Polarisation of both fluorescence, with $\tau = 10^{-8}$ sec, and phosphorescence with $\tau = 10^{-2}$ sec are included.

With "small" molecules in solvents of low viscosity, the polarisation of fluorescence ($\tau = 10^{-8}$ sec) varies from low values to high values over the range + 20 to − 100°C and this temperature range is thus suitable (e.g. with ethanolic solutions) for determining V. With polymeric molecules it is desirable to increase the temperature range by taking measurements above room temperature.

The method has mainly been employed for determining the size of large molecules, such as those of proteins. The usual method is to treat the protein with a fluorescent reagent that will react irreversibly with it to produce a fluorescent polymer molecule. The reagents most frequently used are derived from common fluorescent compounds by the introduction of an isocyanate, isothiocyanate or sulphonyl chloride group. For example protein conjugates have been prepared from the isocyanates of fluorescein, rhodamine B and anthracene, amongst other compounds. The fluorescent isocyanates react irreversibly with free amino or sulphhydryl groups of the protein. The

TABLE 43

VALUES OF POLARISATION AS A FUNCTION OF VISCOSITY AND TEMPERATURE

(It is assumed that $\beta = 0$ and that unpolarised exciting light is used, i.e.: $1/p = 3 + (3.333\mathbf{R}\mathbf{T}\tau/V\eta))$

Solvent	Temp. (°C)	η/T	Polarisation (p)			
			small mol., V = 500		large mol., V = 5 × 10^4	
			$\tau = 10^{-8}$	$\tau = 10^{-2}$	$\tau = 10^{-8}$	$\tau = 10^{-2}$
Water	20	3.4×10^{-5}	0.006	< 0.001	0.215	< 0.001
Ethanol	− 98	2.5×10^{-3}	0.191	< 0.001	0.331	< 0.001
Glycerol	20	5.1×10^{-2}	0.322	< 0.001	0.333	0.001
Glycerol	0	0.44	0.332	< 0.001	0.333	0.008
Glycerol	− 42	2.9×10^{2}	0.333	0.045	0.333	0.313

reagent chosen and the conditions of reaction must obviously be such that they affect the physical structure and size of the protein molecule to the minimum extent. The lifetimes of the fluorescent protein conjugates are usually close to those of the simple fluorescent molecules. For details of the applications of fluorescent protein conjugates the reader is referred to the review by Steiner and Edelhoch[308].

Comparatively little use has been made of the application of rotational depolarisation of phosphorescence. Because of the much greater lifetime of phosphorescence, the viscosity at which depolarisation can be observed is much greater than with fluorescence (compare for example columns 4 and 5 in Table 43). If the lifetime and molar volume of the phosphorescent molecules are known, measurement of depolarisation can in principle be used to determine high viscosity in the range where other methods are difficult. The method is potentially valuable for the investigation of polymeric materials in which a small concentration of phosphorescent material can be dissolved to act as a probe for the measurement of microscopic viscosity. Polarisation of fluorescence is also of value in such investigations, because with some polymers showing high macroscopic viscosity, the local viscosity may be comparatively low, so long as the domains of the polymer chains do not overlap[309].

Because the depolarisation of the fluorescence of a small molecule of short rotational relaxation time in fluid solution is greatly reduced when it becomes bound to a large polymer molecule, measurement of polarisation can be used to determine the fraction of bound molecules in a solution, and hence the equilibrium constant for such a reversible reaction. For example, Velick[310] has measured the equilibrium constants of co-enzyme complexes in this way. The method has also been used to study the equilibrium of the binding of dyestuffs to proteins[311].

2. Determination of Fluorescence Lifetime and Mechanism of Fluorescence Quenching

Collisional quenching of fluorescence causes a reduction in fluorescence yield which follows the Stern–Volmer law (equation 73). However, an identical quenching relationship is observed if the quencher, Q, forms a non-fluorescent complex with the fluorescent molecule, but does not itself quench the excited state of the uncomplexed fluorescer (see Section 2 B 2). The two quenching mechanisms may be distinguished by measuring the lifetime of the emission. With collisional quenching the lifetime is reduced in proportion to the fluorescence intensity, but with quenching by complex formation alone there is no change in the lifetime. Observations of fluorescence polarisation provide a simple indirect method of measuring lifetime and hence of distinguishing between the two mechanisms[312].

3. Determination of Orientation

As discussed in Section 1 D 5 the value of the principal polarisation p_0 in the absence of depolarisation, depends on the angle β between the directions of the transition moments of absorption and emission, in the manner shown by equations 53 and 54. Thus the polarisation of the fluorescence excitation spectrum is expected to show a constant value within any one absorption band but a rapidly changing value in the region of overlap of two absorption bands corresponding to transitions with different values of β. An example of a simple polarisation spectrum is that of cresol, reported by Weber[215]. He found that the polarisation attained a maximum value of about 0.22 for the 275 nm absorption band and fell to -0.065 for the

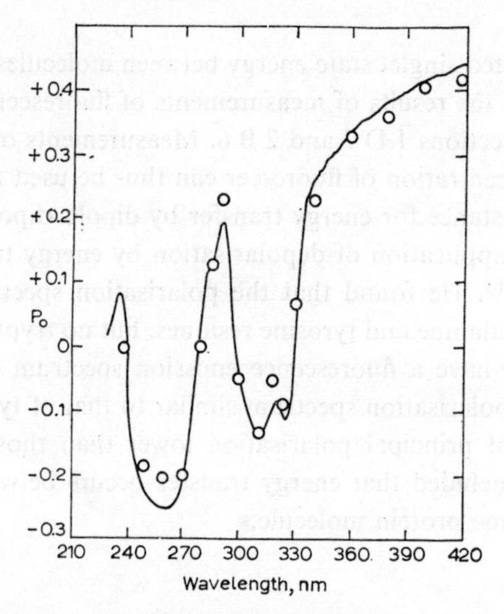

Fig. 143. Polarisation of the fluorescence excitation spectrum of 9-amino-acridine cation in propylene glycol at $-70°C$.
Excitation band width was 3 nm; emission band width was 33 nm (from Ainsworth and Winter[211]).

235 nm band. Using the equation of Jablonski (equation 54) for excitation with natural light, Weber calculated values of $\cos^2 \beta = 0.185$ and 0.78 (the sum of which is close to unity) for the transitions at 275 and 235 nm. He concluded that the two transition moments are at right angles to one another.

Many polarisation spectra are much more complicated than that of cresol, and often show the presence of several transitions whose absorption bands overlap. An example is the spectrum of 9-amino-acridine. Fig. 143 shows the results obtained by Zanker and Wittwer[313] together with the curve obtained by Ainsworth and Winter[211] using their automatic recording polarisation spectrofluorimeter.

The polarisations of the fluorescence excitation, the fluorescence emission, and the phosphorescence spectra of phenanthrene were measured by Azumi and McGlynn[64]. Different vibrations within the same electronic band had different polarisations and Azumi and McGlynn showed how the measurement of polarisation could assist in the analysis of the vibrational structure of the bands.

4. Energy Transfer

Transfer of excited singlet state energy between molecules in solution was first inferred from the results of measurements of fluorescence polarisation, as mentioned in Sections 1 D 7 and 2 B 6. Measurements of polarisation as a function of concentration of fluorescer can thus be used to determine the critical transfer distance for energy transfer by dipole–dipole interaction[66].

A more subtle application of depolarisation by energy transfer has been made by Weber[314]. He found that the polarisation spectrum of proteins containing phenylalanine and tyrosine residues, but no tryptophane residues (i.e. proteins that have a fluorescence emission spectrum characteristic of tyrosine), gave a polarisation spectrum similar to that of tyrosine or cresol, but with values of principal polarisation lower than those of the simple molecules. He concluded that energy transfer occurs between the tyrosine residues in the same protein molecule.

Chapter 5

Application to Analytical Chemistry

A. GENERAL COMMENTS

1. Introduction

The observation of fluorescence as a rough means of identification dates back to the nineteenth century when the appearance of visible fluorescence was often noted along with colour, crystalline form, etc., in describing the properties of a new organic compound. Qualitative observations of visible fluorescence by direct viewing, by photography or by microscopy, has also been in use for a great many years and in the hands of an experienced worker it is still a much used and powerful technique in such fields as criminology, medical and biological examinations, contamination and spoilage of materials, detection of chromatographic zones[315]. Up to about 1945 quantitative fluorimetry had been applied in a variety of fields but had been regarded as unreliable by some workers. This was partly because it had not always been applied with sufficient attention to the basic principles and partly through lack of suitable equipment.

As long ago as 1935 Zscheile[316] used a photoelectric spectrophotometer to measure the fluorescence emission spectra of chlorophylls *a* and *b* and in 1943 he described an improved instrument that employed a rubidium vacuum phototube[317]. Development of the sensitive photomultiplier tube made it possible for photoelectric recording of fluorescence spectra to compete in terms of sensitivity with the less convenient and less precise photographic method, and a variety of instruments were described in the literature during the period up to 1955[318–326]. Since the publication of the paper by Bowman, Caulfield and Udenfriend[127] describing a two-monochromator instrument, equipment has developed to the point where complete fluorescence emission and excitation spectra of small quantities of material can be recorded automatically at the turn of a switch. When properly applied, such photoluminescence measurement is a powerful method of chemical analysis, both qualitative and quantitative, and often at extreme sensitivity. However, like all instrumental methods, its successful application requires the sample to be properly prepared and presented to the instrument, with due attention to the basic principles of the technique. One of the main objects of this chapter is to indicate how the proper conditions may be chosen when designing a method of analysis for a particular purpose, so that the prospective user may avoid the many pitfalls that can arise. Frequent reference will be made to the fundamental principles of photoluminescence dealt with in earlier chapters and the beginner should first acquaint himself with these—in particular the subject matter of chapters 1 and 3.

2. Photoluminescence in Relation to Other Analytical Methods

To see analytical photoluminescence methods in proper prespective it is worthwhile to consider briefly their advantages and limitations in relation to those of other analytical methods. Perhaps the greatest advantage is the possibility of developing methods of high sensitivity—approaching in some cases that of radiochemical methods—and they can thus be applied to the many problems in trace analysis, e.g. the determination of trace contaminants or additives, or the analysis of extremely small specimens for their major constituents. However, in the inorganic field, photoluminescence has to compete with a variety of other powerful methods. Thus spark source mass spectrometry, or emission spectrography, are capable of determining several dozen elements simultaneously at low concentration, and it would clearly be uneconomical to attempt the application of photoluminescence to such problems since, by its very nature, separate procedures would be required for each element, or group of a few elements. On the other hand for certain elements in particular contexts, it is often the most sensitive and convenient method. Each problem must clearly be treated on its merits, having regard to the various analytical facilities at the disposal of the analytical chemist.

In the organic field, photoluminescence has fewer competitors, particularly when small quantities are concerned. Apart from sensitivity, it has the advantage over absorption spectrometry that two, and sometimes four spectra are available as criteria for identification (i.e. excitation and emission spectra of fluorescence and phosphorescence) instead of one. It has the disadvantage that, in order to fluoresce or phosphoresce, a substance must absorb light, but not every absorbing substance emits measurable luminescence. In some applications this can be an advantage since a luminescent substance can often be readily determined without preliminary separation from other substances that are non-luminescent, or luminesce in different spectral regions. In other applications it is often possible with a little chemical ingenuity to convert a non-luminescent substance into a luminescent one. On the other hand luminescence spectra are rarely sufficient to provide complete identification of a previously unknown organic compound, whereas a combination of techniques such as infra-red absorption spectrometry, NMR and mass spectrometry will sometimes give unambiguous identification provided that a sufficiently large quantity of pure specimen is available.

Enough has been said at this stage to warn the beginner against the haphazard application of luminescence methods without reference to the capabilities of other methods. Choice of the correct analytical method for the job is best learnt by experience, but study of the applications of photoluminescence given in this chapter will assist in this respect.

3. Classification of Photoluminescence Methods

Of the various types of photoluminescence described in previous chapters, those that have been most widely applied in chemical analysis, or show the greatest potential for such application, are listed in column 1 of Table 44. Of these, prompt fluorescence and phosphorescence are the most important, and may be applied in two ways. The intrinsic fluorescence or phosphorescence of the substance to be determined may be measured under appropriate conditions, if necessary after preliminary separation from interfering substances. We classify such methods as "natural". If the substance to be determined does not itself luminesce, or if its luminescence appears in a spectral region that is inaccessible owing to instrumental limitations or to masking by the luminescence of other components of the specimen, it may be converted to a suitable luminescent compound by appropriate chemical reactions. We classify these methods as "chemical".

Consideration of the fundamental processes involved in the production of photoluminescence (Chapter 1) indicates that we should search for the phenomenon among those groups of compounds that contain π-electron systems. In the aliphatic series the choice is restricted to highly conjugated compounds such as vitamin A, or to some dicarbonyl compounds such as biacetyl. By far the greatest number of luminescent compounds contain aromatic or heterocyclic ring systems. For prompt fluorescence emission

TABLE

Method	Mode of application	Kind of substance determined	Most useful conditions
Prompt fluorescence	Natural	π^*–π organic Few inorganic	Fluid Glassy Solid Adsorbed
	Chemical	Many types of organic and inorganic	Fluid Glassy Adsorbed
	Quenched	Mainly oxygen Various organic	Fluid Adsorbed
Phosphorescence	Natural	π^*–π organic π^*–n organic Few inorganic	Glassy Solid Adsorbed
	Chemical	Many types in principle	Glassy Adsorbed
	Quenched	Mainly oxygen Various organic	Fluid Adsorbed
Delayed fluorescence	Triplet sensitised	π^*–π organic Rare earths	Fluid
	Quenched	Mainly oxygen	Fluid Adsorbed
Intramolecular energy transfer (fluorescence or phosphorescence)	Chemical	Rare earths	Fluid Glassy

to be appreciable a further condition is also generally required, viz. that the lowest excited singlet state should be π^*–π as indicated in column 3 of Table 44 (see Section 1 B 6). Phosphorescence may be observed when either the π^*–π or the π^*–n excited singlet state has the lowest energy. Few inorganic ions luminesce in solution and frequent use is made of a chemical method by combining the ion with an organic reagent to produce a fluorescent product. Non-fluorescent organic compounds are also converted to fluorescent products by chemical treatment. In both cases the products are nearly always aromatic or heterocyclic, although sometimes the exact nature of the product is not known. Although few *chemical phosphorescence* methods have so far been developed, there is no reason why this should not prove to be a fruitful field for investigation.

THODS OF PHOTOLUMINESCENCE ANALYSIS

Sensitivity	Specificity	Precision
High	Moderate	Very good
Often blank-limited	Good	Moderate
Moderate	Sometimes very good	Low
Often blank-limited	Moderate	Low
Usually blank-limited	Moderate or good	Very good
Usually blank-limited	Moderate or good	Moderate
Usually blank-limited	Moderate or good	Low
Low	Low	Moderate
Moderate	Low	Low
High	Good	Moderate
Moderate	Sometimes very good	Low
Often blank-limited	Moderate	Low
← Few quantitative applications yet made →		
Very high	Low	Moderate
Very high	Low	Low
High	Good	Low
Very high	Low	Moderate
Very high	Low	Low
Moderate	Very good	Very good
Moderate	Very good	Moderate

Valuable methods have been developed based on the measurement of prompt fluorescence under all four conditions considered in column 4 of Table 44. Phosphorescence on the other hand is almost always very weak in fluid media and for analytical purposes it is best applied in glassy or crystalline solution at low temperature—and sometimes even at room temperature—or in the adsorbed state.

In columns 5, 6 and 7 of Table 44 we have made general statements of the sensitivity, specificity and precision to be expected from the different types of method under various conditions. These generalisations must be treated with some caution because they naturally depend on the particular substance being determined. Many of them are governed by instrumental and other factors inherent in the method of measurement. For example, the uniformity of fluid solutions makes for good precision, while freezing at low temperature can introduce errors due to variations in refractive index or even variable transmission of the container if the specimen is immersed in liquid nitrogen. Similarly, frontal illumination of solid solutions introduces error due to geometrical alignment. Again the magnitude of the "blank" is more likely to limit the sensitivity of some methods than others. We shall not discuss columns 5, 6 and 7 further at this stage but the reader may find it useful to refer back to the table when the methods are discussed in more detail later.

In general, quenching methods are of limited applicability because they are somewhat non-specific. They can provide valuable methods in certain cases, e.g. the determination of oxygen, and have therefore been included in the table. The remaining two classes—sensitised delayed fluorescence and intramolecular energy transfer—will be fully dealt with in their respective sections. The second is in reality a special case of a chemical method.

We have deliberately excluded from Table 44 the remaining types of photoluminescence, viz.: sensitised phosphorescence, recombination phosphorescence and recombination delayed fluorescence. This has been done partly because no analytical applications have yet been made of them, and partly because those systems in which they occur can generally be analysed better by the application of one of the other methods.

B. SENSITIVITY

1. Meaning of the Term "Sensitivity"

In photoluminescence analysis the term "sensitivity" may be used in three distinct senses. It may refer to the performance of a particular instrument when used for a given purpose—we shall call this "instrumental sensitivity". One method of defining this is to express the minimum detectable signal-to-noise ratio in terms of the corresponding concentration of some standard substance observed under precisely stated conditions. The term "sensitivity" may refer to the luminescence sensitivity of a particular substance in a stated solvent, independent of the instrument used to measure it. This may be expressed in a variety of ways, but always in terms of basic luminescence parameters of the substance concerned—we therefore call this "absolute sensitivity". Finally we may refer to the overall sensitivity of a particular analytical method. This may depend on both the instrumental sensitivity and the absolute sensitivity of the luminescent substance, but frequently it is limited by the overall luminescence "blank". We shall call this "method sensitivity".

It may seem unnecessary to stress the importance of distinguishing between "instrumental sensitivity" and "method sensitivity", but failure to do so has often led to the publication of extravagant claims about the sensitivity of a particular analytical method. To illustrate the difference we shall quote a simple example. The data in Table 45 refer to comparisons made by Parker and Barnes[131] of the sensitivities of two early instruments—a spectrofluorimeter and a filter fluorimeter. Owing to the much greater light gathering power of the filter system compared with the small monochromator, the *instrumental sensitivity* of the filter fluorimeter (*calculated* as the equivalent concentration of quinine bisulphate) was a hundred times greater than that of the spectrofluorimeter. However the "total fluorescence blank" with the filter instrument was one thousand times greater than its instrumental sensitivity, and the *method sensitivity* for the detection of quinine bisulphate was thus determined by the magnitude and reproducibility of this blank. In this application it was clearly not possible to take advantage of the very high instrumental sensitivity of the filter instrument and the bald statement that "the sensitivity of the filter instrument was 2×10^{-6} µg per ml of quinine bisulphate" could be quite misleading. Indeed, under the conditions quoted in Table 45 it would be better to use the instrument of lower instrumental sensitivity because this gave a smaller overall blank which was

TABLE 45

COMPARISON OF INSTRUMENTAL SENSITIVITIES

Method of isolating fluorescence	Instrumental sensitivity (µg per ml of quinine bisulphate)	Fluorescence blank (equivalent µg per ml of quinine bisulphate)
Small quartz monochromator (30 nm band width)	0.0002	0.0006
Glass filter	2×10^{-6}	0.0023

Note

Flux of exciting light was 3×10^{-9} einstein sec^{-1} at 366 nm. The detector was a 9-stage photomultiplier in both instruments. The instrumental sensitivities correspond to the concentrations of quinine bisulphate that would be required to give a deflection equal to the noise level due to dark current fluctuations etc.

still larger than the instrumental sensitivity. No doubt the blank with the filter instrument could have been reduced, e.g. by more careful choice of filters and/or better reagent purification. We shall come back repeatedly to this important subject of blank in discussing specific methods.

2. Instrumental Sensitivity

The factors governing instrumental sensitivity have been dealt with indirectly in the discussion of the performance of the various components of a spectrofluorimeter, given in Chapter 3. The main factors are:

type of light source.
LGP value of excitation monochromator.
LGP value of analysing monochromator.
wavelength and band width settings of monochromators.
sensitivity of the photomultiplier/amplifier/recorder combination.
time constant.
efficiency of the various entrance and exit slit optics.

We shall assume that the entrance and exit slit optics have been designed to give optimum performance as discussed in Chapter 3. The fluctuation of the recorder pen or indicating meter, when the instrument is operating at high sensitivity, will depend on the time constant of the complete detector system (see Section 3 N 6) and hence the minimum detectable signal due to luminescence will depend on the recording speed required. For chart-recorded spectra a time constant of about 1 second is generally suitable. Most spectra in fluid solution can then be recorded within a few minutes without losing resolution. For resolution of Shpol'skii spectra (see Section 4 F 5) the rate of recording has to be reduced. On the other hand measurement of short lifetimes may require a time constant as short as 0.1 msec or less (Section 3 N 6) and the instrumental sensitivity for this purpose is thus greatly reduced.

The author defines instrumental sensitivity as that concentration of a standard substance required to produce a recorder deflection equal to the overall dark-current fluctuation when the time constant of the detection system is 1 second. Such a statement of instrumental sensitivity must clearly be linked with statements about the conditions under which the test is made. To simplify comparisons between instruments it is useful to consider the complete spectrofluorimeter in two parts—the excitation side and the analysing side. The efficiency of the excitation side may be stated in terms of the light flux reaching the specimen compartment (as measured by the ferrioxalate actinometer—see Section 3 F 2) and the spectral purity of this light. Both these factors will depend on the type of light source and the band width setting of the excitation monochromator. It is obvious that a much greater light intensity of high spectral purity can be obtained by isolating one of the principal lines from a mercury lamp than by isolating a section of the spectrum of a continuous source such as a xenon lamp. This means that emission spectra can be measured at much greater instrumental sensitivity than excitation spectra. It is also obvious that the excitation wavelength will be critical. Thus the extinction coefficient of an anthracene solution is some 80 times greater at 250 nm than at 366 nm and with a high pressure mercury lamp this increase in "absolute sensitivity" of anthracene more than compensates for the lower intensity available at 250 nm. It is not however sufficient to compensate for the much lower intensity from a xenon lamp in this region unless the band width of the excitation monochromator is increased. Likewise on the analysing side the instrumental sensitivity will depend on the band width of the analysing monochromator and the spectral region in which the luminescence appears.

A complete expression of instrumental sensitivity for one particular purpose is given, as an example in Table 46, which refers to the conditions under which the spectra in Fig. 151 of Section 5 C were measured. Such detailed information is rarely quoted in the literature—most published sensitivity data simply quote the "minimum detectable concentration" of quinine bisulphate at stated band widths on the two monochromators.

TABLE 46

METHOD OF EXPRESSING INSTRUMENTAL SENSITIVITY

Light source	200 W extra-high pressure mercury
Excitation monochromator	grating, LGP value 0.26 (with chlorine and OX7 filters)
Wavelength setting	250 nm
Half-band width	9.6 nm
Light flux (without chopper)	1.6×10^{-9} einstein sec^{-1}
Method of illumination	right angle, 1 cm × 1 cm cuvette
Analysing monochromator	grating, LGP value 0.30
Wavelength setting	400 nm
Half-band width	3.3 nm
Photomultiplier	EMI 9558 Q
Time constant	1 second
Instrumental sensitivity	10^{-11} μg per ml of anthracene in ethanol

This is useful for some purposes but is of little help if one is interested in another spectral region entirely.

Although a complete specification of instrumental sensitivity is rarely made, there is one simple test that gives a valuable indication of instrumental performance (sensitivity combined with resolution). This is to measure the Raman spectrum of a suitable solvent in a 1 cm cuvette (see Sections 1 E 7 and 5 C 1). Many current spectrofluorimeters are capable of resolving, or partly resolving, the main Raman band of cyclohexane (wavenumber shift 0.29 μm^{-1}) with excitation in the 350 nm region. A much more critical test is the resolution of the Raman band of carbon tetrachloride (frequency shift 0.07 μm^{-1}) from the scattered exciting light. With a good instrument having a mercury source this band should be at least partly resolved with excitation at 313, 366 and 436 nm (see Fig. 22). Resolution of Raman bands using a continuous source is more difficult and the resolution of the carbon tetrachloride band by this means has not yet been reported. It should be noted that the use of a small cuvette with right angle illumination, although satisfactory for measuring luminescence spectra where inner filter effects have to be kept to a minimum, is far from ideal for the observation of Raman spectra for which the *observable* optical path through the liquid should be as long as possible.

3. Absolute Sensitivity

This refers to a particular substance under stated conditions and is best expressed in a form that is independent of the instrument used to make the measurement. The intensity of prompt fluorescence or phosphorescence emitted by a dilute solution of one substance irradiated by light of a chosen wavelength is proportional to $I_0 \varepsilon \phi c$ (see equation 21). The product $\varepsilon\phi$ is characteristic of the substance at that excitation wavelength and may be used as a measure of the absolute fluorescence or phosphorescence sensitivity of the substance. For most substances, for which ϕ is independent of excitation wavelength, the maximum absolute sensitivity is thus obtained at the peak of the most intense absorption band, although the observed sensitivity with a particular instrument will not necessarily be a maximum at this point because the intensity of exciting light available at some other wavelength may be much greater. To calculate a practical unit of sensitivity Parker and Rees[158] suggested that ε should be expressed as the optical density per cm for a concentration of 1 μg per ml ($= D_\lambda$). The absolute sensitivity ϕD_λ is then expressed in units of ml per μg. Some maximum absolute sensitivity values expressed on this basis are given in column 6 of Table 47. The sensitivity at any other excitation wavelength can then be simply derived from the maximum value by reference to the absorption spectrum (or, more strictly, the corrected excitation spectrum).

When a monochromator, rather than filters, is used to isolate the luminescence, it is usual to select, not the complete emission spectrum, but a narrow band of wavelengths at the peak. The absolute sensitivity will then depend not only on the product ϕD_λ, but also on the effective half-band width, H (= area/peak height) of the spectrum. When comparing one substance with another the factor $\phi D_\lambda/H$ (the fluorescence sensitivity index) is thus also of interest and columns 7 and 8 of Table 47 give this parameter both for the absorption maximum and for the mercury line at 366 nm. It is obvious that the latter gives a much lower absolute sensitivity than the former for these substances, but this wavelength is nevertheless frequently used because it can be readily isolated at high intensity, and the intensity and convenience of using this wavelength, as well as other factors, sometimes outweigh the loss of absolute sensitivity.

Data tabulated in the form shown in Table 47 can be used directly by other workers with different instruments and are particularly useful in discussing the relative merits of different fluorimetric methods. For example, selenium may be determined fluorimetrically by converting it to 4,5-benzopiazselenol[327] or to 3,4-diaminophenylpiazselenol[328]. The fluorescence sensitivity index at 366 nm for the former ($\phi D_{366}/H = 0.12$ in cyclohexane) is more than 20 times greater than for the latter($\phi D_{366}/H = 0.0045$ in toluene) and thus on this basis at least, the former has considerable advantages.

When using absolute fluorescence sensitivity indexes to predict attainable sensitivities with a particular instrument, the characteristics of the latter must also be taken into account. For example, suppose the sensitivity of a spectrofluorimeter to quinine bisulphate with 366 nm excitation is known and the sensitivity for the detection of thionine is required. Reference to column 8 of Table 47 shows that the absolute sensitivity index of thionine is (0.00028)/(0.0073), i.e. 0.038 of that of quinine bisulphate. Suppose that the spectral sensitivity curve of the analysing system was the same as that shown in Fig. 54. The latter shows that the sensitivity at the thionine fluorescence maximum is about 0.6 of that at the quinine bisulphate maximum. Hence the instrumental sensitivity to thionine fluorescence is 0.038×0.6, or 0.023 of its sensitivity to quinine bisulphate with 366 nm excitation. Reference to the

TABLE

ABSOLUTE FLUORESCEN

(D_{max} = optical density per cm for a concentration of 1 μg per ml at the main absorpt

	Absorption maximum (μm⁻¹(nm))	Fluorescence maximum (μm⁻¹(nm))	Half-band width of fluorescence (H)(μm⁻¹)
Quinine bisulphate in 0.1 N sulphuric acid	4.00 (250)	2.17 (461)	0.47
Rhodamine B in ethanol	1.84 (544)	1.75 (571)	0.17
Fluorescein in 0.1 N NaOH	2.04 (490)	1.94 (515)	0.20
Eosin in 0.1 N NaOH	1.93 (518)	1.85 (540)	0.18
Anthracene in ethanol	3.97 (252)	2.50 (400)	0.25
Thionine in 0.1 N H_2SO_4	1.67 (598)	1.61 (621)	0.18

absorption of thionine shows that the sensitivity could be increased more than 100 times by excitation at a wavelength corresponding to the maximum in the absorption spectrum of thionine. The final choice of excitation wavelength would naturally depend on other factors such as the excitation intensities available and the characteristics of other components of the solution.

Use of an absolute sensitivity index in the manner just described requires a knowledge of the spectral sensitivity and relative excitation intensities of the spectrofluorimeter as well as a certain amount of calculation. Many workers avoid this by simply measuring directly the optimum excitation and emission wavelengths with their own instrument and recording the minimum detectable concentrations of a series of substances in the same way. This is of course a perfectly valid procedure so long as the data are subsequently to be used only in connection with the same instrument and the same light source, but it is always worth considering whether some instrumental modification—e.g. substitution of another light source—will give improved performance. In any case, it must always be borne in mind that sensitivity indexes, whether absolute or instrument-related, refer to ideal conditions, i.e. solutions of the pure substance in the absence of all other interfering factors: the actual sensitivity achieved in a practical job of analysis, i.e. the method sensitivity, may be very much lower.

NSITIVITIES
ık; D_{366} = corresponding figure at 366 nm)

ϕ	Fluorescence sensitivity at absorption maximum (ϕD_{max})	Fluorescence sensitivity index	
		At absorption maximum ($\phi D_{max}/H$)	At 366 nm ($\phi D_{366}/H$)
0.55	0.031	0.066	0.0073
0.73	0.16	0.94	0.052
0.85	0.20	1.0	0.015
0.23	0.035	0.19	0.0031
0.30	0.34	1.4	0.012
0.024	0.0056	0.031	0.00028

4. Method Sensitivity

The intensity of fluorescence or phosphorescence emitted by a solute is directly proportional to the intensity of the exciting light and to the overall sensitivity of the detection system. In principle, therefore, it should be possible to improve indefinitely the detection limit of the solute by simply increasing the power or efficiency of the excitation source. Below a certain concentration however, other factors become important and ultimately set a limit beyond which further increase in instrumental sensitivity cannot be usefully employed. Photo-reaction of the solute can be the limiting factor, but more frequently the limit is set by the over-all "blank" value, i.e. by the light reaching the detector from sources other than the luminescence of the solute under investigation. A variety of factors may contribute to the luminescence blank, some instrumental and some inherent in the specimen being analysed. We have grouped these factors under the following headings:

Raman emission from the solvent.
Luminescence of the cuvette and surroundings.
Scattered light.
Luminescent impurities in solvent or reagents.
Luminescence of other components of the specimen.

Each of these will be discussed separately in Section 5 C. Let us consider here in general terms how the presence of a luminescence blank will limit method sensitivity, and consider what types of measurement are likely to be "blank-limited", and what "instrument-limited".

In the presence of a blank that is larger than the electrical signal-to-noise ratio of the detection system itself, the sensitivity of the method will clearly depend on the ability to detect a weak additional signal superimposed on the blank signal. In effect, the blank introduces a new source of "noise" and the magnitude of this "noise" will depend on the precision with which the blank can be reproduced from one test to the next. If for example, the blank is mainly due to Raman emission from the solvent, it may be reproducible to within a few per cent and it may then be realistic to record luminescence levels amounting to as little as 10–20% of that corresponding to the blank. On the other hand, if the method involves a complex series of chemical treatments to separate the material of interest, and/or to produce from it a luminescent compound, the blank may vary considerably from test to test. In these circumstances the minimum meaningful luminescence signal may be several times the mean blank value. In either case it is obviously desirable to design the experiment to reduce the blank to a minimum, when striving for the utmost in sensitivity.

Even with present day sensitive equipment there are occasions when the method sensitivity is still "instrument-limited". This may occur, for example, when one of the monochromators has to be operated with very narrow slits to obtain high resolution, as in the measurement of some Shpol'skii spectra, or when excitation spectra have to be measured over regions where the output from continuous sources is low, e.g. below 250 nm. Methods involving luminescence in the near infra-red region are also likely to be instrument-limited owing to the low sensitivity of detectors in this region.

It should be noted that some methods involving the measurement of a strongly luminescent compound, present as a comparatively major constituent of the specimen, can still be instrument-limited. This happens when the luminescent compound is accompanied by large concentrations of other strongly absorbing substances. To obtain meaningful results it is then necessary to dilute the solution until the total optical density at all wavelengths of interest is small, and the concentration of luminescent substance may then be so low that it is difficult to measure its excitation spectrum in the short-wavelength region.

C. FACTORS CONTRIBUTING TO THE LUMINESCENCE BLANK

1. Raman Spectrum of the Solvent

The nature of the Raman effect was dealt with in Sections 1 E 5–7 where the Raman spectra from some solvents commonly used in spectrofluorimetry were described (see Fig. 22). All solvents containing hydrogen atoms linked to carbon or oxygen show a band shifted about 0.3 μm^{-1} to the long-wavelength side of the exciting light and it is this band that is most likely to cause interference with the measurement of fluorescence spectra. Unfortunately, it is just this type of solvent that is most used, and if the interference cannot be avoided in any other way it is worth considering the use of carbon tetrachloride (at the longer wavelengths above its absorption cut-off) because its Raman band lies so close to the wavelength of the exciting light that it is most unlikely ever to interfere. Chloroform is also worth considering from this point of view because, although it shows a band shifted 0.3 μm^{-1}, the intensity is less than that of water, alcohols or hydrocarbons. For ease of reference, Table 48 gives, for each of five solvents, the position of the main Raman band that is observed when various mercury lines are used for excitation.

The magnitude of the interference caused by the Raman emission from water is illustrated in Fig. 144 by the fluorescence emission spectra of three solutions of quinine bisulphate excited by the 366 nm mercury line. At the low sensitivity used to record the spectrum of the strongest solution (curve B) the scattered light peak is still on scale and the main Raman band produces only a slight inflection on the fluorescence spectrum. When the sensitivity

TABLE 48

POSITIONS OF THE MOST PROMINENT RAMAN BANDS CORRESPONDING TO VARIOUS MERCURY LINES

	Excitation wavelength (nm)				
	248	313	366	405	436
Water	271	350	418	469	511
Ethanol	267	344	410	459	500
Cyclohexane	267	344	409	458	499
Carbon tetrachloride	–	320	376	418	450
Chloroform	–	346	411	461	502

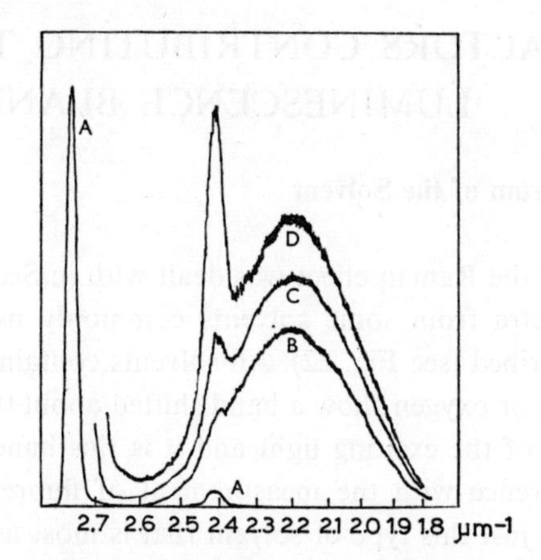

Fig. 144. Fluorescence emission spectrum of quinine bisulphate in 0.1 N sulphuric acid, excited by 366 nm and showing interference by main Raman band of water. A, water alone at low sensitivity; B, 0.1 μg per ml at same sensitivity; C, 0.033 μg per ml at increased sensitivity; D, 0.01 μg per ml at still greater sensitivity (from Parker[73]).

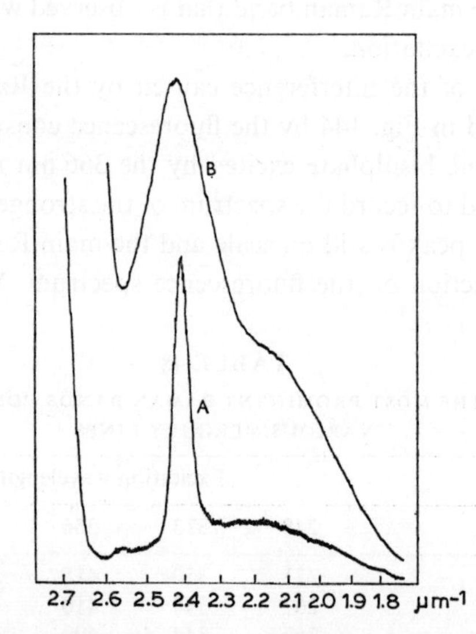

Fig. 145. Effect of slit width on fluorescence emission and Raman emission spectra. 0.001 μg per ml of quinine bisulphate in 0.1 N sulphuric acid excited by 366 nm; A, slit width 0.25 mm; B, slit width 1.75 mm and lower sensitivity (from Parker[73]).

is increased to record the spectra of the more dilute solutions the Raman band becomes increasingly prominent (curves C and D).

The Raman band is so much narrower than the fluorescence band of quinine bisulphate that it is readily identified when narrow slits are used. If however it is attempted to increase the instrumental sensitivity by opening the slits of the analysing monochromator in order to record the spectra of very dilute solutions, the Raman band appears to be considerably broadened and it might be mistaken for part of the structure of the quinine bisulphate band (compare curves A and B in Fig. 145). The interference could be allowed for by measuring the emission from the solvent alone. That the broad band in curve B of Fig. 145 was due to Raman emission could also be confirmed by measuring the fluorescence spectrum with a shorter excitation wavelength—the position of the fluorescence band of the quinine bisulphate would then remain unchanged, but the subsidiary band would move to shorter wavelengths, its wavenumber shift from the exciting light remaining constant. The interference is more troublesome when it is superimposed on a fluorescence spectrum showing vibrational structure. For example, in Fig. 147 the Raman band seriously distorts the fluorescence emission spectrum of the most dilute anthracene solution, which appears to have an additional vibrational band. When fluorescence emission spectra are used to identify substances at very

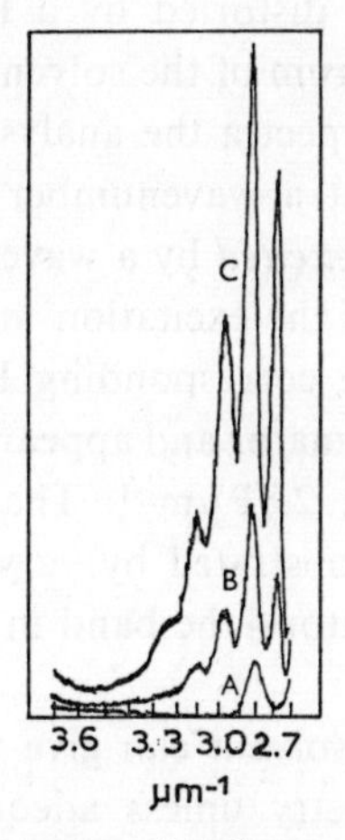

Fig. 146. Fluorescence excitation spectra of anthracene in cyclohexane illustrating interference by Raman spectrum of solvent.
Analysing monochromator set at 397 nm; A, pure solvent; B, 0.007 µg per ml; C, 0.025 µg per ml (from Parker[73]).

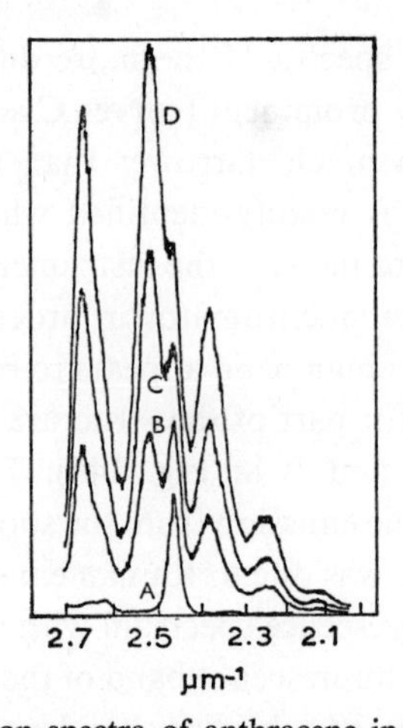

Fig. 147. Fluorescence emission spectra of anthracene in cyclohexane illustrating interference by Raman spectrum of solvent.
Excitation at 366 nm; A, pure solvent; B, 0.025 μg per ml; C, 0.05 μg mer pl; D, 0.075 μg per ml (from Parker[73]).

low concentration it is obviously important to keep the possibility of such interference constantly in mind. This interference appears in the excitation spectrum also, as indicated in Fig. 146. The excitation spectra of the two dilute solutions of anthracene (curves B and C) should be identical with the absorption spectrum but are distorted by a band at 2.81 μm^{-1} which is present in the excitation spectrum of the solvent itself (curve A in Fig. 146). To measure these excitation spectra the analysing monochromator was set at 2.52 μm^{-1} (397 nm), i.e. at a wavenumber corresponding to the main Raman band of cyclohexane *excited* by a wavenumber of 2.81 μm^{-1}. Thus, as the wavenumber setting of the excitation monochromator traversed the region around 2.81 μm^{-1}, the corresponding Raman emission was picked up by the analysing monochromator and appeared in the excitation spectrum as a band with maximum at 2.81 μm^{-1}. That the interference is due to Raman emission can be demonstrated by varying the wavenumber setting of the analysing monochromator—the band in the excitation spectrum will then shift accordingly.

Raman emission from the solvent can give rise to large blank values in high sensitivity filter fluorimetry unless adequate precautions are taken. For example frequent use is made of the 366 nm mercury line for excitation, isolated by means of a "black glass" filter (e.g. Chance OX9A, with long-wavelength cut-off at about 405 nm—see Fig. 66). If carbon tetrachloride is the solvent it is then sufficient to choose a secondary filter having zero transmission below 410 nm. However, with water as solvent, the main Raman band appears at 418 nm, and to avoid a high blank it is then necessary to choose a secondary filter having a short-wavelength cut-off at not less than 430 nm. Similarly, if the mercury line at 405 nm is used for excitation (Raman band for water at 469 nm) the secondary filter must not transmit appreciably below about 480 nm if high blank readings are to be avoided. This means that the choice of secondary filters is sometimes critical, and rejection of part of the required fluorescence band may have to be accepted. The fluorimetric determination of borate ion by reaction with benzoin (see Section 5 G 1) is an example. The fluorescence of the borate–benzoin complex can be readily excited by 366 nm: maximum fluorescence emission occurs at about 480 nm (2.1 μm^{-1}) i.e. well away from the Raman band of the mixed solvent at 409–418 nm (2.39–2.45 μm^{-1}), see Fig. 148. A pale yellow filter cutting off at about 430 nm (2.33 μm^{-1}) is therefore adequate. For various reasons (see Section 5 G 1) it is advantageous to use the 405 nm mercury line for excitation. The borate–benzoin complex absorbs less strongly at this wavelength and hence the intensity of the Raman band relative to the fluorescence is greater. Furthermore the composite Raman band is now situated close to the fluorescence maximum and introduces a considerable distortion in the spectrum (see Fig. 149). The

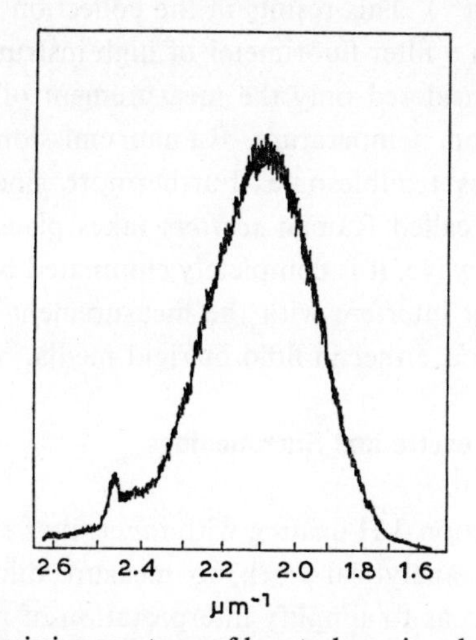

Fig. 148. Fluorescence emission spectrum of borate–benzoin solution in alkaline aqueous ethanol containing 0.004 μg per ml of boron. Exciting wavelength, 366 nm (from Parker[73]).

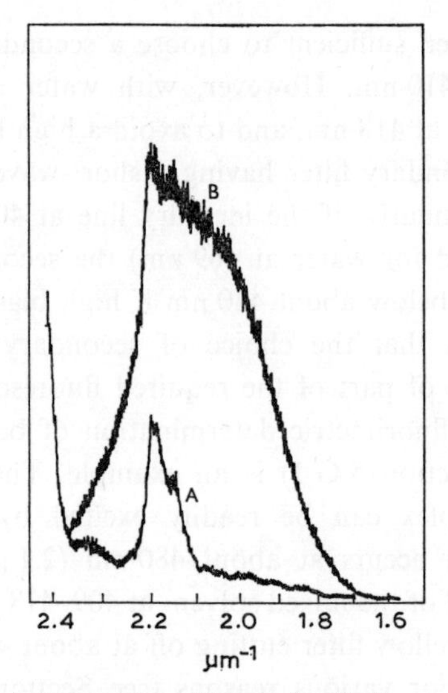

Fig. 149. Fluorescence emission spectra of borate in alkaline aqueous ethanol containing 0.004 µg per ml of boron.
Exciting wavelength, 405 nm; A, before addition of benzoin; B, after addition of benzoin (from Parker[73]).

chosen secondary filter must now have a short wavelength cut-off not less than 480 nm (2.08 μm^{-1}). This results in the collection of less fluorescence, but is acceptable with a filter fluorimeter of high instrumental sensitivity.

So far we have considered only the measurement of the fluorescence of fluid solutions at room temperature. Raman emission from rigid glasses at 77°K is much less troublesome. Furthermore since Raman emission (strictly it should be called Raman *scatter*) takes place within the time of vibration of the light wave, it is completely eliminated by a phosphoroscope and hence it does not interfere with the measurement of phosphorescence or delayed fluorescence, either in fluid or rigid media.

2. Luminescence of Cuvette and Surroundings

As discussed in Section 3 H dealing with inner filter effects, it is desirable, whenever possible in analytical work, to measure dilute solutions by the right angle method so as to simplify interpretation of the results and minimise blank due to scattered light and cuvette luminescence. However, even with the most carefully screened cuvette, in which the analysing monochromator does not view directly the illuminated cuvette faces, sufficient fluorescence from the cuvette faces is scattered by the solvent to give rise to large blank readings. The difficulty is greatest with exciting light of short wavelength. Thus fused quartz fluoresces strongly when exposed to light of wavelength 250 nm and produces a serious interference (see curve A of Fig. 150). For high sensitivity work it is therefore essential to use cuvettes of synthetic silica for which the blank is generally negligible with right angle illumination (see curve B of Fig. 150). Of course, with frontal or in-line illumination the cuvette fluorescence introduces a very much greater blank, and this may be appreciable even with synthetic silica. Similarly the fluorescence of adsorbents such as filter paper, silica gel etc. is often the factor limiting the sensitivity when spots or zones on chromatographic strips or columns are being identified by spectrofluorimetry or spectrophosphorimetry. Special care has to be taken to avoid luminescence blank when making measurements of specimens at liquid nitrogen temperature. If the cuvette is suspended in a fused quartz dewar, the luminescence of the latter can give rise to an enormous blank consisting of both fluorescence and phosphorescence, and it may be necessary to screen the cuvette on all sides

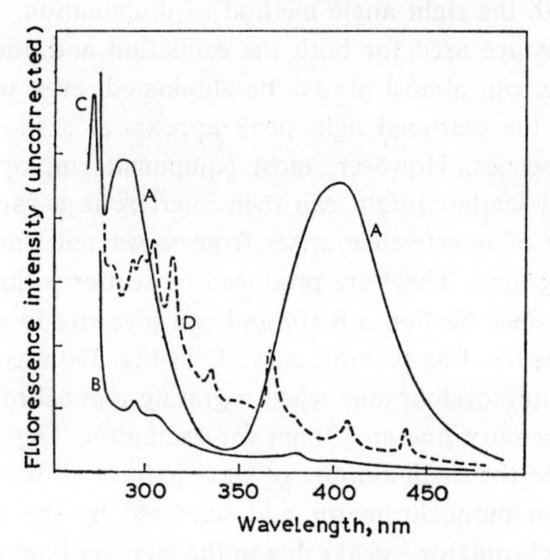

Fig. 150. Illustrating interference from scattered light and fluorescence of fused quartz cuvette.

Specpure cyclohexane excited by light from small grating monochromator set at 250 nm; A, in fused quartz cuvette with filters in beam of exciting light (sensitivity 1000); B, in synthetic silica cuvette with filters (sensitivity 1000); C, as B but sensitivity 500 to show intensity by Raman band; D, in synthetic silica cuvette without filters (sensitivity 1000) (from Parker[210]).

except for small apertures to admit the exciting light and to let out the luminescence.

A particularly insidious form of interference is the artifact produced by phosphorescence of the cuvette (or surroundings) in the measurement of delayed fluorescence (see Section 3 N 8). Here the blank has every appearance of being a true signal—the spectrum is identical with the prompt fluorescence of the substance under investigation. It is however not true delayed fluorescence, but prompt fluorescence excited by *absorption* of the phosphorescence of the container.

Finally we should draw attention to the effect of fluorescence of the entrance slit optics of the analysing monochromator. If these are of fused quartz they can give rise to an objectionable blank when short wavelength excitation is employed with the in-line arrangement, as described in Section 3 H 4. With other arrangements this effect is generally negligible.

3. Scattered Light

Like the interference from cuvette fluorescence, interference from scattered light is least with the right angle method of illumination. If double prism monochromators are used for both the excitation and analysing systems, such interference can almost always be eliminated, even with frontal illumination, since the scattered light peak appears at a shorter wavelength than the fluorescence. However, most equipment employs single monochromators, and scattered light can then interfere in a variety of ways. A common source of interference arises from unwanted wavelengths in the beam of exciting light. These are produced by scatter within the excitation monochromator (see Section 3 B 10) and can give rise to a large blank in high-sensitivity work. For example curve D in Fig. 150 was the "spectrum" obtained from pure cyclohexane when a grating monochromator was used to isolate the mercury line at 250 nm for excitation. The spectrum is due almost entirely to the small amount of stray light of all wavelengths passed by the excitation monochromator and scattered by the solvent into the analysing monochromator—peaks due to the mercury lines at 302, 313, 334, 366, 405 and 436 nm are clearly evident in curve D. This scattered stray light was completely removed by inserting a composite filter consisting of 2 cm of chlorine gas and 2 mm of Chance OX7 glass. This filter absorbed all wavelengths above 300 nm but transmitted freely the required excitation wavelength of 250 nm. The fluorescence emission spectrum of the cyclohexane was then almost completely free of scattered light interference. The intense band at 269 nm (curve C in Fig. 150) is due to the main Raman band of cyclohexane, and gives an indication of the instrumental sensitivity at which the spectra were measured. Another example of the elimination of scattered stray exciting light by the use of filters is given in Section 3 D 2 and Fig. 67. The same section also describes the interference to be expected from stray light from the second order spectrum of a grating monochromator and its removal by means of supplementary filters.

Even when the exciting light is completely monochromatic it is still possible to get interference owing to inefficiency of the analysing monochromator. This interference is especially pronounced in frontal illumination where a comparatively large proportion of the exciting beam is scattered by the specimen and enters the analysing monochromator. A small proportion of this light is then scattered by the optical components of this monochromator so that some of it reaches the photomultiplier at all wavelength settings. Thus in order to work at high sensitivity with frontal illumination it is necessary to insert a short wavelength (and non-fluorescent) cut-off filter between the specimen and the analysing monochromator—or better to use a double monochromator.

4. Luminescence of Impurities in Solvents or Reagents

The ease with which solvents can be freed from fluorescent or phosphorescent impurities varies enormously from one solvent to another. It is of course necessary to purify the solvent only to the degree required for the determination concerned but since stocks of solvent are usually kept to be used for a variety of determinations, some at high sensitivity, it is useful to have available a general test for high purity. A good method for ultra-violet-transmitting solvents is to excite with a high intensity of the 250 nm emission from a high pressure mercury lamp isolated by means of a monochromator and filters as described for curve B in Fig. 150. The main Raman band of most solvents then appears at about 270 nm and the spectral region above 280 nm is free from interference. Nearly all types of fluorescent impurity absorb at 250 nm and will thus be detected. It is of course necessary to check the absorption of the solvent at 250 nm to confirm that high concentrations of absorbing impurities are absent, otherwise a heavily contaminated solvent would not transmit the exciting light to the middle of the cuvette and would appear to be relatively free of fluorescence. In making this test it is always worth recording the main Raman band—at a reduced instrumental sensitivity if necessary. The height of the Raman band acts as an "internal standard" and thus facilitates quantitative comparison between spectra taken at different times, and also provides a check on correct instrumental operation.

Cyclohexane of high purity is available commercially and this can be used without further purification (see curve B in Fig. 150). Laboratory grade absolute ethanol contains a relatively high concentration of fluorescent impurities but these can be largely removed by a single fractionation by the procedure described in Section 3 N 7 (compare curves A and B in Fig. 151). During a certain period the author found that batches of ethanol from one supplier always contained 1–2 parts in 10^{10} of anthracene (see curves A and C in Fig. 151), presumably introduced by adventitious contamination during manufacture or bottling. It was readily removed by fractionation and is mentioned here to illustrate the great *method sensitivity* that can be achieved by fluorescence analysis in favourable circumstances. The fluorescence sensitivity index of anthracene at 250 nm is high, and the fluorescence of purified ethanol was so low that the minimum detectable concentration of anthracene in it was less than 1 part in 10^{11}.

The fluorescence spectra of some specimens of water are shown in Fig. 152.

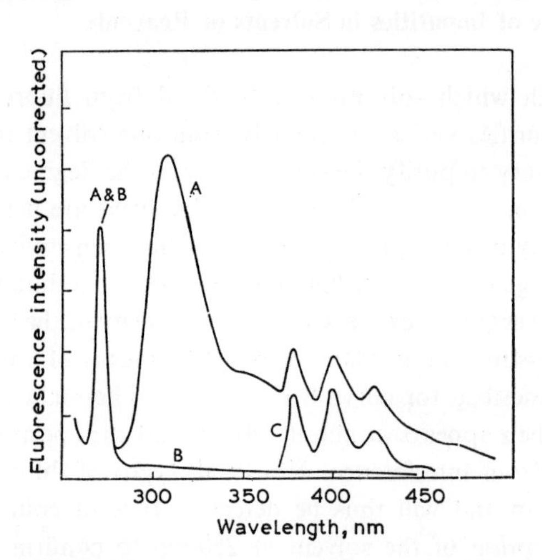

Fig. 151. Fluorescence of impurities in ethanol.
Excitation as for Fig. 150 with filters (sensitivity 625); A, ethanol as received; B, ethanol purified by distillation; C, purified ethanol with 0.0003 μg per ml of anthracene added (from Parker[210]).

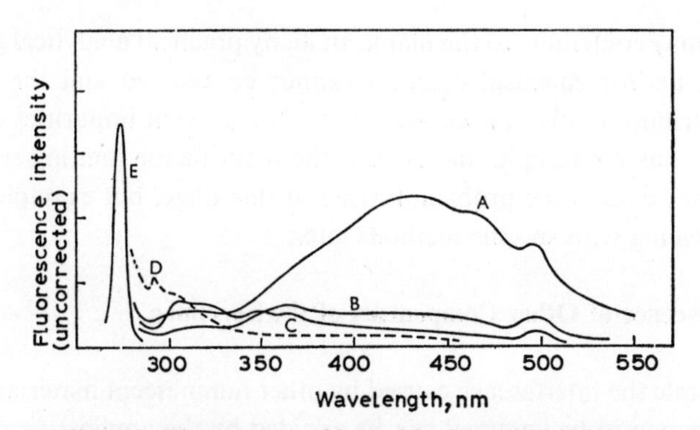

Fig. 152. Fluorescence of impurities in water specimens.
Excitation as for Fig. 150 with filters; A, tap water (sensitivity 2,500); B, twice-distilled water (sensitivity 2,500); C, once-distilled and de-ionised water (sensitivity 2,500); D, once distilled ethanol (sensitivity 2,500); E, as for A, but with sensitivity 625 to show intensity of Raman band (from Parker[210]).

Double distillation of tap water reduces its fluorescence considerably, although at wavelengths greater than 350 nm it is still somewhat more fluorescent than once-distilled ethanol (compare curve B and D in Fig. 152). The difficulty of removing the last traces of fluorescence is probably due to steam volatility of the impurities and preliminary distillation from alkaline permanganate would probably effect a greater degree of purification. De-ionisation does not apparently introduce fluorescent impurities (curve C in Fig. 152) provided that the first runnings from the de-ioniser are rejected.

Ethyl ether can be readily purified, but removal of the last traces of fluorescence from saturated paraffins such as isopentane or n-hexane requires prolonged treatment with oleum. At 77°K the fluorescence and/or phosphorescence of the solvent is generally considerably greater than the fluorescence at room temperature, especially with short-wavelength excitation.

Those methods that require preliminary separation of the luminescent compound before measurement are subject to additional sources of blank. One source is adventitious contamination, which may be serious if a commonly occurring substance such as boron, silicon or lubricating oil is being determined. A second source is impurities in the reagents—this is frequently the main source of blank in inorganic methods. Finally if the substance to be determined has to be reacted with a reagent to produce a luminescent product for measurement, the reagent may decompose slightly to give other luminescent products even in the absence of the substance being determined, and these may contribute to the blank. In many practical analytical problems separation and/or chemical reaction cannot be avoided and the resulting blank contribution often far exceeds that of the solvent impurities discussed in the previous paragraph, and is then the main factor limiting sensitivity. We shall not discuss the problem further at this stage, but examples will be given in dealing with specific methods later.

5. Luminescence of Other Components of the Specimen

In principle the interference caused by other luminescent materials present in the specimen to be analysed can be avoided by the application of appropriate physical or chemical methods to separate the luminescent compound of interest before measuring it. Although preliminary separation is often unavoidable, it is always worth attempting to devise a direct method requiring no pretreatment—or the minimum of pretreatment—of the specimen. Apart from the fact that such direct methods are quicker, they have the advantage that reagent blank is avoided and there is no possibility of loss of material during separation—an important point in trace analysis, or when labile compounds are concerned. Direct methods are perhaps more properly considered along with simultaneous methods, in which several components of a mixture are determined without separation, by the appropriate application of luminescence technique. We shall discuss such methods in a later section but we give here two examples in which the compounds to be determined were present as trace impurities in a strongly fluorescent compound. By appropriate choice of conditions it was possible to determine each of the trace impurities by a direct method in such a way that the fluorescence of the major constituent produced only a small signal at the wavelengths of interest, and this signal could thus be regarded as part of the overall blank controlling the method sensitivity.

The analytical problem we shall describe is the determination of trace amounts of anthracene and fluorene in high-purity phenanthrene. The absorption spectrum of fluorene falls to zero at a wavelength short of 310 nm, whereas phenanthrene absorbs at all wavelengths shorter than 350 nm. There is thus no wavelength at which it is possible to excite the fluorene without also exciting the phenanthrene. However the fluorescence emission spectrum of fluorene (curve E in Fig. 153) is situated in a region where there is substantially no emission from the phenanthrene (curve K in Fig. 153). It is therefore possible to determine fluorene in the presence of phenanthrene

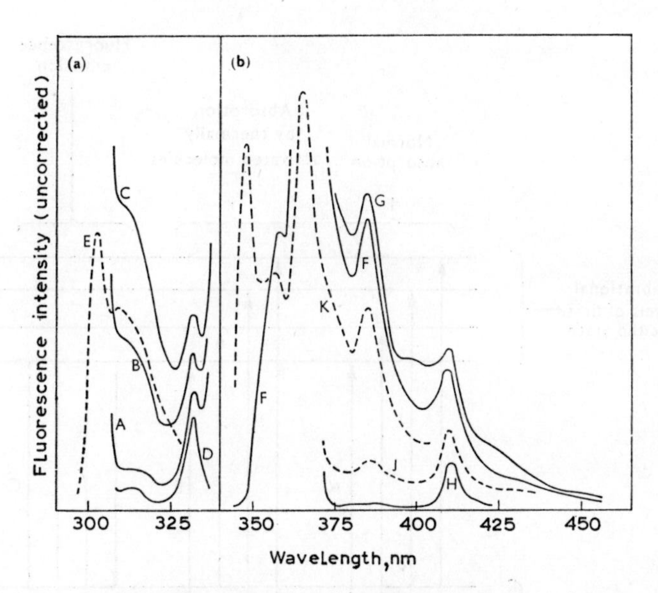

Fig. 153. Determination of fluorene and anthracene in phenanthrene by measurement of fluorescence at room temperature.
(a) Excitation at 302 nm with double monochromator; 100 μg per ml of phenanthrene with A, no addition; B, 0.005 μg per ml of fluorene; C, 0.01 μg per ml fluorene; D, solvent only; E, 1 μg per ml of fluorene alone, excited by 250 nm. (b) Excitation at 366 nm with double monochromator; F, 1% w/v of phenanthrene in ethanol; G, as for F, with 0.01 μg per ml of anthracene added; H, solvent only; J, as for F, but at − 90°C; K, 10 μg per ml of phenanthrene excited at 302 nm (from Parker[210]).

provided that the concentration of the latter is limited to keep inner filter effects within reasonable bounds. The optimum wavelength is that at which the ratio of the extinction coefficients of fluorene to phenanthrene is a maximum: this corresponds approximately to the mercury line at 302 nm. The concentration of the phenanthrene has to be limited to about 100 μg per ml (optical density per cm of 0.2 at 302 nm). The chosen excitation wavelength lies at the edge of the fluorene emission spectrum and scattered exciting light interferes with the measurement of dilute solutions of fluorene at either of the fluorescence maxima. By using a double monochromator the intensity of scattered light is negligible at 312 nm and the fluorescence due to low concentrations of fluorene then appears as an inflection at 312 nm on the side of the scattered light curve (see curves A, B and C in Fig. 153). The main contribution to the blank is the Raman emission from the solvent (curve D in Fig. 153). The additional signal due to the fluorene in the

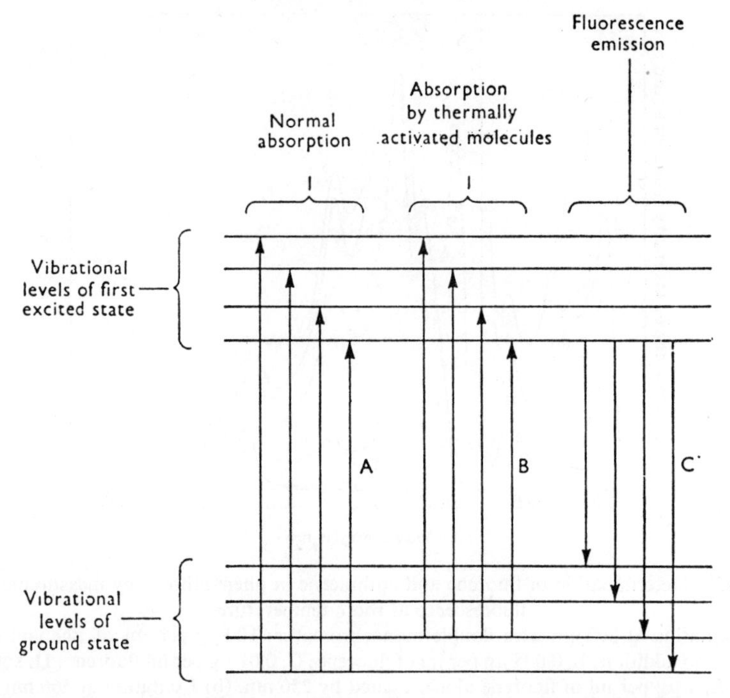

Fig. 154. Illustrating excitation in the anti-Stokes region (from Parker[210]).

specimen (curve A) corresponds to about 10 parts per million of fluorene in the phenanthrene specimen and this is about the minimum content that could be detected with confidence. Note that in this example the method sensitivity is limited, not by the fluorescence of the phenanthrene, but by its *absorption*. The actual concentration of fluorene in the solution is only 0.001 μg per ml but the concentration of phenanthrene has to be limited to 100 μg per ml.

The determination of trace quantities of anthracene in phenanthrene appears, at first sight, to present little problem. Anthracene absorbs the mercury line at 366 nm, which is situated well outside the normal absorption limit of phenanthrene. Thus, with a double monochromator to produce pure 366 nm radiation, the phenanthrene would be expected to produce no fluorescence whatsoever. In fact it shows a weak fluorescence (curve F in Fig. 153), part of which is situated at shorter wavelengths than the exciting light (although most of the short-wavelength part of the spectrum is not recorded in Fig. 153 owing to self-absorption). This effect is known as anti-

Stokes excitation and is readily explained by reference to the energy-level diagram shown in Fig. 154. Although at room temperature most of the phenanthrene molecules are in the lowest vibrational level of the ground state and give rise to an absorption band corresponding to transitions A in Fig. 154, a small proportion of them are thermally activated to an upper vibrational level of the ground state and give rise to absorption corresponding to the transitions B in Fig. 154. The absorption transition of *lowest* energy in B corresponds to a longer wavelength than the emission transition of *highest* energy in C. Hence it is possible to excite fluorescence from the latter by the absorption of the longer wavelength light corresponding to the former. In the determination of anthracene we are not interested in the wavelength region shorter than 366 nm, but the absorption of 366 nm light by the thermally activated phenanthrene molecules also gives rise to fluorescence corresponding to all the other transitions in C, i.e. it gives rise to the normal fluorescence spectrum of phenanthrene. The latter overlaps the region of anthracene fluorescence and thus increases the blank value for the determination of the latter. Thus the fluorescence obtained from a 1% w/v

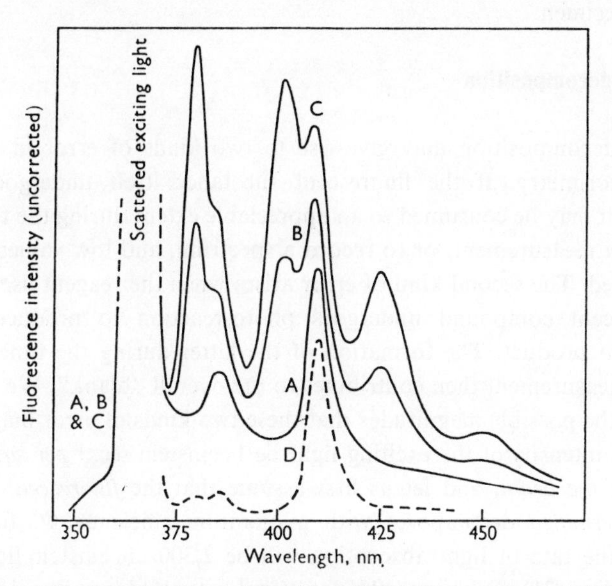

Fig. 155. Determination of anthracene in phenanthrene by measurement of fluorescence at -90°C.

Excitation at 366 nm with double monochromator; 1% w/v of phenanthrene in ethanol with A, no addition; B, 0.01 μg per ml of anthracene added; C, 0.02 μg per ml of anthracene added; D, solvent only (from Parker[210]).

solution of phenanthrene at room temperature is increased only slightly by the addition of 0.01 μg per ml of anthracene (compare curves F and G in Fig. 153). Since the magnitude of the "phenanthrene blank" depends on the concentration of thermally activated phenanthrene molecules, it can be reduced by making the measurement at low temperature. Thus at −90°C the blank value is lower by a factor of about 6 (see curve J of Fig. 153). By working at this temperature it is thus possible to increase the instrumental sensitivity in order to detect a lower concentration of anthracene. The addition of 0.01 μg per ml of anthracene now increases the fluorescence emission considerably (compare curves A and B in Fig. 155) and the main maxima in the anthracene spectrum are now clearly visible. Most of the fluorescence from the pure phenanthrene (curve A in Fig. 155) is now due to contributions from solvent Raman (see curve D) and fluorescence of the phenanthrene itself (see maximum at about 385 nm). By making allowance for these two contributions it can be concluded that the minimum detectable concentration of anthracene in the presence of 1% w/v phenanthrene is about 0.001 μg per ml, or 0.1 parts per million calculated on the phenanthrene specimen.

6. Photo-decomposition

Photo-decomposition may give rise to two kinds of error in analytical spectrofluorimetry. If the fluorescent substance itself undergoes photo-reaction, it may be consumed to an appreciable extent during the time taken to make a measurement, or to record a spectrum, and low values will thus be obtained. The second kind of error arises when the reagent used to form a fluorescent compound undergoes photo-reaction to produce another fluorescent product. The formation of the latter during the time taken to make a measurement then contributes to the overall "blank". We shall now consider the possible magnitudes that these two kinds of error may achieve.

Let the intensity of the exciting light be I_0 einstein sec^{-1} *per sq. cm cross section of the beam*, and let us first assume that the *fluorescent substance being determined* decomposes with a quantum efficiency Φ. In a *dilute* solution the rate of light absorption will be 2,300$\varepsilon c I_0$ einstein litre^{-1} sec^{-1} (see equation 21) and hence the *fractional* rate of decomposition will be given by:

$$\frac{1}{c}\cdot\frac{dc}{dt} = 2300\varepsilon I_0 \Phi \qquad (427)$$

A typical value of ε would be 10^4, and the light isolated from a mercury lamp by a grating monochromator and focussed on the cuvette, might have a cross sectional area of 0.5 cm^2, and hence an intensity of 8.6×10^{-8} einstein sec^{-1} cm^{-2} (see Table 20). In an unfavourable case we might have $\Phi = 0.1$ and hence the fractional rate of decomposition would then be 0.2 per second. That is, in a *still* solution the fluorescence intensity would decrease by about 20% per second. If the solution were stirred and the total volume of the cuvette were ten times the illuminated volume, the rate of decomposition would still be about 2% per second, i.e. sufficient to cause appreciable error during the time taken to make a single reading, or a considerable error during the time taken to record a complete fluorescence spectrum. Clearly, the remedy is to reduce the intensity of the exciting light and increase the sensitivity of the detector system.

Let us now consider the second case, in which an excess of reagent R at concentration c_R is added to the solution of the compound to be determined, A, present at a concentration c_A, to produce a fluorescent product, B:

$$A + R \rightarrow B \text{ (fluorescent)} \qquad (428)$$

Let us further assume that the reagent undergoes photo-decomposition at a quantum efficiency Φ to produce a fluorescent product P. Let ϕ_B and ϕ_P be the fluorescence efficiencies of B and P (assumed to have the same spectrum), and ε_B, ε_R and ε_P the extinction coefficients of B, R and P at the excitation wavelength. The fluorescence signal due to the compound A, after its reaction will be:

$$F_A = 2.3\varepsilon_B c_A I_0 \phi_B k \qquad (429)$$

where k is an instrumental constant. The rate of production of P by photo-decomposition of the reagent will be given by:

$$dc_P/dt = 2300 I_0 \Phi \varepsilon_R c_R \text{ mole litre}^{-1} \text{ sec}^{-1} \qquad (430)$$

and the corresponding rate of increase of fluorescence signal due to this photo-decomposition will be:

$$dF_P/dt = 2.3\varepsilon_P \phi_P I_0 k \, dc_P/dt \qquad (431)$$

Expressing this as a fraction of the fluorescence due to unit concentration of A, we have:

$$\frac{dF_P/dt}{F_A/c_A} = 1000 I_0 \Phi \cdot \frac{\varepsilon_P \phi_P}{\varepsilon_B \phi_B} \cdot (2.3\varepsilon_R c_R) = 1000 I_0 \Phi \cdot \frac{\varepsilon_P \phi_P}{\varepsilon_B \phi_B} \cdot x \qquad (432)$$

where x is the fraction of light absorbed per cm by the excess of reagent R. Now, substituting some values for an unfavourable situation, we may have, $I_0 = 8.6 \times 10^{-8}$ einstein litre^{-1} sec^{-1}, $\Phi = 1.0$, $\varepsilon_P\phi_P = \varepsilon_B\phi_B$, and $x = 0.01$. We then find:

$$\frac{dF_P/dt}{F_A/c_A} = 8.6 \times 10^{-7} \text{ mole litre}^{-1} \text{ sec}^{-1} \tag{433}$$

This means that in a *still* solution the fluorescence blank would increase by an amount equivalent to a concentration of A of nearly 10^{-6} mole litre^{-1} for every second of exposure to the beam of exciting light. Even if the solution were mixed, and the total volume were ten times greater than the irradiated volume, the continuously rising blank would set a serious limitation on the sensitivity of the method. An example of such an effect is met in the determination of borate by the benzoin method discussed in Section 5 G 1.

D. LUMINESCENCE OF ORGANIC COMPOUNDS IN RELATION TO MOLECULAR STRUCTURE

1. General Rules

Some useful general rules that can be deduced from the fundamental concepts discussed in Chapter 1 are listed below together with some others derived in a more empirical manner:

(a) For excitation of photoluminescence by wavelengths greater than 220 nm a π-electron system is required.

(b) Increased conjugation shifts the longest wavelength absorption band to longer wavelengths. In the absence of special effects (e.g. hydrogen bonding, electrolytic dissociation) the fluorescence emission spectrum is situated immediately to the long-wavelength side of the absorption limit. The phosphorescence spectrum is situated to the long-wavelength side of the fluorescence spectrum.

(c) High efficiency of prompt fluorescence in fluid solution is usually associated with compounds having π^*–π as the lowest excited singlet state. These compounds frequently also show intense phosphorescence in rigid media.

(d) Systems having π^*–n as the lowest excited singlet state show very low efficiencies of prompt fluorescence but high efficiencies of phosphorescence in rigid media.
(e) Compounds having a_π–l as the lowest excited singlet state can have high prompt fluorescence efficiencies, and high phosphorescence efficiencies at low temperature.
(f) Charge transfer states can give rise to prompt fluorescence, particularly at low temperature, as well as phosphorescence.
(g) Substitution of ortho–para directing groups does not interfere with the ability to fluoresce and may increase it. Most meta-directing groups except CN reduce the prompt fluorescence efficiency[329].
(h) Increase in molecular rigidity increases prompt fluorescence efficiency in fluid solution and favours resolution of vibrational structure.
(j) Substitution of side groups often shifts the spectra to longer wavelengths and reduces vibrational structure. The effect is least with alkyl groups. Many groups introduce new types of excited state (see (c), (d) and (e) above).
(k) Substitution of heavy atoms reduces fluorescence efficiency.

Clearly in complex compounds the effect of one rule may nullify that of another. There are also many exceptions to some of the rules and it has to be emphasised therefore that they must be applied with caution. In the following sections we illustrate the rules with examples and indicate some of the exceptions.

2. Degree of π-Electron Conjugation (rules (a) and (b))

That π-electrons are required to produce photoluminescence follows mainly from two facts. Firstly few organic compounds devoid of π-electrons absorb appreciably above 220 nm, and of course absorption is a prerequisite of photoluminescence. Secondly, aliphatic compounds such as halogen derivatives which do absorb in the quartz UV region, generally undergo photo-dissociation, e.g. ethyl iodide. Many aliphatic compounds that do contain π-electron systems (e.g. aldehydes, ketones, acids, amides) have π^*–n lowest excited singlet states and by rule (d) are only weakly fluorescent. Biacetyl is an exception, but even with this the efficiency of fluorescence is considerably less than that of phosphorescence. Phosphorescence of other aliphatic compounds in solution has been little investigated. Many absorbing aliphatic compounds contain electron attracting (m-directing) groups such as nitro, and by rule (g) would not be expected to fluoresce. Alkyl cyanides are transparent in the quartz ultra-violet region. In the aliphatic series therefore, one has to look for fluorescence among the compounds containing highly conjugated carbon–carbon double bonds, such as Vitamin A, which in ethanol has an absorption maximum at about 3.0 μm^{-1} and fluoresces in the blue-green region (about 2.0 μm^{-1})[330].

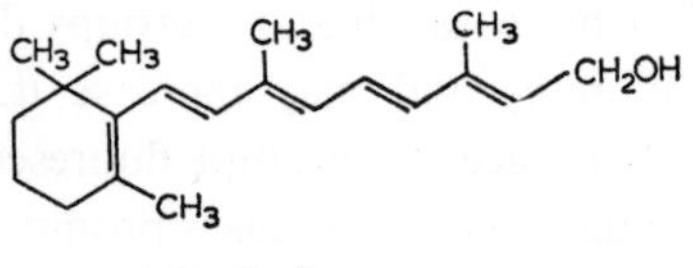

CH_3 CH_3 CH_3 CH_3 CH_2OH CH_3

Vitamin A_1

The effect of length of conjugated chain on the energy of the lowest excited singlet and triplet states is well illustrated by the series benzene, naphthalene, anthracene, naphthacene, all of which are fluorescent:

BENZENE (3)
S_1 3.80 μm^{-1}
T_1 2.94 μm^{-1}

NAPHTHALENE (5)
S_1 3.18 μm^{-1}
T_1 2.13 μm^{-1}

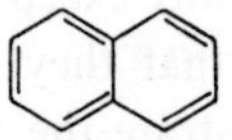

ANTHRACENE (7)
S_1 2.66 μm^{-1}
T_1 1.47 μm^{-1}

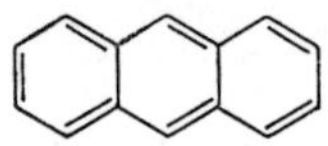

NAPHTHACENE (9)
S_1 2.11 μm^{-1}
T_1 1.03 μm^{-1}

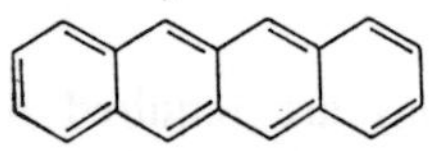

As the number of linearly conjugated (i.e. peripheral) double bonds increases from 3 to 9 the fluorescence and the phosphorescence shift from the ultra-violet (benzene) to the visible and near infra-red (naphthacene). Note that the triplet–singlet splitting remains approximately constant in absolute terms (about 1.0 μm^{-1}) but increases considerably as a proportion of the singlet energy.

If the conjugated chain contains a "cross-link", i.e. if it is not entirely peripheral, the energies are increased. This may be seen by comparing phenanthrene with anthracene, and benzanthracene with naphthacene:

Phenanthrene (7)
S_1 2.87 μm^{-1}
T_1 2.16 μm^{-1}

1,2-Benzanthracene (9)
S_1 2.61 μm^{-1}
T_1 1.65 μm^{-1}

An analogous effect is to be found among the selenodiazols[327]. Thus 3,4-benzo-piazselenol has one of its double bonds in the "cross-linked" position and shows a weak fluorescence at about 2.5 μm^{-1} (see Fig. 106) but the 4,5-isomer, in which all double bonds are peripheral, shows a strong fluorescence at about 1.9 μm^{-1} (see Fig. 186):

3,4-Benzo-piazselenol
S_1 ∽ 2.5 μm^{-1}

4,5-Benzo-piazselenol
S_1 ∽ 1.9 μm^{-1}

3. Nature of the Lowest Excited Singlet State (rules (c), (d), (e) and (f))

Unsubstituted aromatic hydrocarbons all have lowest excited π^*–π states —they always show some fluorescence, often very intense, and generally phosphorescence also in rigid media (see Fig. 156). π^*–n states may be introduced by hetero-ring atoms or by substituent groups. An example of the former is pyridine, which is non-fluorescent, compared with benzene which gives both fluorescence and phosphorescence. The series naphthalene, quinoline, indole, illustrates several points (see also Section 1 B 6 and 4 F 3):

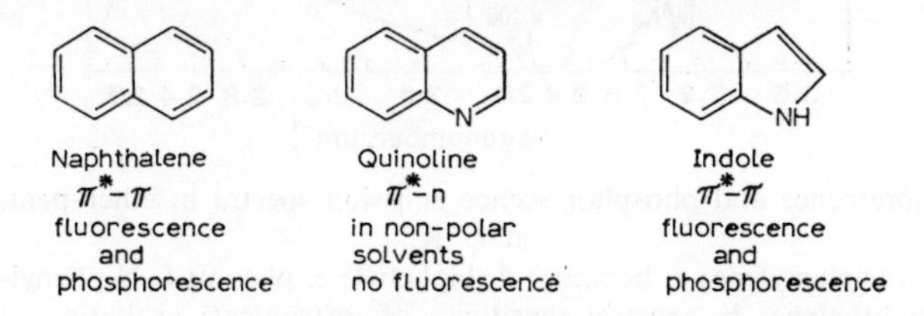

Thus naphthalene has a π^*–π lowest excited singlet state and gives both fluorescence and phosphorescence. Quinoline in non-polar solvents has π^*–n lowest and is substantially non-fluorescent. The spatial arrangement of the orbital of the lone pair of electrons in indole is such that it conjugates with the aromatic π-electrons in the ground state and cannot give rise to a π^*–n transition. This compound therefore has π^*–π lowest and is again both fluorescent and phosphorescent. In polar solvents the positions of the π^*–π and π^*–n levels of quinoline are reversed and it therefore gives fluorescence emission under these conditions. Substitution of other active groups in pyridine and quinoline produces changes in the nature of the energy levels and can thus give rise to fluorescence, e.g. in 3-hydroxy-pyridine.

The commonest example of a substituent group introducing a lowest π^*–n level is carbonyl and this is responsible for the fact that simple (and otherwise unsubstituted) aromatic aldehydes, ketones, carboxylic acids, esters and amides are non-fluorescent (or very weakly fluorescent), but generally phosphorescent in rigid media (see Section 1 B 6 and Fig. 157). As the complexity of the aromatic system increases the separation between

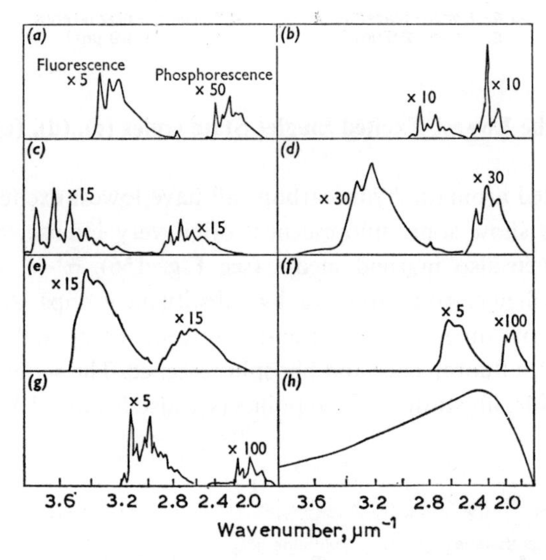

Fig. 156. Fluorescence and phosphorescence emission spectra in ether–pentane–ethanol at 77°K.

a, fluorene; b, triphenylene; c, benzene; d, biphenyl; e, phenol; f, N-phenyl-2-naphthyl-amine; g, naphthalene; h, spectral sensitivity of instrument; excitation at 4.03 μm^{-1} (248 nm); quartz analysing spectrometer with E.M.I. 6256 photomultiplier; hand-band width at 2.5 μm^{-1} was 0.019 μm^{-1}; the left hand curves refer to the fluorescence region measured with the choppers in-phase; the right hand curves are phosphorescence, measured with the choppers out-of-phase (from Parker and Hatchard[203]).

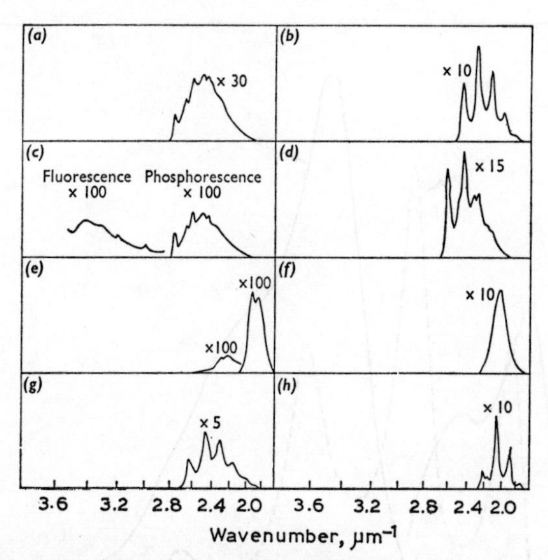

Fig. 157. Phosphorescence emission spectra in ether–pentane–ethanol at 77°K. a, benzoic acid; b, benzophenone; c, acetanilide; d, benzoin; e, 4-nitro-N-ethylaniline; f, benzil; g, acetophenone; h, anthraquinone; conditions as for Fig. 156 (from Parker and Hatchard [203]).

the π^*–π and π^*–n levels decreases and ultimately is reversed. With compounds occupying intermediate positions (e.g. pyrene 3-aldehyde) the nature of the lowest excited state depends on the solvent, and fluorescence increases as the polarity of the solvent increases (see Section 4 F 3). Introduction of a second substituent completely alters the nature and distribution of the excited states and if the second substituent is an electron donating group the CT state is often the lowest excited state. This has a short lifetime and can give rise to both fluorescence and phosphorescence (see Section 1 B 6). Thus benzophenone is non-fluorescent but strongly phosphorescent at 77°K, while 4,4′-tetramethyldiaminobenzophenone gives rise to both fluorescence and phosphorescence under these conditions (see Fig. 158).

Substituents having lone pair electrons that are not capable of giving rise to true π^*–n states (e.g. those classified as a_π–l by Kasha—see Section 1 B 6) are electron donating groups and frequently give rise to both fluorescence and phosphorescence. Thus most phenols and aromatic amines are fluorescent in fluid solution. These groups have the power of nullifying the effect of π^*–n substituents as described in the previous paragraph. Non-fluorescent π^*–n heterocyclics also become fluorescent when substituted with these groups, e.g., hydroxypyridine.

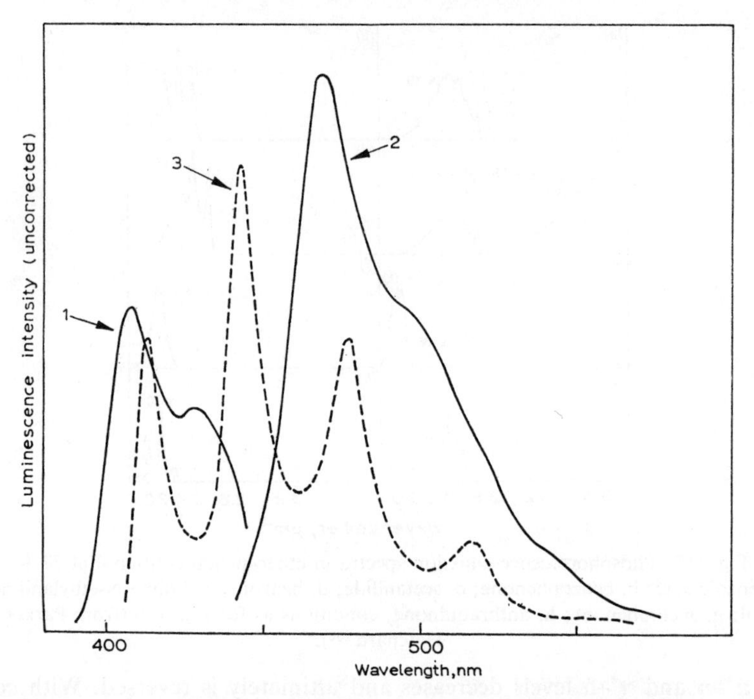

Fig. 158. Comparison of photoluminescence of compounds forming CT and π^*–n states. 1,2, fluorescence (× 6) and phosphorescence (× 2) of 2.5×10^{-6} M 4,4′-tetramethyldiaminobenzophenone in ether–pentane–ethanol at 77°K; 3, phosphorescence (× 1) of 2.5×10^{-6} M benzophenone (no fluorescence); excitation at 250 nm; band width of analysing monochromator was 3.3 nm.

4. Classification of Substituent Groups (rule (g))

Williams and co-workers[329, 331, 332] have made extensive studies of the effect of substituents on the prompt fluorescence of aromatic and heterocyclic compounds in fluid solution. They found that generally ortho–para directing substituents tend to enhance the fluorescence but meta-directing groups prevent fluorescence, and that the former can re-instate the fluorescence when present together with the latter. Many of the examples obeying these rules can be explained in terms of the fundamental factors considered in the previous section. Thus π^*–n substituents such as carbonyl, ester, carboxylic acid, are meta-directing groups while a_π–l groups are ortho–para directing:

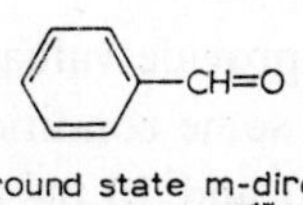

-CH=O

Ground state m-directing
excited state π^*-n
non-fluorescent

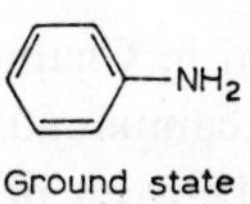

-NH_2

Ground state o,p-directing
excited state a_π-l
fluorescent

The meta-directing CN group is an exception to the Williams rule—CN-substituted aromatics are frequently fluorescent. Presumably the π^*–n state has a higher energy than the π^*–π in these compounds although the reason for this is not clear. Some other exceptions to the rule should be noted. Thus the ionised phenolic group is ortho–para directing but according to Williams[329] the phenolate ion is non-fluorescent. This fact is applied to distinguish between phenol and anisole:

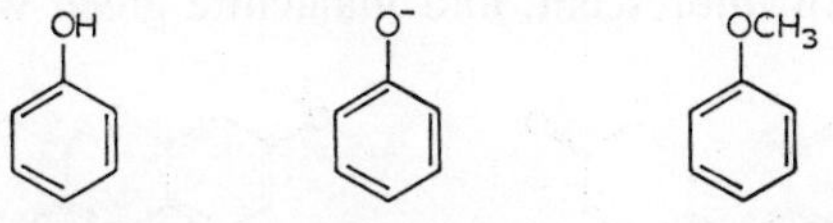

OH

Fluorescent
(neutral
solution)

O^-

non-Fluorescent
(alkaline
solution)

OCH_3

Fluorescent
in neutral
and alkaline
solution

On the other hand the β-naphtholate ion is fluorescent (see Section 4 C). Again, although NH_3^+ is a meta-directing group and, in accordance with this, aniline hydrochloride is said to be non-fluorescent[329], the hydrochlorides of the naphthylamines are fluorescent (see Section 4 C). Some of these discrepancies may be explained by the fact that as the size of the aromatic or heterocyclic system increases the influence of meta-directing groups decreases. For example some nitro-stilbene derivatives are fluorescent in fluid solution e.g., 4-dimethylamino-4′-nitrostilbene:

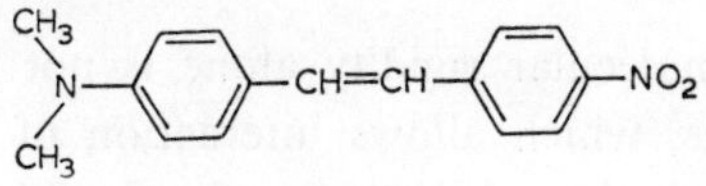

$(CH_3)_2N$–C_6H_4–CH=CH–C_6H_4–NO_2

a compound recommended by Lippert and co-workers[163] for correcting fluorescence emission spectra. The fluorescence is probably due at least in part to the presence of the amino group since the simpler compound meta-nitro-dimethylaniline also fluoresces[163].

Superimposed on the rules given in Section 5 D 1 we naturally have to consider the effects of internal and external hydrogen bonding and of electrolytic dissociation in the ground and excited states, the basic principles of which were discussed in Chapter 4. These provide valuable methods of distinguishing between compounds that under some conditions give similar spectra. For example, between pH values of 3 and 10 m- and p-hydroxy-benzoic acids are non-fluorescent but the o-compound gives rise to fluorescence (internal hydrogen bonding), while at a pH value of 14 all three acids are said to fluoresce[332].

5. Position and Vibrational Resolution of Fluorescence Emission Spectra (rules (h) and (j))

It is often stated that increase in molecular rigidity increases fluorescence efficiency in fluid solution—the most frequently quoted comparisons are phenolphthalein with fluorescein, and malachite green with rhodamine B:

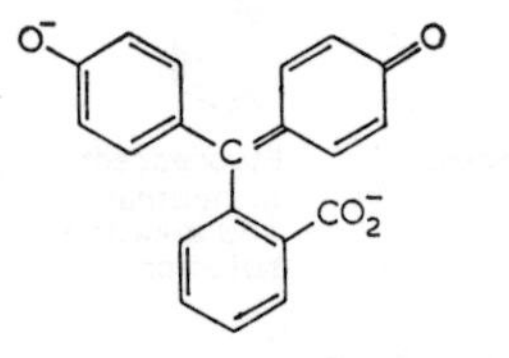

Phenolphthalein (non-fluorescent)

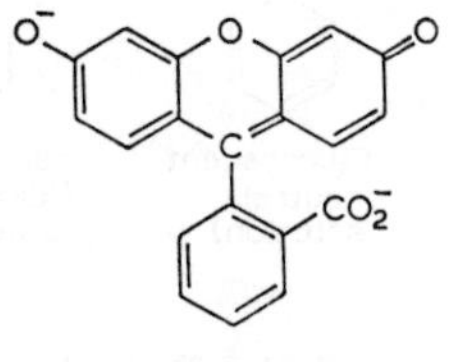

Fluorescein (fluorescent)

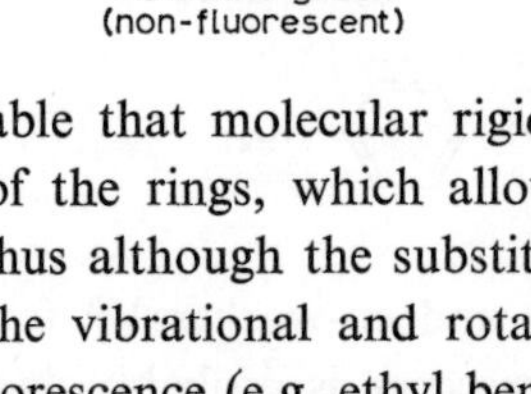

Malachite green (non-fluorescent)

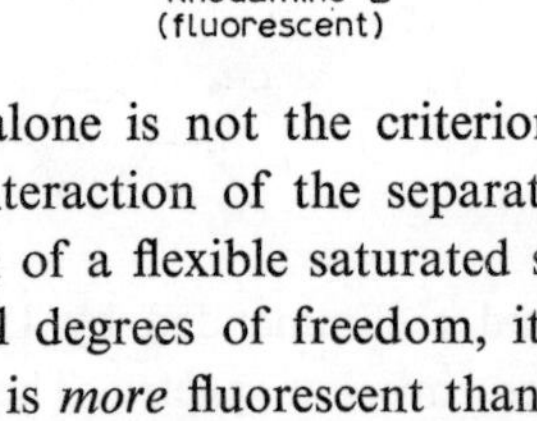

Rhodamine B (fluorescent)

It is probable that molecular rigidity alone is not the criterion, but co-planarity of the rings, which allows interaction of the separate electron systems. Thus although the substitution of a flexible saturated side group increases the vibrational and rotational degrees of freedom, it does not prevent fluorescence (e.g. ethyl benzene is *more* fluorescent than benzene). However such substitution does generally reduce the resolution of vibrational structure in the fluorescence spectrum—compare benzene and toluene in Fig. 159. Similarly, increased rigidity due to the formation of an additional ring system often increases vibrational resolution—compare phenyl naphthalene with benzfluorene; phenyl-β-naphthylamine with benz-

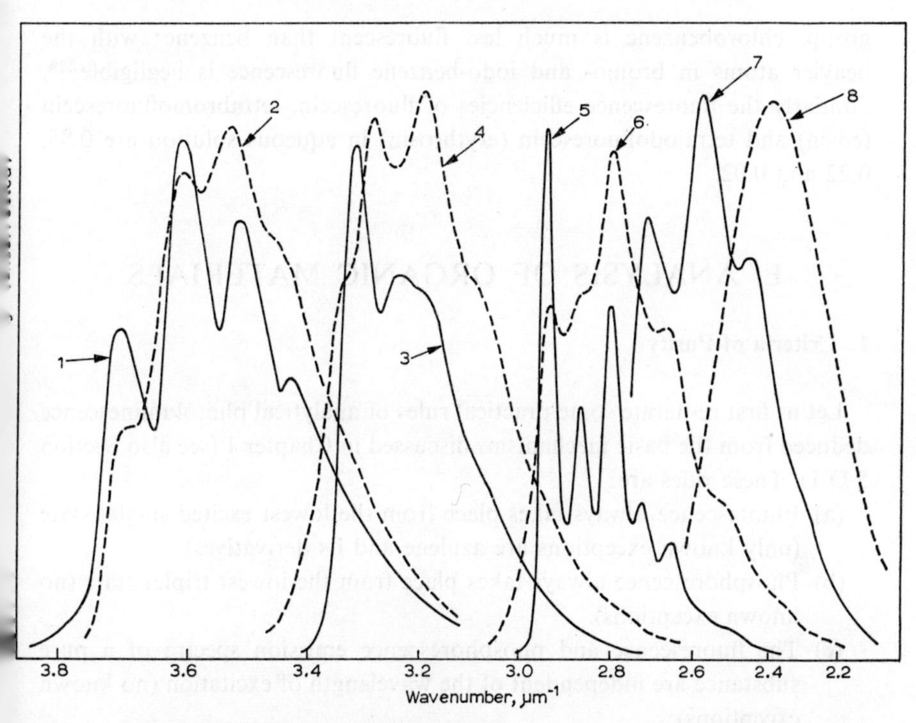

Fig. 159. Illustrating effects of molecular rigidity and substitution on the vibrational resolution of fluorescence emission spectra in ethanol at room temperature. 1,2, benzene and toluene; 3,4, fluorene and diphenyl; 5,6, 2,3-benzfluorene and phenyl naphthalene; 7,8, 3,4-benzcarbazole and phenyl-2-naphthylamine.

carbazole in Fig. 159. Surprisingly, fluorene shows no more resolution than diphenyl in fluid solution at room temperature (Fig. 159) although its spectrum is better resolved in rigid medium (curves a and d in Fig. 156). The presence of "active" substituent groups frequently removes the vibrational structure almost completely and shifts the spectrum to longer wavelengths, for example phenol and aniline compared with benzene, or methoxy naphthalene compared with naphthalene.

6. Heavy Atom Substitution (rule k)

Substitution of heavy atoms increases the rate of intersystem crossing from S_1 to T_1 owing to increased spin–orbit coupling, and hence decreases fluorescence efficiency. Thus, although chlorine is an ortho–para directing group, chlorobenzene is much less fluorescent than benzene: with the heavier atoms in bromo- and iodo-benzene fluorescence is negligible[329]. Similarly the fluorescence efficiencies of fluorescein, tetrabromofluorescein (eosin) and tetraiodofluorescein (erythrosin) in aqueous solution are 0.85, 0.22 and 0.02.

E. ANALYSIS OF ORGANIC MATERIALS

1. Criteria of Purity

Let us first re-iterate some practical rules of analytical photoluminescence deduced from the basic mechanisms discussed in Chapter 1 (see also Section 5 D 1). These rules are:

(a) Fluorescence always takes place from the lowest excited singlet state (only known exceptions are azulene and its derivatives).

(b) Phosphorescence always takes place from the lowest triplet state (no known exceptions).

(c) The fluorescence and phosphorescence emission spectra of a pure substance are independent of the wavelength of excitation (no known exceptions).

(d) The corrected fluorescence and phosphorescence excitation spectra are identical with or closely similar to, the absorption spectrum (true for all substances except those showing intramolecular energy transfer or decomposition at very short wavelengths).

(e) The fluorescence emission band is situated just to the long-wavelength side of the longest wavelength absorption band (many exceptions).

(f) The phosphorescence band is situated at longer wavelengths than the fluorescence band (no known exceptions).

All these rules except (e) may be applied as stringent tests of purity of a fluorescent or phosphorescent compound, whether it be a standard specimen, or a fraction separated from a mixture in the course of an analysis. They must of course be applied to dilute solutions in which inner filter effects are negligible. Thus if a substance is found to give a fluorescence emission with maximum at shorter wavelengths than its long wavelength absorption band, the presence of an impurity can be concluded (rule (a)). Similarly if a phosphorescence emission is found at shorter wavelengths than a fluorescence emission the presence of more than one substance can be deduced (rules (b) and (f)). A frequent problem that arises is the presence of a very weak "tail" at the long wavelength limit of absorption. Sometimes this is due to a transition of low probability, but it is often found to be due to traces of impurity and this can be decided by measuring the excitation spectrum (fluorescence or phosphorescence). If the "tail" is due to impurity the excitation spectrum will be lower than the absorption spectrum in this region (rule (d)). It may happen that the absorption by the impurity is small at all wavelengths relative to that of the main component, and within experimental error rule (d) will then be obeyed. If, however, the impurity is fluorescent or phosphorescent it will make its presence felt by the appearance of a second emission band, the relative intensity of which will vary with the wavelength of excitation (rule (c)).

In applying these rules care must be taken to check whether non-conformance is due to impurity or to the fact that the substance exists in two forms in solution. The commonest example of this is ionisation to give acid and basic forms, which are distinct species with their own absorption and emission spectra. Such substances can be made to conform by measuring in sufficiently acid or alkaline solution (see Section 4 C). The substance may isomerise in solution, e.g. cis and trans stilbenes. The substance is then no longer pure in the strict sense. Such cases can generally be identified from the known structure of the substance under investigation.

The exceptions to rule (e) require some special comment. The fluorescence emission band may be situated at a much longer wavelength than the longest wavelength absorption band. This may be due to fluorescent impurity in a non-fluorescent compound, but frequently such large Stokes' shifts are due to hydrogen bonding or other solvent effects (see Section 4 F). Such substances will still conform to the other rules provided that they do not exist in more than one form in solution. Electrolytic dissociation in the excited state (Section 4 C) can also give rise to apparently large Stokes' shifts. Thus, the pH value of a solution containing an organic acid (e.g. β-naphthol) may have been adjusted to a value at which only the acid form is present, as indicated by the absorption spectrum. The emission spectrum may still be due to the basic form, appearing at much longer wavelengths than the absorption spectrum, or it may show bands due to both forms. The remedy here is to work at sufficiently high or low pH values that dissociation in the excited state is negligible. The appearance of new emission bands due to excited dimer formation is generally not troublesome in the measurement of prompt fluorescence for analytical purposes because one normally operates at low concentrations at which dimer formation is negligible. It must however be taken into account when measuring long-lived emission spectra in fluid solution. The analytical applications of the latter are discussed in Section 5 E 4.

Unexpected emission bands observed in glassy solvents at low temperature must be subjected to careful scrutiny to ensure that they are not due to crystallisation of a part of the solute during the process of cooling. The presence of both dissolved and crystalline material is equivalent to the presence of two different substances since the crystals will in general have different fluorescence and phosphorescence spectra. In particular, the crystals of some compounds will exhibit prompt and/or delayed emission from the excited dimer which may be mistaken for an impurity.

2. Direct Analysis of Mixtures by Fluorescence Measurement

The tests for purity described above are in effect methods for the analysis of mixtures in which one component is present in very much greater concentration than the others. The practical rules may be applied to devise methods for the analysis of mixtures—whether they be of this type, or consist of several major constituents. The first step is to measure the absorption spectra and the fluorescence and/or phosphorescence emission spectra of the pure components. If the fluorescence or phosphorescence emission spectrum of one component is well separated from the corresponding emission spectra of the other components there will obviously be no difficulty in analysing for this component. If the emission spectra overlap, then inspection of the absorption spectra will indicate the optimum wavelengths for excitation of each component having regard to the absorption spectra of the others. By excitation at appropriate absorption maxima it is thus frequently possible to obtain reasonably pure emission spectra although the other components may be excited to some extent.

Let us consider first an example of a three-component mixture that can be simply analysed, viz. a mixture of fluorene, phenanthrene and anthracene, first investigated by Thommes and Leininger[333]. A special case of this in which fluorene and anthracene were present in very small concentration, was discussed in Section 5 C 5. From that discussion it will be apparent that when all three compounds are present as major constituents their separate direct determination presents no problems. This follows from the fact that they have well-separated long-wavelength absorption limits. Thus anthracene can be excited at a wavelength (e.g. 366 nm) not absorbed by the other two compounds. The phenanthrene can be excited at a wavelength (e.g. 313 nm) not absorbed by the fluorene, and although the anthracene is also excited, it is possible to find a wavelength within the phenanthrene emission spectrum (e.g. 350 nm) at which the anthracene does not emit (see Fig. 153 curve K). Similarly when the fluorene is excited by a shorter wavelength still, its fluorescence can be measured at, say 315 nm, where neither the phenanthrene nor the anthracene emit (see Fig. 153 curve E). This principle is applicable to a wide variety of mixtures. Notice that only the emission spectrum of the component absorbing at the longest wavelength can be obtained undistorted by the others. For example, Sawicki and co-workers[334] have measured the emission spectrum of perylene excited at 430 nm in the presence of 25 other aromatic hydrocarbons. By careful choice of excitation wavelength the spectrum of another component absorbing at a shorter wavelength can also be obtained almost undistorted. An example is the mixture benz(a)pyrene and perylene, which is difficult to separate chromatographically. The absorption of perylene at 295 nm is low relative to that of benz(a)pyrene and by exciting at this wavelength a substantially undistorted emission spectrum of benz(a)pyrene is obtained in the presence of an equal concentration of perylene[334]. If the mixture is such that a suitable excitation wavelength cannot be found, the pH of the solution may be varied, or the mixture submitted to chemical treatment designed to convert the substance to be determined into a compound having an absorption spectrum shifted to longer wavelengths, i.e. out of range of interference by other components. For example, in concentrated sulphuric acid the long-wavelength excitation limit of benz(a)pyrene is shifted from about 400 nm to 520 nm with emission at 546 nm. By this means it can be determined in the presence of 50 other aromatic hydrocarbons[334].

In this and the following section we have dwelt at length on the analysis of mixtures of unsubstituted aromatic and heterocyclic compounds because their absorption and emission spectra are very characteristic and thus provide graphic examples of the capabilities of photoluminescence measurement. The application of the same principles is implicit in the other applications discussed in later sections where the spectra of many compounds are much less characteristic. When the emission spectrum consists of a single broad band, the reliability of identification can be increased considerably by measuring the excitation spectrum. The greater specificity of the latter follows from the fact that the absorption spectra of fluorescent organic

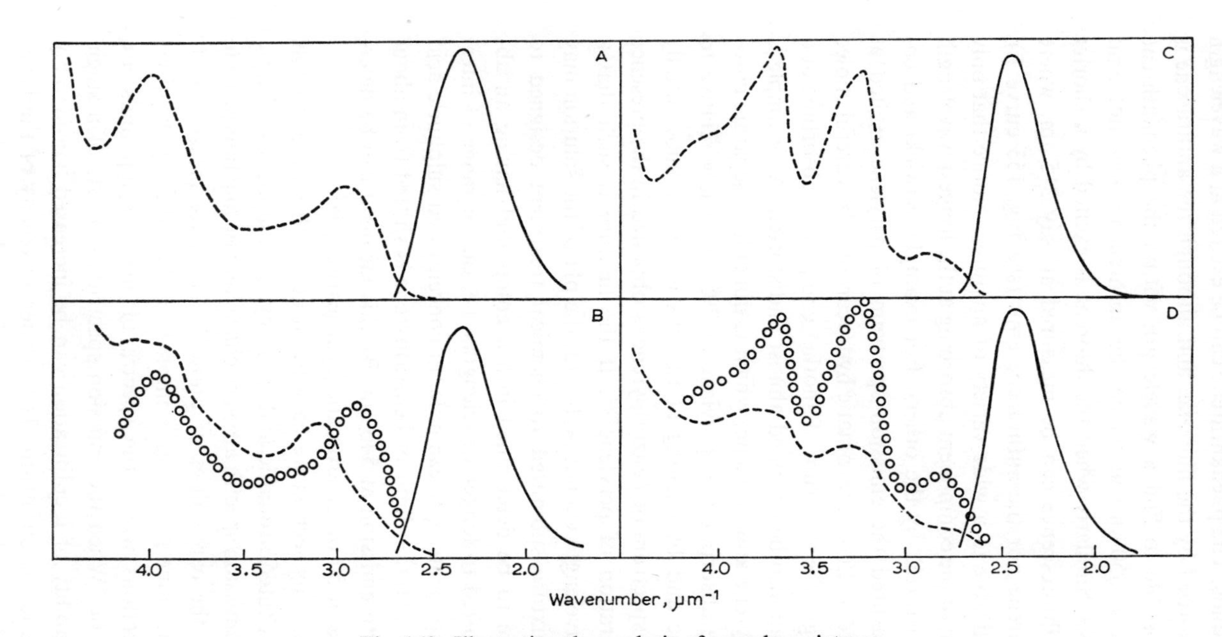

Fig. 160. Illustrating the analysis of complex mixtures.
Full curves are fluorescence emission spectra; dotted curves are absorption spectra; circles are approximate corrected fluorescence excitation spectra; A, pure N-phenyl-α-naphthylamine; B, extract of rubber containing N-phenyl-α-naphthylamine; C, pure N-phenyl-β-naphthylamine; D, extract of rubber containing N-phenyl-β-naphthylamine (from Parker and Barnes[131]).

compounds frequently contain at least two maxima. In effect, the corrected excitation spectrum provides a means of measuring the absorption spectrum of a luminescent compound in the presence of large quantities of other absorbing materials (see Section 1 B 4). This was pointed out as long ago as 1957 by Parker and Barnes[131] and we shall describe two examples taken from this early work, viz. the determination of the antioxidants α- and β-phenylnaphthylamines in vulcanised rubber. These compounds show strong fluorescence bands with maxima at 2.34 and 2.42 μm^{-1} (427 and 413 nm) —see Fig. 160, curves (a) and (c)— and can be observed directly by excitation of a diluted extract of the rubbers (Fig. 160 curves (b) and (d)). Their identity can be confirmed by measuring the corresponding excitation spectra which are very similar to the absorption spectra (with modern equipment they are found to be nearly identical with the absorption spectra). In contrast, direct absorption measurements on the rubber extracts give little clue as to the identity of the antioxidants because the absorptions of the latter are masked by the strong absorptions due to pine tar and other ingredients of the rubbers.

3. Direct Analysis of Mixtures by Phosphorescence Measurement

Frequently the measurement of phosphorescence provides a solution to the analysis of mixtures when the measurement of fluorescence does not. Consider first the analysis of specimens from various positions on a zone refined column of fluorene[210]. Pure fluorene ceases to absorb at a wavelength short of 313 nm and should therefore show neither fluorescence nor phosphorescence when excited at this wavelength. The phosphorescence spectrum of the unzoned material showed a complex system of bands (curve A in Fig. 161), of which five corresponded in wavelength to five of the six bands (marked C in Fig. 161), in the phosphorescence spectrum of carbazole (curve C). A specimen taken from near the top of the zone refined column showed these bands at much greater intensity, while specimens from lower down the column showed practically no indication of the carbazole bands (curves A, B and C in Fig. 162). Clearly carbazole segregates from fluorene against the zoning direction and can be completely removed. The concentrations of carbazole in the fluorene specimens were found to be 30 p.p.m. (curve A in Fig. 161), 190 p.p.m. (curve B in Fig. 161) and less than 2 p.p.m. (curves A, B and C in Fig. 162). By increasing either the fluorene concentration, the intensity of the exciting light or the sensitivity of the detection system, considerably lower concentrations of carbazole could have been detected. Note that a double monochromator had to be used to isolate the exciting light sufficiently pure to avoid exciting the fluorene itself. Note also that the *prompt fluorescence* of the carbazole in the original specimen was swamped by the fluorescence of a second impurity, and could not therefore have been used for its determination. The prompt fluorescence of the second impurity was associated with the phosphorescence bands marked U in Figs. 161 and 162. It segregated downwards on zone refining and was ultimately completely removed by rezoning the top section of the column. It appeared to be a naphthalene derivative. It is noted that Kanda and co-workers[335] determined the content of dibenzofuran in commercial fluorene by photographic recording of the phosphorescence spectrum excited with wavelengths that also excited the phosphorescence of the fluorene. Inter-

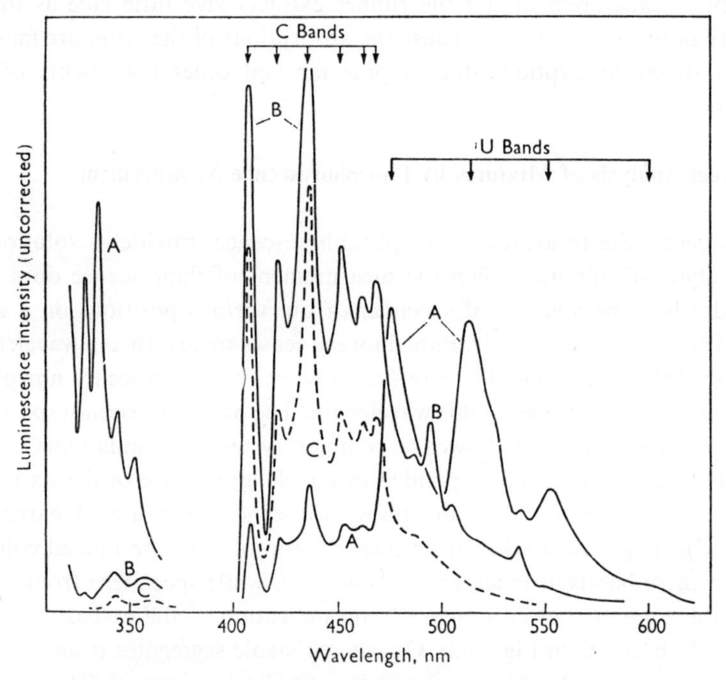

Fig. 161. Fluorescence and phosphorescence of impurities in fluorene. Excitation in ethanol at 77°K with 313 nm isolated by double monochromator; phosphorescence spectra (right hand curves) at 410 times greater sensitivity than fluorescence spectra (left hand curves); fluorene specimens (3×10^{-2} M) from A, original material; B, specimen situated 4 cm from top of zone-refined column; C, 4×10^{-6} M carbazole alone (from Parker[210]).

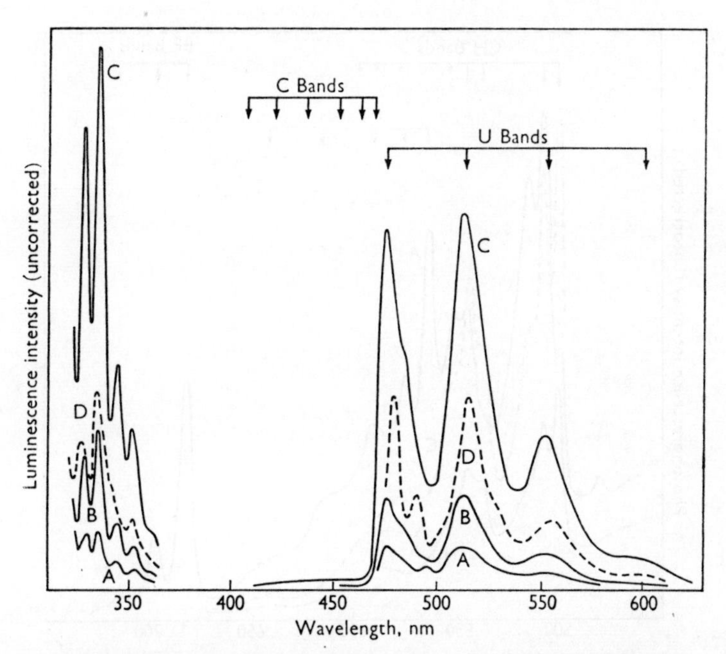

Fig. 162. Fluorescence and phosphorescence of impurities in fluorene. Excitation in ethanol at 77°K with 313 nm isolated by double monochromator; phosphorescence spectra (right hand curves) at 410 times greater sensitivity than fluorescence spectra (left hand curves); A, B, C, 3×10^{-2} M fluorene from specimens situated at 15, 30 and 45 cm from top of zone-refined column; D, 10^{-3} M β-methylnaphthalene (from Parker[210]).

ference from the latter was avoided by measuring the short-wavelength emission band of the dibenzofuran, which appears at a shorter wavelength than the fluorene phosphorescence. They were able to observe the 0–0 band of dibenzofuran at a concentration of 10^{-6}M.

A somewhat different principle has been employed for the detection of trace impurities in zone-refined benz(a)pyrene[210]. This compound absorbs at all wavelengths shorter than 405 nm and emits intense fluorescence covering the region 400–500 nm. The detection of impurities in it by fluorescence measurement is thus limited to those compounds that absorb at wavelengths longer than 405 nm, or emit at wavelengths shorter than 400 nm. The measurement of phosphorescence is very apt because the phosphorescence of benz(a)pyrene (marked BP in Figs. 163 and 164) is situated in the red and near infra-red regions, and thus the whole of the spectral region shorter than 680 nm is left clear for the observation of

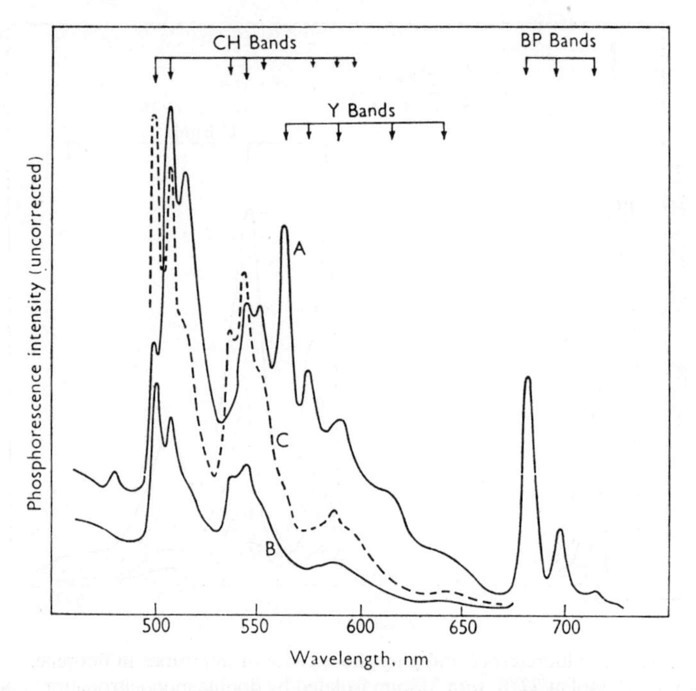

Fig. 163. Phosphorescent impurities in benz(a)pyrene.
Excitation of 10^{-4} M benz(a)pyrene in ethanol at 77°K by 313 nm; A, untreated material; B, C, specimens situated at 4 and 0.5 cm from the top of zone-refined column (from Parker[210]).

impurity phosphorescence. Further, the phosphorescence of the benz(a)-pyrene itself may be used as an internal standard. In the 2 mm thick layer of frozen solution, the benz(a)pyrene (10^{-4}M) absorbed about 15% at 313 nm and 65% at 366 nm and its absorption did not therefore prevent the impurities being excited. The recrystallised but unzoned material gave different spectra with the two wavelengths of excitation (curves A in Figs. 163 and 164), indicating that more than one phosphorescent impurity was present. One band system, marked Y in Figs. 163 and 164, was present in both spectra and the impurity responsible segregated rapidly towards the bottom of the column during zone refining (compare curves B with curves A in Figs. 163 and 164). Of the compounds responsible for the remaining bands in Fig. 163, some segregated downwards and were removed. One set of bands persisted in all sections of the column and these were found to be due to chrysene (bands marked CH in Fig. 163). By the method of standard additions the chrysene content was found to be as much as 0.03% even after many hundreds of zone passes. Muel and Hubert-Habart[336] used a similar principle to monitor the purification of the benz(a)pyrene specimen that they prepared for measurement of its complete phosphorescence emission spectrum.

For further examples of the analysis of mixtures we shall refer to the work of Zander[337] who has made a study of the application of photoluminescence analysis to coal-tar chemistry[338–341]. Pyrene is less phosphorescent than some of its impurities and hence the conditions are favourable for its study by phosphorescence analysis. Fig. 165 shows the detection of three major impurities in technical pyrene. Thus excitation at 360 nm within the strong absorption band of fluoranthene gave the almost undistorted phosphorescence spectrum of this impurity (curve 1). Identity was confirmed by the measurement of the lifetime of the phosphorescence. Excitation at 290 nm gave a different spectrum (curve 3), two of the main bands of which corresponded to those of benzo[b]naphtho[1,2-d]furan (curve 4). Finally excitation at 340 nm

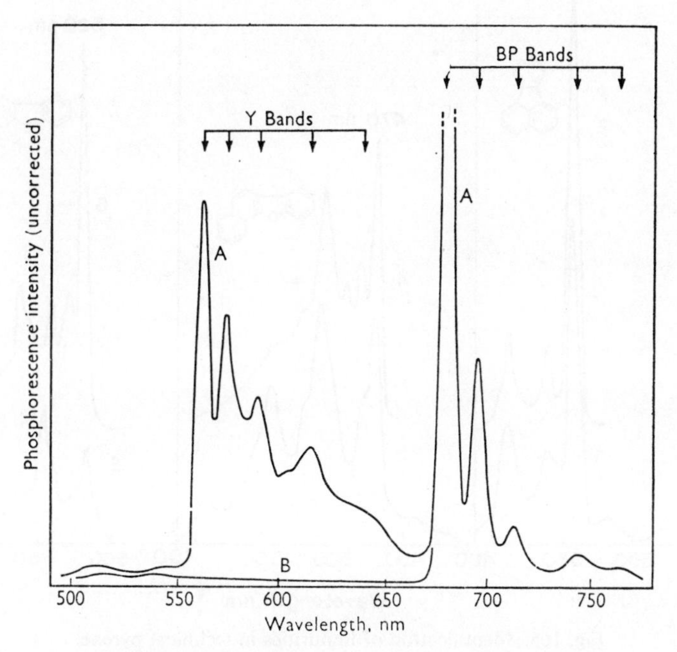

Fig. 164. Phosphorescent impurities in benz(a)pyrene.
Excitation of 10^{-4} M benz(a)pyrene in ethanol at 77°K by 366 nm; A, untreated material; B, specimen situated 4 cm from top of zone-refined column (from Parker[210]).

gave yet a third phosphorescence spectrum (curve 5), part of which is due to fluoranthene, but the short wavelength section was tentatively attributed to benzo[b]naphtho[2,3-d]furan. Zander also pointed out that phosphorescence measurement is ideal for the determination of carbazole and phenanthrene in anthracene (non-phosphorescent), and he gave calibration graphs covering the range 0.1–2%. He also showed that phosphorescence measurement provided a simple and reliable direct method for determining the strongly phosphorescent triphenylene in coal-tar fractions containing 2% of this substance in the presence of a variety of other 5-ring compounds (see Fig. 166). He used the intense sharp triphenylene emission at 463 nm for identification and determination. Excitation at the triphenylene absorption maximum at 258 nm was suitable. He avoided inner filter effects due to other components by working at concentrations below 10^{-5}M. Although the tar fractions contained many other phosphorescent compounds, none of their spectra overlapped 463 nm seriously.

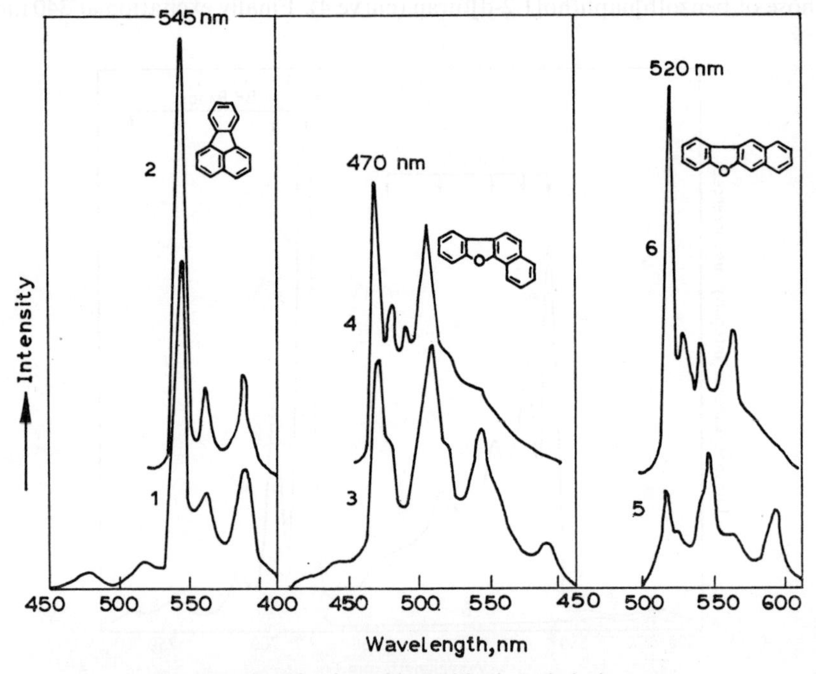

Fig. 165. Identification of impurities in technical pyrene.
1, 3, 5, phosphorescence spectra of technical pyrene in ether–pentane–ethanol at 77°K by excitation with 360 nm, 290 nm and 340 nm; 2, 4, 6, pure specimens of fluoranthene, benzo[b]naphtho-[1, 2-d]furan and benzo[b]naphtho[2, 3-d]furan (from Zander[337]).

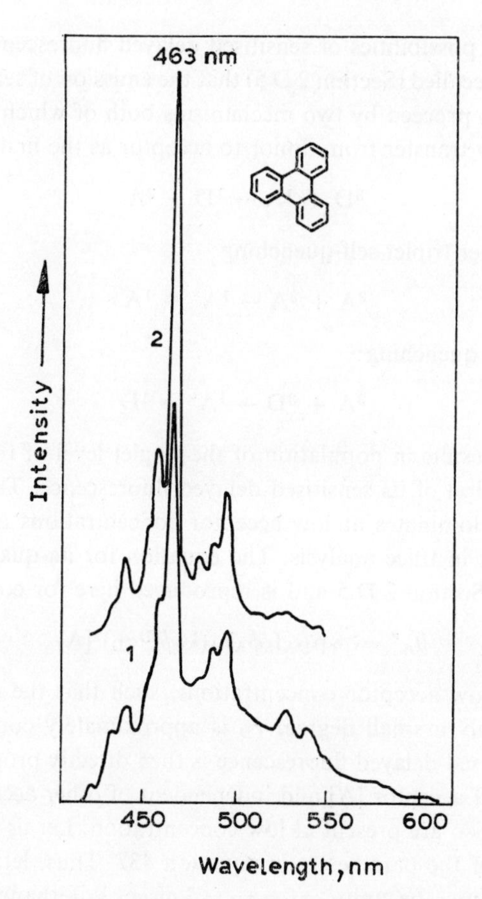

Fig. 166. Determination of triphenylene in tar fractions by phosphorescence measurement in ether–pentane–ethanol at 77°K.
1, tar fraction containing triphenylene; 2, pure triphenylene (from Zander[337]).

4. Direct Analysis of Mixtures by the Measurement of Sensitised Delayed Fluorescence

Although direct spectrofluorimetric or spectrophosphorimetric methods for the analysis of mixtures can sometimes be devised, their application is not by any means universal. In view of the hazards (high reagent blank, adventitious contamination and loss of material) associated with separation procedures, it is worth considering any other measurement that can distinguish between the components of mixtures of related organic compounds. We shall now discuss the possibilities of sensitised delayed fluorescence for this purpose. It will be recalled (Section 2 D 5) that the emission of sensitised delayed fluorescence can proceed by two mechanisms both of which require triplet-to-singlet energy transfer from donor to acceptor as the first stage, viz.

$$^3D + {}^1A \rightarrow {}^1D + {}^3A \tag{434}$$

followed by either triplet self-quenching

$$^3A + {}^3A \rightarrow {}^1A^* + {}^1A \tag{435}$$

or mixed triplet quenching:

$$^3A + {}^3D \rightarrow {}^1A^* + {}^1D \tag{436}$$

each of which results in population of the singlet level of the acceptor and hence the emission of its sensitised delayed fluorescence. The mixed-triplet mechanism predominates at low acceptor concentrations and is therefore of most interest in trace analysis. The equation for its quantum efficiency was derived in Section 2 D 5 and is reproduced here for convenience, viz.,

$$\theta_A'' = p_e p_{DA} I_a \phi_A \tau_A (k_c \phi_t^D \tau_D)^2 [A] \tag{437}$$

Note that for low acceptor concentrations, such that the donor triplet is quenched to only a small degree, τ_D is approximately constant. The efficiency of sensitised delayed fluorescence is then directly proportional to the concentration of acceptor [A] and independent of other acceptors provided that the latter also are present at low concentration. Let us substitute some typical values of the parameters in equation 437. Thus, let the probability factors p_e and p_{DA} be unity, $\tau_A = \tau_D = 5$ msec, k_c (ethanol) $= 0.57 \times 10^{10}$ litre mole^{-1} sec^{-1} at 22°C, $I_a = 5 \times 10^{-6}$ einstein litre^{-1} sec^{-1}, $\phi_t^D = 0.1$ and $\phi_A = 0.5$. With an acceptor concentration of 10^{-8}M, the efficiency of its sensitised delayed fluorescence, θ_A'', will be 10^{-3}, i.e. well within the capabilities of measurement by a good spectrophosphorimeter. Further, for concentrations of 10^{-8}M and below, θ_A'' will be approximately proportional to [A], and all acceptors present will exhibit their own sensitised delayed fluorescence spectrum independently of the presence of the others. Of course, one acceptor present at high concentration will quench the whole of the donor triplet and the sensitised delayed fluorescence of the other acceptors present at much lower concentration will not then appear. At -75°C in ethanol the situation is even more favourable because although k_c is reduced by a factor of about 20 (see Fig. 24), many triplet lifetimes increase by a

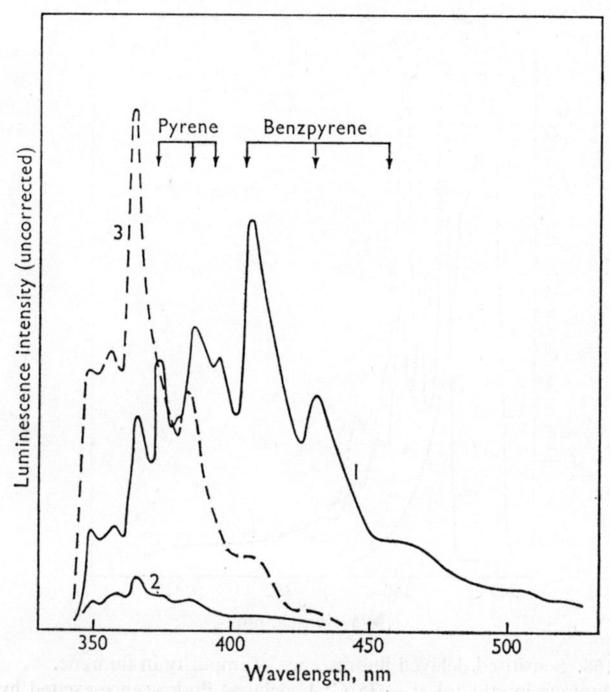

Fig. 167. Delayed fluorescence of benz(a)pyrene and pyrene sensitised by phenanthrene. Solution contained 10^{-3} M phenanthrene, 10^{-9} M benz(a)pyrene and 10^{-9} M pyrene in ethanol at 22°C; 1, delayed fluorescence excited by 313 nm at a rate of absorption of 10^{-6} einstein litre^{-1} sec^{-1}; 2, as for 1, but after 30 min irradiation; 3, prompt fluorescence at 600 times lower sensitivity (from Parker, Hatchard and Joyce[125]).

factor of 10–20 and θ_A'' for a given acceptor concentration is often some 10–20 times greater than at room temperature.

The observation of such sensitised delayed fluorescence has proved to be particularly valuable as a criterion of purity of aromatic hydrocarbons having relatively high triplet levels[111, 125]. The procedure simply involves the measurement of the delayed fluorescence spectrum of a deaerated and relatively concentrated solution of the specimen using a high intensity of exciting light absorbed by the host material. If the spectrum of delayed fluorescence differs from that of the prompt fluorescence the presence of one or more impurities should be suspected. Due account must be taken of the possibility of excited dimer emission, from the host material, e.g. from pyrene, but this consists of a broad structureless band whereas the sensitised delayed fluorescence spectra from aromatic hydrocarbon impu-

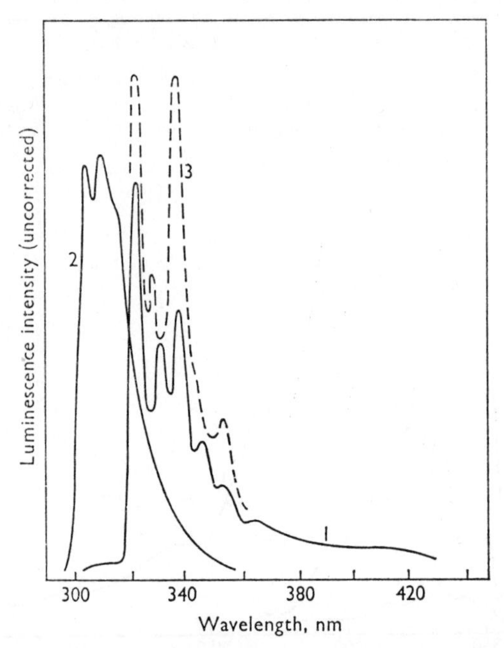

Fig. 168. Sensitised delayed fluorescence of impurity in fluorene.
5×10^{-5} M fluorene in ethanol at -75°C; 1, delayed fluorescence excited by 302 nm at a rate of light absorption of 0.9×10^{-6} einstein litre^{-1} sec^{-1}; 2, prompt fluorescence at 2000 times lower sensitivity; 3, prompt fluorescence of acenaphthene solution (from Parker, Hatchard and Joyce[125]).

rities nearly always show some vibrational structure. Examples of this test are shown in Figs. 167, 168 and 169. In the first[125], 1 p.p.m. of pyrene and benz(a)pyrene are detected in a phenanthrene specimen. The second shows the detection of an unsuspected impurity in a recrystallised and zone refined specimen of fluorene[125] and the third shows the enormous intensity of sensitised delayed fluorescence produced by an impurity in a twice recrystallised specimen of chrysene[104].

It should be noted that the process of sensitisation involves the transfer to the acceptor of a proportion of the energy of the exciting light that is large relative to the concentration of acceptor, and photo-decomposition, even at a small quantum efficiency, will result in the quite rapid disappearance of the acceptor from the irradiated solution. Thus, attempts to increase the sensitivity of the method by using extremely high intensities of exciting light may result in a considerable degree of photo-decomposition during the few

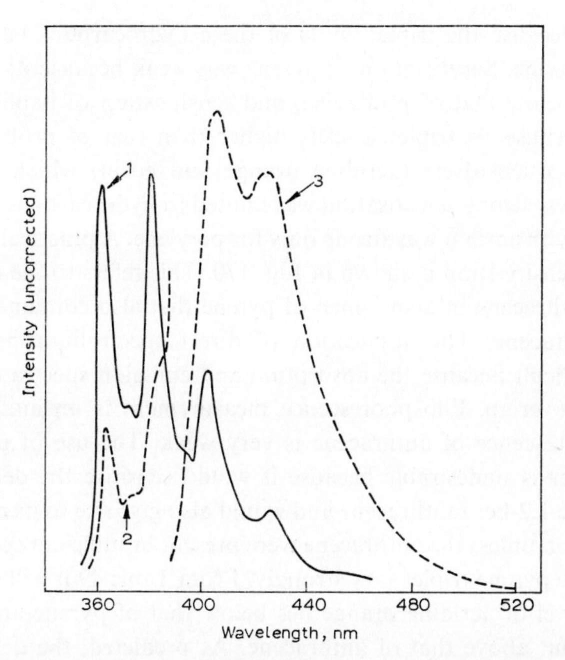

Fig. 169. Sensitised delayed fluorescence of impurity in chrysene. 2.6×10^{-5} M chrysene in ethanol at 21°C; 1, prompt fluorescence at sensitivity 3; 2, delayed fluorescence at sensitivity 400 showing weak band due to chrysene; 3, delayed fluorescence at sensitivity 25, showing intense bands due to impurity. Rate of light absorption was 2×10^{-6} einstein litre^{-1} sec^{-1} (from Parker and Joyce[104]).

minutes required to record the delayed fluorescence spectrum. This is illustrated in Fig. 167: on irradiation of the original solution (represented by curve 1) with a very high intensity, the peak emission of the benz(a)-pyrene was observed to decrease quite rapidly. The original intensity could at first be restored by shaking the cell to expose fresh solution to the light beam. After prolonged irradiation with shaking, almost the whole of the pyrene and benz(a)pyrene was decomposed as indicated by curve 2 of Fig. 167.

The measurement of sensitised anti-Stokes delayed fluorescence (see Section 2 D 6) has interesting possibilities. Thus Parker, Hatchard and Joyce[110, 125] have shown that by careful choice of a dyestuff as triplet-sensitiser it is possible to sensitise selectively the delayed fluorescence from certain components of a mixture. Some results with six aromatic hydrocarbons and three sensitisers are shown in Table 49. Thus, using proflavine, sensitisation of benzanthracene, anthracene, benz(a)pyrene and perylene was strong because the triplet levels of these hydrocarbons lie well below that of proflavine. Sensitisation of pyrene was weak because its triplet level is only just below that of proflavine, and sensitisation of naphthalene was negligible because its triplet level is higher than that of proflavine. With the other two sensitisers (acridine orange and eosin) which have lower triplet energies, strong sensitisation was limited to hydrocarbons lower down the list, and with eosin it was strong only for perylene. A practical application of selective sensitisation is shown in Fig. 170. This refers to the detection of 0.01% of anthracene in a specimen of pyrene that also contained 0.01% of 1,2-benzanthracene. The application of direct spectrofluorimetry to this system is difficult because the absorption and emission spectra of the three compounds overlap. Phosphorescence measurement is unsuitable because the phosphorescence of anthracene is very weak. The use of pyrene itself as a sensitiser is undesirable because it would sensitise the delayed fluorescence of the 1,2-benzanthracene and would also give rise to its own delayed dimer emission unless the anthracene were present in sufficient concentration to quench the pyrene triplet very strongly. From Table 49 it will be seen that the triplet level of acridine orange lies below that of pyrene and 1,2-benzanthracene but above that of anthracene. As predicted, the delayed fluorescence from the mixture sensitised by acridine orange to 436 nm showed

TABLE 49

SELECTIVE SENSITISATION OF DELAYED FLUORESCENCE AT 22°C

(From Parker, Hatchard and Joyce[125])

Compound	Triplet level (μm^{-1})	$\theta_A \times 10^4$, with sensitisation by: Proflavine hydrochloride	Acridine orange hydrochloride	Eosin
Naphthalene	2.12	< 0.02	< 0.02	< 0.02
Proflavine hydrochloride	1.71	—	—	—
Pyrene	1.69	0.2	< 0.02	< 0.02
1,2-Benzanthracene	1.65	0.9	< 0.02	< 0.02
Acridine orange hydrochloride	1.60	—	—	—
Anthracene	1.48	2	3	0.02
3,4-Benzpyrene	1.47	6	5	0.2
Eosin	1.42	—	—	—
Perylene	1.26	0.8	2	5

Note

Acceptor concentrations were 10^{-6} M. Rates of absorption of light by the donors were 6×10^{-6}, 5.5×10^{-6} and 11×10^{-6} einstein per litre per second at 436, 436 and 546 nm for proflavine hydrochloride, acridine orange hydrochloride and eosin, respectively.

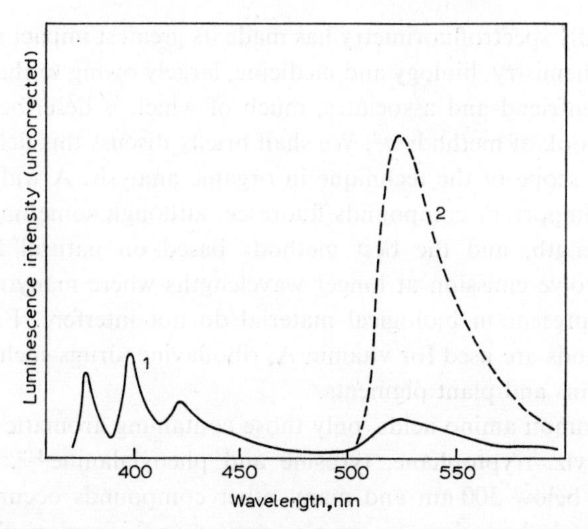

Fig. 170. Detection of anthracene in impure pyrene by selective sensitisation of delayed fluorescence.
Solution contained 10^{-2} M pyrene, 10^{-6} M 1,2-benzanthracene and 10^{-6} M anthracene with 1.7×10^{-5} M acridine orange hydrochloride as sensitiser at 22°C; 1, delayed fluorescence excited by 436 nm at a rate of absorption of 6×10^{-6} einstein litre^{-1} sec^{-1}; 2, prompt fluorescence at 4000 times lower sensitivity (from Parker, Hatchard and Joyce[125]).

bands due to anthracene, but none due to pyrene or 1,2-benzanthracene.

Like all analytical methods, selective sensitisation has its limitations. In particular a pure sensitised delayed fluorescence spectrum can only be obtained from that component of a mixture having the lowest triplet energy. Sensitisation at higher triplet energies will give mixed spectra due to several components.

5. Other Methods Utilising Prompt Fluorescence in Fluid Solution

This group includes by far the largest number of photoluminescence methods so far published and covers such fields as biology and medicine, food and drugs, plant biochemistry, agriculture etc., and natural products such as petroleum, as well as the fields of coal-tar products and air pollution, of which examples have been given in the previous sections. We shall have space to describe only a few methods that exemplify particularly the problems involved in the determination of trace amounts of complex substances that, in general, give rise to relatively diffuse spectra.

Undoubtedly spectrofluorimetry has made its greatest impact so far in the fields of biochemistry, biology and medicine, largely owing to the pioneering work of Udenfriend and associates, much of which is described in an excellent text-book of methods[342]. We shall briefly discuss this field mainly to illustrate the scope of the technique in organic analysis. A wide variety of biologically important compounds fluoresce, although sometimes at rather short wavelength, and the best methods based on natural fluorescence generally involve emission at longer wavelengths where many of the other compounds present in biological material do not interfere. For example, natural methods are used for vitamin A, riboflavine, drugs such as quinine, and porphyrins and plant pigments.

Of the common amino acids, only those containing aromatic systems are fluorescent, viz. tryptophane, tyrosine and phenylalanine[343]. Even these absorb only below 300 nm and many other compounds occurring in, for example, protein hydrolysates, are also excited in this region. Waalkes and Udenfriend[344] therefore developed a chemical method for tyrosine (reaction with α-nitroso-β-naphthol in the presence of nitric and nitrous acids) giving a fluorescent product that could be excited in the visible region at 460 nm. The method could then be applied for example to the determination of tyrosine in plasma or tissue in a comparatively simple procedure that did not require complete separation of the tyrosine. This principle of shifting the fluorescence characteristics to longer wavelengths is very frequently applied in biochemical work to avoid interference by the many other substances present, and to provide more specific identification.

A considerable amount of chemical ingenuity has gone into the development of some methods. An attractive example is the application of the trihydroxyindole reaction (a ring-closure reaction) to the simultaneous determination of epinephrine and norepinephrine[342]. The reaction involves the following stages:

oxidation
pH 6.5 or 3.5

OH^-

Trihydroxy-indole (fluorescent)

Epinephrine R = CH_3
Norepinephrine R = H

Both amines react at a pH value of 6.5 but only epinephrine at a pH value of 3.5. Many other ring-closure reactions are used, e.g. the reaction of hexoses such as glucose with 5-hydroxy-1-tetralone in sulphuric acid to yield the fluorescent benzonaphthenedione[345]:

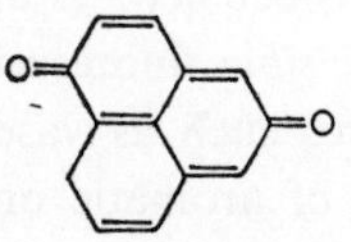

or the reaction of polycarboxylic acids with resorcinol, e.g. malic acid gives umbelliferone-4-carboxylic acid:

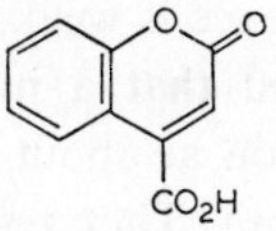

CO_2H

A great many condensation reactions in sulphuric acid, with or without the addition of another reagent, give rise to fluorescent products—frequently of unknown structure—many of which have been applied for analytical purposes. Enough has been said to indicate that absence of natural fluorescence should certainly not exclude fluorimetric methods from consideration when devising new analytical procedures for a particular problem in organic analysis.

We shall quote one more example of a chemical method from another field, viz. analysis of air pollutants, for which Sawicki and co-workers have devised comprehensive systems of analysis based on chromatography coupled with photoluminescence measurements (see for example references 346 and 347). One compound of interest in these studies is 1,4-cyclohexanedione. This was reacted with o-phthalaldehyde in sulphuric acid to give the di-cation of pentacenequinone:

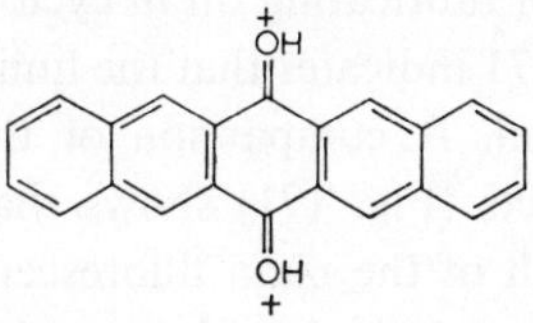

$\overset{+}{O}H$

$\underset{+}{O}H$

This emitted fluorescence at 650 nm with a characteristic excitation spectrum having peaks at 350, 575 and 625 nm, and inflections at 382 and 540 nm. The detection limit was 10^{-6}g[348].

Let us now turn to a quite different field. Apart from its application to the determination of specific substances, spectrofluorimetry can be used as an empirical or "fingerprint" method for the identification of complex mixtures of natural or artificial origin, or for comparing different fractions from a natural product. We shall consider here its application to the examination of petroleum fractions. The *visible* fluorescence of petroleum oils and of complex hydrocarbon mixtures such as vaseline, wax etc. has long been known, and is due to traces of aromatic or heterocyclic compounds absorbing in the near ultra-violet or violet regions. Von Eisenbrand[349, 350] has investigated such emission from liquid paraffin and from yellow vaseline and ceresin excited by mercury light at 366 nm. Per unit concentration of total hydrocarbon this fluorescence is weak, and relatively uncharacteristic. Parker and Barnes[351] observed that a much more intense fluorescence situated in the ultra-violet region at about 360 nm was excited by light of wavelengths shorter than 300 nm. They tested a variety of different types of motor-car engine oil and found substantially the same fluorescence emission and excitation spectra from all. They concluded that the emission was not due to oil additives but to the aromatics present in this particular fraction of the petroleum itself. They applied the method to the determination of oil mist in the low pressure air from compressors. About 50 litres of air was passed through small discs of filter paper. The filter was eluted with a small volume of cyclohexane and the fluorescence intensity measured. To avoid errors due to variation in the aromatic content of lube oils they used a sample of oil from the sump of the compressor to prepare standard solutions for comparison. The sensitivity was limited by the blank arising from the filter paper and cyclohexane. This corresponded to about 0.6 μg of oil per ml of cyclohexane or 0.06 μg of oil per litre of air.

By using a high intensity of exciting light at 250 nm and a large grating monochromator to analyse the fluorescence emission, it is possible to detect very low concentrations of lubricating oil in cyclohexane. Thus comparison of curves 2 and 5 in Fig. 171 indicates that the limit is here set by the solvent blank at about 0.02 μg/ml. A comparison of the spectra obtained from various petroleum products (Fig. 171) shows that the lighter the fraction the shorter the wavelength of the main fluorescence emission. Thus motor spirit appears to contain compounds of benzenoid type. In diesel fuel there is a large proportion of emission in the naphthalene region, while in lubricating oil the maximum emission appears in the phenanthrene region at about 360 nm. Crude oil contains in addition more complex compounds

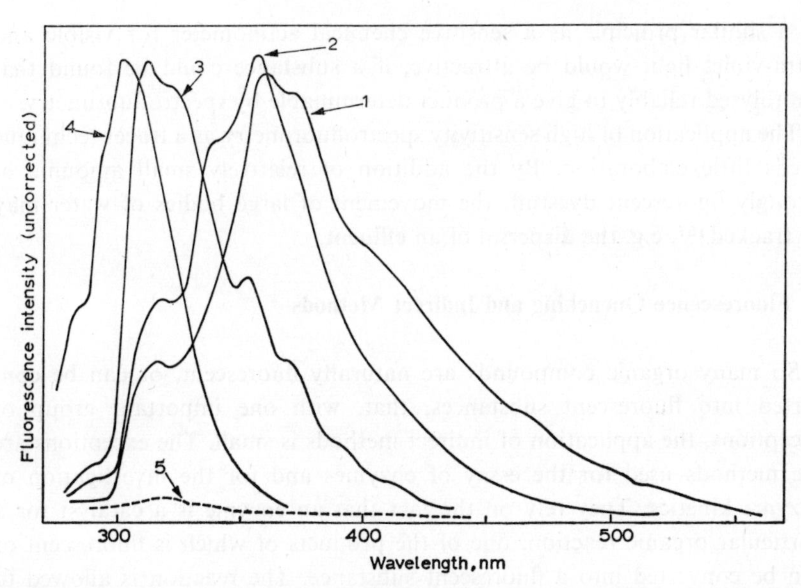

Fig. 171. Fluorescence of cyclohexane solutions of petroleum products at room temperature.
Excitation at 250 nm isolated by grating monochromator with OX7 and chlorine gas filters; band width of analysing monochromator was 3.3 nm; 1, crude oil, 2 μg per ml, × 2.5; 2, automobile engine lubricant, 20 μg per ml, × 1; 3, automobile diesel fuel, 20 μg per ml, × 1; 4, automobile petrol, 20 μg per ml, × 2; 5, cyclohexane alone, × 15.

emitting at longer wavelengths. There is little doubt that photoluminescence would be a powerful tool for the investigation of the aromatic and heterocyclic compounds present in crude oils, especially after preliminary separation by the various forms of chromatography. Low temperature fluorescence and phosphorescence as described in Section 5 E 7 also provide characteristic tests.

A novel application of spectrofluorimetry in fluid solution has been described by Armstrong and Grant[352] who devised a highly sensitive chemical dosimeter for ionising radiation based on the radiolysis of aqueous solutions of calcium benzoate. They determined the salicylic acid produced in the radiolysis by exciting the irradiated solution at 290 nm and measuring the fluorescence at 400 nm. Diphenyl is also a radiation product but does not interfere. They found the fluorescence intensity to be a linear function of the concentration of salicylic acid which was itself proportional to the dose of X- or gamma-radiation over the range 5 to 100 rads. The application of a similar principle as a sensitive chemical actinometer for visible and ultra-violet light would be attractive, if a substance could be found that photolysed reliably to give a product determinable by spectrofluorimetry.

The application of high sensitivity spectrofluorimetry as a tracer technique needs little elaboration. By the addition of relatively small amounts of strongly fluorescent dyestuff, the movement of large bodies of water may be tracked[157], e.g. the dispersal of an effluent.

6. Fluorescence Quenching and Indirect Methods

So many organic compounds are naturally fluorescent, or can be converted into fluorescent substances, that, with one important group of exceptions, the application of indirect methods is small. The exceptions are the methods used for the essay of enzymes and for the investigation of enzyme kinetics. They rely on the fact that an enzyme is a catalyst for a particular organic reaction, one of the products of which is fluorescent or can be converted into a fluorescent substance. The reaction is allowed to proceed under appropriate conditions and the rate of formation of product is a measure of the concentration of enzyme. The sensitivity of spectrofluorimetric methods makes possible the determination of exceedingly low concentrations of enzymes, or conversely of the substrates on which they act[342].

The deliberate application of fluorescence quenching is rarely employed in organic analysis but one can conceive situations where it might be used to increase selectivity. For example saturation of a cyclohexane solution of pyrene with air quenches the fluorescence by a factor of about 20 while that of anthracene and many other aromatic hydrocarbons is reduced by only a small amount. Thus measurement of the aerated solution would minimise the interference by the pyrene with the measurement of the anthracene: deaeration could then be employed to enhance the pyrene emission for its measurement. The use of quenching for the determination of molecular oxygen is discussed in Section 5 G 3.

7. Other Methods Based on Fluorescence and Phosphorescence in Glassy Media

The extreme sensitivity and discrimination that can be obtained by the use of phosphorescence at 77°K in the analysis of some aromatic hydrocarbons and heterocyclics was discussed in some detail in Section 5 E 3, with special reference to the work of Zander[337] and Parker[210]. We shall now describe its application to other types of compound. To a large extent spectrophosphorimetry and spectrofluorimetry are complementary techniques since absence of fluorescence in an absorbing molecule generally indicates a high triplet formation efficiency and, provided that radiationless processes at 77°K are not too rapid, this means high phosphorescence efficiency under these conditions. As an incidental outcome of their investigation of the triplet state in rigid media Lewis and Kasha[30] suggested that the phosphorescence of organic molecules could be used for their identification. This suggestion was taken up by Keirs and co-workers[353] who pointed out that, as the phosphorescence spectra from over 200 compounds in rigid media at low temperature had at that time already been studied by the photochemists, the field was a fertile one for exploitation by the analytical chemists. They devised tentative schemes for the analysis of mixtures, e.g. of benzaldehyde, benzophenone and 4-nitrodiphenyl in ether–pentane–ethanol (5:5:2) at 77°K, using filtered or unfiltered light from a mercury lamp for excitation, a Becquerel-type phosphoroscope, and a spectrograph for recording the emission spectra. They also suggested that phosphorescence lifetimes, or careful control of the exciting wavelength, might be used as a means of discriminating between compounds having similar phosphorescence emission spectra. Freed and Salmre[354] used phosphorescence spectra to distinguish between some indole derivatives of importance in the study of the central nervous system. They used a methanol–ethanol mixture (9:1) at 77°K and found that the sensitivity was limited by the phosphorescence of impurities in the solvents and of the fused silica containers. Papers by Holzbecher[355] and by Parker and Hatchard[203] discussed the possibilities of phosphorescence measurement as an analytical technique. In the latter, Parker and Hatchard showed that it was possible to obtain high sensitivities with monochromatic exciting light isolated from a monochromator (see for example Fig. 172) and demonstrated the possibility of recording corrected phosphorescence excitation spectra (see Fig. 173). They recommended the use of separately-driven choppers (see Section 3 N 1) so that the spectra of (phosphorescence + fluorescence), and phosphorescence only, could be conveniently measured in succession.

Winefordner and co-workers have described a variety of methods for the determination of compounds of pharmacological interest. Winefordner and Latz[356] determined aspirin in 0.4 ml of blood serum or plasma by chloroform extraction, evaporation and measurement of the phosphorescence in ether–pentane–ethanol (5:5:2) at 77°K using "white" light from a xenon arc for excitation, and measurement of the phosphorescence at 410 nm. They were able to determine 10–1000 μg of aspirin per ml of serum or plasma without interference—the total blank corresponded to about 5 μg of aspirin per ml of serum. Winefordner and Tin[357] gave excitation and emission wavelengths for 22 compounds in ethanol at 77°K, with limits of detection ranging from 1.2 μg/ml for lidocaine to 0.005 μg/ml for benzoic acid. They developed methods for the determination of procaine, phenobarbital, cocaine and chlorpromazine in blood serum, and cocaine and atropine in urine[358]. These were based on extraction with chloroform or ether, evaporation and measurement in ether–pentane–ethanol. They pointed out[357] that while analysis of some mixtures could be made directly by appropriate choice of excitation and emission wavelengths, others required separation. Winefordner and Moye[359] described a method for the determination of nicotine, nornicotine and anabasine in tobacco, based on

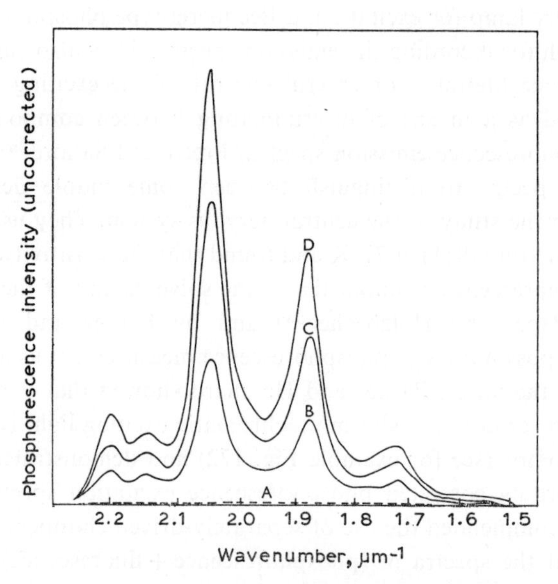

Fig. 172. Phosphorescence emission spectra of dilute solutions of anthraquinone in ether–pentane–ethanol at 77°K.
Excitation at 3.19 μm^{-1} (313 nm); quartz analysing monochromator with 9558Q photomultiplier; half-band width was 0.024 μm^{-1} at 2.5 μm^{-1}; A, solvent only; B, 0.03 μg per ml; C, 0.06 μg per ml; D, 0.09 μg per ml.

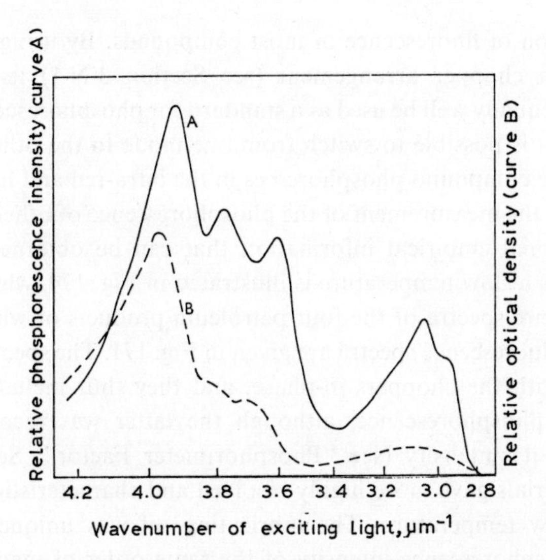

Fig. 173. Phosphorescence excitation spectrum of anthraquinone in ether–pentane–ethanol at 77°K.
A, excitation spectrum; B, absorption spectrum in ethanol at 295°K (from Parker and Hatchard[203]).

chloroform extraction, from alkaline solution, separation by thin-layer chromatography and phosphorimetric measurement at 77°K in ethanol containing 2% H_2SO_4. All three compounds gave similar phosphorescence excitation (270 nm) and emission (385 nm) spectra.

To avoid the errors that can arise either by slight variation in the position of the cuvette or by light scatter in the liquid nitrogen, or in the specimen itself if the solvent cracks, Parker and Hatchard[90] have found it advantageous to add an internal standard that fluoresces or phosphoresces in a region that does not overlap that of the compounds to be determined. A very suitable standard for use in solvent mixtures containing ethanol is the hydrochloride of the oxidised form of diaminophenoxazine:

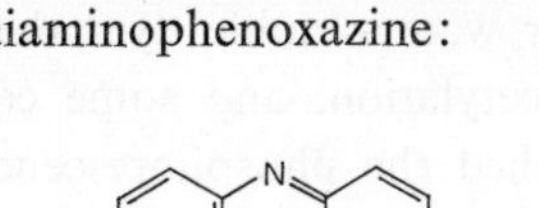

$$H_2N\text{–}[\text{phenoxazine}]\text{=}\overset{+}{N}H_2$$

This compound is the oxygen analogue of thionine—it has a greater fluorescence efficiency than the latter, with maximum at 600 nm, i.e. well removed from the region of fluorescence of most compounds. By using the synchronously driven chopper arrangement (see Section 3 N 1) its fluorescence intensity can equally well be used as a standard for phosphorescence measurements, since it is possible to switch from one mode to the other at the flick of a lever. The compound phosphoresces in the infra-red and hence does not interfere with the measurement of the phosphorescence of other compounds.

The additional empirical information that can be obtained by making measurements at low temperature is illustrated in Fig. 174, which shows the low temperature spectra of the four petroleum products of which the room temperature fluorescence spectra are given in Fig. 171. The spectra in Fig. 174 were taken with the choppers in-phase, and they thus include both fluorescence and phosphorescence, although the latter was recorded at only one third of its intensity (see "Phosphorimeter Factor", Section 3 N 2). All four materials gave more highly resolved and characteristic fluorescence spectra at low temperature. The lubricating oil was unique in having a recorded phosphorescence intensity of the same order of magnitude as the fluorescence. The phosphorescence spectra were measured free of fluorescence by running the choppers out-of-phase. The phosphorescence spectra all showed some structure and thus provided additional criteria for identification purposes. The wavelength regions in which the phosphorescence emissions were observed are indicated in the legend to Fig. 174.

Although nearly all analytical applications of phosphorescence have been made at low temperature, it is worth considering measurements in rigid transparent organic glasses (i.e. "plastics") at room temperature. For example Oster and co-workers[360] found that many aromatic hydrocarbons and dyestuffs exhibited phosphorescence under these conditions. They used three methods to introduce the solute: casting from a solvent; melting with the solute; polymerisation of the monomer in which the solute was dissolved. For aromatic hydrocarbons they used polystyrene, polymethylmethacrylate, polyvinylacetate, cellulose acetate, ethyl or methyl cellulose and a polycarbonate. For water-soluble dyes they used polyvinylalcohols of different degrees of acetylation, and some cellulose derivatives. They found that oxygen quenched the phosphorescence and they made use of the quenching to determine the rate of diffusion of oxygen into the plastic. The lifetime of the phosphorescence generally depended on the nature of the plastic as well as on the temperature and they concluded that triplet quenching was influenced not only by micro-Brownian movement of the polymer segments, but also by specific interaction with the polymeric matrix.

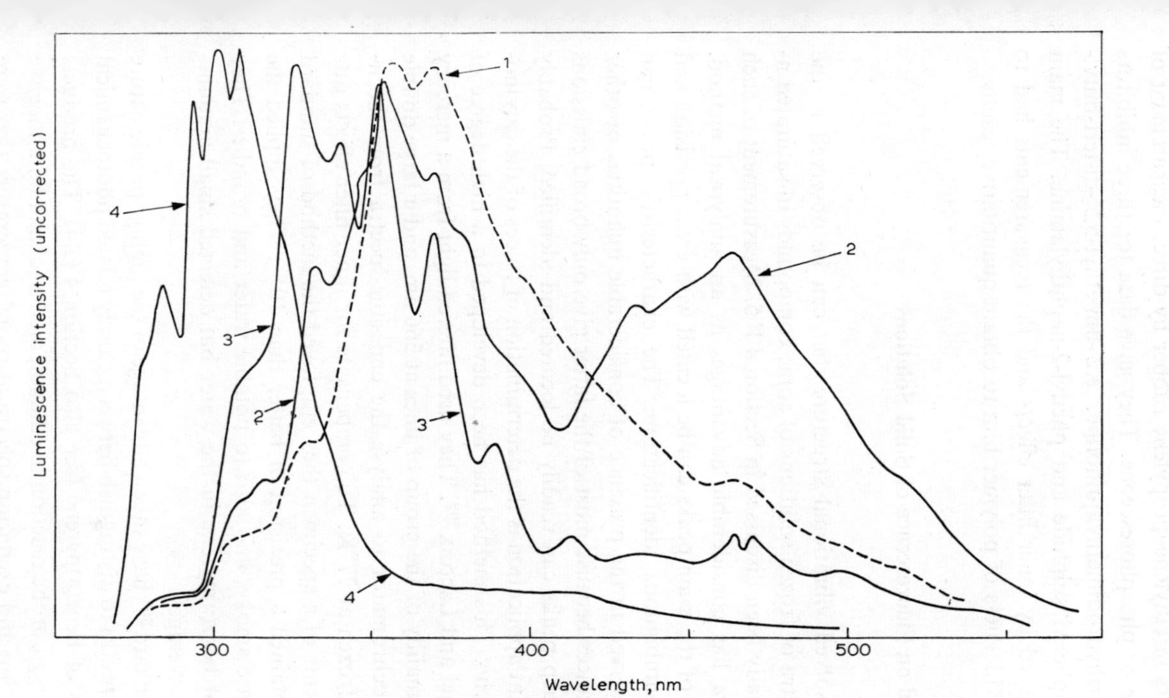

Fig. 174. Fluorescence and phosphorescence of petroleum products in 3:1:1 ether–pentane–ethanol at 77°K. Excitation at 250 nm isolated by grating monochromator with OX7 and chlorine gas filters; band width of analysing monochromator was 3.3 nm; 1, crude oil (curve shows mainly fluorescence; phosphorescence appeared between 420 and 570 nm with maximum intensity about one quarter that of fluorescence); 2, automobile engine lubricant (part of curve above 420 nm is mainly phosphorescence; the true relative intensity is 3 times that indicated because of the phosphorimeter factor); 3, automobile diesel fuel (part of curve above 420 nm is mainly phosphorescence; see remarks on curve 2); 4, automobile petrol (curve shows mainly fluorescence; phosphorescence appeared between 360 and 480 nm with maximum intensity about one seventh that of fluorescence).

Drushel and Sommers[361] have described methods for the determination of some inhibitors in ethylene–propylene rubber by direct measurement of their fluorescence or phosphorescence. They gave data for three inhibitors viz. polymeric trihydrodimethylquinoline, 2,2′-dimethyl-5,5′-di-tert-butyl-4,4′-dihydroxy-diphenyl sulphide and phenyl-2-naphthylamine. The main difficulty was caused by inner filter effects and the measurements had to be corrected for thickness of polymer film to obtain quantitative results.

8. Methods Based on Fluorescence of Solid Solutions

The highly resolved vibrational structure that can be observed in the fluorescence spectra of frozen solutions of some compounds in saturated n-paraffins has already been discussed in Section 4 F 5. Measurement of such Shpol′skii spectra has considerable advantages as an analytical method. The wavelengths of the main peaks can be located with great precision and thus provide unambiguous identification. The characteristic peaks can frequently be observed in the presence of considerable quantities of other fluorescent substances because most of the latter give only broad emissions on which the sharp peaks can readily be located and identified. Probably the most important application is the determination of traces of the carcinogenic benz(a)pyrene. This method has been developed to a high degree of sensitivity by Muel and Lacroix[292]. They used filtered light from a mercury lamp, consisting mainly of the group of lines at 366 nm, and a large double glass prism monochromator to analyse the emission spectra from the n-octane solutions frozen at 77°K. To compensate for inner filter effects due to other components of a specimen they employed the method of standard additions and obtained a precision of better than 10%. They applied the method to cigarette smoke, and also to potable water and brandies. They found no trace of benz(a)pyrene in the water but detected small amounts in the alcoholic beverages.

Parker and Hatchard[307] have taken advantage of the highly specific nature of the Shpol′skii spectra to distinguish between closely related photochemical reaction products of benz(a)pyrene (see also Section 4 G 4). The products obtained by carrying out the reaction in ethanol, methanol and butanol gave fluorescence emission and excitation spectra at room temperature that were indistinguishable from one another. The Shpol′skii spectra of the products were similar in general form to that of benz(a)pyrene itself (compare curves

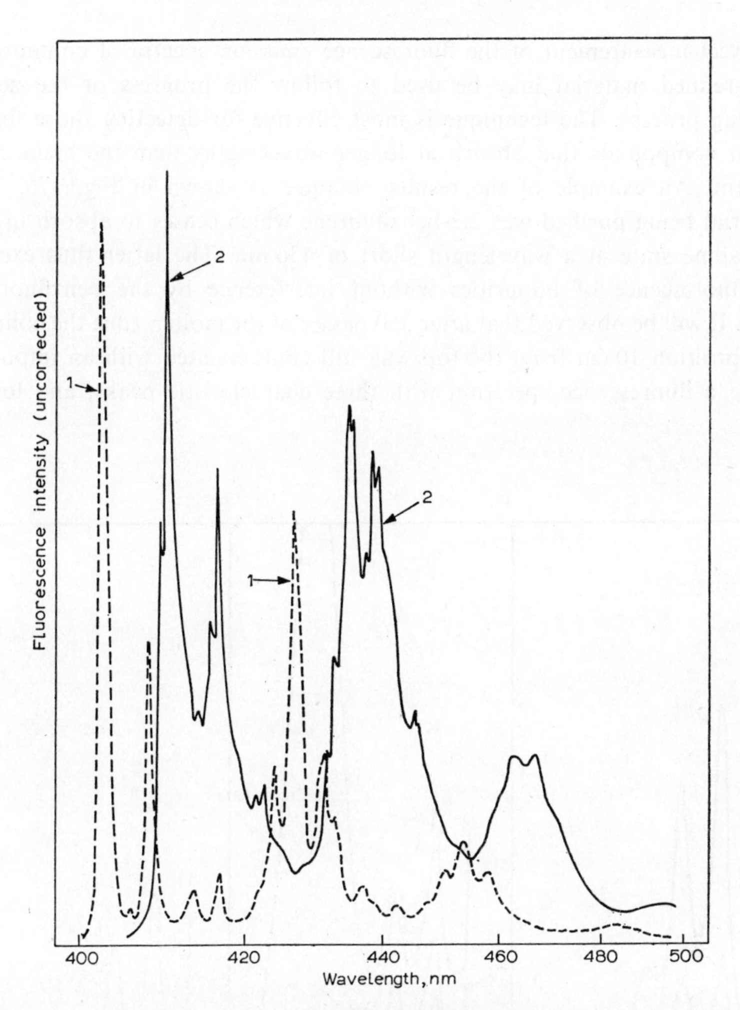

Fig. 175. Application of Shpol'skii spectra.
Prompt fluorescence in 75% cyclohexane–25% n-octane at 77°K excited by 366 nm; band width 0.2 nm at 400 nm; 1, benz(a)pyrene; 2, photo-reaction product from ethanol (from Parker and Hatchard[307]).

1 and 2 in Fig. 175) but shifted to longer wavelengths. The spectra of the three products also showed differences in the wavelengths and relative intensities of the sharp peaks. It would be difficult to obtain such characteristic fingerprints by other methods when less than 1 microgram of specimen is available.

Direct measurement of the fluorescence emission spectra of columns of zone-refined material may be used to follow the progress of the zone-refining process. The technique is most effective for detecting those fluorescent compounds that absorb at longer wavelengths than the main constituent. An example of the results obtained is shown in Fig. 176. The material being purified was 2,3-benzfluorene which ceases to absorb in the crystalline state at a wavelength short of 436 nm. The latter thus excites the fluorescence of impurities without interference by the benzfluorene itself. It will be observed that after 350 passes of the molten zone the column at a position 10 cm from the top was still contaminated with an impurity giving a fluorescence spectrum with three characteristic peaks, and lower

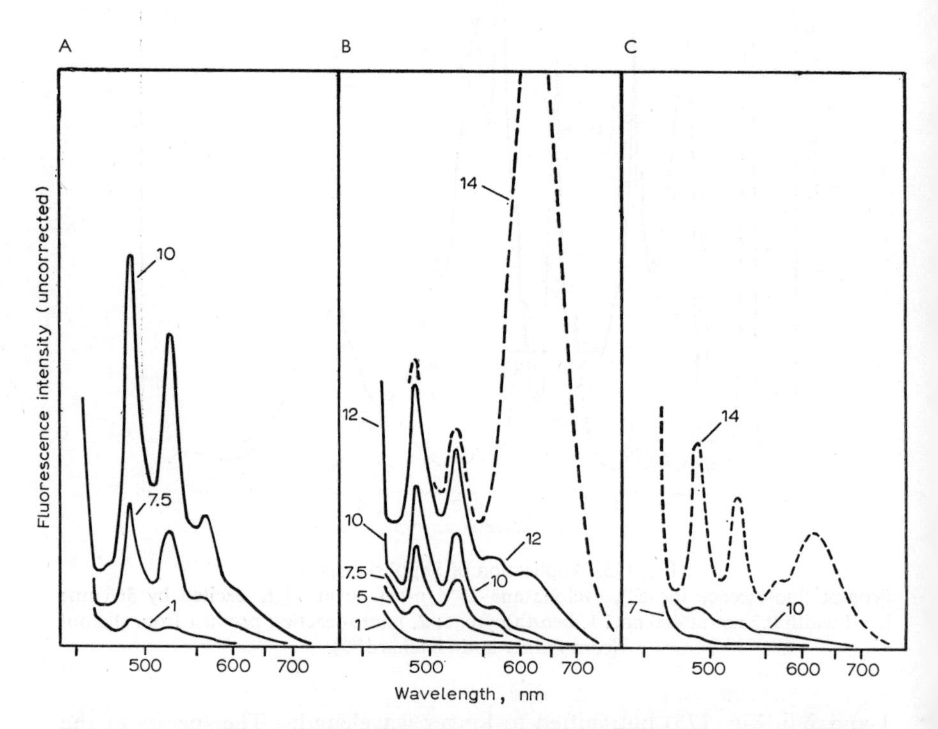

Fig. 176. Zone-refining of 2,3-benzfluorene monitored by fluorescence measurement. Excitation at 436 nm isolated by double prism monochromator; double silica prism analysing monochromator; frontal illumination with equipment shown in Fig. 88; distance from top of column in cm is indicated against each curve; total length of column 17 cm; A, after 350 passes of molten zone; B, after 530 passes at 3 times greater sensitivity; C, after 730 passes at same sensitivity as B.

sections of the column (not shown) were very heavily contaminated. After 530 zone passes a fair degree of purification had occurred and after 730 zone passes the top 10 cm were substantially free of fluorescent impurity. The spectra shown in Fig. 176 represent very low levels of fluorescence intensity and to avoid interference from scattered exciting light, double monochromators were used to isolate the exciting light and to analyse the fluorescence. Although the impurity responsible for the three-peaked spectrum was coloured yellow it could not be detected by its colour in the parts of the column represented by the curves in the left hand section of Fig. 176.

9. Fluorescence and Phosphorescence of Adsorbates

Sawicki and Stanley[362] have described a series of fluorescence spot tests on filter paper for glyoxal, pyruvaldehyde, salicylaldehyde and other aromatic aldehydes, using salicylal-hydrazone or 2-hydroxy-1-naphthalhydrazone as reagents. The visual detection limits varied from 0.01 μg for pyruvaldehyde to 10 μg for salicylaldehyde. Sawicki and Johnson[363] reported visual spot tests for a wide range of compounds likely to be met in air-pollution studies, including polycyclic aromatic hydrocarbons and their amino, nitro and aldehyde derivatives; acridine, benzo- and dibenzo-acridines; benzo- and dibenzo-carbazoles; 4-aminoazobenzene derivatives; phenols, aza heterocyclic compounds and pesticides such as phenothiazine or parathion. The tests were carried out on glass fibre or cellulose paper, on thin-layer alumina chromatograms or on cellulose acetate chromatograms. The visual fluorescence or phosphorescence from the spots was observed at liquid nitrogen temperature. Characteristic colours were obtained with dry spots, or spots wet with solvent. Different colours were obtained after adding a drop of sulphuric or trifluoracetic acids, or 29% methanolic tetraethyl ammonium hydroxide. They tabulated the results obtained with over 80 compounds and suggested that measurement of emission spectra from such spots would be worth while. In another paper Bender, Sawicki and Wilson[364] have described the fluorescent detection and spectrofluorimetric characterisation of carbazoles separated by thin-layer chromatography, and Sawicki, Elbert and Stanley[365] have used selective quenching agents to distinguish between fluorescent spots on a thin-layer chromatogram.

F. LUMINESCENCE OF INORGANIC AND INORGANIC–ORGANIC COMPOUNDS IN RELATION TO STRUCTURE

1. Inorganic Compounds

Of those metallic ions that, in aqueous solution, absorb in the quartz ultra-violet region, most do not give rise to luminescence. Light absorption may cause photoreaction by charge transfer reactions such as:

$$FeCl^{2+} \xrightarrow{h\nu} Fe^{2+} + Cl\cdot \qquad (438)$$

The lack of fluorescence from some anions such as nitrate is also due to photo-decomposition. In the coloured complexes of some transition elements the absorbed energy is degraded via lower excited states arising from the presence of partly filled *d* orbitals. In the rare earth series it is the $4f$ shell that is partly filled, and electrons occupying $4f$ levels can be raised by light absorption into unoccupied $4f$ levels. The $4f$ levels are well shielded from external influences by the outermost electrons that occupy the $5s$ and $5p$ orbitals in the trivalent ions. As a result, radiationless deactivation is small and in crystal phosphors all rare earths that contain between 2 and 12 *f*-electrons give rise to line emission, viz., Pr, Nd, Sm, Eu, Gd, Tb, Dy, Ho, Er, Tm. In fluid solution line emission is said to be restricted to the five ions in the middle of the series, viz., samarium, europium, gadolinium, terbium and dysprosium[126]. The rare earth absorption spectra are complex and emission may take place from more than one energy level. The simple salts (e.g. chlorides, sulphates) of the five ions that luminesce in solution show line absorption which in aqueous medium is weak and difficult to excite effectively at low concentration. Terbium chloride may be excited by the pressure-broadened mercury lines at 366 nm and concentrations down to about 10^{-5}M can be detected with a sensitive spectrofluorimeter. Samarium, europium and dysprosium chlorides are less strongly excited by this group of wavelengths (see Fig. 177 and Table 52 of Section 5 G). With shorter wavelength excitation solutions of gadolinium chloride give a line emission at 310 nm (Fig. 177). The absorption and luminescence intensity may be increased by the formation of inorganic complexes. Thus Taketatsu and co-workers[366] have measured the emission spectra of terbium and europium carbonates in aqueous potassium carbonate solution. When

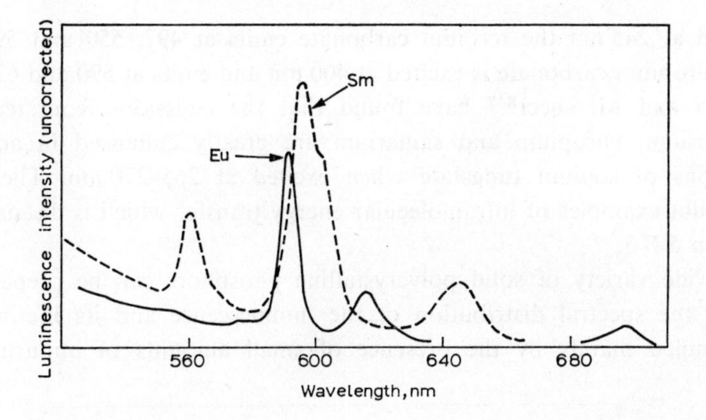

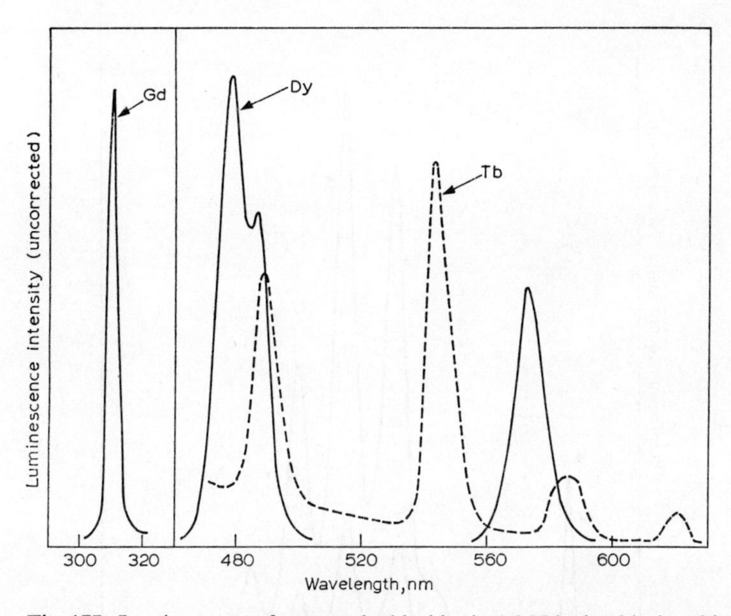

Fig. 177. Luminescence of rare earth chlorides in 1.0 N hydrochloric acid. Excitation at 250 nm (Gd) and 366 nm (Sm, Eu, Tb, Dy) using 200 W extra-high pressure mercury lamp and grating monochromator at band width 9.6 nm; band width of analysing monochromator was 3.3 nm; concentrations and relative sensitivities at which curves were measured were: Gd, 4×10^{-2} M at 4.5; Sm, 4×10^{-2} M at 20; Eu, 4×10^{-2} M at 2; Tb, 10^{-3} M at 4; Dy, 4×10^{-2} M at 1; Raman band at 418 nm was about 3 times as intense as the Tb line at 479 nm.

excited at 245 nm the terbium carbonate emits at 497, 550 and 593 nm. The europium carbonate is excited at 400 nm and emits at 590 and 620 nm. Alberti and Massucci[367] have found that the emissions from terbium, dysprosium, europium and samarium are greatly enhanced in aqueous solutions of sodium tungstate when excited at 265–270 nm. These are no doubt examples of intramolecular energy transfer which is discussed in Section 5 F 3.

A wide variety of solid polycrystalline phosphors can be prepared in which the spectral distribution of the luminescence and its lifetime are determined mainly by the presence of small amounts of impurities or

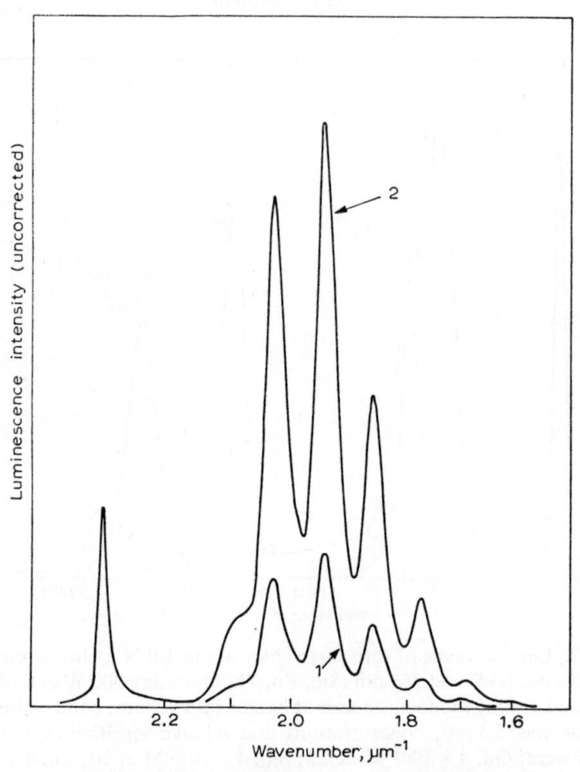

Fig. 178. Effect of sulphuric acid concentration on the intensity of the luminescence of uranyl ion.
Excitation at 436 nm isolated from 200 W extra-high pressure mercury lamp by quartz prism double monochromator; double silica prism analysing monochromator with band width 0.008 μm^{-1} at 2.0 μm^{-1}; 1, 2, uranyl sulphate (5×10^{-3} M) at 20°C in 0.1 N and 5 N sulphuric acid.

"activators". Polycrystalline base materials include zinc, cadmium, calcium and strontium sulphides, potassium chloride, zinc selenide, calcium and magnesium tungstates, beryllium, zinc and cadmium silicates, and many others. Impurity activators include copper, silver, manganese, antimony, thallium, lead, rare earths, bismuth and uranium. For details of the methods of preparation of such phosphors and the theories of their luminescence the reader is referred elsewhere[126, 368, 369]. Reference will be made later to their analytical application.

Apart from the rare earths, photoluminescence in fluid solution has been observed from the inorganic ions of very few elements. The photoluminescence of uranium in solution is restricted to those compounds containing the uranyl ion, UO_2^{2+}. All uranyl salts have absorption and emission spectra of the same type[126], although they differ in detail. Similar spectra are observed in solution, in the pure crystalline state and in solid solutions in, for example, sodium fluoride. The spectra show a vibrational band structure corresponding to the vibrational modes of the O–U–O species (see Fig. 178). The electronic transition is partly forbidden and in the crystalline salts the luminescence has a lifetime of about 10^{-4} sec, but it nevertheless has a high efficiency in many salts. The luminescence efficiency of uranyl salts depends on the nature and concentration of the anion, and on the solvent. It seems that the emission originates in a variety of uranyl complexes. The effect is illustrated in Fig. 178 which shows the increased luminescence intensity obtained at high concentrations of sulphuric acid.

Thallous ion emits fluorescence in sodium chloride solution[370] and cerium and tin in sulphuric acid[371, 372].

2. Organo-Metallic Complexes

The simplest type of fluorescent organo-metallic complex is that in which an ionisable fluorescent compound combines with a metallic salt to form a complex ion. Thus the orange-fluorescent rhodamine B combines with tervalent thallium chloride or bromide to form an orange-red fluorescing complex which is soluble in benzene[373]. The thallium is considered to be combined as the "salt" of rhodamine B cation with $TlBr_4^-$.

A large proportion of the fluorescent organo-metallic compounds are to be found among the metal chelates. These consist frequently of a single metal ion joined to one or more molecules of a polydentate organic compound to give a rigid molecule containing several fused ring systems in which the metal atom is incorporated. Since the structure of such complexes is largely organic in nature, we may expect their photoluminescence to obey rules similar to those outlined in Section 1 B 6. Provided that the metal atom does not contain lower unfilled atomic orbitals (i.e. provided that it is not a transition element or a rare earth) this does indeed seem to be the case—the metal atom then behaves like an inert or "saturated organic atom" and forms a part of the overall ring system of the molecule. Sometimes the organic reagent itself is non-fluorescent. Many reagents have flexible molecules or contain groups capable to giving π^*–n states and the development of fluorescence on chelation with a metal atom is no doubt then due to the formation of a rigid molecule and/or to lowering of the π^*–π energy level relative to the π^*–n, or possibly to the activation of an excited state of charge transfer type. If all n-electrons in the reagent are coordinated with the metal atom in the chelate, the latter will not have a π^*–n excited state, and this may be the reason for the formation of fluorescent chelates from some non-fluorescent reagents. The metal chelates often absorb at longer wavelengths than the parent reagent, thus facilitating their excitation in the presence of an excess of the latter. As judged by the intense oxygen quenching, the luminescence of the borate–benzoin complex (see Section 5 G 1) has an exceptionally long lifetime and in this chelate the lowest excited state may be π^*–n. It is possible that a variety of other metal chelates, at present thought to be non-luminescent, may in fact have relatively long-lived singlet states for the same reason, and their fluorescence may have been obscured by oxygen quenching. In general one would expect chelates having π^*–n excited singlet states to exhibit intense phosphorescence in rigid media at low temperature owing to the increased efficiency of intersystem crossing to the triplet state.

Space does not permit a complete review of all fluorescent chelates but it will be instructive to give some examples of the commoner types. Comprehensive literature references to luminescence and its analytical applications may be found in the reviews by White and co-workers[374]. Organic molecules capable of forming chelates contain at least two functional groups, each of which can combine with the metal ion by donating a pair of electrons. Some groups may lose a proton in this process of co-ordination, e.g. CO_2H, OH, NH_2, and some may not, e.g. O, OR, N, etc. These functional groups must be so arranged that on co-ordination with the metal atom they form a ring system having at least 4 members—by far the majority form five or six membered rings. The size of the metal ion can also be important—with small ions steric hindrance may prevent chelation with some reagents. For example substitution of a methyl group in the 2-position of 8-hydroxy-quinoline prevents the formation of a tris-complex with Al^{3+}, but not with Cr^{3+} or Fe^{3+}[375].

Examples of chelating agents and fluorescent chelates derived from them are shown in Fig. 179. The structures of some chelates are well established. For example 8-hydroxy-quinoline (C in Fig. 179) in its ionised form (Q^-) combines with metal ions via its oxygen and nitrogen atoms to form the

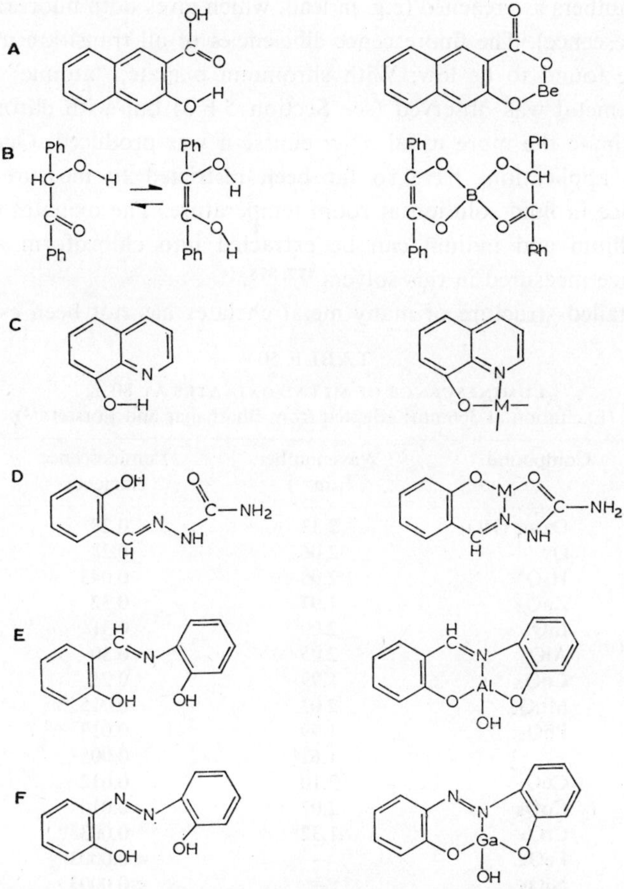

Fig. 179. Some chelating agents and fluorescent chelates.
A, 2-hydroxy-3-naphthoic acid; B, benzoin; C, 8-hydroxyquinoline; D, salicylaldehyde semicarbazone; E, salicylidine-2-aminophenol; F, 2,2′-dihydroxyazobenzene.

normal oxinates. These are neutral; thus divalent metal oxinates have the formula MQ_2, the trivalent MQ_3 and the tetravalent MQ_4. Many oxinates are only very weakly fluorescent, e.g. those of transition metals such as iron, cobalt and nickel. Bhatnagar and Forster[376] have measured the luminescence efficiencies of a variety of oxinates and substituted oxinates at 80°K. Some of their results for unsubstituted oxinates are shown in Table 50. Except for very high Z numbers, the fluorescence efficiencies with closed shell metal ions are approximately the same as that of the oxine anion. Thus the heavy atom effect does not appear to operate, at least until high atomic numbers are reached (e.g. in lead, which gives both fluorescence and phosphorescence). The fluorescence efficiencies of all transition metal chelates were found to be low. With chromium oxinate, "atomic" emission from the metal was observed (see Section 5 F 3) but with chromium dibromo-oxinate the more usual π^*–π emission was produced. Quantitative analytical applications have so far been restricted to measurements of fluorescence in fluid solution at room temperature. The oxinates of aluminium, gallium and indium can be extracted into chloroform and their fluorescence measured in this solvent[377, 378].

The detailed structure of many metal chelates has not been established

TABLE 50

LUMINESCENCE OF METAL OXINATES AT 80°K
(Excitation at 366 nm; adapted from Bhatnagar and Forster[376])

Compound	Wavenumber (μm^{-1})	Luminescence efficiency
Oxine (HQ)	2.33	0.28
Q^-	2.06	0.28
H_2Q^+	2.08	0.043
ZnQ_2	1.97	0.32
InQ_3	2.00	0.31
AlQ_3	2.05	0.30
CdQ_2	1.99	0.21
MnQ_2	2.02	0.015
PbQ_2	1.99	0.013
	1.62[a]	0.005[a]
CuQ_2	2.10	0.012
CoQ_3	2.07	0.010
CrQ_3	1.32[b]	0.0033[b]
FeQ_3	—	< 0.0003
NiQ_2	—	< 0.0003

[a] Phosphorescence.
[b] Metal ion emission.

with certainty. Thus the aluminium and beryllium complexes of 2-hydroxy-3-naphthoic acid contain one molecule of reagent per metal atom[379] and a likely method of chelation is that shown in Fig. 179 (A). However in aqueous solution the beryllium is expected to have two further water molecules co-ordinated and the aluminium three water molecules and an OH^- group. The same principle applies to the chelates of salicylaldehyde semicarbazone, salicylidene-2-aminophenol and 2,2′-dihydroxyazobenzene and its derivatives (Fig. 179 (D), (E) and (F)). Salicylidene-2-aminophenol gives fluorescent complexes with a variety of metal ions[380–382]. It is a particularly sensitive test for aluminium, and Dagnall and co-workers found that the fluorescent aluminium complex contained one molecule of reagent per atom of aluminium[383]. The monomer would therefore be expected to have the structure shown in Fig. 179 (E) but with two molecules of water completing the co-ordination of the aluminium. They suggested that it probably consisted of a low order polynuclear species, e.g. a dimer linked by elimination of water between hydroxyl groups. Similar effects may occur in the trivalent metal chelates of 2,2′-dihydroxyazobenzene and its derivatives[381] which should also have one OH^- ion attached (see Fig. 179 (F)). The trivalent metal chelates of salicylaldehyde semicarbazone (Fig. 179 (D)) should have two attached OH^- groups (not shown in Fig. 179) and should therefore be capable of forming oxygen bridged polymers. The scandium chelate provides a very sensitive fluorimetric method for scandium[384] and it has been suggested that it is present as a polymeric species[385].

The last paper quoted above gives an example of each of three methods by which fluorescence measurements can be used to determine the mole ratio of metal to chelating agent in a chelate. The first of these is the well known method of continuous variations[386] in which the fluorescence intensity is measured of a series of solutions each containing the same total concentration of (metal ion + reagent) but varying ratios of metal ion to reagent concentration. The maximum fluorescence is observed at the point where the ratio [metal ion]/[reagent] in the solution is the same as that in the complex. In the second method[387] the fluorescence is measured of a series of solutions all containing the same concentration of metal ion, but with varying concentrations of reagent. The curve obtained changes slope at the point where the ratio [metal ion]/[reagent] in the solution is the same as that in the chelate. The sharpness of the change over decreases as the dissociation constant of the chelate increases. In the third method[388] two fluorescence–concentration curves are plotted. In the first, a constant concentration c of the metal ion is reacted with varying smaller concentrations of the reagent, and in the second, a concentration c, of reagent is reacted with varying smaller concentrations of metal ion. Both curves are linear and the ratio of their slopes is equal to the mole ratio of metal ion to reagent in the chelate.

3. Intramolecular Energy Transfer

If the metal ion forming part of a chelate does not contain partly unfilled inner shells of electrons, the excited states produced by promotion of one of the electrons of the metallic ion to an upper orbital have energies much greater than the π^*–π and π^*–n singlet and triplet states corresponding to the promotion of one of the electrons in the organic part of the molecule. As a result therefore the metal ion behaves for all practical purposes as a "saturated organic atom" and serves merely as one of the links holding the various parts of the molecule together. In considering the emission spectra, only the π^*–π, π^*–n and other states characteristic of the organic part of the chelate are relevant, and although these states are naturally modified to some extent by the process of chelation, as described in the previous section, the spectra are still diffuse.

If, however, the chelate is formed from a transition metal or a rare earth, a series of new energy levels is introduced corresponding to transitions of electrons between the various *d* or *f* levels in the metallic atom. With many compounds, some of these levels are situated at energies lower than the singlet, or even the triplet π^*–π and π^*–n levels and must therefore be taken into account when considering the luminescence emission.

The three general cases to be considered are shown in Fig. 180. At first sight the introduction of the metal ion merely introduces a third *manifold* of levels in addition to the singlet and triplet manifolds. Clearly, if the lowest level of the *atomic* manifold is situated higher than either the lowest singlet or triplet level (A in Fig. 180) we should expect $S_1 \rightarrow S_0$ and $T_1 \rightarrow S_0$ to be unaffected. If on the other hand there is an atomic level situated lower than either the singlet or the triplet level (C in Fig. 180) we should anticipate that *intersystem crossing* to the "*a*" level would occur and that emission from the "*a*" level would be observed. This process is generally called intramolecular energy transfer (see also Section 2 B 7). It was first demonstrated by Weissman[389–391], who found that solutions of europium benzoylacetonate and europium salicylaldehyde emit strongly the luminescence charac-

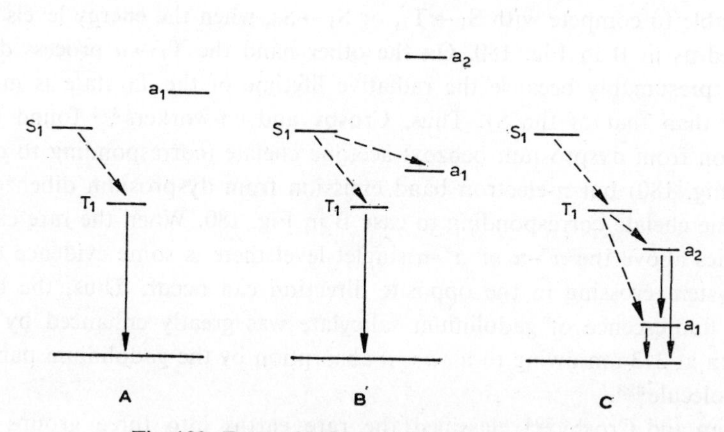

Fig. 180. Energy level manifolds of metal chelates.
A, transfer to atomic levels impossible; B, fluorescence quenched (complexes of transition metals with low-lying *d* levels); C, transfer from triplet level and line emission observed (some rare earth chelates).

teristic of the europium ion when excited with light absorbed in the π^* bands of the chelates.

Although in principle we should expect this process to occur with transition metals, the *d* electrons of the latter are not well shielded from external influences and the luminescence is strongly quenched. In other cases (as also with the 4 *f* levels of some rare earths) very low-lying *d* levels are present and the emission, if it occurs, is in the infra-red region. As a result many transition metal complexes—even those formed from fluorescent reagents— are weakly or non-fluorescent. Some do however give line emission, e.g. the oxine chelate of chromium which emits at 1.32 μm^{-1} in rigid medium at 80°K (see Table 50). In contrast, the 4 *f* levels of the rare earths are well shielded from external influences and this has three consequences. First, the luminescence corresponding to the $a \rightarrow a$ transitions is not so easily quenched. Second, it retains much of the nature of an atomic emission and thus consists of a narrow line. Third, emissions may occur *from* higher "*a*" levels than "a_1", and they may terminate at higher levels than the ground state. Hence, several emission lines may be observed from one rare earth. If the *f* levels are so well shielded from external influences, it may be asked why they can be populated by intersystem crossing from π^*–π or π^*–n levels. The answer seems to be that the probability of such intersystem crossing is indeed low. Thus the process $S_1 \rightarrow a$ does not seem to be able to compete with $S_1 \rightarrow T_1$, or $S_1 \rightarrow S_0$, when the energy levels are situated as in B in Fig. 180. On the other hand the $T_1 \rightarrow a$ process does occur, presumably because the radiative lifetime of the T_1 state is much longer than that of the S_1. Thus, Crosby and co-workers[392] found line emission from dysprosium benzoyl acetone chelate (corresponding to case C in Fig. 180) but π-electron band emission from dysprosium dibenzoylmethane chelate corresponding to case B in Fig. 180. When the rare earth level lies above the π^*–π or π^*–n singlet level there is some evidence that intersystem crossing in the opposite direction can occur. Thus, the blue violet fluorescence of gadolinium salicylate was greatly enhanced by excitation at 313 nm owing to an $a \leftarrow a$ absorption by the gadolinium part of the molecule[393].

Whan and Crosby[394] classified the rare earths into three groups according to the luminescence of their chelates with benzoyl acetone and dibenzoylmethane viz.:

Gd, La, Lu	high yields of π^* fluorescence and/or phosphorescence ($\phi_f + \phi_p$)
Pr, Nd, Ho, Er, Tm, Yb	lower values of ($\phi_f + \phi_p$)
Eu, Tb, Dy, Sm	very low values of ($\phi_f + \phi_p$): strong line emission

The first group do not contain $4f$ levels below the triplet level, and hence no intersystem crossing from the triplet level is possible. The second group possess low-lying $4f$ states via which the triplet energy is degraded. They found a rough correlation between the number of $4f$ levels to which transfer could occur and the efficiency of luminescence quenching. Absence of line emission in these rare earths was attributed to radiationless deactivation via vibronic coupling to the ligands and hence to the surrounding solvent. Those members of this group that might otherwise be expected to emit in the visible region contain a large number of closely spaced levels through which this radiationless deactivation can operate, whereas the members of the third group have some prominent gaps separating the emitting levels from lower levels.

Bhaumik and Telk[395] have measured the luminescence efficiencies and lifetimes of the line emissions from several chelates of terbium, europium and samarium in a variety of fluid solvents at room temperature. Their data illustrate several interesting points. First, the luminescence efficiencies are high in some compounds and the intersystem crossing to the rare earth levels must be efficient to compete with the high rate of the radiationless conversion processes undergone by triplet states in fluid solution. Second, dissolved oxygen has a negligible effect in spite of the relatively long luminescence lifetimes. (Once the rare earth excited state has been reached, the relative stability and long lifetime of this state can perhaps be understood in terms of the shielding provided for the $4f$ orbitals by the outer electrons.) Third, the luminescence efficiency of some chelates is quite strongly dependent on solvent—this may be due to chemical changes.

4. Intermolecular Energy Transfer

Energy transfer may take place from a triplet excited organic molecule to a rare earth compound by a diffusion-controlled process, in a manner analogous to the triplet-to-singlet and triplet-to-triplet energy transfer processes discussed in Sections 2 C 4 and 2 D 2. Thus, El-Sayed and Bhaumik[396] found evidence that the benzophenone triplet can transfer its energy to europium chelates by a diffusion-controlled process, and Matovich and Suzuki[397] have reported that, in acetophenone and other aromatic ketones, rare earth salts can be excited via the solvent. Ballard and Edwards[398] studied the concentration dependence of the emission spectra of solutions of the nitrates of Sm^{3+}, Dy^{3+}, Tb^{3+} and Eu^{3+} in acetophenone. They interpreted their results in terms of diffusion-controlled intermolecular energy transfer from the acetophenone triplet to the rare earth ion. Heller and Wasserman[399] found that the luminescent levels of Tb^{3+} and/or Eu^{3+} could be sensitised by diffusion-controlled transfer of energy from the triplet states of 21 aromatic aldehydes and ketones in acetic acid solution at room temperature. Winefordner and McCarthy[400] have made use of the same principle to sensitise the luminescence of europium, terbium, samarium and dysprosium in acetic anhydride (see Section 5 G 2).

5. Organic Compounds Containing Non-Metals

In principle it is possible to utilise the fluorescence of any organic compound to determine one of the elements it contains. Thus *in principle* it is possible to determine chlorine by converting it to the fluorescent 9-chloranthracene, or nitrogen by converting it to (say) β-naphthylamine. In practice, of course, the series of reactions required to perform such conversions are so complex that they are of no use as general analytical methods. For the latter, simple reactions that proceed with uniform yield (preferably 100%) are preferred and it is worth looking at the whole field of organic reaction mechanisms to decide what types of reaction may be suitable. Ring-closure reactions that result in the formation of additional heterocyclic ring systems are likely candidates. Such reactions may not provide methods for the determination of an element in all its forms, but only for one specific compound of it. For example, the reaction of hydrazine with o-phthalaldehyde—or possibly with a more complex o-dialdehyde—might be used to determine nitrogen in this particular form:

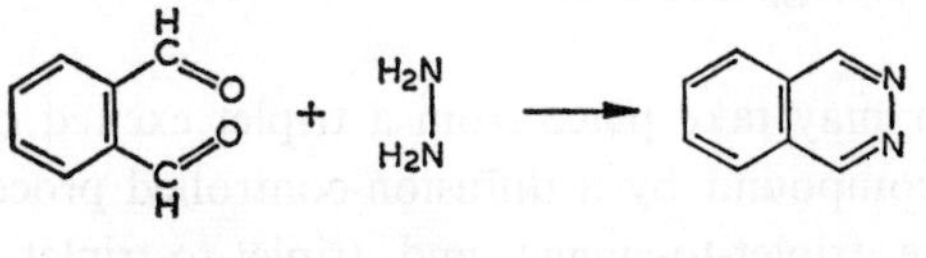

A fluorimetric method for the determination of nitrogen combined as cyanide has been reported by Hanker and co-workers[401]. The hydrogen cyanide was converted to cyanogen chloride by reaction with chloramine T. The cyanogen chloride was then reacted with nicotinamide giving ultimately a product with a strong fluorescence in alkaline medium.

Ring-closure methods for the determination of sulphur and selenium in all their usual inorganic forms have been described by Parker and co-workers[327, 402]. Sulphate, sulphite etc. is distilled as H_2S from hypophosphorous acid–HI mixture. The H_2S is trapped on a small filter paper moistened with zinc acetate and the resulting zinc sulphide converted to the red fluorescent dyestuff thionine by reaction with an acid solution of ferric sulphate and p-phenylene diamine:

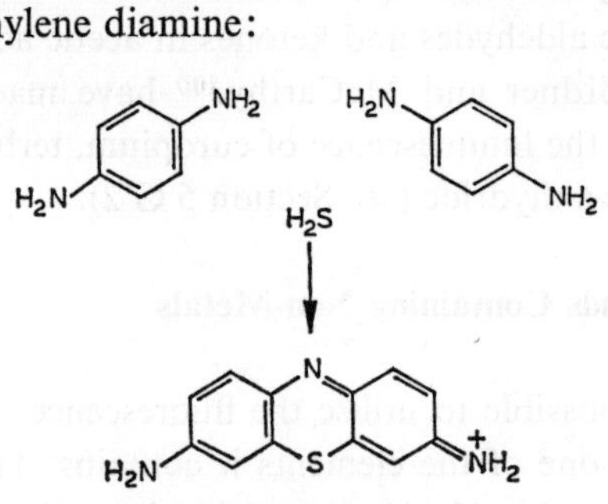

The thionine is extracted into butanol for fluorescence measurement. Selenium, as selenous acid in aqueous solution, is reacted with 2,3-diaminonaphthalene to give the strongly fluorescent 4,5-benzo-piazselenol:

$$NH_2,\ NH_2 \xrightarrow{H_2SeO_3} N,\ Se,\ N$$

G. ANALYSIS FOR ELEMENTS AND INORGANIC COMPOUNDS

1. Application of Prompt Fluorescence in Fluid Solution

There are very few viable methods based on the *natural* luminescence of simple inorganic ions in solution. The luminescence from the simple salts of the rare earth ions is very weak, and that of the uranyl ion appears to be due to inorganic complexes of one sort or another, and hence is not strictly "natural". Similarly the fluorescence of thallous ion in sodium chloride solution[370] is probably due to chloride complexes. Sill and Peterson[371] made a critical study of the fluorescence of uranium using 254 nm for excitation, with visual observation of the green fluorescence. The fluorescence in dilute sulphuric acid was enhanced considerably by the addition of hydrofluoric or phosphoric acids, and they recommended a concentration of about 0.5% w/v of the latter. They were thus able to detect 10 μg of uranium in 200 ml of solution. Only tin (yellowish green) and cerium (white) gave fluorescence under these conditions. Strongly absorbing ions act as inner filters and reduce the fluorescence intensity. They must therefore be removed, or the solution diluted to bring the inner filter effect to within acceptable limits. Sill and Peterson recommended that the results be checked by the method of standard additions. They gave procedures for the determination of traces of uranium in refractory materials. The sensitivity and specificity obtainable with a spectrofluorimeter is illustrated by the spectra shown in Fig. 181, which also shows the enhancement by the addition of phosphoric acid. With the laboratory grade reagents used, the sensitivity for uranium was limited by the reagent blank to about 10^{-7}M (0.02 μg per ml). With purified reagents about 0.002 μg per ml of uranium could be detected and specifically identified by the characteristic vibrational structure.

The fluorescence emission spectra of metal chelates are generally diffuse and not in themselves sufficient to provide unambiguous identification. For the latter reliance is placed on the specificity of the particular reagent under the chosen conditions of test, and frequently it is also necessary to carry out separation from interfering elements. This is not to say that such fluorescent chelates have not proved valuable for the determination of a variety of metals, e.g. aluminium, gallium, beryllium, zirconium, thorium, germanium, magnesium, zinc, tungsten, tin, thallium, vanadium, ruthenium etc., see reviews by White[374]. To illustrate the principles we shall discuss one example—the well known fluorimetric determination of aluminium ions with 8-hydroxyquinoline. In principle this is simple: the reaction is carried out at a pH value of 5–6 and the aluminium oxinate extracted into chloroform for the measurement of its fluorescence. Gallium and indium also give oxinates having fluorescence bands that overlap that of aluminium oxinate (see Fig. 182), and Collat and Rogers[377] devised a method for the simultaneous

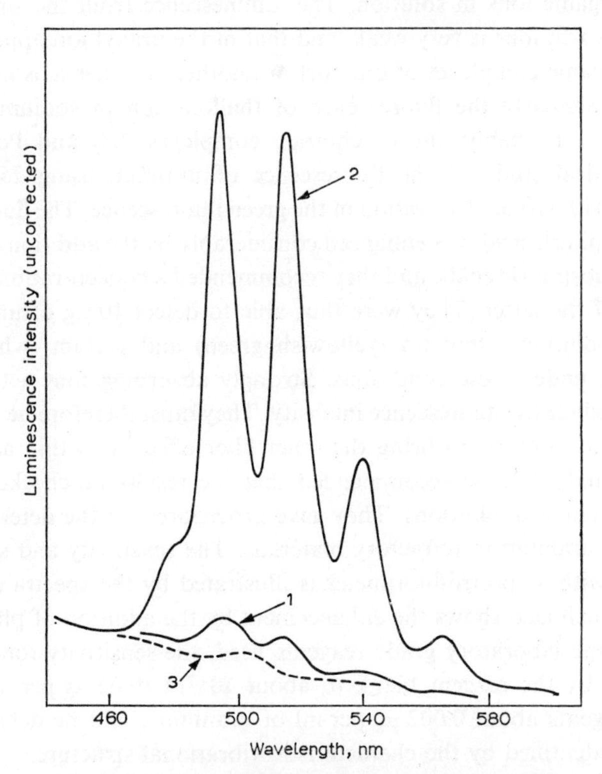

Fig. 181. Spectrofluorimetric determination of uranium.
Excitation at 250 nm isolated by grating monochromator with chlorine and OX7 filters; band width of analysing monochromator was 6.6 nm; 1, 10^{-6} M uranyl sulphate in 0.1 N sulphuric acid; 2, as for 1, with the addition of 0.5% w/v orthophosphoric acid; 3, as for 2, but without uranyl sulphate.

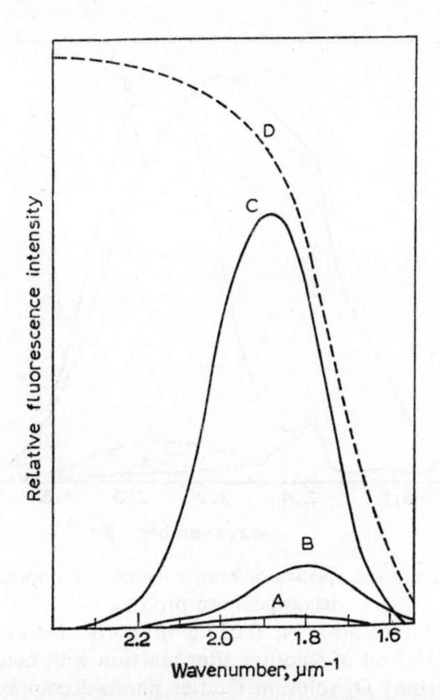

Fig. 182. Spectrofluorimetric determination of aluminium. Uncorrected fluorescence emission spectra of oxinates obtained by extraction of solutions containing 2.5 μg of metal; A, indium; B, gallium; C, aluminium; D, spectral sensitivity curve of instrument (from Rees[378]).

determination of gallium and aluminium in admixture. This was based on the fact that the relative excitation efficiencies of the two oxinates at 366 nm and 436 nm are different (i.e. the ratios of their extinction coefficients at these two wavelengths are different). Unfortunately the differences are not sufficient to allow a very precise determination, and in particular it is not possible to determine small concentrations of one element in the presence of much larger concentrations of the other by this method.

The work of Rees[378] is a good example of how careful purification of reagents and attention to other sources of contamination can result in a highly sensitive fluorimetric method. The problem was the determination of trace amounts of aluminium in high-purity synthetic silica. To minimise reagent blank and adventitious contamination, chemical treatment was kept as simple as possible. Gallium and indium were not separated, but by excitation at 405 nm and measurement near the maximum of the fluorescence emission spectrum of aluminium oxinate, the gallium and the indium gave readings corresponding to only 15% and 3% of their weight of aluminium.

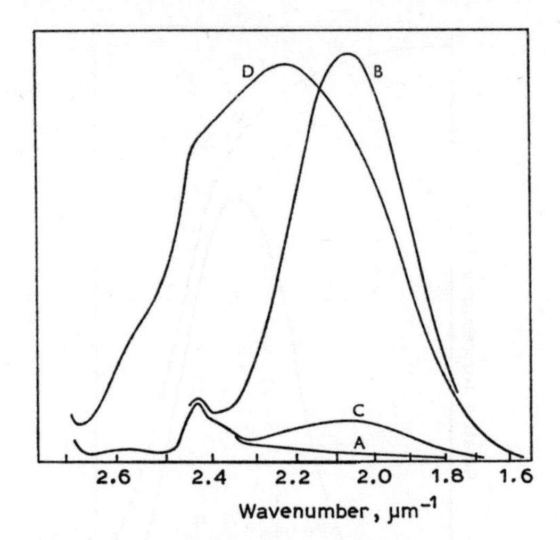

Fig. 183. Fluorescence emission spectra of borate–benzoin compound and benzoin photo-decomposition product.
Excitation at 2.73 μm^{-1} (366 nm); A, 0.07 μg of boron before addition of benzoin; B, 0.07 μg of boron in 16.5 ml of solution after reaction with benzoin; C, reagent blank after reaction with benzoin; D, solution C after photo-decomposition; spectra are uncorrected; spectral sensitivity curve as for curve D in Fig. 182 (from Parker and Barnes[404]).

One of the main sources of blank was found to be the hydrofluoric acid used to dissolve the specimen, and reagent grade material (0.2 p.p.m. aluminium) was distilled in an all-platinum apparatus. By purification of all other reagents, and exclusion of all but silica or platinum apparatus, Rees was able to reduce the overall blank of the method to 0.02 p.p.m. of aluminium calculated on the silica specimen. Lower blanks than this for ubiquitous elements such as aluminium, silicon, boron or magnesium, are very difficult to achieve when a refractory or metallic specimen has to be dissolved.

The question of reagent blank has been one of the main problems facing the analytical chemist in the semiconductor industry where impurities at the parts per thousand million level are of interest. We choose two further examples to illustrate the problems involved in devising fluorimetric methods for this type of analysis. First the determination of traces of boron in high purity silicon. The fluorimetric reaction of borate ion with benzoin was first applied by White and co-workers[403] to the determination of boron in the range 0.1–10 μg. Subsequent work by Parker and Barnes[131, 404] showed that the method was capable of considerably greater sensitivity but that a variety of factors had to be controlled. The reaction was carried out in aqueous ethanol containing sodium carbonate. The 366 nm mercury line excites the fluorescence efficiently, but causes rapid decomposition of the benzoin to a product having a fluorescence band overlapping that of the borate–benzoin complex, as shown in Fig. 183. The rate of decomposition produced by the 405 nm mercury line was less relative to the fluorescence of the boron compound, and the use of 405 nm also had the advantage that inner filter effects by the excess benzoin reagent, and by other likely absorbing impurities, were less. It had the disadvantage that the Raman band overlapped the fluorescence of the borate–benzoin compound but this source of interference was eliminated by appropriate choice of the secondary filter (see Fig. 184).

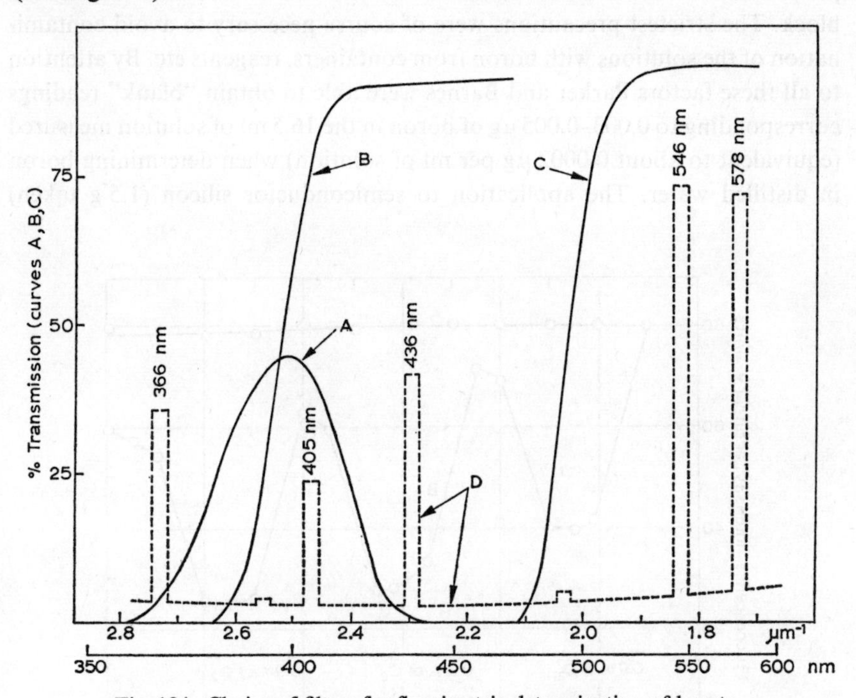

Fig. 184. Choice of filters for fluorimetric determination of borate.
A, 1 cm of a 0.75% w/v solution of iodine in carbon tetrachloride and 1 cm of a solution containing 100 g of $Cu(NO_3)_2.3H_2O$ in 60 ml of N nitric acid; B, 1.5 cm of a 1% w/v aqueous solution of sodium nitrite; C, 1.5 cm of a 1% w/v solution of potassium chromate in 0.05% w/v sodium hydroxide; D, approximate spectral distribution of light from 250 W compact source mercury lamp (quanta); the combination of A and B was used to isolate the 405 nm mercury line for excitation; filter C was used to reject scattered exciting light and solvent Raman while passing sufficient of the fluorescence (from Parker and Barnes[404]).

Oxygen even at low concentration, produced three undesirable effects, namely (i) reversible quenching of the borate–benzoin fluorescence and slow irreversible decomposition (see Fig. 185), (ii) oxidation of the benzoin to produce absorbing products, and (iii) acceleration of the photo-decomposition of the benzoin. To obtain a reproducibly low blank value it was necessary to de-oxygenate completely both the alkaline borate solution and the ethanolic benzoin solution, using the apparatus shown in Fig. 86. The benzoin had to be purified by several recrystallisations in the dark room and its solutions stored in the dark. The borate–benzoin equilibrium had a temperature coefficient that resulted in a decrease in fluorescence of 3–4% per °C and the cuvette thus had to be situated in a constant temperature block. The strictest precautions were of course necessary to avoid contamination of the solutions with boron from containers, reagents etc. By attention to all these factors Parker and Barnes were able to obtain "blank" readings corresponding to 0.003–0.005 μg of boron in the 16.5 ml of solution measured (equivalent to about 0.0002 μg per ml of solution) when determining boron in distilled water. The application to semiconductor silicon (1.5 g taken)

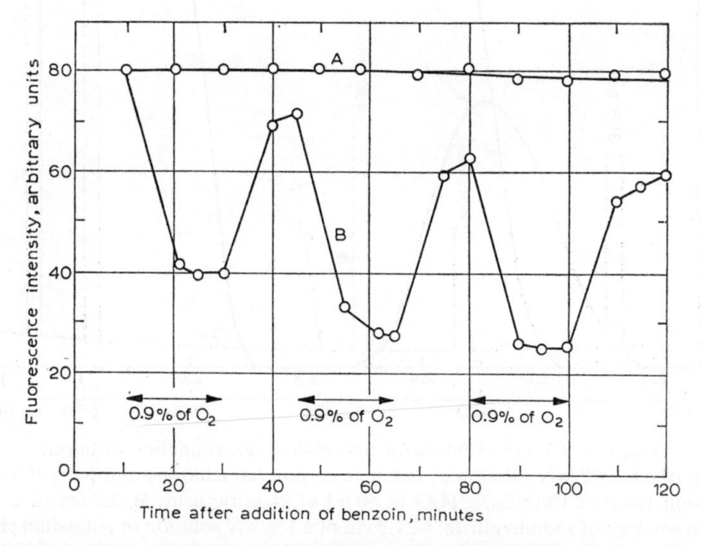

Fig. 185. Reversible quenching and irreversible decomposition of borate–benzoin solution induced by the presence of oxygen.

Excitation at 405 nm of a solution containing 0.018 μg per ml of boron after reaction with benzoin; A, with pure nitrogen passing; B, with pure nitrogen passing, but with 0.9% v/v of oxygen periodically injected into the gas stream (from Parker and Barnes[131]).

naturally involved additional reagents, and pretreatment consisting of hydrothermal decomposition under high pressure[405] and vacuum distillation as ethyl borate. The overall blank was thus increased from 0.005 to about 0.02 μg. They estimated that detection was reasonably reliable down to about 0.03 p.p.m. This may be compared with the detection limit of 0.002 p.p.m. for the determination of borate in water where no chemical pretreatment is required. Incidentally it may be noted that once the details of a fluorimetric method involving diffuse emission spectra have been worked out, its routine application can be relegated to a simple filter instrument—as in this case.

Finally, let us consider a sensitive method for a non-metal, selenium. This is a less widely distributed element and it does not present the difficulty of adventitious contamination met in the previous method. The limiting factor still turns out to be reagent blank, but the "blank" fluorescence is now due to other factors than selenium contamination. This method also demonstrates how an organic reagent can be "tailor-made" to suit the purpose required. The reagent originally used was 3,3′-diaminobenzidine proposed by Hoste and Gillis[406, 407] as a colorimetric reagent, and applied as a fluorimetric reagent by Parker and Harvey[328] who showed that reaction with selenous acid can produce two piazselenols, the relative proportions of which depend on the relative concentrations of selenous acid and reagent:

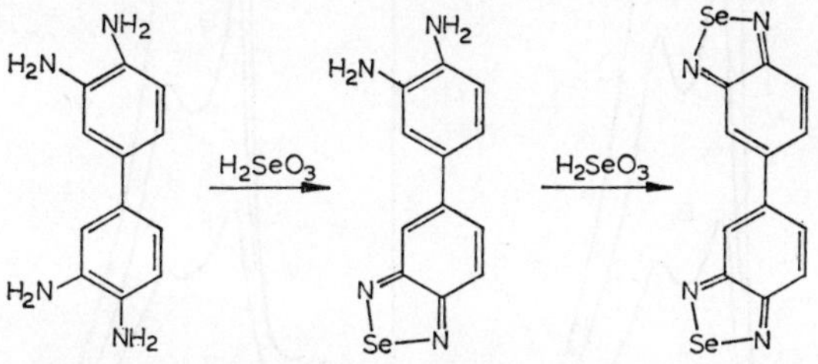

With a large excess of 3,3′-diaminobenzidine as used in the analytical procedure, only the mono-selenium compound is formed. Since this contains two free amino groups it is strongly basic and can only be extracted from aqueous solution into an organic solvent by increasing the pH value of the solution. This complicates the analytical procedure in the presence of metals whose hydroxides are precipitated in neutral or alkaline solution, and necessitates their removal or the addition of large amounts of complexing agents which have to be purified to a high degree. Clearly, reagents containing only one pair of o-diamino groups would not suffer from this disadvantage, and in fact it was found[327] that 2,3-diaminonaphthalene was almost ideal: the corresponding 4,5-benzpiazselenol (see Section 5 D 2) could be extracted into dekalin or cyclohexane from 0.1 N HCl; it was strongly excited by the 366 nm mercury line and gave a characteristic fluorescence emission spectrum (see Fig. 186); and its fluorescence sensitivity index, $\phi D/H$, was 0.1 at 366 nm compared with 0.005 for the compound formed from 3,3′-diaminobenzidine at its optimum mercury line (436 nm). The factor limiting sensitivity was the blank fluorescence from trace impurities in the 2,3-diaminonaphthalene. By recrystallisation of the reagent and carrying out all manipulations in the dark room, this was kept to the equivalent of 0.002 μg of selenium in the total extract. Reagents used to dissolve 1 g of arsenic or gallium arsenide increased this value by only 0.006 μg.

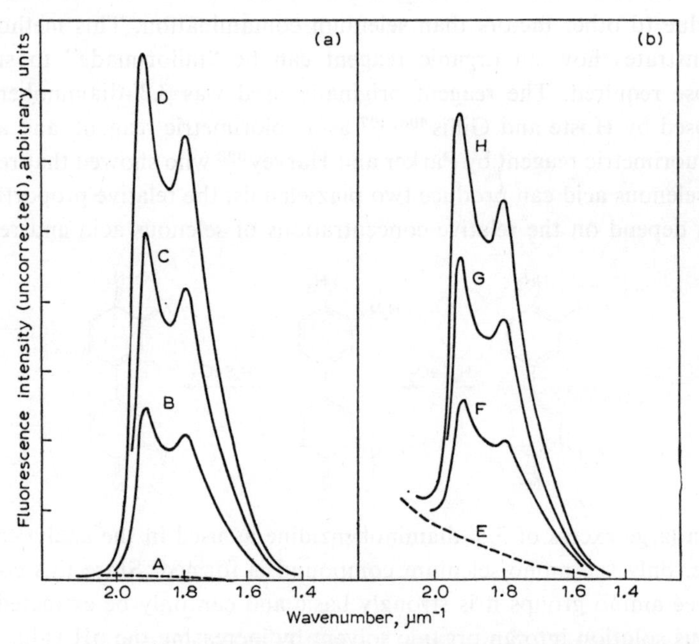

Fig. 186. Spectrofluorimetric determination of selenium with 2,3-diaminonaphthalene. Fluorescence emission spectra of 4,5-benzopiazselenol in aerated dekalin; excitation at 2.73 μm^{-1} (366 nm); half-band width of analysing monochromator 0.025 μm^{-1} at 2.5 μm^{-1}; A, B, C, D, pure compound at concentrations 0.0, 0.0053, 0.0106 and 0.0159 μg per ml; E, F, G, H, solutions derived from selenous acid by reaction with 2,3-diaminonaphthalene and extraction with dekalin, and containing selenium equivalent to 0.0, 0.0014, 0.0028, 0.0042 μg per ml (from Parker and Harvey[327]).

2. Methods Based on Intra- and Intermolecular Energy Transfer

In contrast to the broad emission bands observed from most metallic chelates, the emissions that can be evoked from some rare earth ions consist of one or more narrow lines which provide a means of specific identification. Furthermore, the lifetime of the emissions is in many cases sufficiently long to be resolved by a mechanical phosphoroscope, even in oxygenated fluid solution, and thus in principle all other fluorescence, scattered light, Raman emission and phosphorescence of organic compounds can be eliminated. The application of the phosphoroscope has not so far been reported, but a variety of methods employing spectrofluorimeters have appeared. Thus, Kononenko and co-workers[408] used 1,10-phenanthroline with salicylic acid for the extraction and fluorimetric determination of Eu and Tb, and 1,10-phenanthroline with thenoyltrifluoracetone[409] for the determination of Eu and Sm.

Ballard and Edwards[410] made use of the complexes of Eu and Sm with thenoyl trifluoracetone and trioctyl phosphate. They excited the complexes with 366 nm radiation and found quantum efficiencies of 0.67 for the europium emission at 613 nm and 0.0093 for the samarium emission at 597 nm. They calculated the corresponding absolute sensitivities (ϕD_{366}) to be 0.114 and 0.0016. These results may be compared with the corresponding values for some strongly fluorescent organic compounds given in Table 47. Ballard and Edwards applied their method to the determination of europium and samarium in granite at the 1–10 p.p.m. level.

Taketatsu and co-workers[366] determined Tb and Eu as their carbonate complexes, and Alberti and co-workers employed tungstate[367, 411] or oxalate[412] to enhance the luminescence emission from solutions of Sm, Eu, Tb and Dy. Dagnall and co-workers[413] described a sensitive method for terbium based on the formation of a 1:1:1 complex with ethylenediaminetetracetic acid (EDTA) and sulphosalicylic acid. They concluded that the terbium line emissions were produced by intramolecular energy transfer from the energy level associated with the sulphosalicylic acid part of the molecule, and that the EDTA protects the terbium ion and the coordinated sulphosalicylic acid from collisional interference by other rare earth ions or solvent molecules.

McCarthy and Winefordner[400] measured the ratio of the intensity of the rare earth line to that of the background for the rare earth emissions in the presence of a variety of aromatic carbonyl compounds in de-oxygenated

TABLE 51

LINE TO BACKGROUND RATIOS FOR 10^{-2} M RARE EARTH SOLUTIONS IN ACETIC ANHYDRIDE CONTAINING CARBONYL COMPOUNDS
(Adapted from McCarthy and Winefordner[400])

Rare earth	Carbonyl compound (10^{-2} M)	Ratio
Sm	Veratraldehyde	33
Eu	o-Tolualdehyde	28
Tb	4,4′-Dimethoxybenzophenone	1100
Dy	Anisaldehyde	19

TABLE 52

METHODS FOR DETERMINING RARE EARTHS

Reagent	Rare earth	Emission wavelength observed	Limiting sensitivity (M)	Limitation
HCl (see Fig. 177)	Sm	560, 594, 642	$\sim 5 \times 10^{-3}$	Blank
	Eu	591, 616, 697	$\sim 5 \times 10^{-4}$	limited
	Gd	310	$\sim 10^{-4}$	(impurities
	Tb	490, 544, 588, 622	$\sim 10^{-5}$	etc.)
	Dy	478, 487, 572	$\sim 10^{-4}$	
Thenoyl trifluor-acetone and trioctyl phosphate[410]	Sm	562, 597, 642	7×10^{-6}	Instrument
	Eu	613	3×10^{-8}	limited
Carbonate[366]	Eu	590, 620	3×10^{-5}	Probably
	Tb	497, 550, 593	2×10^{-6}	instrument limited
Tungstate[367, 411]	Sm	560, 595, 640	3×10^{-6}	Instrument
	Eu	590, 615, 650	3×10^{-8}	limited
	Tb	490, 545, 590, 620	6×10^{-7}	
	Dy	480, 570	6×10^{-8}	
EDTA and sulphosalicylic acid[413]	Tb	485, 545 575	2×10^{-8}	Blank limited
Carbonyl compounds in acetic anhydride[400]	Sm	~ 550, 588	2×10^{-5}	Blank
	Eu	~ 573, 607	5×10^{-6}	limited
	Tb	~ 491, 541, 591	10^{-8}	
	Dy	~ 477, 578	8×10^{-7}	

acetic anhydride. The reagents giving the greatest ratios are shown in Table 51. McCarthy and Winefordner concluded that the increased emission in the presence of the sensitisers was due to *intermolecular* transfer from the triplet states of the sensitisers to the rare earth ions. They found that extraneous ions reduced the intensity of emission, and this effect was attributed to collisional competition between the foreign ion and the rare earth. They therefore recommended that a standard addition technique should be employed when the system was used for analytical purposes.

The summary of data in Table 52 gives some idea of the increased sensitivity that may be obtained by sensitisation, either by the intra- or intermolecular process. The limiting sensitivity values given for the chloride solutions are for excitation at 366 nm, except for Gd which was excited at 250 nm. The values are only very approximate and are estimated on the assumption that a signal equal to 1/10th of the background could be detected. The background was due to traces of fluorescent impurities in the solutions measured and could of course vary very considerably from sample to sample. For this reason the direct measurement of chloride solutions is not recommended, quite apart from the disadvantage of low sensitivity.

Finally application of computers to rare earth analysis should be mentioed. Stanley and co-workers[414] have investigated the possibilities of general methods for the analysis of mixtures of those rare earths giving measurable sensitised luminescence emission, using the computerised technique of successive spectrum stripping.

3. Quenching Methods

The measurement of luminescence quenching as a general method for the determination of the concentration of inorganic quenchers has not found wide application because it is not very specific, and better methods for the determination of most inorganic substances can be found. There is, however, one substance, namely molecular oxygen, for which the measurement of luminescence quenching can provide a useful method covering a wide range of concentrations. The method is particularly apt for the determination of oxygen in inert gas streams such as nitrogen, argon, etc.—a requirement of considerable industrial importance. We shall therefore outline the possibilities of the method for this purpose.

The Stern–Volmer equation (Section 2 B 2) may be re-written in the form:

$$\text{Percent quenching} = 100(1 - F/F_0) = 100\left(1 - \frac{1}{1+x}\right) \qquad (439)$$

where:

$$x = k_Q \tau_0 S p_{O_2} \qquad (440)$$

S is the solubility of oxygen in the solution and p_{O_2} the partial pressure of oxygen in equilibrium with it. Clearly for a given solubility and partial pressure the percent quenching may be varied over a wide range of values by varying the viscosity of the solution or the lifetime of the luminescing substance. Whatever values are chosen, the quenching curve will have the same shape, determined by equation 439, and we can therefore construct a general quenching curve, which can be made to suit the chosen system by simply varying the horizontal scale, as shown in Fig. 187.

In principle it is possible to calculate the calibration curve from equations

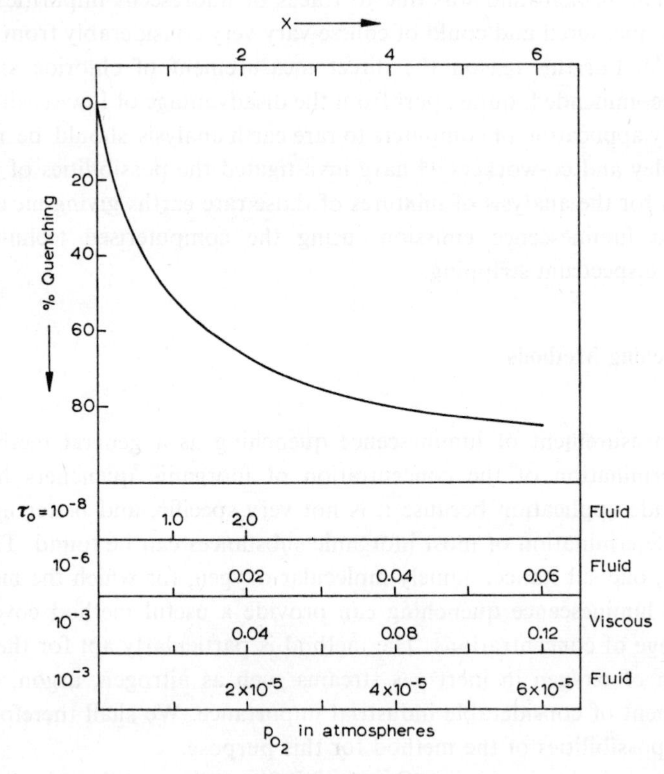

Fig. 187. Calibration curve for determination of oxygen by luminescence quenching.

TABLE 53

SYSTEMS CHOSEN FOR DETERMINATION OF OXYGEN
(Solubility of oxygen = $S = 5 \times 10^{-3}$ mole litre^{-1} atmosphere^{-1}; $k_Q = 2 \times 10^{10}$ for the fluid solvent and 1×10^{7} for the viscous solvent)

Solvent	Luminescence lifetime (sec)	$p_{O_2}/x = 1/(k_Q \tau_0 S)$
Fluid	10^{-8}	1.0
Fluid	10^{-6}	0.01
Viscous	10^{-3}	0.02
Fluid	10^{-3}	10^{-5}

439 and 440, but in practice the values of these parameters (particularly k_Q) are not known precisely, and empirical calibration at several known partial pressures of oxygen is necessary. To assist in choosing a system suitable for a particular range of partial pressures we shall construct scales for four hypothetical systems. We shall choose two solvents, both having $S = 5 \times 10^{-3}$ mole litre^{-1} atmosphere^{-1}. One is a typical fluid solvent with $k_c = 0.5 \times 10^{10}$ litre mole^{-1} sec^{-1}, and one a viscous solvent (e.g. glycerol) with $k_c = 4 \times 10^{6}$ litre mole^{-1} sec^{-1} at room temperature (see Table 53). Now oxygen quenching generally has a rate constant some 3–4 times greater than that calculated for diffusion-controlled reactions [76] (equation 80), and for convenience therefore we shall assume that k_Q has the values 2×10^{10} and 1×10^{7} litre mole^{-1} sec^{-1} for the fluid and viscous solvents. For luminescence having the lifetimes shown in column 2 of Table 53 we then find that unit value of x (50% quenching) corresponds to the oxygen partial pressures shown in column 3, and the four lower horizontal scales for p_{O_2} in Fig. 187 can thus be inserted. A compound having a normal fluorescence lifetime, e.g. fluorescein, in a fluid solvent, is thus suitable for high oxygen partial pressures up to 2 atmospheres or more. A compound such as pyrene with an exceptionally long-lived fluorescence ($\sim 10^{-6}$ sec) is suitable for the range 0–5% oxygen in a gas at atmospheric pressure. A somewhat similar range may be achieved in the viscous solvent with phosphorescence or E-type delayed fluorescence having a lifetime of about 1 msec (e.g. eosin). The same compound in fluid solution gives a much more sensitive range with the capability of detecting about 1 p.p.m. of oxygen ($\sim$ 10% quenching). Luminescence of longer lifetime will give a corresponding increase in sensitivity. There is also the possibility of interpolating between the last two ranges by adjusting the viscosity with mixed solvents. Calibration of the low range presents some difficulty. One method is to inject known small amounts of oxygen by

passing the oxygen-free gas stream through the anode compartment of an electrolysis cell. An alternative is to de-oxygenate the solution under high vacuum as described in Section 3 J 2, including in the cell a short length of sealed capillary tube containing a known amount of air. The luminescence intensity is measured before and after breaking the capillary tube, and the oxygen content calculated from the relative volumes of gas space and liquid in the cell. In measuring delayed fluorescence using a spectrophosphorimeter, the prompt fluorescence may be used as an internal standard. The main problem that arises in this method of oxygen measurement is photodecomposition, and some systems have to be frequently renewed.

The delayed fluorescence emissions of many dyestuffs adsorbed on silica gel have lifetimes considerably greater than 1 msec and are very strongly quenched by molecular oxygen. For example, 50% quenching of trypaflavine delayed fluorescence occurs at an oxygen pressure of about 0.5×10^{-7} atmosphere[415–417]. The full unquenched emission from such a system is best achieved by pumping down to high-vacuum. Calibration at such low oxygen concentrations presents some difficulty. As an alternative to intensity measurements, the lifetime of the delayed fluorescence may be used as a measure of oxygen content (see equation 71). This applies also to the measurement of delayed fluorescence or phosphorescence in solution where the lifetimes may be in the 1–10 msec range.

Finally it should be remembered that odd-electron molecules such as nitric oxide also produce strong quenching of triplet states.

4. Indirect Methods

There are some inorganic ions for which conversion to a fluorescent compound has proved to be difficult or impossible. Recourse then has to be made to indirect methods, which may be of two kinds. The ion may be allowed to react with a fluorescent substance to form non-fluorescent products, and the decrease in fluorescence intensity is then taken as a measure of the concentration of reacting ion. Such methods are sometimes referred to as "quenching methods", but the process is not fluorescence quenching in the strict sense defined in section 3 H 1. One of the ions for which no direct fluorimetric methods have yet been devised is fluoride and it is fortunate therefore that this ion is particularly suitable for indirect methods. Many tests depend on the formation of unionised or insoluble complexes between aluminium and fluoride. Aluminium is reacted with an organic reagent to produce a fluorescent complex, and in the presence of fluoride the fluorescence intensity is reduced. Reagents that have been employed are the aluminium complexes of quercetin[418], morin or 8-hydroxyquinoline[419], the dyestuff alizarin garnet[420], for which phosphates were found to have no effect. A method based on the use of a thorium–cochineal complex has also been described[418]. Paper strips impregnated with magnesium oxinate are used for the determination of fluoride contamination in air[421].

In the second kind of indirect method the ion to be determined is reacted with a non-fluorescent chelate to release a product that is itself fluorescent or can be converted to a fluorescent substance. Palladium, being a transition element, forms a non-fluorescent (and water soluble) chelate with 8-hydroxyquinoline-5-sulphonate. Palladium also forms a complex ion with cyanide, and in the presence of the latter the equivalent amount of the organic palladium complex is dissociated. The released oxine sulphonate is then combined with magnesium to form the fluorescent chelate of the latter[422]:

$$\underset{\text{(non-fluorescent)}}{Pd(OxS)_2} + 4CN^- \rightarrow 2OxS + Pd(CN)_4^{2-} \tag{441}$$

$$2OxS + Mg^{2+} \rightarrow \underset{\text{(fluorescent)}}{Mg(OxS)_2} \tag{442}$$

Similar reactions take place with sulphide and thiocyanate ions. A method depending on the release of a fluorescent dye from a thorium chelate by sulphate, fluoride and phosphate ions has also been reported[423]. It may be noted that a more specific method for phosphate has been reported by Minns[424]. The phosphate is converted to phosphomolybdic acid, the latter reacted with thiamine in isobutanol to produce the fluorescent thiochrome.

5. Inorganic Substances in Crystalline or Glassy Media

The luminescence of polycrystalline phosphors induced by a variety of inorganic "activators" was mentioned in Section 5 E 1. It has been used for the determination of rare earths, e.g. in calcium tungstate. The luminescence of many inorganic phosphors is very susceptible to impurity "quenching" and this has been applied as an analytical method. For example, chromium has been determined by its quenching action on the luminescence of a uranium-activated sodium fluoride bead after it had been dipped in a chromium solution and fused[425], and traces of copper, silver, mercury and platinum have been detected by their effects on the luminescence of a silver-activated cadmium sulphide phosphor[426]. The most widely applied analytical application of inorganic phosphors is the determination of uranium by observing the characteristic fluorescence produced when traces of the element are fused with sodium fluoride. This was first developed by Hernegger and Karlik[427] and has been reviewed by Rodden[428]; the minimum detectable quantity is 10^{-11}g.

In view of the difficulties associated with the blank arising from reagents required to dissolve transparent materials such as silica or crystalline alumina, the application of direct excitation methods is attractive. Thus chromium has been determined in ruby by means of its luminescence[429] and the luminescing agents in some crystalline laser materials can be determined directly, once standard specimens have been analysed by other methods. Observation of the luminescence of fused quartz and synthetic silica can

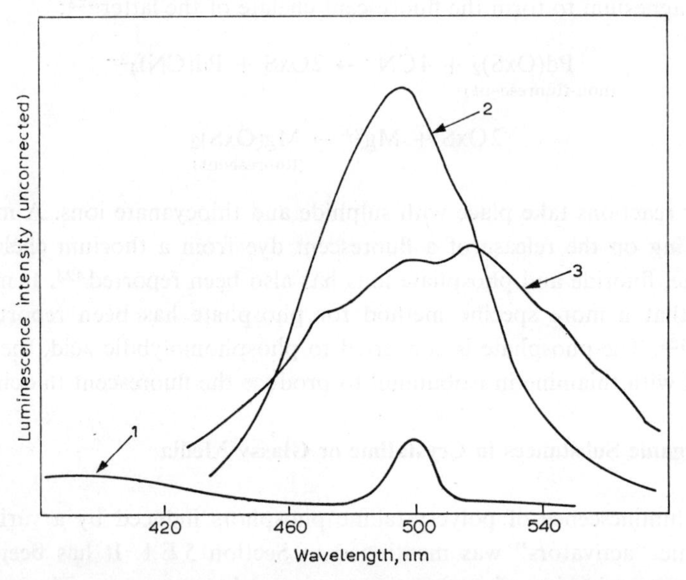

Fig. 188. Effect of copper content on the luminescence of synthetic silica. Excitation at 250 nm isolated from mercury lamp by grating monochromator with OX7 and chlorine gas filters; band width of analysing monochromator was 3.3 nm; 1, pure specimen; 2, specimen deliberately doped and containing 0.13 p.p.m. of copper; 3, specimen containing 0.15 p.p.m. of copper as adventitious contamination; relative sensitivities at which the spectra were recorded were 70, 1, 8; the Raman band from ethanol measured at the same sensitivity as curve 3 had a peak height about 50% greater than the maximum of curve 3; the maximum at 500 nm in curve 1 is second order scattered exciting light.

provide a sensitive indication of the presence of certain impurities. Much of the blue-violet emission from fused quartz excited by light of wavelength 250 nm seems to be due to the presence of traces of aluminium (see Fig. 85). However, the quantitative interpretation of such luminescence requires considerable care because the intensity and spectral distribution of the luminescence depends on the previous heat treatment of the material. Presumably the impurity element can enter the matrix in different sites. Thus the introduction of copper into synthetic silica gives rise to a green luminescence, but the intensity and spectral distribution of the latter can vary considerably for different specimens having the same copper content, as indicated in Fig. 188.

H. FUTURE DEVELOPMENTS

Of the many experimental techniques of photoluminescence, the measurement of prompt fluorescence in fluid solution at room temperature is still the most widely applied. It is simple and rapid, and equipment now available will operate at high efficiency and sensitivity over wide regions of the spectrum. Probably the most important improvement would be the invention of a light source giving a sufficiently intense continuum below 250 nm to permit the measurement of excitation spectra of very dilute solutions in this spectral region. The additional advantages of fluorescence and phosphorescence measurement at low temperature are becoming more widely recognised and applications in this field will undoubtedly expand rapidly in the coming years. Application of long-lived luminescence in fluid solution is still in its infancy and trace analysis by the measurement of sensitised delayed fluorescence, depending as it does on the low concentrations required to quench long-lived triplet molecules, has great potentialities. Increase in phosphorimeter resolution time, or development of other simple equipment for resolving luminescence having a lifetime in the range 10^{-6}–10^{-4} sec should provide additional scope, e.g. for the application of the annihilation-controlled delayed fluorescence from relatively short-lived triplets.

The measurement of luminescence polarisation, until recently applied by a limited number of workers for special purposes, is expected to be more widely adopted, e.g., for distinguishing between overlapping spectral bands, or for reducing the effects of scattered light, as well as for the investigation of the more subtle effects connected with energy transfer and molecular rotation. We may expect the discovery of unusual media or other conditions giving extra discrimination in particular situations. We have seen how the use of certain crystalline solvents at low temperature can, with certain substances, give extraordinarily sharp emission spectra, and there are no doubt many other special conditions waiting to be discovered, e.g. the use of unusual solvents or mixtures of solvents, rigid polymers or perhaps adsorption on a particular type of surface, or on a particular kind of polymer molecule in solution (see for example the work of Oster and associates [430–432] with adsorbed dyestuffs). In addition there are several well-known phenomena that have been little exploited analytically. One example is the external heavy atom effect, which has recently been applied by Zander in connection with the measurement of both phosphorescence [433] and fluorescence [434]. Another is the application of photochemical reactions to produce new luminescent species. Unusual catalytic effects, particularly the use of enzymes, also provide a lively field for further investigation.

Complete analysis of a complex mixture of components having overlapping luminescence excitation and emission spectra can in principle be achieved by measuring the emission spectra of the mixture excited by a large number of wavelengths. The application of such measurements to the enumeration of the components of complex mixtures has been discussed by Weber [435] who sets down the results as an $m \times n$ matrix, in which the m columns are determined by the wavelengths of excitation and the n rows by the wavelengths of observation. The matrix elements are numbers proportional to the response of the detector. In general, the rank of the matrix gives the number of components in the system having distinct absorption and fluorescence spectra. Weber applied the method to one-and two-component mixtures. With the use of a computer, the method should prove to be a powerful tool for the investigation of complex mixtures of unknown luminescent components. The consecutive recording of a complete series of spectra is however a tedious task and an alternative approach to this problem has been described by Schachter and Haenni [436]. In this method the excitation and emission monochromators are scanned simultaneously and the results are presented on an oscilloscope screen. The X axis of the display represents the emission wavelength, the Y axis the excitation wavelength, and the beam trigger is arranged to operate at a preselected level of emission intensity to produce a "fluorograph" or contour at that intensity level. A combination of fluorographs corresponding to several intensity levels are then assembled into a three-dimensional model, or "stereofluorograph" which contains all details of the three spectral parameters. Schachter and Haenni applied the technique to polynuclear aromatic hydrocarbons and proposed its use for finger printing pure compounds. A valuable development would be the direct coupling of the output from this device to a computer so that the intensities corresponding to a large number of excitation/emission wavelength combinations could be recorded and analysed rapidly by Weber's method. Investigation of complex mixtures would then be greatly facilitated.

The analytical chemist is required to analyse ever smaller specimens. In the organic field the biochemist wishes to determine the composition of parts of living cells, and the application of micro-fluorescence or phosphorescence spectrometry shows great promise. For example, Olson[437] has described an apparatus consisting essentially of a combination of a microscope with a spectrofluorimeter, and has applied it to the study of the photochemistry of living cells and the spectral changes undergone during irradiation. Similar equipment has been used by Loeser[438] to measure chlorpromazine fluorescence spectra in living single cells and to study the fluorescence of acridine orange, reserpine, tetracycline, hormones, etc. in microscopic specimens. Similar techniques could be adapted to measure very small volumes of solution, or powder specimens of both organic or inorganic origin.

In work on the micro scale, the laser, as an excitation source, clearly has the advantage of providing a high flux of exciting light on a very small area. Photochemical decomposition will no doubt present problems, but these can sometimes be minimised by working in rigid medium at low temperature. On the somewhat larger scale, the laser can be used to obtain light fluxes sufficiently great to produce annihilation-controlled delayed fluorescence, even from short-lived triplets. High power lasers also provide a useful method of exciting phosphorescence and delayed fluorescence by triplet–singlet absorption (see for example references 439 and 440). Lasers giving some wavelengths in the ultra-violet region are already available and with the present rapid advance in laser technology we can expect their widespread use as excitation sources in many areas of photoluminescence measurement.

Definitions of Symbols

Roman

A	Proportionality constant.
^1A, ^1B, etc.	Molecules A, B etc. in their singlet ground states.
$^1A^*$, $^1B^*$, etc.	Molecules A, B, etc. in their first excited singlet states.
^3A, ^3B, etc.	Molecules A, B, etc. in their lowest triplet states.
A_{nm}	Einstein transition probability for spontaneous emission.
A_u	Area of light source.
A_v	Area of image of light source.
$\mathbf{A}$	Frequency factor.
a	Molecular radius.
B, $B_{\bar{\nu}}$	Relative band width at constant slit width.
B_{mn}	Einstein transition probability for absorption.
B_{nm}	Einstein constant for stimulated emission.
c	Concentration.
c_h	"Half-value" concentration.
$\mathbf{c}$	Velocity of light *in vacuo*.
D, D_λ	Optical density.
$\mathbf{D}_A$, $\mathbf{D}_B$, etc.	Diffusion coefficients.
d	Height of dispersing element.
d, d_1, d_2	Depths at which fluorescence is viewed.
E	Quantity of energy.
ΔE	Energy difference.
E_R	Repulsion energy between ground state molecules.
E_s	Energy of first excited singlet state.
$dE/d\lambda$	Energy emitted per sec within unit wavelength interval.
$\mathbf{E}$	Electric vector.
F	Aperture of monochromator.
F	Observed intensity of fluorescence.
f	Activity coefficient.
f	Focal length of monochromator.

f'	Fraction of compound present in its basic form.
f_D	Fluorescence emission spectrum of donor normalised to unity.
f_D^i, f_M^i	Intensities of dimer and monomer emission at isosbestic wavelength.
G	Gibbs free energy.
g_l, g_u	Multiplicities of ground and excited states.
H	Half-band width of emission spectrum.
H	Heat content (enthalpy).
$-\Delta H$, $-\Delta H^*$	Heats of formation in ground and excited states.
$-\Delta H_A$	Heat of formation of excited dimer.
h	Planck's constant (6.624×10^{-27} erg sec).
h	Height of slit.
I	Intensity of light.
I_A	Rate of light absorption (quanta sec^{-1}) in a volume l cc, where l is the optical depth.
I_a	Rate of light absorption (einstein $litre^{-1}$ sec^{-1}).
I_{DF}	Total rate of emission of delayed fluorescence.
I_{DF}', I_{DF}''	Rates of emission of delayed fluorescence by interaction of like and unlike triplets.
I_p	Rate of emission of phosphorescence.
$I_{\parallel}$, $I_{\perp}$	Intensity of vertically and horizontally polarised light.
I_λ	Intensity of a continuous source at wavelength λ (einstein sec^{-1} nm^{-1} per unit solid angle).
i_x, i_y	Photomultiplier output with polar oriented to pass vertically or horizontally polarised light.
i_a, i_d	AC and DC components.
J	Intensity of monochromatic line from a discontinuous source (einstein sec^{-1} per unit solid angle).
K, K*	Equilibrium constants in ground and excited states.
K	Proportionality constant.
K′ and K″	Constants relating to dimer/monomer emission ratio.
(K′)*	$= k_1/k_2$.
K_1, K_2, K_3	Constants relating to dimer/monomer emission ratio.
K_c	Equilibrium constant for complex formation.
K	Stern–Volmer quenching constant.
k	Effective molecular cross section for light absorption.
k	Proportionality constant.

k	Boltzmann's constant (1.3805×10^{-16} erg deg^{-1}).
k (with subscript)	See list of rate constants below.
L, $L_{\bar{\nu}}$	Transmission factor of a monochromator.
^1L_a, ^1L_b	Lowest excited singlet states of aromatic hydrocarbon (Platt).
LGP	Light gathering power of a monochromator.
l	Optical depth.
M	Molarity.
m	Linear dispersion of monochromator (mm nm^{-1}).
m_f, m_s	Degrees of modulation of fluorescence and exciting light.
N	Avogadro's number (6.023×10^{23} mole^{-1}).
N	Pre-exponential factor.
n	Number of molecules.
n	Refractive index.
n	Lone pair orbital.
P	Dipole moment.
P, $P_{\bar{\nu}}$	Photomultiplier response per quantum.
P_{SS}, P_{TS}	Observed phosphorescence intensities by excitation in the $S_1 \leftarrow S_0$ and $T_1 \leftarrow S_0$ absorption bands.
p	Degree of polarisation.
p′	Observed degree of polarisation.
p_e	Apparent degree of polarisation.
p_L	Degree of polarisation of exciting beam.
p_0	Principal polarisation.
p_{O_2}	Partial pressure of oxygen.
p (with subscript)	Probability factor—see list below.
Q	Quencher of excited singlet state.
Q	Rate of emission of fluorescence.
$d\mathbf{Q}/d\bar{\nu}$	Einsteins emitted per second within unit wavenumber interval.
q, q_1, q_2	Quenchers of triplet state.
R	Recorder chart reading.
R	Ratio of rates of second order to first order triplet decay.
R_c	Critical distance for energy transfer.
R_M, R_D	Total rates of population of excited monomeric and dimeric states.
$R_{\bar{\nu}}$	Instrument reading when set to wavenumber $\bar{\nu}$.
R	Gas constant.

$R_{\parallel}$, $R_{\perp}$	Observed readings of vertical and horizontal components of fluorescence with vertically polarised exciting light.
$R_{\parallel}'$, $R_{\perp}'$	Corresponding readings with horizontally polarised exciting light.
r	Molecular radius.
S	Entropy.
S	Solubility.
ΔS^*	Entropy of formation of excited dimer.
S_0, S_1 etc.	Ground and excited singlet states.
S_λ	Spectral sensitivity of monochromator/photomultiplier combination expressed in relative units of wavelength interval per quantum.
$S_{\bar{\nu}}$	As for S_λ but in relative units of wavenumber interval per quantum.
s	Height of light source.
T	Lowest triplet state.
T	Relative transmission of monochromator for vertically and horizontally polarised light.
T_1, T_2 etc.	First, second, etc. triplet states.
T_λ	Light flux passing monochromator from monochromatic line at wavelength λ (einstein sec^{-1}).
$T_{\Delta\lambda}$	Light flux passing monochromator set with band width $\Delta\lambda$ from a continuous source (einstein sec^{-1}).
T	Absolute temperature.
t	Time.
u	Object distance.
u_ν	Radiation density.
V	Molar volume.
v	Image distance.
w	Instrumental constant.
w, w_1, w_2	Slit widths of monochromator.
x	Fraction of light absorbed per cm.
x	Fraction of compound that dissociates in the excited state before fluorescing.

First Order Rate Constants

k_1	Reaction of excited acid with water.
k_3	Reaction of excited base with water.

k_d	Excited dimer dissociation.
k_e	Thermal activation followed by intersystem crossing $S_1 \leftarrow T_1$.
k_f	Emission of fluorescence.
k_f	Fluorescence of acid.
k_f'	Fluorescence of base.
k_f'	Fluorescence of dimer.
k_g	Intersystem crossing $S_1 \rightarrow T_1$.
k_g'	Intersystem crossing $S_n \rightarrow T_n$.
k_{ga}	Intersystem crossing $T_1^* \rightarrow S_1$.
k_h, k_j	Composite constants for first order triplet decay (with superscripts as necessary to indicate compound concerned)
k_h^0	Limiting value of k_h at low temperature.
k_m	Intersystem crossing $T_1 \rightarrow S_0$.
k_n	Internal conversion $S_1 \rightarrow S_0$.
k_n'	Internal conversion $S_n \rightarrow S_{n-1}$ or $T_n \rightarrow T_{n-1}$.
k_n'	De-activation of excited dimer by non-radiative processes.
k_0	Radiationless de-activation of excited singlet state of acid.
k_0'	Radiationless de-activation of excited singlet state of base.
k_p	Emission of phosphorescence.
k_r	Reaction of product of triplet–triplet annihilation to give products other than excited singlet monomer.
k_s	Reaction of product of triplet–triplet annihilation to give singlet excited monomer.
k_v	Reaction of triplet with solvent.

Second Order Rate Constants

k_2	Reaction of excited base with H_3O^+.
k_4	Reaction of excited acid with OH^-.
k_A	Association of excited and ground state monomer.
k_a	Reaction of excited base with acid.
k_a	Composite constant for second order triplet decay.
k_b	Reaction of excited acid with base.
k_c	Diffusion-controlled rate constant calculated from $k_c = 8\mathbf{RT}/3000\,\eta$.
k_G	Quenching of triplet by ground state.
k_Q	Quenching of singlet by solute Q.

k_q, k_q'	Quenching of triplet by solute q.
k_X	Formation of product X by triplet–triplet annihilation.

Probability Factors

p_{AA}	Probability that an encounter between two triplets of compound A will ultimately lead to the formation of $^1A^*$.
p_a	Probability that the annihilation of two like triplets will ultimately lead to the formation of an excited singlet molecule.
p_c	Probability that an encounter between two like triplets will ultimately give rise to an excited singlet molecule.
p_{DA}	Probability that an encounter between triplets of compounds D and A will ultimately lead to the formation of $^1A^*$.
p_e	Probability that an encounter between ^1A and ^3D will lead to the formation of ^3A.
p_e'	Proportion of quenching encounters between ^1A and ^3D that lead to the formation of ^3A.

Greek

α	Electrical polarisability.
α	Angular dispersion of monochromator.
α	Angle between directions of emitted ray and incident ray.
α	Birk's constant (ratio of rates of formation of excited dimer and excited monomer by triplet–triplet interaction).
β	Angle between absorption and emission oscillators.
γ	Inclination to the vertical of plane of polarisation of the horizontal ray emitted at right angles to the direction of propagation of the exciting light.
δ	Depolarisation ratio.
ε, ε'	Molecular extinction coefficient.
ε_{SS}, ε_{TS}	Molecular extinction coefficients within the $S_1 \leftarrow S_0$ and $T_1 \leftarrow S_0$ absorption bands.
ξ	Phase of fluorescence relative to that of the exciting light.
η	Viscosity.
$1/\eta_0$	"Limiting value" of fluidity at high temperature.
θ	Angle between transition moment and electric vector.

θ	Total efficiency of P-type delayed fluorescence.
θ_A, etc.	Efficiency of P-type delayed fluorescence of A, etc.
θ'	Efficiency of delayed fluorescence produced by interaction between like triplets.
θ''	Efficiency of delayed fluorescence produced by interaction between unlike triplets.
θ_1	Solid angle subtended by lens at light source.
θ_2	Solid angle subtended by collimator at slit.
κ	Orientation factor.
λ	Wavelength.
ν	Frequency.
$\bar{\nu}$	Wavenumber.
π	Mathematical constant.
π	Type of orbital.
ρ	Rotational relaxation time.
σ	Type of orbital.
σ	Encounter distance.
$\sigma_{\bar{\nu}}$	Reflectivity.
τ, τ'	Lifetime.
τ_A, τ^A	Triplet lifetime of compound A.
τ_D	Lifetime of excited dimer at infinite dilution.
τ_{DF}	Lifetime of delayed fluorescence.
$\tau_M{}^0$	Lifetime of monomer at infinite dilution.
τ_0	Lifetime in absence of added quencher.
τ_R	Radiative lifetime of triplet state.
τ_r	Radiative lifetime of singlet state.
τ_{SS}, τ_{TS}	Lifetimes of delayed fluorescence excited by absorption in the $S_1 \leftarrow S_0$ and $T_1 \leftarrow S_0$ absorption bands.
τ_t	Lifetime of triplet.
Φ	Quantum efficiency of photochemical reaction.
Φ_D, Φ_M	Total efficiencies of luminescence of dimer and monomer.
ϕ	Azimuthal angle between molecular axis and direction of observation.
ϕ	Efficiency of prompt fluorescence or phosphorescence.
ϕ, ϕ'	Fluorescence efficiencies of acidic and basic forms.
ϕ_A, etc.	Fluorescence efficiency of A, etc.
ϕ_e	Efficiency of E-type delayed fluorescence.

ϕ_f	Efficiency of prompt fluorescence.
ϕ_f^0	Efficiency of prompt fluorescence in absence of added quencher.
ϕ_0, ϕ_0'	Fluorescence efficiencies of acidic and basic forms in strongly acid and strongly alkaline solutions respectively.
ϕ_p	Efficiency of phosphorescence.
ϕ_t	Triplet formation efficiency.
ω	Rate of modulation.

Some Textbooks on Photoluminescence and Related Subjects

I. B. Berlman, *Handbook of Fluorescence Spectra of Aromatic Molecules*, Academic Press, New York, 1965.

E. J. Bowen, *The Chemical Aspects of Light*, Clarendon Press, Oxford, 1946.

E. J. Bowen and F. Wokes, *Fluorescence of Solutions*, Longmans, Green & Co., London, 1953.

J. G. Calvert and J. N. Pitts, Jr., *Photochemistry*, John Wiley and Sons, Inc., New York, 1966.

F. Daniels (ed.), *Photochemistry in the Liquid and Solid States*, John Wiley and Sons, Inc., New York, 1960.

C. Ellis and A. A. Wells, *The Chemical Action of Ultraviolet Rays*, Reinhold Publishing Corporation, New York, 1941.

Th. Förster, *Fluoreszenz Organischer Verbindungen*, Vandenhoeck and Ruprecht, Göttingen, 1951.

G. F. J. Garlick, *Luminescence*, Handbuch der Physik, Vol. 26, Springer-Verlag, Berlin, 1958.

D. M. Hercules (ed.), *Fluorescence and Phosphorescence Analysis*, Interscience Publishers, Inc., New York, 1966.

H. Kallman and G. M. Spruch (eds.), *Luminescence of Organic and Inorganic Materials*, John Wiley and Sons, Inc., New York, 1962.

M. A. Konstantinova-Schlezinger (ed.), *Fluorimetric Aanlysis*, Israel Program for Scientific Translations, Jerusalem, 1965 (from Russian).

H. W. Leverenz, *Luminescence of Solids*, Chapman and Hall Ltd., London, 1950.

W. D. McElroy and B. Glass (eds.), *Light and Life*, The John Hopkins Press, 1961.

J. N. Murrell, *The Theory of the Electronic Spectra of Organic Molecules*, Methuen and Co., Ltd., London, 1963.

W. A. Noyes, Jr., G. S. Hammond and J. N. Pitts, Jr., *Advances in Photochemistry*, Interscience Publishers, Inc., New York, Vol. 1 (1963), Vol. 2 (1964), Vol. 3 (1964), Vol. 4 (1966).

R. A. Passwater, *Guide to the Fluorescence Literature*, Plenum Press, New York, 1967.

P. Pringsheim, *Fluorescence and Phosphorescence*, Interscience Publishers, Inc., New York, 1949.

J. A. Radley and J. Grant, *Fluorescence Analysis in Ultra-violet Light*, 4th edition, Chapman and Hall Ltd., London, 1954.

C. N. R. Rao, *Ultra-violet and Visible Spectroscopy*, Butterworths, London, 1961.

C. Reid, *Excited States in Chemistry and Biology*, Butterworths Scientific Publications, London, 1957.

S. Udenfriend, *Fluorescence Assay in Biology and Medicine*, Academic Press, New York and London, 1962.

C. E. White and A. Weissler, in L. Meites (ed.), *Handbook of Analytical Chemistry*, McGraw Hill, 1964.

A. B. Zahlan (ed.), *The Triplet State*, Beirut Symposium, Cambridge University Press, Cambridge, 1967.

Reactivity of the Photoexcited Organic Molecule, Proc. 13th Conf. on Chemistry, Brussels, 1965, Interscience Publishers, Inc., New York, 1967.

References

1

1 E. LIPPERT, W. LÜDER and F. MOLL, *Spectrochim. Acta*, 10 (1959) 858.
2 E. LIPPERT, *Z. Naturforsch.*, 10a (1955) 541; *Z. Elektrochem.*, 61 (1957) 962.
3 N. MATAGA, Y. KAIFU and M. KOIZUMI, *Bull. Chem. Soc. Japan*, 28 (1955) 690; 29 (1956) 465.
4 H. K. HUGHES, *Anal. Chem.*, 24 (1952) 1349; 33 (1961) 1968.
5 E. J. BOWEN and F. WOKES, *Fluorescence of Solutions*, Longmans Green & Co., London, 1953.
6 TH. FÖRSTER, *Fluoreszenz Organischer Verbindungen*, Vandenhoeck and Ruprecht, Göttingen, 1951.
7 J. B. BIRKS and D. J. DYSON, *Proc. Roy. Soc. (London) A*, 275 (1963) 135.
8 J. N. MURRELL, *The Theory of the Electronic Spectra of Organic Molecules*, Methuen and Co. Ltd., London, 1963.
9 G. PORTER, in *Reactivity of the Photoexcited Organic Molecule*, Proc. 13th Conf. on Chemistry, Brussels, 1965, Interscience Publishers, Inc., New York, 1967.
10 M. KASHA, *Discussions Faraday Soc.*, No. 9 (1950) 14.
11 M. KASHA, in W. D. MCELROY and B. GLASS (eds.), *Light and Life*, The John Hopkins Press, 1961.
12 G. PORTER and P. SUPPAN, *Trans. Faraday Soc.*, 61 (1965) 1664.
13 C. A. PARKER, unpublished work.
14 R. S. MULLIKEN, *J. Phys. Chem.*, 56 (1952) 801.
15 G. BRIEGLEB and J. CZEKALLA, *Angew. Chem.*, 72 (1960) 401.
16 D. CZEKALLA and K. O. MEYER, *Z. Physik. Chem. (Frankfurt)*, 27 (1961) 185; G. D. SHORT and C. A. PARKER, *Spectrochim. Acta*, 23 A (1967) 2487.
17 J. SIDMAN, *Chem. Rev.*, 58 (1958) 689.
18 R. BAUER and A. BACZYNSKI, *Bull. Acad. Polon. Sci.*, 6 (1958) 113.
19 C. A. PARKER and C. G. HATCHARD, *Trans. Faraday Soc.*, 57 (1961) 1894.
20 E. C. LIM, J. D. LAPOSA and J. M. H. YU, *J. Mol. Spectry.*, 19 (1966) 412.
21 M. F. O'DWYER, M. A. EL-BAYOUMI and S. J. STRICKLER, *J. Chem. Phys.*, 36 (1962) 1395.
22 S. KATO, S. LIPSKY and C. L. BRAUN, *J. Chem. Phys.*, 37 (1962) 190.
23 E. J. BOWEN and J. SAHU, *J. Phys. Chem.*, 63 (1959) 4.
24 E. J. BOWEN and D. SEAMAN, in H. KALLMAN and G. M. SPRUCH (eds.), *Luminescence of Organic and Inorganic Materials*, John Wiley and Sons, Inc., New York, 1962.
25 F. PERRIN, *Ann. Phys. (Paris)*, 12 (1929) 169.
26 A. JABLONSKI, *Nature*, 131 (1933) 839.
27 A. JABLONSKI, *Z. Physik*, 94 (1935) 38.
28 W. L. LEVSHIN and L. A. VINOKUROV, *Physik. Z. Sowjet U.*, 10 (1936) 10.
29 G. N. LEWIS, D. LIPKIN and T. T. MAGEL, *J. Am. Chem. Soc.*, 63 (1941) 3005.
30 G. N. LEWIS and M. KASHA, *J. Am. Chem. Soc.*, 66 (1944) 2100.
31 S. BOUDIN, *J. Chim. Phys.*, 27 (1930) 285.
32 H. KAUTSKY, *Chem. Ber.*, 68 (1935) 152.
33 H. L. J. BÄCKSTRÖM and K. SANDROS, *Acta Chem. Scand.*, 12 (1958) 823.
34 C. A. PARKER and C. G. HATCHARD, *Trans. Faraday Soc.*, 59 (1963) 284.
35 C. A. PARKER and C. G. HATCHARD, *J. Phys. Chem.*, 66 (1962) 2506.
36 C. A. PARKER, C. G. HATCHARD and T. A. JOYCE, *J. Mol. Spectry.*, 14 (1964) 311.

37 P. P. Dikun, *Zh. Eksperim. i Teor. Fiz.*, 20 (1950) 193.
38 R. Williams, *J. Chem. Phys.*, 28 (1958) 577.
39 B. Stevens, E. Hutton and G. Porter, *Nature*, 185 (1960) 917.
40 Th. Förster and K. Kasper, *Z. Elektrochem.*, 59 (1955) 977.
41 B. Stevens and E. Hutton, *Nature*, 186 (1960) 1045.
42 B. Stevens and E. Hutton, *Spectrochim. Acta*, 18 (1962) 425.
43 C. A. Parker and C. G. Hatchard, *Nature*, 190 (1961) 165.
44 J. B. Birks and I. H. Munro, in H. Kallman and G. M. Spruch (eds.), *Luminescence of Organic and Inorganic Materials*, John Wiley and Sons, Inc., New York, 1962.
45 C. A. Parker and C. G. Hatchard, *Proc. Chem. Soc.*, (1962) 147.
46 C. A. Parker and C. G. Hatchard, *Proc. Roy. Soc. (London) A*, 269 (1962) 574.
47 B. Stevens, M. S. Walker and E. Hutton, *Proc. Chem. Soc.*, (1963) 62.
48 G. N. Lewis and D. Lipkin, *J. Am. Chem. Soc.*, 64 (1942) 2801.
49 P. Debye and J. O. Edwards, *J. Chem. Phys.*, 20 (1952) 236.
50 H. Linschitz, M. G. Berry and D. Schweitzer, *J. Am. Chem. Soc.*, 76 (1954) 5833.
51 W. C. Meyer and A. C. Albrecht, *J. Phys. Chem.*, 66 (1962) 1168.
52 E. Dolan and A. C. Albrecht, *J. Chem. Phys.*, 37 (1962) 1149; 38 (1963) 567.
53 W. M. McClain and A. C. Albrecht, *J. Chem. Phys.*, 43 (1965) 465.
54 E. C. Lim and G. W. Swenson, *J. Chem. Phys.*, 36 (1962) 118; 39 (1963) 2768.
55 E. C. Lim and W. Y. Wen, *J. Chem. Phys.*, 39 (1963) 847.
56 E. C. Lim, C. P. Lazzara, M. Y. Yang and G. W. Swenson, *J. Chem. Phys.*, 43 (1965) 970.
57 K. D. Cadogan and A. C. Albrecht, *J. Chem. Phys.*, 43 (1965) 2550.
58 B. Stevens and M. S. Walker, *Chem. Commun.*, (1965) 8.
59 C. A. Parker and T. A. Joyce, *J. Chem. Soc. A*, (1966) 821.
60 W. A. Gibbons, G. Porter and M. I. Savadatti, *Nature*, 206 (1965) 1355.
61 B. Brocklehurst, W. A. Gibbons, F. T. Lang, G. Porter and M. I. Savadatti, *Trans. Faraday Soc.*, 62 (1966) 1793.
62 F. Perrin, *J. Phys. Radium*, 7 (1926) 390.
63 A. Jablonski, *Z. Physik*, 96 (1935) 236.
64 T. Azumi and S. P. McGlynn, *J. Chem. Phys.*, 37 (1962) 2413.
65 A. Jablonski, *Acta Phys. Polon.*, 10 (1950) 33, 193.
66 G. Weber, *Trans. Faraday Soc.*, 50 (1954) 552.
67 P. P. Feofilov and B. J. Sveshnikoff, *J. Phys. USSR*, 3 (1940) 493.
68 J. R. Partington, *An Advanced Treatise on Physical Chemistry*, Vol. 4, Longmans, Green and Co., London, 1953.
69 G. Herzberg, *Molecular Spectra and Molecular Structure. II. Infra-red and Raman Spectra of Polyatomic Molecules*, D. Van Nostrand Co. Inc., New York, 1959.
70 J. H. Hibben, *The Raman Effect and Its Chemical Applications*, Reinhold Publishing Corp., Inc., New York, 1939.
71 J. Eisenbrand and H. Picher, *Z. Elektrochem.*, 50 (1944) 72.
72 J. Eisenbrand, *Optik*, 11 (1954) 557.
73 C. A. Parker, *Analyst*, 84 (1959) 446.
74 J. M. Price, M. Kaihara and H. K. Howerton, *Appl. Opt.*, 1 (1962) 521.
75 O. Stern and M. Volmer, *Phys. Z.*, 20 (1919) 183.
76 E. J. Bowen, *Trans. Faraday Soc.*, 50 (1954) 97.
77 C. D. Hodgman (ed.), *Handbook of Chemistry and Physics*, 13th edition, Chemical Rubber Publishing Co., Cleveland, 1948.
78 B. Williamson and V. K. de la Mer, *J. Am. Chem. Soc.*, 70 (1948) 717.
79 P. Debye, *Trans. Electrochem. Soc.*, 82 (1942) 265.

80 M. Smoluchowski, *Z. Physik. Chem. (Leipzig)*, 92 (1917) 129.
81 G. G. Stokes (1850), *Mathematical and Physical Papers*, Vol. III, Cambridge University Press, London, 1903, pp. 1 and 55.
82 A. Einstein, *Ann. Physik*, 17 (1905) 549; 19 (1906) 371; *Z. Elektrochem.*, 14 (1908) 235.
83 A. D. Osborne and G. Porter, *Proc. Roy. Soc. (London) A*, 284 (1965) 9.
84 C. A. Parker, *Trans. Faraday Soc.*, 60 (1964) 1998.
85 C. Tanaka, J. Tanaka, E. Hutton and B. Stevens, *Nature*, 198 (1963) 1192.
86 C. A. Parker, *Nature*, 200 (1963) 331.
87 R. M. Noyes, in G. Porter and B. Stevens (eds.), *Progress in Reaction Kinetics*, Vol. 1, Pergamon Press, Oxford, 1961.
88 M. Beer and H. C. Longuet–Higgins, *J. Chem. Phys.*, 23 (1955) 1390.
89 G. Viswanath and M. Kasha, *J. Chem. Phys.*, 24 (1956) 574.
90 C. A. Parker and C. G. Hatchard, unpublished work.
91 J. W. Hilpern, *Trans. Faraday Soc.*, 61 (1965) 605; J. B. Birks, J. M. De C. Conte and G. Walker, *Phys. Letters*, 19 (1965) 125.
92 R. M. Hochstrasser and R. D. McAlpine, *J. Chem. Phys.*, 44 (1966) 3325; I. H. Munro, T. D. S. Hamilton, J. P. Ray and G. F. Moore, *Phys. Letters*, 20 (1966) 386.
93 E. Gaviola and P. Pringsheim, *Z. Physik*, 24 (1924) 24.
94 F. Weigert and G. Käppler, *Z. Physik*, 25 (1924) 99.
95 J. Perrin, *in 2eme Conseil de Chimie Solvay, Brussels*, 1924, Gauthier-Villars, Paris, 1925.
96 J. Perrin, *Compt. Rend.*, 184 (1927) 1097.
97 Th. Förster, *Ann. Physik*, series 6, 2 (1948) 55.
98 Th. Förster, *Discussions Faraday Soc.*, 27 (1959) 7.
99 J. G. Calvert and J. N. Pitts, Jr., *Photochemistry*, John Wiley and Sons, Inc., New York, 1966.
100 E. J. Bowen and R. Livingston, *J. Am. Chem. Soc.*, 76 (1954) 6300.
101 J. T. Dubois and B. Stevens, in H. P. Kallman and G. M. Spruch (eds.), *Luminescence of Organic and Inorganic Materials*, John Wiley and Sons Inc., New York, 1962.
102 G. Weber and F. W. J. Teale, *Trans. Faraday Soc.*, 54 (1958) 640.
103 T. Medinger and F. Wilkinson, *Trans. Faraday Soc.*, 61 (1965) 620.
104 C. A. Parker and T. A. Joyce, *Trans. Faraday Soc.*, 62 (1966) 2785.
105 A. A. Lamola and G. S. Hammond, *J. Chem. Phys.*, 43 (1965) 2129.
106 A. Terenin and V. Ermolaev, *Trans. Faraday Soc.*, 52 (1956) 1042.
107 H. L. J. Bäckström and K. Sandros, *Acta Chem. Scand.*, 14 (1960) 48.
108 G. Porter and F. Wilkinson, *Proc. Roy. Soc. (London) A*, 264 (1961) 1.
109 K. Sandros and H. L. J. Bäckström, *Acta Chem. Scand.*, 16 (1962) 958.
110 C. A. Parker, C. G. Hatchard and T. A. Joyce, *Nature*, 205 (1965) 1282.
111 C. A. Parker, in W. A. Noyes, Jr., G. S. Hammond and J. N. Pitts, Jr., (eds.), *Advances in Photochemistry*, Vol. 2, Interscience Publishers, New York, 1964.
112 C. A. Parker, *Spectrochim. Acta*, 19 (1963) 989.
113 C. A. Parker and T. A. Joyce, *Chem. Commun.*, (1966) 108.
114 C. A. Parker and T. A. Joyce, *Photochem. Photobiol.*, 6 (1967) 395.
115 B. Stevens and M. S. Walker, *Proc. Roy. Soc. (London) A*, 281 (1964) 420.
116 S. Czarnecki, *Bull. Acad. Polon. Sci.*, 9 (1961) 561.
117 B. Muel, *Compt. Rend.*, 255 (1962) 3149.
118 T. Azumi and S. P. McGlynn, *J. Chem. Phys.*, 38 (1963) 2773.
119 T. Azumi and S. P. McGlynn, *J. Chem. Phys.*, 39 (1963) 1186.
120 F. Dupuy, *Thesis*, University of Bordeaux, 1964.

121 T. Azumi and S. P. McGlynn, *J. Chem. Phys.*, 39 (1963) 3533; F. J. Smith and S. P. McGlynn, *J. Chem. Phys.*, 42 (1965) 4308.
122 S. P. McGlynn, M. R. Padhye and M. Kasha, *J. Chem. Phys.*, 23 (1955) 593.
123 C. A. Parker, *Proc. Roy. Soc.* (*London*) *A*, 276 (1963) 125.
124 C. A. Parker and C. G. Hatchard, *Proc. Chem. Soc.*, (1962) 386.
125 C. A. Parker, C. G. Hatchard and T. A. Joyce, *Analyst*, 90 (1965) 1.
126 P. Pringsheim, *Fluorescence and Phosphorescence*, Interscience Publishers Ltd., London, 1949.
127 R. L. Bowman, P. A. Caulfield and S. Udenfriend, *Science*, 122 (1955) 32.
128 G. Weber and F. W. J. Teale, *Biochem. J.*, 65 (1957) 476.
129 F. W. J. Teale, *Biochem. J.*, 76 (1960) 381.
130 C. E. White, D. E. Hoffman and J. S. Magee, *Spectrochim. Acta*, 9 (1957) 105.
131 C. A. Parker and W. J. Barnes, *Analyst*, 82 (1957) 606.
132 H. E. Johns and A. M. Ralph, *Photochem. Photobiol.*, 4 (1965) 673 and 693.
133 W. Elenbaas and J. Riemens, *Philips Techn. Rev.*, 11 (1950) 299.
134 E. H. Nelson, *Phot. J.*, 89B (1949) 54.
135 C. W. Sill, *Anal. Chem.*, 33 (1961) 1584.
136 E. J. Bowen, *The Chemical Aspects of Light*, Clarendon Press, Oxford, 1946.
137 M. Kasha, *J. Opt. Soc. Am.*, 38 (1948) 929.
138 J. Sharpe, *Electronic Technology*, 38 (1961) 196, 248.
139 G. R. Harrison and P. A. Leighton, *Phys. Rev.*, 88 (1931) 899.
140 E. J. Bowen, *Proc. Roy. Soc.* (*London*) *A*, 154 (1936) 349.
141 W. H. Melhuish, *New Zealand J. Sci. Technol.*, 37, 2B (1955) 142; *J. Opt. Soc. Am.*, 52 (1962) 1256.
142 G. Weber and F. W. J. Teale, *Trans. Faraday Soc.*, 53 (1957) 646.
143 C. A. Parker, *Nature*, 182 (1958) 1002.
144 C. A. Parker, *Proc. Roy. Soc.* (*London*) *A*, 220 (1953) 104.
145 C. A. Parker, *Trans. Faraday Soc.*, 50 (1954) 1213.
146 C. G. Hatchard and C. A. Parker, *Proc. Roy. Soc.* (*London*) *A*, 235 (1956) 518.
147 C. A. Parker and C. G. Hatchard, *J. Phys. Chem.*, 63 (1959) 22.
148 J. Lee and H. H. Seliger, *J. Chem. Phys.*, 40 (1964) 519.
149 J. H. Baxendale and N. K. Bridge, *J. Phys. Chem.*, 59 (1955) 783.
150 C. A. Parker and W. T. Rees, *Analyst*, 87 (1962) 83.
151 Y. Hirshberg and E. Fischer, *Rev. Sci. Instr.*, 30 (1959) 197.
152 R. J. Argauer and C. E. White, *Anal. Chem.*, 36 (1964) 368.
153 W. Slavin, R. W. Mooney and D. T. Palumbo, *J. Opt. Soc. Am.*, 51 (1961) 93.
154 F. R. Lipsett, *J. Opt. Soc. Am.*, 49 (1959) 673.
155 G. K. Turner, *Science*, 146 (1964) 183.
156 W. H. Melhuish and R. H. Murashige, *Rev. Sci. Instr.*, 33 (1962) 1213.
157 C. W. Sill, *Anal. Chem.*, 33 (1961) 1579.
158 C. A. Parker and W. T. Rees, *Analyst*, 85 (1960) 587.
159 F. Benford, G. P. Lloyd and S. Schwarz, *J. Opt. Soc. Am.*, 38 (1948) 445 and 964.
160 C. A. Parker, *Anal. Chem.*, 34 (1962) 502.
161 H. C. Borresen and C. A. Parker, *Anal. Chem.*, 38 (1966) 1073.
162 W. H. Melhuish, *J. Phys. Chem.*, 64 (1960) 762.
163 R. Lippert, W. Nagele, I. Seibold-Blankenstein, W. Steiger and W. Voss, *Z. Anal. Chem.*, 170 (1959) 1.
164 J. H. Chapman, Th. Förster, G. Kortüm, E. Lippert, W. H. Melhuish, G. Nebbia and C. A. Parker, *Photoelec. Spectrometry Group Bull.*, No. 14 (1962) 378; *Appl. Spectry.*, 17 (1963) 171; *Z. Anal. Chem.*, 197 (1963) 431.

165 H. V. Drushel, A. L. Sommers and R. C. Cox, *Anal. Chem.*, 35 (1963) 2166.
166 P. Rosen and G. M. Edelman, *Rev. Sci. Instr.*, 36 (1965) 809.
167 T.D. S. Hamilton, *J. Sci. Instr.*, 43 (1966) 49.
168 G. Kortüm and W. Hess, *Z. Physik. Chem. (Frankfurt)*, 19 (1959) 142.
169 R. M. Hill, R. Oechsli and E. Marg, *J. Opt. Soc. Am.*, 51 (1961) 1139.
170 M. I. Mekshenkov and A. P. Andreitsev, *Biophysics (USSR)*, 6 (1961) 88.
171 F. R. Lipsett, *Photoelec. Spectrometry Group Bull.*, No. 16 (1965) 474.
172 S. I. Vavilov, *Z. Physik*, 22 (1924) 266.
173 L. S. Forster and R. Livingston, *J. Chem. Phys.*, 20 (1952) 1315.
174 E. J. Bowen and J. W. Sawtell, *Trans. Faraday Soc.*, 33 (1937) 1425.
175 W. H. Melhuish, *J. Phys. Chem.*, 65 (1961) 229.
176 E. H. Gilmore, G. E. Gibson and D. S. McClure, *J. Chem. Phys.*, 20 (1952) 829.
177 I. B. Berlman, *Handbook of Fluorescence Spectra of Aromatic Molecules*, Academic Press, New York, 1965.
178 E. J. Bowen and R. J. Cook, *J. Chem. Soc.*, (1953) 3059.
179 C. A. Parker and T. A. Joyce, *Chem. Commun.*, (1967) 744.
180 H. Hellstrom, *Arkiv Kemi Mineral. Geol.*, 12A (1937) 17.
181 J. Umberger and V. LaMer, *J. Am. Chem. Soc.*, 67 (1945) 1099.
182 E. J. Bowen, in W. A. Noyes, Jr., G. S. Hammond and J. N. Pitts, Jr. (eds.), *Advances in Photochemistry*, Vol. 1, Interscience Publishers, New York, 1963.
183 E. J. Bowen and A. H. Williams, *Trans. Faraday Soc.*, 35 (1939) 765.
184 J. J. Hermans and S. Levinson, *J. Opt. Soc. Am.*, 41 (1951) 460.
185 J. B. Birks and W. A. Little, *Proc. Phys. Soc. (London) A*, 66 (1953) 921.
186 E. Bailey and G. Rollefson, *J. Chem. Phys.*, 21 (1953) 1315.
187 B. D. Venetta, *Rev. Sci. Instr.*, 30 (1959) 450.
188 S. H. Liebson, *Nucleonics*, 10 (1952) 41.
189 A. Scharmann, *Z. Naturforsch.*, 11A (1956) 398.
190 A. Schmillen, *Z. Physik*, 135 (1953) 294.
191 J. B. Birks and D. J. Dyson, *J. Sci. Instr.*, 38 (1961) 282.
192 J. B. Birks, D. J. Dyson and I. H. Munro, *Proc. Roy. Soc. (London) A*, 275 (1963) 575.
193 J. H. Malmberg, *Rev. Sci. Instr.*, 28 (1957) 1027.
194 S. S. Brody, *Rev. Sci. Instr.*, 28 (1957) 1020.
195 J. B. Birks, T. A. King and I. H. Munro, *Proc. Phys. Soc. (London)*, 80 (1962) 355.
196 C. F. Hendee and W. B. Brown, *Philips Tech. Rev.*, 19 (1957) 50.
197 R. G. Bennett, *Rev. Sci. Instr.*, 31 (1960) 1275.
198 N. Mataga, M. Tomura and H. Nishimura, *Mol. Phys.*, 9 (1965) 367.
199 J. T. Dubois and M. Cox, *J. Chem. Phys.*, 38 (1963) 2536.
200 J. T. Dubois and R. L. Van Hemert, *J. Chem. Phys.*, 40 (1964) 923.
201 T. V. Ivanova, P. I. Kudriashov and V. I. Sveshnikov, *Dokl. Akad. Nauk SSSR*, 138 (1961) 572.
202 E. Becquerel, *Ann. Chim. Phys. (Paris)*, 27 (1871) 539.
203 C. A. Parker and C. G. Hatchard, *Analyst*, 87 (1962) 664.
204 J. F. Coetzee, G. P. Cunningham, D. K. McGuire and G. R. Padmanabhan, *Anal. Chem.*, 34 (1962) 1139.
205 D. L. Maricle, *Anal. Chem.*, 35 (1963) 683.
206 D. L. Maricle and W. G. Hodgson, *Anal. Chem.*, 37 (1965) 1562.
207 J. D. Winefordner and P. A. St. John, *Anal. Chem.*, 35 (1963) 2211.
208 H. Greenspan and E. Fischer, *J. Phys. Chem.*, 69 (1965) 2466.
209 W. G. Pfann, *Zone Melting*, 2nd edition, John Wiley & Sons, Inc., New York, 1966.

210 C. A. PARKER, in P. W. SHALLIS (ed.), *Proc. SAC Conf., Nottingham*, W. Heffer & Sons Ltd., Cambridge, 1965.
211 S. AINSWORTH and E. WINTER, *Appl. Opt.*, 3 (1964) 371.
212 J. JOHNSON and E. G. RICHARDS, *Arch. Biochem. Biophys.*, 97 (1962) 250.
213 G. WEBER, *Biochem. J.*, 51 (1952) 155.
214 G. WEBER, *J. Opt. Soc. Am.*, 46 (1956) 962.
215 G. WEBER, *Biochem. J.*, 75 (1960) 335.
216 P. G. BOWERS and G. PORTER, *Proc. Roy. Soc. (London) A*, 296 (1967) 435.
217 H. LEONHARDT and A. WELLER, in H. P. KALLMANN and G. M. SPRUCH (eds.), *Luminescence of Organic and Inorganic Materials*, John Wiley & Sons, New York, London, 1962.
218 M. KASHA, *J. Chem. Phys.*, 20 (1952) 71.
219 T. MEDINGER and F. WILKINSON, *Trans. Faraday Soc.*, 62 (1966) 1785.
220 C. A. PARKER and T. A. JOYCE, *Chem. Commun.*, (1966) 234.
221 R. F. BORKMAN and D. R. KEARNS, *Chem. Commun.*, (1966) 446.
222 G. PORTER and F. WILKINSON, *Trans. Faraday Soc.*, 57 (1961) 1686.
223 R. LIVINGSTON and P. J. MCCARTIN, *J. Phys. Chem.*, 67 (1963) 2511.
224 G. PORTER and L. J. STIEF, *Nature*, 195 (1962) 991.
225 W. G. HERKSTROETER, A. A. LAMOLA and G. S. HAMMOND, *J. Am. Chem. Soc.*, 86 (1964) 4537.
226 D. S. MCCLURE, *J. Chem. Phys.*, 17 (1949) 905.
227 G. JACKSON and G. PORTER, *Proc. Roy. Soc. (London) A*, 260 (1961) 13.
228 A. TERENIN and V. L. ERMOLAEV, *J. Chim. Phys.*, 55 (1958) 698.
229 C. A. PARKER and T. A. JOYCE, *Nature*, 210 (1966) 701.
230 G. G. HALL, *Proc. Roy. Soc. (London) A*, 213 (1952) 113.
231 S. P. MCGLYNN, T. AZUMI and M. KASHA, *J. Chem. Phys.*, 40 (1964) 507.
232 D. F. EVANS, *J. Chem. Soc.*, (1957) 1351.
233 W. ROTHMAN, A. CASE and D. R. KEARNS, *J. Chem. Phys.*, 43 (1965) 1067.
234 P. AVAKIAN, E. ABRAMSON, R. G. KEPLER and J. C. CARIS, *J. Chem. Phys.*, 39 (1963) 1127.
235 K. WEBER, *Z. Physik. Chem. (Leipzig)*, B 15 (1931) 18.
236 TH. FÖRSTER, *Z. Elektrochem.*, 54 (1950) 42.
237 TH. FÖRSTER, *Z. Elektrochem.*, 54 (1950) 531.
238 TH. FÖRSTER, in F. DANIELS (ed.), *Photochemistry in the Liquid and Solid States*, John Wiley & Sons Inc., New York, 1960.
239 W. BARTOK, P. J. LUCCHESI and N. S. SNIDER, *J. Am. Chem. Soc.*, 84 (1962) 1842.
240 A. WELLER, *Z. Physik. Chem. (Frankfurt)*, 17 (1958) 224.
241 A. WELLER, in G. PORTER and B. STEVENS (eds.), *Progress in Reaction Kinetics*, Vol. 1, Pergamon Press, London, 1961.
242 L. D. DERKACHEVA, *Opt. Spectry. (USSR)*, 9 (1960) 110.
243 A. WELLER, *Z. Elektrochem.*, 61 (1957) 956.
244 A. WELLER, *Z. Elektrochem.*, 56 (1952) 662.
245 A. WELLER, *Z. Elektrochem.*, 58 (1954) 849.
246 A. WELLER, *Z. Physik. Chem. (Frankfurt)*, 3 (1955) 238.
247 A. WELLER, *Z. Elektrochem.*, 60 (1956) 1144.
248 J. B. BIRKS, J. H. APPLEYARD and R. POPE, *Photochem. Photobiol.*, 2 (1963) 493.
249 B. STEVENS and M. I. BAN, *Trans. Faraday Soc.*, 60 (1964) 1515.
250 TH. FÖRSTER and K. KASPER, *Z. Physik. Chem. (Frankfurt)*, 1 (1954) 275.
251 K. KASPER, *Z. Physik. Chem. (Frankfurt)*, 12 (1959) 52.
252 E. DÖLLER and TH. FÖRSTER, *Z. Physik. Chem. (Frankfurt)*, 34 (1962) 132.

253 J. B. Birks, M. D. Lumb and I. H. Munro, *Proc. Roy. Soc. (London) A*, 280 (1964) 289.
254 F. Bandow, *Z. Physik. Chem. (Leipzig)*, 196 (1951) 329.
255 E. Döller, *Z. Physik. Chem. (Frankfurt)*, 34 (1962) 151.
256 I. B. Berlman and A. Weinreb, *Mol. Phys.*, 5 (1962) 313.
257 E. Döller and Th. Förster, *Z. Physik. Chem. (Frankfurt)*, 31 (1962) 274.
258 J. B. Birks and L. G. Christophorou, *Spectrochim. Acta*, 19 (1963) 401.
259 J. B. Birks and J. B. Aladekomo, *Spectrochim. Acta*, 20 (1964) 15.
260 J. B. Aladekomo and J. B. Birks, *Proc. Roy. Soc. (London) A*, 284 (1965) 551.
261 J. B. Birks and T. A. King, *Proc. Roy. Soc. (London) A*, 291 (1966) 244.
262 J. B. Birks and L. G. Christophorou, *Proc. Roy. Soc. (London) A*, 274 (1963) 552.
263 J. B. Birks and L. G. Christophorou, *Proc. Roy. Soc. (London) A*, 277 (1964) 571.
264 T. V. Ivanova, G. A. Mokeeva and B. Ya. Sveshnikov, *Opt. Spectry. (USSR)*, 12 (1962) 325.
265 J. B. Birks, C. L. Braga and M. D. Lumb, *Proc. Roy. Soc. (London) A*, 283 (1965) 83.
266 E. J. Bowen and D. W. Tanner, *Trans. Faraday Soc.*, 51 (1955) 475.
267 J. B. Birks and L. G. Christophorou, *Nature*, 197 (1963) 1064.
268 J. B. Birks and J. B. Aladekomo, *Photochem. Photobiol.*, 2 (1963) 415.
269 R. L. Barnes and J. B. Birks, *Proc. Roy. Soc. (London) A*, 291 (1966) 570.
270 T. D. S. Hamilton, *Photochem. Photobiol.*, 3 (1964) 153.
271 J. B. Birks and L. G. Christophorou, *Nature*, 196 (1962) 33.
272 J. R. Platt, *J. Chem. Phys.*, 17 (1949) 484.
273 F. Hirayama, *J. Chem. Phys.*, 42 (1965) 3163.
274 S. S. Yanari, F. A. Bovey and R. Lumry, *Nature*, 200 (1963) 242.
275 E. A. Chandross, *J. Chem. Phys.*, 43 (1965) 4175.
276 E. A. Chandross, J. W. Longworth and R. E. Visco, *J. Am. Chem. Soc.*, 87 (1965) 3259.
277 J. B. Birks, G. F. Moore and I. H. Munro, *Spectrochim. Acta*, 22 (1966) 323.
278 C. A. Parker, *Spectrochim. Acta*, 22 (1966) 1677.
279 J. B. Birks, *J. Phys. Chem.*, 67 (1963) 1299; 68 (1964) 439.
280 C. A. Parker and G. D. Short, *Trans. Faraday Soc.*, 63 (1967) 2618.
281 S. Nagakura and H. Baba, *J. Am. Chem. Soc.*, 74 (1952) 5693.
282 R. Livingston, W. F. Watson and J. McArdle, *J. Am. Chem. Soc.*, 71 (1949) 1542.
283 N. Mataga, Y. Kaifu and M. Koizumi, *Bull. Chem. Soc. Japan*, 29 (1956) 373.
284 K. Bredereck, Th. Förster and H. G. Oesterlin, in H. P. Kallman and G. M. Spruch (eds.), *Luminescence of Organic and Inorganic Materials*, John Wiley & Sons, New York, 1962.
285 E. V. Shpol'skii, *Soviet Phys. Usp.*, 3 (1960) 372.
286 E. V. Shpol'skii, A. A. Il'ina and L. A. Klimova, *Dokl. Akad. Nauk SSSR*, 87 (1952) 935.
287 E. V. Shpol'skii and L. A. Klimova, *Bull. Acad. Sci. SSSR Ser. Fiz.*, 18 (1954) 357.
288 E. J. Bowen and B. Brocklehurst, *J. Chem. Soc.*, (1954) 3875.
289 E. J. Bowen and B. Brocklehurst, *J. Chem. Soc.*, (1955) 4320.
290 E. V. Shpol'skii, *Soviet Phys. Usp.*, 2 (1959) 378.
291 P. P. Dikun, *Vopr. Onkol.*, 5 (1959) 672.
292 B. Muel and G. Lacroix, *Bull. Soc. Chim. France*, (1960) 2139.
293 R. I. Personov, *J. Anal. Chem. USSR*, 17 (1962) 503.
294 P. A. Teplyakov, *Opt. Spectry. (USSR)*, 15 (1963) 350.
295 G. V. Gobov, *Opt. Spectry. (USSR)*, 15 (1963) 194.
296 Z. S. Ruzevich, *Opt. Spectry. (USSR)*, 15 (1963) 191.

297 L. A. Klimova, *Opt. Spectry.* (*USSR*), 15 (1963) 185.
298 D. N. Shigorin, N. A. Shcheglova and R. N. Nurmukhametov, *Bull. Acad. Sci. SSSR*, 23 (1959) 39.
299 N. A. Shcheglova, D. N. Shigorin and N. S. Dokunikhin, *Russ. J. Phys. Chem.*, 38 (1964) 1067.
300 L. Grajcar and S. Leach, *J. Chim. Phys.*, (1964) 1523.
301 C. A. Parker, C. G. Hatchard and B. Webb, unpublished work.
302 S. Leach, R. Lopez-Delgado and L. Grajcar, *J. Chim. Phys.*, (1966) 194.
303 C. G. Hatchard and C. A. Parker, *Trans. Faraday Soc.*, 57 (1961) 1093.
304 C. A. Parker and W. T. Rees, *Nature*, 184 (1959) 1223.
305 C. A. Parker and W. T. Rees, *J. Chim. Phys.*, (1959) 761.
306 S. Leach, in H. P. Kallman and G. M. Spruch (eds.), *Luminescence of Organic and Inorganic Materials*, John Wiley & Sons Inc., New York, 1962.
307 C. A. Parker and C. G. Hatchard, *Photochem. Photobiol.*, 5 (1966) 699.
308 R. F. Steiner and H. Edelhoch, *Chem. Rev.*, 62 (1962) 457.
309 G. Oster and Y. Nishijima, *Fortschr. Hochpolymer. Forsch.*, 3 (1964) 313.
310 S. F. Velick, *J. Biol. Chem.*, 233 (1958) 1455.
311 D. J. B. Lawrence, *Biochem. J.*, 51 (1952) 168.
312 G. Weber, *Trans. Faraday Soc.*, 44 (1948) 185.
313 V. Zanker and A. Wittwer, *Z. Physik. Chem.* (*Frankfurt*), 24 (1960) 183.
314 G. Weber, *Biochem. J.*, 75 (1960) 345.
315 J. A. Radley and J. Grant, *Fluorescence Analysis in Ultra-violet Light*, 4th edition, Chapman and Hall Ltd., London, 1954.
316 F. P. Zscheile, *Protoplasma*, 22 (1935) 513.
317 F. P. Zscheile and D. G. Harris, *J. Phys. Chem.*, 47 (1943) 623.
318 R. A. Burdett and L. C. Jones, *J. Opt. Soc. Am.*, 37 (1947) 554.
319 F. J. Studer, *J. Opt. Soc. Am.*, 38 (1948) 467.
320 W. Priestley, *Anal. Chem.*, 22 (1950) 509.
321 A. J. Derr, *J. Opt. Soc. Am.*, 41 (1951) 872.
322 J. L. Laner and E. J. Rosenbaum, *J. Opt. Soc. Am.*, 41 (1951) 450.
323 E. J. Bowen, *Photoelec. Spectrometry Group Bull.*, No. 6 (1953) 124.
324 F. B. Huke, R. H. Heidel and R. H. Fassel, *J. Opt. Soc. Am.*, 43 (1953) 400.
325 J. S. McAnally, *Anal. Chem.*, 26 (1954) 1526.
326 A. G. Gornall and H. Kalant, *Anal. Chem.*, 27 (1955) 474.
327 C. A. Parker and L. G. Harvey, *Analyst*, 87 (1962) 558.
328 C. A. Parker and L. G. Harvey, *Analyst*, 86 (1961) 54.
329 R. T. Williams and J. W. Bridges, *J. Clin. Pathol.*, 17 (1964) 371.
330 W. A. Hagins and W. H. Jennings, *Discussions Faraday Soc.*, No. 27 (1959) 180.
331 R. T. Williams, *J. Roy. Inst. Chem.*, 83 (1959) 611.
332 A. Rosen and R. T. Williams, *Photoelec. Spectrometry Group Bull.*, No. 13 (1961) 339.
333 G. A. Thommes and E. Leininger, *Talanta*, 7 (1961) 181.
334 E. Sawicki, T. R. Hauser and T. W. Stanley, *Intern. J. Air Pollution*, 2 (1960) 253.
335 Y. Kanda, R. Shimada, K. Hanada and S. Kajigaeshi, *Spectrochim. Acta*, 17 (1961) 1268.
336 B. Muel and M. Hubert-Habart, J. *Chim. Phys.*, (1958) 377.
337 M. Zander, *Angew. Chem., Intern. Ed. Engl.*, 4 (1965) 930.
338 M. Zander, *Erdoel Kohle, Erdgas, Petrochemie*, 15 (1962) 362.
339 M. Zander, *Chem. Ber.*, 97 (1964) 2695.
340 M. Zander and W. H. Franke, *Chem. Ber.*, 98 (1965) 588.
341 H. D. Sauerland and M. Zander, *Erdoel, Kohle, Erdgas, Petrochemie*, 19 (1966) 502.

342 S. Udenfriend, *Fluorescence Assay in Biology and Medicine*, Academic Press, New York and London, 1962.
343 F. W. J. Teale and G. Weber, *Biochem. J.*, 65 (1956) 476.
344 T. P. Waalkes and S. Udenfriend, *J. Lab. Clin. Med.*, 50 (1957) 733.
345 T. Momose and Y. Ohkura, *Talanta*, 3 (1959) 155.
346 E. Sawicki, W. Elbert, T. W. Stanley, T. R. Hauser and F. T. Fox, *Anal. Chem.*, 32 (1960) 810.
347 E. Sawicki, T. W. Stanley and J. Noe, *Anal. Chem.*, 32 (1960) 816.
348 E. Sawicki and H. Johnson, *Anal. Chim. Acta*, 34 (1966) 381.
349 J. von Eisenbrand, *Deut. Lebensm. Rundschau*, 58 (1962) 230 and 319.
350 J. von Eisenbrand, *Deut. Apotheker-Z.*, 103 (1963) 623.
351 C. A. Parker and W. J. Barnes, *Analyst*, 85 (1960) 3.
352 W. A. Armstrong and D. W. Grant, *Nature*, 182 (1958) 747.
353 R. J. Keirs, R. D. Britt and W. E. Wentworth, *Anal. Chem.*, 29 (1957) 202.
354 S. Freed and W. Salmre, *Science*, 128 (1958) 1341.
355 Z. Holzbecher, *Chem. Listy*, 53 (1959) 713.
356 J. D. Winefordner and H. W. Latz, *Anal. Chem.*, 35 (1963) 1517.
357 J. D. Winefordner and M. Tin, *Anal. Chim. Acta*, 31 (1964) 239.
358 J. D. Winefordner and M. Tin, *Anal. Chim. Acta*, 32 (1965) 64.
359 J. D. Winefordner and H. A. Moye, *Anal. Chim. Acta*, 32 (1965) 278.
360 G. Oster, N. Geacintov and A. V. Khan, *Nature*, 196 (1962) 1089.
361 H. V. Drushel and A. L. Sommers, *Anal. Chem.*, 36 (1964) 836.
362. E. Sawicki and T. W. Stanley, *Chemist-Analyst*, 49 (1960) 107.
363 E. Sawicki and H. Johnson, *Microchem. J.*, 8 (1964) 85.
364 D. F. Bender, E. Sawicki and R. M. Wilson, *Anal. Chem.*, 36 (1964) 1011.
365 E. Sawicki, W. C. Elbert and T. W. Stanley, *J. Chromatog.*, 17 (1965) 120.
366 T. Taketatsu, M. A. Carey and C. V. Banks, *Talanta*, 13 (1966) 1081.
367 G. Alberti and M. A. Massucci, *Anal. Chem.*, 38 (1966) 214.
368 H. W. Leverenz, *Luminescence of Solids*, Chapman & Hall Ltd., London, 1950.
369 G. F. J. Garlick, *Luminescence*, Handbuch der Physik, Vol. 26, Springer-Verlag, Berlin, 1958.
370 C. W. Sill and H. E. Peterson, *Anal. Chem.*, 21 (1949) 1266.
371 C. W. Sill and H. E. Peterson, *Anal. Chem.*, 19 (1947) 646.
372 K. F. Gudymenko, M. U. Belyi and M. A. Skachko, *Zavodsk. Lab.*, 24 (1958) 1066.
373 F. Feigl, V. Gentil and D. Goldstein, *Anal. Chim. Acta*, 9 (1953) 393.
374 C. E. White, *Anal. Chem.*, 21 (1949) 104; 22 (1950) 69; 24 (1952) 85; 26 (1954) 129; 28 (1956) 621; 30 (1958) 729; 32 (1960) 47 R. (With A. Weissler), *Anal. Chem.*, 34 (1962) 81 R; 36 (1964) 116 R; 38 (1966) 155 R.
375 H. Irving, E. J. Butler and M. F. Ring, *J. Chem. Soc.*, (1949) 1489.
376 D. C. Bhatnagar and L. S. Forster, *Spectrochim. Acta*, 21 (1965) 1803.
377 J. W. Collat and L. B. Rogers, *Anal. Chem.*, 27 (1955) 961.
378 W. T. Rees, *Analyst*, 87 (1962) 202.
379 G. F. Kirkbright, T. S. West and C. Woodward, *Anal. Chem.*, 37 (1965) 137.
380 Z. Holzbecher, *Chem. Listy*, 47 (1953) 1023; 49 (1955) 1030.
381 D. C. Freeman and C. E. White, *J. Am. Chem. Soc.*, 78 (1956) 2678.
382 E. A. Bozhevol′nov and G. V. Serebryakova, *Opt. Spectry.* (*USSR*), 13 (1962) 216.
383 R. M. Dagnall, R. Smith and T. S. West, *Talanta*, 13 (1966) 609.
384 I. M. Korenman and V. S. Efimychev, *J. Anal. Chem. USSR*, 17 (1962) 426.
385 G. F. Kirkbright, T. S. West and C. Woodward, in P. W. Shallis (ed.), *Proc. SAC Conf., Nottingham*, W. Heffer & Sons Ltd., Cambridge, 1965.

386 P. Job, *Ann. Chim. (Paris)*, 9 (1928) 113.
387 J. H. Yoe and A. L. Jones, *Ind. Eng. Chem., Anal. Ed.*, 16 (1944) 111.
388 A. E. Harvey and D. L. Manning, *J. Am. Chem. Soc.*, 72 (1950) 4488; 74 (1952) 4744.
389 S. I. Weissman, *J. Chem. Phys.*, 10 (1942) 214.
390 P. Yuster and S. I. Weissman, *J. Chem. Phys.*, 17 (1949) 1182.
391 S. I. Weissman, *J. Chem. Phys.*, 18 (1950) 1258.
392 G. A. Crosby, R. E. Whan and R. M. Alire, *J. Chem. Phys.*, 34 (1961) 743.
393 R. Tomaschek, *Reichsber. Physik*, 1 (1944) 139.
394 R. E. Whan and G. A. Crosby, *J. Mol. Spectry.*, 8 (1962) 315.
395 M. L. Bhaumik and C. L. Telk, *J. Opt. Soc. Am.*, 54 (1964) 1211.
396 M. A. El-Sayed and M. L. Bhaumik, *J. Chem. Phys.*, 39 (1963) 2391.
397 E. Matovich and C. K. Suzuki, *J. Chem. Phys.*, 39 (1963) 1442.
398 R. E. Ballard and J. W. Edwards, *Spectrochim. Acta*, 21 (1965) 1353.
399 A. Heller and E. Wasserman, *J. Chem. Phys.*, 42 (1965) 949.
400 W. J. McCarthy and J. D. Winefordner, *Anal. Chem.*, 38 (1966) 848.
401 J. S. Hanker, R. M. Gamson and H. Klapper, *Anal. Chem.*, 29 (1957) 879.
402 C. A. Parker and W. T. Rees, in J. P. Cali (ed.), *Trace Analysis of Semiconductor Materials*, Pergamon Press, Oxford, 1964.
403 C. E. White, A. Weissler and D. Busker, *Anal. Chem.*, 19 (1947) 802.
404 C. A. Parker and W. J. Barnes, *Analyst*, 85 (1960) 828.
405 C. L. Luke and S. S. Flaschen, *Anal. Chem.*, 30 (1958) 1406.
406 J. Hoste, *Anal. Chim. Acta*, 2 (1948) 402.
407 J. Hoste and J. Gillis, *Anal. Chim. Acta*, 12 (1955) 158.
408 L. I. Kononenko, R. S. Lauer and N. S. Poluektov, *Zh. Analit. Khim.*, 18 (1963) 1468.
409 L. I. Kononenko, N. S. Poluektov and M. P. Nikonova, *Zavodsk. Lab.*, 30 (1964) 779.
410 R. E. Ballard and J. W. Edwards, in P. W. Shallis (ed.), *Proc. SAC Conf., Nottingham*, W. Heffer & Sons Ltd., Cambridge, 1965.
411 G. Alberti and M. A. Massucci, *Gazz. Chim. Ital.*, 95 (1965) 997 and 1006.
412 G. Alberti, R. Bruno, M. A. Massucci and A. Saini, *Gazz. Chim. Ital.*, 95 (1965) 1021.
413 R. M. Dagnall, R. Smith and T. S. West, *Analyst*, 92 (1967) 358.
414 E. C. Stanley, B. I. Kinneberg and L. P. Varga, *Anal. Chem.*, 38 (1966) 1362.
415 H. Kautsky, A. Hirsch and W. Baumeister, *Chem. Ber.*, 64 (1931) 2053.
416 J. Franck and P. Pringsheim, *J. Chem. Phys.*, 11 (1943) 21.
417 H. Kautsky and G. O. Müller, *Z. Naturforsch.*, 2A (1947) 167.
418 A. K. Babko and P. V. Khodulina, *Ukr. Khim. Zh.*, 17 (1951) 191.
419 H. H. Willard and C. A. Horton, *Anal. Chem.*, 24 (1952) 862.
420 W. A. Powell and J. H. Saylor, *Anal. Chem.*, 25 (1953) 960.
421 S. W. Chaikin and C. I. Glassbrook, *Research for Industry*, 5 (1953) 2.
422 J. S. Hanker, A. Gelberg and B. Witten, *Anal. Chem.*, 30 (1958) 93.
423 V. A. Nazarenko and M. B. Shustova, *Zavodsk. Lab.*, 24 (1958) 1344.
424 R. E. Minns, *Meeting on Analysis of Semiconductors*, Society for Analytical Chemistry, London, 1960. See also J. P. Cali, *Trace Analysis of Semiconductor Materials*, Pergamon Press, Oxford, London, 1964, p. 243.
425 I. E. Starik, E. F. Starik and G. V. Kostyrev, *Tr. Radievogo Inst. Akad. Nauk SSSR*, 7 (1956) 111.
426 A. Bradley and N. V. Sutton, *Anal. Chem.*, 31 (1959) 1554.
427 F. Hernegger and B. Karlik, *Sitz.-ber. Akad. Wiss. Wien, Math.-Naturw. Kl.*, IIA, 144 (1935) 217.

428 C. J. RODDEN, *Anal. Chem.*, 21 (1949) 327.
429 P. P. FEOFILOV and L. A. KUZNETSOVA, *Inzh. Fiz. Zh.*, 1 (1958) 46.
430 G. OSTER, *Trans. Faraday Soc.*, 47 (1951) 660; *J. Polymer Sci.*, 16 (1955) 235.
431 G. OSTER and J. S. BELLIN, *J. Am. Chem. Soc.*, 79 (1957) 294.
432 N. WOTHERSPOON and G. OSTER, *J. Am. Chem. Soc.*, 79 (1957) 3992.
433 M. ZANDER, *Z. Anal. Chem.*, 226 (1967) 251; 227 (1967) 331.
434 M. ZANDER, *Z. Anal. Chem.*, 229 (1967) 352.
435 G. WEBER, *Nature*, 190 (1961) 27.
436 M. M. SCHACHTER and E. O. HAENNI, *Anal. Chem.*, 36 (1964) 2045.
437 R. A. OLSON, *Rev. Sci. Instr.*, 31 (1960) 844.
438 C. N. LOESER, *Rev. Sci. Instr.*, 37 (1966) 237.
439 R. G. KEPLER, J. C. CARIS, P. AVAKIAN and E. ABRAMSON, *Phys. Rev. Letters*, 10 (1963) 400.
440 A. TERENIN, G. KOBYSHEV and G. LIALIN, *Photochem. Photobiol.*, 5 (1966) 689.
441 C. A. PARKER, *Chemistry in Britain*, 2 (1966) 160.
442 S. IWATA, J. TANAKA and S. NAGAKURA, in A. B. ZAHLAN (ed.) *The Triplet State*, Beirut Symposium, Cambridge University Press, Cambridge, 1967.
443 J. B. BIRKS in A. B. ZAHLAN (ed.), *The Triplet State*, Beirut Symposium, Cambridge University Press, Cambridge, 1967.

Index of Tables

Table No.		Page No.
1	Approximate sizes of quanta	3
2	Nomenclature and symbols of absorption spectrometry	17
3	Error due to inner filter effect	20
4	Raman bands observed in selected solvents	66
5	Diffusion-controlled rate constants	75
6	Phosphorescence efficiencies and lifetimes of eosin	89
7	Systems investigated by Terenin and Ermolaev	92
8	Effect of light intensity on the delayed fluorescence of anthracene solutions	101
9	Effect of solvent on the delayed fluorescence of phenanthrene at 77°K	107
10	Molecular parameters for various amounts of deviation from square law delayed fluorescence	112
11	Classification of photoluminescence of organic compounds	126
12	Comparison of monochromators	147
13	Band widths expressed in wavelength and wavenumber	154
14	Dimensions of sources	168
15	Flux from 200 W extra-high pressure mercury lamp through grating monochromator	168
16	Outputs of lines from 25 W cadmium and zinc lamps	180
17	Relative intensities of lines from 2½ kW high pressure mercury–cadmium lamp	181
18	Minimum useful height and width of source	182
19	Light fluxes through quartz prism monochromator	183
20	Light flux from mercury lines isolated by monochromators or filters	183
21	Filters for isolating mercury lines	189
22	Quantum efficiency of the ferrioxalate actinometer	210
23	Reflectivity of magnesium oxide	253
24	Spectral sensitivity in the ultra-violet region	255
25	Fluorescence quantum efficiencies	266
26	Fluorescence and phosphorescence efficiencies at 77°K	281
27	Viscosity of low temperature glasses	290
28	Fluorescence quenching and increase of triplet formation caused by bromobenzene	307
29	Efficiencies of triplet formation and of monomer and dimer fluorescence	308
30	Triplet formation efficiencies from delayed fluorescence measurements	310
31	Sensitised delayed fluorescence of perylene	312
32	Triplet formation efficiency of perylene	313
33	Triplet data in ethanol at room temperature	315
34	Triplet lifetimes of eosin	319
35	Intersystem crossing rates for eosin	321
36	Triplet energies from phosphorescence measurements	324

Table No.		*Page No.*
37	Selection of triplet energies	325
38	Acid dissociation constants for the excited singlet state	333
39	Acid dissociation constants in the triplet and singlet states	343
40	Enthalpies and entropies of photoassociation	356
41	Dimer formation	358
42	Dimer and monomer emission bands of naphthalene in ethanol	368
43	Polarisation as a function of viscosity and temperature	393
44	Methods of photoluminescence analysis	400
45	Comparison of instrumental sensitivities	403
46	Method of expressing instrumental sensitivity	405
47	Absolute fluorescence sensitivities	408
48	Solvent Raman bands corresponding to mercury lines	411
49	Selective sensitisation of delayed fluorescence	454
50	Luminescence of metal oxinates at 80°K	476
51	Rare earths in acetic anhydride containing carbonyl compounds	492
52	Methods for determining rare earths	492
53	Systems chosen for determination of oxygen	495

Index of Figures

Figure No.		*Page No.*
1	Transitions giving rise to absorption and fluorescence emission spectra	6
2	Absorption and fluorescence emission spectra of anthracene and quinine bisulphate	8
3	The Franck–Condon Principle	11
4	Change of solvation after excitation or emission	14
5	Derivation of Beer–Lambert Law	16
6	Relationship between fluorescence excitation spectrum and absorption spectrum	22
7	Method of calculating radiative lifetimes	24
8	Estimating the extent of the first absorption band	26
9	Types of electronic transition	29
10	Interaction of lone pair with π-orbital	31
11	Illustrating solvent shifts	34
12	Triplet and singlet energy levels and intersystem crossing	37
13	Intersystem crossing	39
14	Relative spectral positions of absorption, fluorescence and phosphorescence spectra	43
15	Directions of electric vector in polarised and unpolarised light	52
16	Absorption of plane polarised light	52
17	Emission of polarised light from oriented molecules	54
18	Absorption and emission from randomly oriented molecules	56
19	Absorption bands and polarisation of fluorescence	58
20	Light scattering by spherical particles	62
21	Light scattering by non-spherical particles	63
22	Raman spectra of solvents	67
23	Reciprocal rates of transitions	69
24	Diffusion-controlled rate constant in ethanol	75
25	Absorption and fluorescence emission spectra of azulene	78
26	Fluorescence from upper excited singlet states	80
27	Phosphorescence and delayed fluorescence of phenanthrene in ethanol	90
28	Variation of phosphorescence and delayed fluorescence of phenanthrene with temperature	91
29	Dependence of triplet-to-singlet transfer efficiency on triplet levels	95
30	Spectra of long-lived emission from eosin solutions	98
31	E-type delayed fluorescence as a function of temperature	99
32	Decay of delayed fluorescence in rigid medium	106
33	Growth of delayed fluorescence in rigid medium	106
34	Mixed first and second order decay curves	112
35	Efficiency of delayed fluorescence with mixed first and second order decay	114
36	Sensitised delayed fluorescence of anthracene	115
37	Sensitised delayed fluorescence of naphthacene	117

Figure No.		*Page No.*
38	Dependence of delayed fluorescence efficiencies on naphthacene concentration	118
39	Delayed fluorescence efficiencies for a typical donor–acceptor system	120
40	Mechanism of sensitised anti-Stokes delayed fluorescence	122
41	Sensitised anti-Stokes delayed fluorescence of naphthalene	123
42	Sensitised anti-Stokes delayed fluorescence of perylene	124
43	Mutual sensitisation of delayed fluorescence	125
44	General purpose spectrofluorimeter	130
45	Working principle of a monochromator	132
46	Spectral distribution of light passing monochromator from continuous source	135
47	Images of entrance slit produced by spectral lines	137
48	Variation of relative intensities of lines and continuum with slit width	138
49	Arrangements for determining linear dispersion of monochromator	140
50	Method of measuring linear dispersion	141
51	LGP values of monochromators (wavelength basis)	148
52	LGP values of monochromators (wavenumber basis)	155
53	Method of measuring line to background intensity	160
54	Spectral sensitivity curves	161
55	Spectrum of 6 W low pressure mercury lamp	162
56	Spectrum of 125 W medium pressure mercury lamp	163
57	Spectrum of extra-high pressure mercury lamp	164
58	Spectrum of 350 W xenon lamp	165
59	Spectrum of 100 W tungsten lamp	166
60	Spectral distribution of radiation from black body	170
61	Circuits for operating xenon and mercury lamps	173
62	Transitions giving rise to some intense mercury lines	175
63	Some typical lamp forms	176
64	Two kinds of water-cooled lamp housing	185
65	Transmission curves of short-wavelength cut-off filters	187
66	Transmission curves of band-pass filters	188
67	Use of filters to improve monochromator performance	190
68	The thermopile	195
69	The vacuum photocell	198
70	Spectral response of various photocathodes	199
71	The photomultiplier	201
72	Electrostatic dynode systems of photomultipliers	202
73	Sensitivity–voltage calibration for photomultiplier	203
74	The fluorescence quantum counter	205
75	Light absorption by recommended ferrioxalate solutions	211
76	Principle of potentiometer recorder in single beam operation	216
77	Arrangement for double beam recording	218
78	Methods for illumination and viewing	221
79	Effect of excessive absorption of the exciting light on the excitation spectrum	223
80	Effect of a second absorbing solute on the excitation spectrum	224
81	Effect of a second absorbing solute on the emission spectrum (right angle illumination)	225
82	Effect of a second absorbing solute on the emission spectrum (frontal illumination)	228

Figure No.		Page No.
83	Effect of self-absorption on the emission spectrum	229
84	Self-absorption and concentration quenching with the in-line arrangement	231
85	Photoluminescence spectra of fused quartz and synthetic silica	235
86	Cell for reaction and measurement in the absence of oxygen	236
87	General purpose cell compartment	238
88	Cell compartment for frontal illumination	238
89	Vacuum system for de-aeration of solutions	240
90	Cell for vacuum de-aeration of solutions	240
91	Dewar vessel for measurements at high or low temperatures	242
92	Optical absorption cell for use with gaseous heating or cooling agents	244
93	Apparatus for frontal illumination at low temperature	245
94	Manual correction of excitation spectrum	248
95	Comparison of absorption and excitation spectra	248
96	Arrangement for recording corrected excitation spectrum	249
97	Distortion of uncorrected excitation spectrum	251
98	Calculation of sensitivity curve for the visible region	254
99	Calibration in the ultra-violet region	254
100	Spectral sensitivity curve of grating monochromator with 9558Q photomultiplier	258
101	Direct recording of corrected emission spectra	259
102	Hamilton's function board and tapped helipot	260
103	Diagram of spectrophosphorimeter	273
104	Chopper phase and phosphorescence intensity	274
105	Fluorescence and phosphorescence emission spectra of impure phenanthrene	276
106	Luminescence of 3,4-benzopiazselenol	277
107	Delayed fluorescence of anthracene in ethanol	280
108	Decay of luminescence with time	284
109	Circuit for measuring lifetimes	285
110	Sensitised luminescence of impurities in ethanol	288
111	Polarising elements	294
112	Measurement of polarisation of fluorescence	295
113	Section drawing of polariscope	296
114	Arrangement for determining the polarisation of the fluorescence excitation spectrum	299
115	Recording polarisation spectrofluorimeter	301
116	Perylene delayed fluorescence sensitised by acenaphthene	311
117	Triplet decay rates as a function of temperature	317
118	Fluorescence emission spectra as a function of pH value	330
119	Energy levels of acidic and basic forms	332
120	Variation of fluorescence intensity with pH value	335
121	Molecular structures of photodimers	345
122	Potential energy diagram for dimer	346
123	Change in fluorescence spectrum of pyrene with concentration	350
124	Monomer and dimer fluorescence intensities after flash excitation	351
125	Temperature dependence of fluorescence efficiencies of pyrene monomer and dimer	352
126	Dimer/monomer emission ratio in prompt fluorescence	354
127	Fluorescence spectra of fluoronaphthalene at various temperatures	355

Figure No.		*Page No.*
128	Method of deriving fluorescence spectrum of excited dimer	357
129	Absorption spectrum of pyrene in ethanol	359
130	Energy levels of monomer and excited dimer	360
131	Prompt fluorescence of pyrene in ethanol	364
132	Delayed fluorescence of pyrene in ethanol	365
133	Mechanism for prompt and delayed fluorescence of pyrene	366
134	Prompt and delayed fluorescence of naphthalene in ethanol	367
135	Dimer/monomer emission ratio in delayed fluorescence	369
136	Ground and excited states of pyrene-3-aldehyde	377
137	Effect of temperature on vibrational bands in pyrene spectrum	380
138	Shpol'skii spectrum of pyrene (silica prism)	383
139	Shpol'skii spectrum of pyrene (glass prism)	384
140	Shpol'skii spectrum of 3-ethylpyrene	385
141	Fluorescence emission spectra of thionine solution photolysed at 77°K	389
142	Photoreaction of benz(a)pyrene in presence of polymer	391
143	Polarisation spectrum of 9-aminoacridine cation	395
144	Interference by main Raman band of water	412
145	Effect of slit width on recorded Raman spectrum	412
146	Interference of Raman band with excitation spectrum	413
147	Interference of Raman band with structured emission spectrum	414
148	Emission from borate–benzoin solution with 366nm excitation	415
149	Emission from borate–benzoin solution with 405nm excitation	416
150	Interference by scattered light and cuvette fluorescence	417
151	Fluorescence of impurities in ethanol	420
152	Fluorescence of impurities in water	421
153	Determination of fluorene in phenanthrene	423
154	Excitation in the anti-Stokes region	424
155	Determination of anthracene in phenanthrene	425
156	Fluorescence and phosphorescence emission spectra at 77°K	432
157	Fluorescence and phosphorescence emission spectra at 77°K	433
158	Comparison of photoluminescence due to CT and π*-n states	434
159	Effects of molecular rigidity and substitution on vibrational resolution of fluorescence emission spectra	437
160	Illustrating analysis of complex mixtures	442
161	Fluorescence and phosphorescence of impurities in fluorene	444
162	Fluorescence and phosphorescence of impurities in fluorene	445
163	Phosphorescent impurities in benz(a)pyrene	446
164	Phosphorescent impurities in benz(a)pyrene	447
165	Identification of impurities in technical pyrene	448
166	Determination of triphenylene in tar fractions	449
167	Sensitised delayed fluorescence of benz(a)pyrene and pyrene	451
168	Sensitised delayed fluorescence of impurity in fluorene	452
169	Sensitised delayed fluorescence of impurity in chrysene	453
170	Application of selective sensitisation	455
171	Fluorescence of petroleum products	459
172	Phosphorescence emission spectra of anthraquinone	462
173	Phosphorescence excitation spectrum of anthraquinone	463
174	Fluorescence and phosphorescence of petroleum products	465

Figure No.		Page No.
175	Application of Shpol'skii spectra	467
176	Zone refining monitored by fluorescence measurement	468
177	Luminescence of rare earth chlorides	471
178	Luminescence of uranyl ion in sulphuric acid	472
179	Some chelating agents and fluorescent chelates	475
180	Energy level manifolds of metal chelates	479
181	Spectrofluorimetric determination of uranium	484
182	Spectrofluorimetric determination of aluminium	485
183	Fluorescence of benzoin photodecomposition product	486
184	Choice of filters for fluorimetric determination of borate	487
185	Reversible quenching and irreversible decomposition induced by oxygen	488
186	Spectrofluorimetric determination of selenium	490
187	Determination of oxygen by luminescence quenching	494
188	Effect of copper content on the luminescence of synthetic silica	498

Subject Index

Note: A page number in italics followed by the letter T or F refers to a table or figure to be found on that page.

absorbance, definition of, 17, *17T*
absorption of light:
  conditions for, 6
  law of, 15, *16F*
  rate of, 282
  time required for, 11, 70
absorption spectrometry, nomenclature, *17T*
absorption spectrum:
  definition of, 18
  examples of, *8F*, *22F*, *43F*, *78F*, *359F*
  integration of, 25, *24F*, *26F*
  method of plotting, 18, *8F*
absorptivity, definition of, *17T*
acceptor:
  groups, 32
  triplet, see triplet acceptor
acenaphthene, 103, *266T*, *310T*, *315T*, *356T*, *311F*, *317F*, *452F*
acetanilide, *281T*, *433F*
acetic anhydride, 481, 493, *492T*
acetonaphthone, *324T*
acetophenone, 481, *92T*, *281T*, *324T*, *433F*
acetylpropionyl, *324T*
acridine, 341, 376, *333T*, *343T*
acridine, detection of, 469
acridine orange, 124, 322, 326, 454, 501, *266T*, *315T*, *454T*, *455F*
acridinium ion, *333T*
acriflavine, 50
actinometer:
  chemical, advantages of, 208
  ferrioxalate, 208
activators for inorganic phosphors, 473
aesculin, 205
air pollutants, analysis of, 457, 458, 469
alcohols, aliphatic, light absorption by, 30
alizarin garnet, 497
alkyl halides, light absorption by, 30
aluminium, 475, 476, 477, 485, 496, 497, *476T*, *475F*, *485F*
amine glasses, 49
amines:
  aliphatic, light absorption by, 30
  pK* values of, 340
amino acids, 456
aminoacridine, 396, *395F*
aminoanthraquinone, 382
aminoazobenzenes, detection of, 469
aminopyrene sulphonate, 329
amplification of photomultiplier output, 215
anabasine, determination of, 462
analysis (see also under name of material):
  classification of methods of, *400T*
  empirical methods of, 458, 464
  historical, 397
analysis, inorganic:
  energy transfer methods, 491
  fluorescence methods in fluid solution, 483
  in crystalline and glassy media, 497
  indirect methods, 496
  of mixtures, 484
  quenching methods, 493
analysis, organic:
  chemical fluorescence methods, 455
  criteria of purity, 438
  direct delayed fluorescence methods, 449
  direct fluorescence methods, 422, 440, 500
  direct phosphorescence methods, 443
  external heavy atom effect in, 500
  fluorescence methods in crystalline or glassy solution, 460, 466
  fluorescence methods in fluid solution, 419, 440, 455
  indirect fluorescence methods, 460
  of adsorbates, 469
  phosphorescence methods, 443, 460
  quenching methods, 460

aniline, 31, 375, 435, 437, *325T*, *31F*
anisaldehyde, *492T*
anisil, 94
anisole, 435, *325T*
annihilation, see energy transfer, triplet-to-triplet
anthanthrene, *358T*
anthracene, 10, 47, 114, 117, 130, 223, 231, 268, 280, 326, 360, 362, 390, 420, 430, *101T*, *266T*, *281T*, *307T*, *310T*, *312T*, *315T*, *358T*, *408T*, *454T*, *8F*, *115F*, *117F*, *118F*, *125F*, *224F*, *231F*, *280F*, *413F*, *414F*, *423F*, *425F*
anthracene:
  analysis of, 448
  determination of, 420, 424, 440, 454, 460, *423F*, *425F*, *455F*
  vapour, delayed fluorescence of, 46
anthracene isocyanate, 393
anthranilic acid, *266T*
anthraquinone, 382, *281T*, *324T*, *433F*, *462F*, *463F*
anti-Stokes delayed fluorescence, 122, 449
anti-Stokes fluorescence, 10, 424, *423F*, *424F*
apparatus, 128
artifacts, 220, 291
aspirin, determination of, 461
atomic manifold, 478
atropine, determination of, 462
azaheterocyclic compounds, detection of, 469
azulene, 79, 381, 438, *78F*

band width:
  magnitude of, 134
  wavelength–wavenumber conversion, *154T*
beam splitter, 206
Beer–Lambert Law, *16F*
benz-, see also benzo-
benzacridine, detection of, 469
benzaldehyde, 93, 435, 461, *92T*, *324T*
benzanthracene, 103, 125, 250, 315, 373, 431, *266T*, *315T*, *356T*, *358T*, *454T*, *22F*, *125F*, *317F*, *357F*
benzanthracene, determination of, 453
benzanthracene derivatives, *358T*
benzcarbazole, 436, *437F*
benzcarbazole, detection of, 469
benzene, 223, 430, 431, 436, *75T*, *266T*, *281T*, *325T*, *358T*, *224F*, *432F*, *437F*
benzene, purification of, 289
benzfluorene, 436, 468, *437F*, *468F*
benzil, 94, *281T*, *324T*, *433F*
benzo-, see also benz-
benzoate ion, 459, *325T*
benzoic acid, *281T*, *325T*, *433F*
benzoic acid, determination of, 462
benzoin, 237, 474, 486, *92T*, *281T*, *415F*, *416F*, *433F*, *475F*, *486F*, *488F*
benzonaphthenedione, 457
benzonaphthofuran, determination of, 447, *448F*
benzophenone, 33, 79, 94, 433, 461, 481, *92T*, *281T*, *324T*, *34F*, *433F*, *434F*
benzopiazselenol, 407, 431, 482, 490, *277F*
benzoylacetone, 478, 480
benzoylfluorene, *324T*
benzperylene, 381, *358T*
benzpyrene, 104, 356, 381, 384, 390, *266T*, *315T*, *358T*, *454T*, *369F*, *467F*
benzpyrene:
  analysis of, 445, *446F*, *447F*
  determination of, 441, 452, 453, 466, *451F*
  photochemical products from, 466, *467F*
benzyl radical, 383, 390
beryllium, 477, 484
biacetyl, see diacetyl
biochemistry and biology, fluorescence analysis in, 455
bioluminescence, 1
biphenylene, see diphenylene
biphenyl, see diphenyl
biphotonic reactions, 49, 50, 51
black body radiation, 169, *170F*
blank:
  as a limitation of sensitivity, 403, 409
  due to anti-Stokes fluorescence, 425
  due to impurities in reagents, 419, 485, 486, 489
  due to impurities in solvents, 419
  due to luminescence of cuvette, 416
  due to other components of specimen, 422, 485
  due to photodecomposition, 426, 487
  due to Raman scatter, 411, 487
  due to scattered light, 418
  effect of specimen arrangement on, 233

blood serum, analysis of, 461
borate, determination of, 237, 415, 486
borate–benzoin complex, 474, 487, *415F*, *416F*, *486F*, *488F*
boron in semiconductor materials, 486
brandy, benz(a)pyrene in, 466
bromobenzene, 438, *307T*
bromobutane, 289
bromonaphthalene, *325T*
bromopyrene, 356
butyrophenone, *324T*

cadmium, *476T*
cadmium lamps, 179, *180T*
cage effect, 74
calibration of spectrofluorimeter:
  intensity scales, 219
  spectral sensitivity, 247, 252
carbazole, 387, *92T*, *324T*
carbazole, determination of, 443, 448, *444F*
carbon tetrachloride, 227, *66T*
carbonyl group:
  orbitals of, 30
  vibrational bands, 382
carboxylic acids, pK* values of, 340
ceresin, 458
cerium, 473, 483
charge transfer complexes, 32
charge transfer reactions, 470
charge transfer states, 32, 429, 433
charge transfer transitions, 33
chelates, 473, 483, *475F*
chelates, analysis of, 477
chemiluminescence, 1
chloramine-T, 482
chloroanthraquinone, 382
chlorobenzene, 438
chloroform, *66T*
chloronaphthalene, *92T*, *325T*
chlorophylls, 313, 314, 319, 324, 325, 376, *266T*, *315T*
chloropyrene, 356
chlorpromazine, 501
chlorpromazine, determination of, 462
choppers for phosphorimeter:
  adjustment of, 278
  construction of, 274
chromatograms, examination of, 239, 469
chromium, 475, 476, 479, 497, 498, *476T*
chrysene, 104, *266T*, *310T*, *315T*
chrysene:
  analysis of, 452, *453F*
  determination of, 447, *446F*
cigarette smoke, benz(a)pyrene in, 466
coal tar fractions, analysis of, 447, *448F*, *449F*
cobalt, 476, *476T*
cocaine, determination of, 462
cochineal, 497
co-enzyme complexes, 394
collimator, 131
colour temperature, 171
computers, application in photoluminescence work, 259, 493, 500
copper, 497, *476T*, *498F*
coronene, 381, *325T*
correction of spectra, 246, 252
cresol, 395
critical transfer distance, 84
crystal lattice, interaction of solute with, 381
crystalline matrix, photolysis in, 382
crystallisation, interference by, 440
crystal phosphors, inorganic, 470, 472, 497
cuvette, luminescence of, 416, *417F*
cyanide, determination of, 482, 497
cyanobenzophenone, *324T*
cyanogen chloride, 482
cyanopyrene, 356
cyclohexane:
  crystalline forms of, 386
  diffusion-controlled rate constants in, *75T*
  Raman bands of, *66T*
  purity of, 289, 420
cyclohexanedione, determination of, 457

dark current:
  cause of, 201, 204
  compensation of, 218
de-aeration, methods of, 237, 239
decay curves, *106F*, *112F*, *284F*
delayed fluorescence:
  annihilation controlled, 108
  artifacts and trivial effects in, 291
  as a criterion of purity, 451
  definition of, 42
  effect of viscosity on, 379, *107T*
  efficiency, determination of, 279

delayed fluorescence (*continued*)
E-type, see E-type delayed fluorescence
kinds of, 42, *126T*
mutual sensitisation of, 122, *125F*
photochemical reactions and, 287, 390
P-type, see P-type delayed fluorescence
quenching of, by oxygen, 495
recombination, 43, 50, *126T*
sensitised, see sensitised delayed fluorescence
spectra, examples of, *90F*, *98F*, *115F*, *117F*, *123F*, *124F*, *125F*, *280F*, *288F*, *311F*, *365F*, *367F*, *451F*, *452F*, *453F*, *455F*
depolarisation:
by energy transfer, 60, 83, 396
by re-emission, 299
by rotation, 59
measurement of, 295
depolariser, 296
diacetyl, 45, 93, 271, *324T*, *325T*
diacetylbenzene, *324T*
diaminobenzidine, 489
diaminonaphthalene, 482, *490F*
diaminophenoxazine, 463
diaminophenylpiazselenol, 407, 489
dianthracene, 345, 362, 390, *345F*
dianthrylethane, 82, 379
dibenzacridine, detection of, 469
dibenzanthracene, *325T*, *358T*
dibenzcarbazole, detection of, 469
dibenzofuran, determination of, 444
dibenzothiophene, *324T*
dibenzoylbenzene, *324T*
dibenzoylmethane, 480
dibenzpyrene, *358T*
dichloroanthracene, 381
dichlorobenzophenone, *324T*
diesel fuel, 458, *459F*, *465F*
diffusion-controlled processes, 74, 271, 378, 495
diffusion-controlled rate constants, *75T*, *75F*
diffusion rate of oxygen, determination of, 464
diffusive transport, 107
dihydroxyazobenzene, 477, *475F*
dihydroxyazonaphthalene sulphonic acid, 249
dimer, see excited dimer and photodimer
dimers, types of, 344
dimethoxybenzophenone, *492T*
dimethylaminonaphthalene-:
benzyl sulphonamide, 85
phenyl sulphonamide, 85
sulphonamide, 86
sulphonate, 205
dimethylaminonitrostilbene, 435
dimethylanthracene, 361, 373, *266T*, *315T*, *358T*
dimethyldibutyldihydroxydiphenyl sulphide, 466
dimethylnaphthylamine, 331, *325T*, *343T*
dimethylnaphthylammonium ion, *325T*
dinaphthacene, *345F*
dipentacene, *345F*
diphenyl, 437, 459, *92T*, *281T*, *432F*, *437F*
diphenylamine, 48, 387, *92T*
diphenylamine radical ion, 48
diphenylanthracene, 268, 361, *266T*, *307T*, *315T*, *358T*
diphenylene, 81
diphenylene, oxide, *324T*
diphenylnitrogen radical, 48
diphenyloxazole, *358T*
diphenylpolyenes, 381
diphenylpropane, 361
diphenylpropanone, *324T*
dipole moment:
induced, 13
of excited molecule, 15
permanent, 13
solvation and, 374
dipole strength of transition, 27
dispersing element, 131
dispersion:
angular, 133
linear, 133, 139, *140F*, *141F*
dissociation constants in ground and excited states, *333T*, *343T*
donor, see triplet donor
donor groups, 32
d-orbitals, 470, 478
dysprosium, 470, 472, 480, 481, *492T*, *471F*
dysprosium, determination of, 491, *492T*

Einstein relations, 27
einstein, unit of light quanta, 2
electrochemiluminescence, 362, 373
electroluminescence, 1

electron spin, 36
electronic excitation, 7
emission of light, see light
emission spectrum:
  automatic recording of, 258
  correction of, 252
  distortion of, 225, 227, 230
  plotting of, 18, 258
encounter rate constant, 74, 271, *75T*, *75F*
energy level diagram, 7, 36, *6F*, *37F*
energy levels:
  and the wave equation, 5
  of acidic and basic forms, *332F*
energy, molecular, kinds of, 7
energy transfer:
  depolarisation by, 60, 83, 396
  intramolecular, 85, 478
  long range, 60, 83, 104, 107, 370, 396
  singlet-to-singlet, 83, 85, 271, 305
  to metal ion, 481
  triplet-to-singlet, 92, *95F*
  triplet-to-triplet, 99, 101, 104, 107, 115, 321, 370
energy, units of, *3T*
enthalpy, 343, 345, 353, *356T*
entrance slit, illumination of, 142
entrance slit optics, fluorescence of, 232, 418
entropy, 353, *356T*
enzymes, assay of, 460
eosin, 45, 89, 97, 122, 124, 268, 320, 326, 378, 438, 454, *89T*, *266T*, *315T*, *319T*, *321T*, *408T*, *454T*, *98F*, *99F*, *124F*
epinephrine, determination of, 456
equilibrium constants:
  from polarisation measurements, 394
  of excited states, 340
erbium, 480
erythrosin, 438
ethanol, 288, 420, *66T*, *75T*, *393T*, *420F*
ethers, aliphatic, 30
ethylbenzene, 436
ethylene, 30
ethylenediaminetetraacetic acid, 491
ethyl ether, 289
ethylphenylglyoxalate, *324T*
ethylpyrene, 386, *385F*
E-type delayed fluorescence:
  and intersystem crossing, 320
  characteristics of, *126T*
E-type delayed fluorescence (*continued*)
  definition of, 46
  efficiency of, 97
  historical development of, 44
  kinetics of, 97, *99F*
  lifetime of, 98
  sensitised, 127
europium, 470, 472, 478, 480, *492T*, *471F*
europium:
  benzoylacetonate, 478
  chelates, 480
  determination of, 491, *492T*
  salicylaldehyde, 478
  trisdibenzoylmethide, 86
excimer, definition of, 47, 48
excimers, see excited dimer
exciplex, 359
excitation:
  by light absorption, 6
  -emission characteristics, automatic recording of, 500
  spectrum, automatic recording of, 249
  spectrum, correction of, 246
  spectrum, definition of, 21
  spectrum, distortion of, 223, 227, 230
  spectrum, triplet-singlet, 326
  spectrum, use for identification, 441
  thermal, see thermal activation
  wavelength, choice of, 440
excited complex, 340, 359
excited dimer:
  delayed fluorescence of, 47, 102, 363, *368T*
  delayed fluorescence spectra of, *365F*, *367F*
  effect of temperature on, 351, 367, 369
  formation, factors affecting, 359
  formation, kinetics of, 348
  general mechanism for excitation of, 371
  hypothetical long-lived, 47, 104
  in rigid medium, 362
  intramolecular, 361
  lifetime of, 350
  list of, *358T*
  of pyrene, 59, 347, *350F*, *351F*, *352F*, *354F*
  prompt fluorescence of, 346, *368T*
  prompt fluorescence spectra of, *346F*, *355F*, *357F*, *364F*, *367F*
  stabilisation of, 359

excited dimer (*continued*)
stability of, 346
thermodynamic data for, 351, *356T*
types of, 344
excited state (see also under sigma*, pi*, singlet, triplet and charge transfer states):
dissociation constants in, *333T*, *343T*
equilibria in, 328
isomerisation in, 342
kinds of, 28, 36
kinetics of protolytic reactions in, 338, 341
exit slit optics, 151
experimental methods, 128
extinction coefficient, 17, *17T*
eyes, protection of, 184

ferrioxalate actinometer:
characteristics of, 209
experimental procedure for, 210
light absorption by, *211F*
mechanism of, 209
quantum efficiency of, *210T*
filters:
band pass, 187
cut off, 186
fluorescence of, 150
for exciting light, 188
for filter fluorimetry, 192, 487, *487F*
for luminescence, 191
to eliminate Raman bands, 415, 487
to improve monochromator performance, *190F*
to isolate mercury lines, *189T*
to remove scattered light, 418
transmission curves of, 186, *187F*, *188F*, *487F*
fingerprinting, 501
flavone, *324T*
f-levels, 470, 478
flexible molecules, 379
fluoranthene, 103, 318, *266T*, *315T*, *317F*
fluoranthene, determination of, 447, *448F*
fluorene, 282, 437, *266T*, *281T*, *324T*, *325T*, *432F*, *437F*
fluorene:
analysis of, 443, 452, *444F*, *452F*
determination of, 422, 440, *423F*
fluorenone, *324T*
fluorescein, 44, 83, 84, 207, 228, 247, 250, 268, 436, 438, 495, *266T*, *408T*, *229F*, *248F*
fluorescein–boric acid system, 44
fluorescein isocyanate, 393
fluorescence:
analysis, see analysis
anti-Stokes, 10, 425, *423F*, *424F*
blank, see blank
decay curve, 270, *351F*
definition of, 42
delayed, see delayed fluorescence
efficiency, see below
emission, kinetics of, 22
emission spectrum, see below
excitation spectrum, see below
from upper excited states, 78, *78F*, *80F*
intensity, 19
intensity measurements, sensitivity of, 21
lifetime of, 70, 71, 269, 303, 394
molecular structure and, 30, 32, 33, 376, 428, 470
origin of, 7
pH dependence of, 328, 334, *330F*, *335F*
polarisation of, 53, 55, 392
prompt, characteristics of, *126T*
prompt, definition of, 42
quenching, see below
Raman spectrum as standard in measurement of, 406, 419, 420
rotational depolarisation of, 59
self-absorption of, 225, 227, 230, 264, *229F*
sensitised, 84
sensitivity index, 407, *408T*
solvent dependence of, 373
spot tests, 469
standard substances, 265
temperature dependence of, 41, 81, 488
tracing, 460
viscosity dependence of, 379
fluorescence efficiency:
determination of, 260
expressions for, 72
low temperature determination of, 281
of flexible molecules, 82
significance of, 9
temperature dependence of, 81
triplet formation and, 82, 88, *310T*
values of, 268, *266T*, *281T*, *308T*, *310T*, *315T*, *476T*

fluorescence emission spectrum:
automatic recording of, 258, *259F*
correction of, 252, *254F*
examples of, *8F*, *78F*, *225F*, *228F*, *229F*, *231F*, *235F*, *276F*, *277F*, *311F*, *330F*, *355F*, *357F*, *364F*, *412F*, *413F*, *414F*, *415F*, *416F*, *417F*, *420F*, *421F*, *423F*, *425F*, *432F*, *433F*, *434F*, *437F*, *442F*, *444F*, *445F*, *452F*, *453F*, *455F*, *459F*, *465F*, *467F*, *468F*, *471F*, *472F*, *484F*, *485F*, *486F*, *490F*, *498F*
measurement of, 129
mirror image relationship of, 10, *8F*, *43F*
plotting of, 18, 258, *8F*
fluorescence excitation spectrum:
automatic recording of, 249, *249F*
correction of, 246, *248F*
definition of, 21
examples of, *22F*, *223F*, *224F*, *248F*, *251F*, *442F*
measurement of, 129
significance of, *22F*
fluorescence quenching:
analysis by, 460, 469, 493
by charge transfer, 305
by chemical reaction, 305
by collision, 72
by complex formation, 73
by energy transfer, 85, 305
by heavy atoms, 306, *307T*
by oxygen, 493, *494F*
self-, 73, 84, 232, 347, *231F*
study of photochemical reactions by, 387
fluoride, determination of, 496
fluorimeter, definition of, 128
fluorograph, 500
fluorometer:
definition of, 128
types of, 269
fluoronaphthalene, *325T*, *356T*
fluorophotometer, definition of, 128
forbidden transitions, 38, 174, 177, 473
Franck–Condon Principle, 11, *11F*
frequency, 2
frontal illumination, 226, 237

gadolinium, 470, 480, *492T*, *471F*
gallium, 476, 484, *485F*
geometry of specimen compartment, 220, 233
germanium, 484
Glan–Thompson prism, 293, *294F*
glassy solvents, 289, 388, 460, *290T*
glucose, determination of, 457
glycerol, 289, *75T*, *89T*, *393T*, *98F*, *99F*
glyoxal, detection of, 469
granite, analysis of, 491
grating (see also under monochromators):
anomalies of, 257
blaze of, 149
dispersion produced by, 133
efficiency of, 149
spectral orders, interference by, 150
growth curves, *106F*, *351F*

half-band width, 135
half-value concentration, 349
heat content, see enthalpy
heavy atom effect, 306, 373, 437, 500, *307T*
heptane, 381
hexane, 381, 421
hexanol, 382
hexaphenylethane, 48
holmium, 470, 480
hormones, 501
Hund's rule, 37
hydrazine, 482
hydrogen abstraction, 94
hydrogen bonding, 35, 374, 382
hydroxy-:
-benzaldehyde, *92T*
-benzoic acid, 436
-naphthalhydrazone, 469
-naphthoic acid, 477, *475F*
-pyrene, 375
-pyrene sulphonate, 329, *333T*
-pyridine, 432
-quinoline, 475, 484, 497, *476T*, *475F*, *485F*
-quinoline sulphonate, 497
-tetralone, 457
-tryptamine creatinine sulphate, 250, *251F*

illumination, methods of, *221F*
impurities:
quenching by, 286, 378
removal of, 286
incandescence, 1
incident radiant power, *17T*
indium, 476, 485, *476T*, *485F*

indole, 431
infra-red light, excitation by, 51
inner filter effects:
  distinguished from quenching, 220
  in rigid media, 388
  theoretical, due to absorption of exciting light, 20, *20T*
  with frontal illumination, 226, *223F*, *228F*, *229F*
  with in-line illumination, 229, *231F*
  with right-angle illumination, 222, 487, *223F*, *224F*, *225F*
inorganic compounds (see also under analysis, and name of element or compound):
  determination of, 483
  photoluminescence of, 470
internal conversion:
  definition of, 9
  rate of, 70
internal standard:
  for low temperature measurements, 463
  Raman spectrum as, 406, 419, 420
intersystem crossing:
  description of, 38, *37F*, *39F*
  rate of, 70, 304, 313, 320, *321T*
  to atomic levels, 86, 478, 480
  via upper triplet states, 41, 82
iodobenzene, 438
iodonaphthalene, 76, *325T*
ionising radiation, dosimeter for, 459
ionising solvents, purification of, 289
iron, 208, 470, 475, *476T*
isopentane, purification of, 289

Jablonski diagram, 44, *37F*

kinetics:
  of diffusion-controlled processes, 74, 107
  of E-type delayed fluorescence, 97
  of excited dimer processes, 348, 363, 369, 371
  of phosphorescence, 87, 97
  of prompt fluorescence, 23, 71, 81, 303
  of protolytic processes, 328, 338, 341
  of P-type delayed fluorescence, 99, 103, 108
  of sensitised delayed fluorescence, 114, 122
  of singlet state processes, 303
kinetics (*continued*)
  of triplet state processes, 313
  survey of processes, 69, 303

lamps, see light sources
lanthanum, 480
laser:
  as an excitation source, 501
  principle of, 28
laser materials, analysis of, 498
lead, 476, *476T*
lidocaine, determination of, 462
lifetime, see under type of luminescence or excited state
lifetime, measurement:
  by flash excitation, 270, 283
  by polarisation measurements, 394
  interpretation of decay curves, 283
  of fluorescence, 269
  of phosphorescence and delayed fluorescence, 282, *285F*
lifetime, radiative:
  calculation of, 23, *24F*, *26F*
  definition of, 23
  of singlets and triplets, 38
  of singlet state, 72, 303
  of triplet state, 88, 313
light:
  absorption and emission, conditions, 6
  absorption, law of, 15, *16F*
  emission, spontaneous and stimulated, 27
  fluxes, 182, *168T*, *183T*
  gathering power, 144, 154
  intensity, see below
  leakage past choppers, 279
  nature of, 2
  scattering, 60 (see also scattered light)
  sources, see below
  velocity of, 2
light intensity:
  measurement of, 194, 197, 200, 204, 208
  modulation of, 269
  monitoring of, 206, *205F*
light sources:
  area of, 142
  brightness of, 167
  compensation for fluctuation of, 217
  dimensions of, *168T*, *182T*
  focussing of, 144
  housings for, *185F*

light sources (*continued*)
measurement of spectral distribution of, 158
minimum usable size of, 179, *182T*
operation of, 171, 172, 174, 177, 178, *173F*
precautions in the use of, 184
rate of emission from, 143
spectral distribution of, *168T*, *180T*, *181T*, *162F*, *163F*, *164F*, *165F*, *166F*
types of, *176F*
line emission from solutions, 380, 470, 478, 490, *476T*, *492T*, *383F*, *384F*, *467F*
line to background ratio, *138F*, *160F*
liquid paraffin, quenching rates in, 76
lone pair orbital, 30, *31F*
lubricating oil, determination of, 458, 464
luminescence:
definition of, 1
long-lived, measurement of, 272, 282
polarisation of, 53, 55 (see also under polarisation)
lutetium, 480

magnesium, 484, 497
magnesium oxide, reflectivity of, *253T*
malachite green, 436
malic acid, determination of, 457
manganese, *476T*
medicine, fluorescence analysis in, 455
mercury, detection of, 497
mercury–cadmium lamp, 179
mercury lamp:
extra high pressure, 169, *168T*, *164F*
high pressure, 178
low pressure, 174, *162F*
medium pressure, 176, *163F*
precautions in the use of, 184
mercury lines, isolation of, 174, 188
mercury vapour:
radiative transitions in, *175F*
self-absorption by, 176
mesitylene, *358T*
metal-organic compounds, 473
metastable state, 44
methoxy-:
-anthraquinone, 382
-benzoic acid, 342
-benzoic ester, 342
-naphthalene, 437, *266T*, *310T*, *315T*
methyl-:
-acridinium chloride, *266T*
-anthracene, 461, *266T*, *307T*, *315T*, *358T*
-anthraquinone, 382
-benzanthracene, 373
-cyclohexane, 290
-naphthalene, *92T*, *325T*, *356T*, *445F*
-naphthylamine, 331
-pyrene, 356
-salicylate, 342
Michler's ketone, see tetramethyldiaminobenzophenone
microspectrofluorimetry, 501
mixtures, enumeration of components in, 500
modulation of exciting light, 269
molar volume, determination of, 392
molecular absorptivity, *17T*
molecular extinction coefficient, *17T*
molecular structure:
luminescence and, 428, 470
pK* value and, 340
monitor, *205F*
monochromators:
aperture of, 143
choice of, 156
comparison of, 147, *147T*
double, 151, 156
exit beam geometry and purity from, 151, *135F*
focal length of, 143
grating inefficiencies in, 149
light flux passed by, 145, 146
light gathering power of, 144, 151, 156, *147T*, *148F*, *155F*
linear dispersion of, *140F*, *141F*
precautions in the use of, 157
principle of, 131, *132F*
resolving power of, 152
scattered light in, 150
spectral sensitivity curves of, *161F*, *258F*
morin, 497
motor spirit, 458, *459F*, *465F*
multiplicity, 36
mutual sensitisation, 122

naphthacene, 117, 121, 227, 390, 430, *325T*, *358T*, *117F*, *118F*, *228F*
naphthacenequinone, *324T*

naphthaldehyde, *324T*
naphthalene, 92, 104, 323, 356, 366, 430, 431, 454, *92T, 266T, 281T, 310T, 315T, 324T, 358T, 368T, 454T, 123F, 367F, 432F*
naphthalene derivatives, *358T*
naphthalene diol, *333T*
naphthoate ion, *325T*
naphthoflavone, *324T*
naphthoic acid, 340, *325T, 333T, 343T*
naphthol, 331, 339, 375, *325T, 333T, 343T, 330F*
naphtholate ion, 435, *325T*
naphthol sulphonate, *333T*
naphthylamine, 329, 435, *266T, 325T, 333T, 343T, 33F*
naphthylamine sulphonate, 329
naphthylammonium ion, *325T*
naphthylphenylketone, *324T*
neodymium, 226, 480, *225F*
nickel, 476, *476T*
Nicol prism, 293, *294F*
nicotinamide, 482
nicotine, determination of, 462
nitrate, 470
nitrodimethylaniline, 435
nitrodiphenyl, 461
nitroethylaniline, *281T, 433F*
nitronaphthalene, *325T*
nitrosonaphthol, 456
noise level:
  due to blank, 410
  instrumental, 217, 286
non-metals, 481
norepinephrine, determination of, 456
nornicotine, determination of, 462
n–pi*, see pi*–n

octane, 381
oils, petroleum, determination of, 458, 464
0–0 transition, 10, 13, 374, *14F*
optical density:
  definition of, 17, *17T*
  measurement of, 263, 264
optical depth, *17T*
orbitals, types of, 28, *29F, 31F*
organic compounds (see also under analysis, and name of compound):
  analysis of, 438
  photoluminescence of, 428
orientation of transition moment, 395
oscillator strength, 27
oxine, see hydroxyquinoline
oxygen:
  determination of, *495T*
  quenching by, 38, 236, 379, 464, 488, *495T, 488F, 494F*
  removal of, 237, 239, 287
ozone, photochemical formation of, 174, 185

palladium, 497
paraffins, purification of, 289
parathion, detection of, 469
Pauli exclusion principle, 36
pentacene, *358T*
pentacenequinone, 457
pentane, 289
perylene, 50, 122, 268, 291, 310, 312, *266T, 312T, 315T, 358T, 454T, 124F, 311F*
perylene, determination of, 441, 453
perylene vapour, 46
pesticides, detection of, 469
petroleum ether, 289
petroleum fractions, analysis of, 458, 464, *459F, 465F*
pharmacological analysis, 461
phase of choppers, 278
phase of fluorescence, 269
phenanthrene, 46, 91, 99, 103, 104, 105, 114, 122, 275, 287, 363, 381, 396, 431, *107T, 266T, 281T, 310T, 315T, 324T, 90F, 91F, 106F, 115F, 123F, 276F*
phenanthrene, analysis of, 422, 452, *423F, 451F*
phenanthrene, determination of, 440, 448
phenanthrene vapour, 46
phenanthroline, 209, 491
phenobarbital, determination of, 462
phenol, 49, 375, 433, 435, *266T, 281T, 325T, 333T, 432F*
phenolphthalein, 436
phenols:
  detection of, 469
  pK* values of, 340, *333T, 343T*
phenothiazine, detection of, 469
phenyl-:
  -alanine, 396, 456
  -anthracene, *266T, 307T, 315T*
  -anthraquinone, 382

phenyl-(continued)
-glyoxal, *324T*
-naphthalene, 436, *437F*
-naphthylamine, 227, 436, *281T*, *228F*, *432F*, *437F*, *442F*
-naphthylamines, determination of, 443, 466
phenylenediamine, 482
phosphate, determination of, 497
phosphorescence:
alpha, 44, 46
analysis, see analysis
beta, 44, 46
characteristics of, *126T*
definition of, 38, 42
efficiency, 87, 279
efficiency, values of, *89T*, *281T*, *476T*
emission spectrum, *43F*
emission spectrum, examples of, *90F*, *98F*, *276F*, *277F*, *432F*, *433F*, *434F*, *444F*, *445F*, *446F*, *447F*, *448F*, *449F*, *462F*, *465F*
excitation spectrum, 326, 461, *463F*
fluid solution, 45
historical development of, 44
lifetime of, 38, 90, *317F*
lifetime, values of, *89T*, *92T*, *281T*
molecular structure and, 428
polarisation of, 53, 393, 396
quenching of, 90, 93, 495
rate of, 70
recombination, 43, 50, 51, *126T*
sensitised, 92, *92T*, *126T*
spot tests, 469
temperature dependence of, 89, *90F*, *91F*
viscosity dependence of, 378
phosphorimeter:
definition of, 129
factors, 275
types of, 272
phosphors, inorganic, 472
photo-association, see excited dimer, photodimer
photocells, types of, 193
photocells, vacuum, 197, *198F*, *199F*
photochemical products, analysis of, 386, 466
photochemical reactions:
interference by, 426, 452, 487
sensitised, 287, 390
photochemical reactions (continued)
studied by photoluminescence, 328, 386, 466, *389F*, *391F*
photochemistry, laws of, 4
photo-detectors, types of, 193
photodimer, 345, 360, 390, *358T*, *345F*
photodissociation, 48
photographic plates, fluorescent coatings for, 204
photo-ionisation, 48
photoluminescence (see also under fluorescence, phosphorescence, delayed fluorescence):
analysis, 399, 438, 483, *400T* (see also under analysis)
decay of, *126T*
definition of, 1
intensity dependence of, *126T*
kinds of, 126, *126T*
molecular structure and, 428, 470
photochemical reactions studied by, 328, 386, 466, *389F*, *391F*
recombination, 48
temperature dependence of, *126T*
photomultiplier tubes:
measurement of output from, 215
operation of, 203, *201F*
principle of, 200, *201F*
pulsing of, 270
sensitivity adjustment of, 203, 219, *203F*
spectral sensitivity of, *199F*
types of, *202F*
photo-oxidation, 48, *488F*
phthalaldehyde, 457, 482
pH value, effect on fluorescence, 328, 334, *335F*
piazselenols, 407, 431, 482, 489, *277F*, *490F*
pi-electron conjugation, 429
pi-electrons, fluorescence and, 30, *31F*, *377F*
pi*–n transitions, 31, 33, 38, 376, 429, 432, 474, 478, *377F*
pi*–pi transitions, 30, 376, 428, 432, 474, 476, 478, *377F*
pK values of ground and excited states, 330, 340, *333T*, *343T*
Planck's constant, 2
Planck's law, 169
platinum, 497
Platt classification, 359

polarisation:
  degree of, 53, 297
  energy transfer and, 396
  filter fluorimeters, 295
  measurement of, 292, 299, *295F*, *296F*, *299F*, *301F*
  measurements, application of, 392
  principal, 55
  spectra, 57, 298, *58F*, *395F*
  spectrophotometers, 299
  temperature and viscosity dependence, *393T*
polariscope, *296F*
polarised light:
  absorption of, 52, *52F*, *56F*
  description of, 51, *52F*
  emission of, *54F*, *56F*
  production of, 293
polariser, 52, 293
polarising films, 295
polarising units, 267, *294F*
polymeric materials as solvents, 464
polymer molecules, dyestuffs adsorbed on, 500
polyphenyls, 381
polystyrene, 361
polyvinylpyridine-n-butyl bromide, 390
porphyrins, 456
potassium ferrioxalate as a chemical actinometer, 209
potential energy diagram, 11, 39, *11F*, *39F*, *346F*
praseodymium, 480
precautions, 157, 184
pressure-broadened lines, 176
prisms, dispersion of, 133, 147
probability (see also "Definition of Symbols"):
  of triplet-to-triplet energy transfer, 102, 108, 115, *312T*, *315T*
  of triplet-to-singlet energy transfer, 115
procaine, determination of, 462
proflavine, 124, 320, 322, 326, 390, 453, *266T*, *315T*, *454T*, *125F*
propanol, 289
propiophenone, *324T*
propylene glycol, purification of, 289
protein:
  binding of dyestuffs, 393
  conjugates, 393
protein (*continued*)
  fluorescence polarisation of, 396
  size of, 393
protolytic equilibria:
  in the singlet state, 328, 342, *333T*, *343T*
  in the triplet state, 343, *343T*
P-type delayed fluorescence:
  by triplet–singlet absorption, 327
  characteristics of, *126T*
  definition of, 48
  description of, 99
  deviation from square law, 110, *114F*
  historical development of, 46
  in rigid media, 103, *107T*, *106F*
  lifetime of, 103, *317F*
  non-exponential decay of, 105, 110, *106F*, *112F*
  of excited dimers, 363, 366, 369
  sensitised, see sensitised delayed fluorescence
  spectra, see delayed fluorescence spectra
purity, tests of, 286, 438, 451
pyrene, 103, 104, 287, 308, 322, 347, 363, 369, 373, 379, 380, 381, 382, 385, 495, *266T*, *308T*, *310T*, *315T*, *324T*, *356T*, *358T*, *454T*, *125F*, *288F*, *317F*, *350F*, *351F*, *352F*, *354F*, *364F*, *365F*, *366F*, *369F*, *380F*, *383F*, *384F*
pyrene:
  aldehyde, 376, 433
  analysis of, 447, 454, *448F*, *455F*
  derivatives, *358T*
  determination of, 452, 460, *451F*
  sulphonate, 356
  vapour, 46
pyridine, 31, 341, 432, *31F*
pyrrole, 31, *31F*
pyruvaldehyde, detection of, 469

quantum counters, 204, *205F*
quantum efficiency:
  of delayed fluorescence, see delayed fluorescence efficiency
  of phosphorescence, see phosphorescence efficiency
  of photochemical reactions, 5, 388
  of photoluminescence, 5
  of prompt fluorescence, see fluorescence efficiency
quantum, size of, *3T*

Quantum Theory, 2
quartz:
  analysis of, 485, 498
  cuvette, effect of luminescence of, 416, *417F*
  luminescence of, 232, 235, 498, *235F*, *417F*
quasi-linear spectra, 381
quenching agents, selective, use in analysis, 469
quenching of luminescence (see also under kind of luminescence):
  approximate calculation of, 77, 96
  by impurity, 286, 378
  by oxygen, 38, 236, 379, 460, 464, 488
  determination of oxygen by, 493
  differentiation from inner filter effects, 221
  photochemical reactions studied by, 387
  representation as a pseudo first order process, 71
quercetin, 497
quinine bisulphate, 84, 223, 226, 227, 247, 265, 403, 407, 413, *266T*, *408T*, *8F*, *138F*, *223F*, *225F*, *248F*, *412F*
quinine sulphatoperiodide, 295
quinoline, 376, 432, *325T*, *343T*
quinolinium ion, *325T*
quintuplet state, 363, 371

radiative lifetime, see lifetime
radicals, production of, 48, 51, 390
radioluminescence, 1
Raman effect, 64
Raman scatter, polarisation of, 65
Raman spectrum:
  as a test of instrumental sensitivity, 406
  contribution to blank, 411, 487
  in filter fluorimetry, 414
  of cyclohexane, interference by, 419, *417F*
  of ethanol, interference by, *413F*, *414F*, *415F*, *416F*, *420F*
  of solvents, 66, 411, *66T*, *411T*, *67F*
  of water, interference by, 411, *412F*, *421F*
  recognition of, 139, 413
rare earths, 86, 470, 473, 479, *492T*, *471F*
rare earths, determination of, 491, 497, *492T*
Rayleigh scattering, 61
reciprocal centimetre, 19
reciprocal micrometre (= reciprocal micron), 19, 153
recorder:
  double beam operation of, 217, *218F*
  oscilloscope type, 217, 283, *285F*
  potentiometer type, 216, *216F*
  single beam operation of, 216
reflectivity of magnesium oxide, *253T*
refractive index, correction for, 263
relaxation, 14, 374, *14F*
reserpine, 501
resolution of vibrational bands, 379, 382, 436
resonance radiation, 176
resorcinol, 457
rhodamine B, 205, 250, 268, 436, 473, *266T*, *408T*
rhodamine B isocyanate, 393
riboflavine, 456
rigid molecules, 380, 436
rigid solvents, 289, 388, 466, *290T*
ring closure reactions in analysis, 456
Rochon prism, 268, *294F*
rotational depolarisation, measurement of, 295, 392
rotational relaxation time, 59, 392
rubber, anti-oxidants in, 443, *442F*
rubber, synthetic, inhibitors in, 466
rubrene, *266T*
ruby, analysis of, 498
ruthenium, 484

salicylaldehyde, 469, 478
salicylaldehyde semicarbazone, 477, *475F*
salicylalhydrazone, 469
salicylic acid, 342, 491
salicylic acid, determination of, 459
salicylidine–aminophenol, 477, *475F*
samarium, 470, 472, 480, 481, *492T*, *471F*
samarium, determination of, 491, *492T*
saturated compounds, light absorption by, 29
scandium, 477
scattered light:
  contribution to blank, 418
  depolarisation of, 62, 63
  from "particles", *62F*, *63F*
  interference by, 150, *417F*
  kinds of, 60

selenium, determination of, 407, 482, 489, *490F*
selenous acid, 482, 489
self absorption, 225, 227, 230, 264, *229F*, *231F*
semithionine, 387
sensitised delayed fluorescence:
  and triplet formation efficiency, 309
  anti-Stokes, 122, 453, *122F*, *123F*, *124F*, *125F*
  direct analysis of mixtures by, 449
  general equations for, 114, *120F*
  in photochemistry, 390, *288F*, *391F*
  mechanism of, 114
  mutual, *125F*
  selective, *454T*, *455F*
  spectra, *115F*, *117F*, *288F*, *311F*, *451F*, *452F*, *453F*, *455F*
sensitised phosphorescence, 92, *92T*, *126T*
sensitised prompt fluorescence, 84
sensitivity:
  absolute, 406, 490, 491, *408T*
  calibration of, 219
  compensation for variations in, 215
  index, 407, 490, *408T*
  instrumental, 402, 403, 404, *403T*, *405T*
  kinds of, 402
  limitations due to blank, 409
  method, 403, 409, 489
  of photoluminescence analysis, 401, *400T*
  optimum instrumental, 157
  optimum wavelengths for, 407
  Raman spectrum as test of, 406
Shpol'skii effect, 379
Shpol'skii spectra, 139, 152, 466, *383F*, *384F*, *385F*, *467F*
sigma*–n transitions, 30
sigma*–sigma transitions, 29
silica:
  analysis of, 485, 498
  fluorescence of, 235, 498, *235F*, *498F*
silicon, analysis of, 486
silver, 497
singlet energy, values of, *92T*, *315T*
singlet state:
  definition of, 36
  kinds of, 28, 431
  lifetime of, 38, 70, 72, 269
  pK values of 334, *333T*, *343T*
singlet state (*continued*)
  quenching by heavy atoms, 305, 373, 437, 500, *307T*
  radiative transition rate, 70, 303
  routes for decay of, 303
singlet-to-singlet energy transfer, see energy transfer
slits, maintenance of, 158
slit width, optimum, 135, 137
Smoluchowski equation, 74
sodium lamp, 179
solutes, purity of (see also purity), 286
solvated electrons, 49
solvation, 13, 374, *14F*
solvent:
  effect of, 15, 33, 373, *34F*, *377F*, *380F*, *383F*, *384F*
  fluorescence of, *417F*, *420F*, *421F*
  for low temperature measurements, see glassy solvents
  purification of, 286
  tests for purity of, 287, 419
sonoluminescence, 1
specificity of photoluminescence analysis, 401, *400T*
specimen:
  arrangement, 233
  compartment for frontal illumination of, 237, *238F*, *245F*
  container for aerated solutions, 234
  de-aeration of, 237, 239, *236F*, *240F*
  general purpose compartment for, 237, *238F*
  illumination of, 151, *221F*
  temperature control of, 241, *242F*, *244F*
spectral distribution:
  data on light sources, *168T*, *180T*, *181T*, *162F*, *163F*, *164F*, *165F*, *166F*
  of a discontinuous source, 139, *138F*, *160F*
  of light from a monochromator, 133, *135F*, *137F*
spectral line to background ratio, 138, 158
spectral purity, 133, 136, *135F*, *137F*
spectral sensitivity factor, 252, *255T*, *98F*, *161F*, *258F*, *280F*
spectrofluorimeter:
  definition of, 128
  general purpose, 129, *130F*
spectrofluorophotometer, 128
spectrometers, 128

spectrophosphorimeter:
definition of, 128
diagram of, *273F*
factor, 275
principle of, *274F*
types of, 272
spectrophotofluorometer, definition of, 128
spin quantum number, 36
square law, 100, *101T*, *107T*, *112T*
standard fluorescent substances, 265
standard for radiation measurements, 194, 208
standard lamp, 195, 252, 253
steady state equations, 71, 87
Stern–Volmer equation, 72
stilbene, 439
Stokes–Einstein equation, 76
Stokes' Law, 10
Stokes shift, 342, 439
stray light, errors caused by, 264
substituent groups, effect on luminescence, 31, 32, 434, 436, 437, *385F*
sulphate, determination of, 482, 497
sulphide, determination of, 482, 497
sulphite, determination of, 482
sulphosalicylic acid, 491
sulphur, determination of, 482

terbium, 470, 472, 480, 481, *492T*, *471F*
terbium determination of, 491, *492T*
tetrabromofluorescein, see eosin
tetracene, see naphthacene
tetracycline, 501
tetraiodofluorescein, see erythrosine
tetramethyldiaminobenzophenone, 32, 33, 433, *324T*, *34F*, *434F*
tetramethylparaphenylenediamine, 50
thallium, 473, 483, 484
thenoyltrifluoracetone, 491
thermal activation, 10, 40, 44, 45, 70, 96, 97, 316, 320, *95F*
thermoluminescence, 1, see also recombination phosphorescence and delayed fluorescence
thermopile:
calibration of, 194, *195F*
construction of, 194
sensitivity of, 196
thiamine, 497
thiochrome, 497
thiocyanate, 497
thionine, 388, 407, 482, *266T*, *408T*
thioxanthone, *324T*
thorium, 484, 497
thulium, 480
time constant of measuring circuit, 284
tin, 473, 483, 484
tobacco, analysis of, 462
tolualdehyde, *492T*
toluene, 436, *266T*, *325T*, *358T*, *437F*
toluidine, 49
transient rate kinetics, 77, 108
transition elements, 470, 476, 479
transitions (see also under singlet state, triplet state, intersystem crossing):
allowed, 27, 174
atomic, *479F*
classification of, 31
electronic, types of, 28, *29F*
energy of, 35
forbidden, 38, 174, 177
intermolecular dipole–dipole, 84
moment of, 52, 57, 395
of complex molecules, *6F*, *37F*
of the mercury atom, *175F*
0–0, 10, 13, 374
probability of, 22, 35, 38
rates of, 70, *69F*
transmission, *17T*
transmittance, *17T*
transmitted radiant power, *17T*
triacetylbenzene, *324T*
triboluminescence, 1
trihydrodimethylquinoline polymer, determination of, 466
trihydroxyindole reaction, 456
triphenylamine, *324T*
triphenylene, 381, *266T*, *281T*, *310T*, *315T*, *324T*, *432F*
triphenylene, determination of, 448, *449F*
triphenylmethylphenyl ketone, *324T*
triphenylmethyl radical, 48, 390
triphenylpentane, 361
triplet acceptors and donors, *92T*, *454T*
triplet energy, see triplet state
triplet formation efficiency:
definition of, 86
determination of, 305, 309, 313
temperature dependence of, 82, 89
values of, *307T*, *308T*, *310T*, *313T*, *315T*

triplet lifetime:
effect of impurities on, 316, 318
effect of solvent on, 320
expression for, 88
measurement of, 282, 313
range of, 70
reason for, 38
temperature dependence of, 315
values of, *312T*, *315T*, *319T*
viscosity dependence of, 315, 378
triplet sensitisers (see also under sensitised delayed fluorescence and triplet state: energy of):
in analysis, 449
list of, *92T*, *454T*
triplet–singlet absorption, 326, 501
triplet–singlet emission, see phosphorescence
triplet–singlet splitting, 100, 430
triplet state:
absorption by, 49, 51
bimolecular decay of, 314
definition of, 36
energy of, 323, *92T*, *315T*, *324T*, *325T*, *454T*
intersystem crossing from, 313, 320
parameters, 313
population of, 41
quenching by ground state, 314
quenching by heavy atoms, 308
quenching by impurities, 96, 286, 378
rate of disappearance of, 313
reaction of, 314, 390
triplet-to-singlet, see energy transfer
triplet–triplet (see also energy transfer, triplet-to-triplet):
annihilation, 107, 108
interaction, 48, 100, 107, 108, 114, 122, 363, 370, 390
trivial effect, 125, 291
trivial process, 83
tryptophane, 396, 456
tungsten, 484
tungsten lamp, spectral distribution of, 170, *166F*
Tyndall effect, 63
Tyndall scattering, 63
tyrosine, 396
tyrosine, determination of, 456

ultra-violet lamps, see light sources
umbelliferone carboxylic acid, 457
units of wavelength, wavenumber and energy, *3T*, *154T*
uranium, 473, 497, *472F*
uranium, determination of, 483, 498, *484F*
urine, analysis of, 462

vacuum system, *240F*
vanadium, 484
vaseline, fluorescence of, 458
vector, electric, *52F*
veratraldehyde, *492T*
vibrational bands, 7, 11, 58, 379, 436, 473
vibrational levels, *6F*
viscosity of solvent, 59, 74, 82, 103, 107, 378, 392, *290T*
vitamin A, 430, 456
volume, molar, determination of, 392

water:
benz(a)pyrene in, 466
purity of, 420
Raman bands of, *66T*
viscosity of, *75T*
wave equation, 5
wavelength:
compared with wavenumber, 153
relationship with frequency, 2
units of, 19, *3T*
wavenumber:
compared with wavelength, 153
definition of, 4
linear drive for, 260
units of, 19, 153, *3T*
wax, fluorescence of, 458

xanthone, *324T*
xenon lamp:
characteristics of, 172
operation of, 172, *173F*
spectral distribution of, 167, *165F*
xylene, *358T*

ytterbium, 480

zinc, 484, *476T*
zinc lamps, 179, *180T*
zirconium, 484
zone refining:
analysis of material from, 443, 445
examination of columns from, 239, 468, *468F*
for purification of solutes, 290